THE CLASSICAL ELECTROMAGNETIC FIELD

THE CLASSICAL ELECTROMAGNETIC FIELD

Leonard Eyges

Dover Publications, Inc.
New York

To my parents,
Edward and Alice Eyges

Published in Canada by General Publishing Company, Ltd., 30 Lesmill Road, Don Mills, Toronto, Ontario.
Published in the United Kingdom by Constable and Company, Ltd.

This Dover edition, first published in 1980, is an unabridged and corrected republication of the work originally published in 1972 by Addison-Wesley Publishing Company, Inc.

International Standard Book Number: 0-486-63947-9
Library of Congress Catalog Card Number: 79-56179

Manufactured in the United States of America
Dover Publications, Inc.
180 Varick Street
New York, N.Y. 10014

PREFACE

This book is intended as a text for a graduate course in electromagnetic theory. I began to write it after teaching such a course at M.I.T. and Northeastern University for several years and being dissatisfied with the existing texts. The reader will judge whether or not I have succeeded in writing a better one, but here I would like to show, at least obliquely, the reasons for my dissatisfaction, by outlining the text and my reasons for writing it as I have done.

The book is divided into two main parts: static fields and time-varying fields. Of course, this division is no novelty; it reflects history and is incorporated in large numbers of textbooks. It has the consequence that the electric and magnetic fields are considered as separate entities until they are amalgamated by special relativity and the Lorentz transformation. Since relativity theory tends these days to be taught earlier and earlier in the physics students' career, the possibility is now open of altering this canonical (static/time-varying) division. After the electrostatic field is discussed, one can, with some confidence, introduce the magnetic field as a result of viewing an electric field from a moving frame. This approach has some merit, and I have not dismissed it lightly. It is more basic, in a way, in that only one field need be postulated; the other can then be derived from it using relativistic transformations which must eventually be introduced anyway. And the symmetry between electric and magnetic fields that is made manifest by relativity is aesthetically very satisfying. But the amount of relativity theory that must be introduced before the magnetic field emerges from the transformed electric field is considerable. I think the danger of the student not seeing the wood for the trees outweighs the advantages of this procedure and so I have not adopted it.

When dealing with static fields, the field concept is, in fact, superfluous. All of electrostatics is comprised (in principle) in Coulomb's law for the force between two charges and in the law of superposition, and all of magnetostatics in Ampère's law for the force between two currents. The split of Coulomb's law into a field produced by one charge, which field then acts on another, is a convenience for static charges, but it is not conceptually necessary. A similar remark applies to the division of Ampère's law into the production of a field by one stationary current and its action on a second. It is only for time-varying fields that the field concept assumes its real importance as a way of preserving the conservation laws of energy and momentum. But since one is forced eventually to introduce the idea of field, it is useful pedagogically to introduce it as soon as possible, i.e., in connection with statics, to promote familiarity and ease with it. Nonetheless, I have also felt it important to clarify the status of the field concept, and so have prefaced the chapters on static fields by one called "Concepts of a Field Theory" which essentially contains an elaboration of the remarks above.

Beyond the static/time-varying division, there is a kind of threefold symmetry. The book treats essentially three kinds of fields: the electric, the magnetic, and their amalgam, the time-varying electromagnetic field. Each of the treatments follows the same pattern, and the shape of this pattern is partly given by the idea of the "summation problem."* By this is meant the problem of evaluating the field from an integral expression over the given, *known* sources, i.e., charges or currents. The summation problem is meant to stand in contrast to boundary-value problems with matter in which matter effectively acts as the source of charges or current that are *not* known, or given, in advance of solving the problem. I have found this division, into a definition of a field and its corresponding summation problem, a sound one pedagogically and am convinced of its usefulness.

Chapter by chapter, the book develops as follows. After the first chapter on concepts of a field theory Chapter 2, on electrostatics, discusses properties of the electric field and of the scalar potential Φ. The chapter ends with the superposition integral for the potential Φ due to an arbitrary continuous or discrete charge distribution.

The problem of actually evaluating this integral for charge distributions of one kind or another is then the summation problem for electrostatics, the subject of Chapter 3. In it are considered distributions which occupy a finite volume and the corresponding multipole expansion; two-dimensional and one-dimensional distributions; surface charges and double layers; and dipolar distributions. One merit of a rather complete discussion of the kinds of distributions possible is that many of them are forced to our attention later. In Green's theorem, for example, the concept of surface charge and double layer enter, but it is very useful to have encountered these before coping with whatever other difficulties Green's theorem may entail. Similarly, the discussion in this chapter of the external field of dipole distributions is immediately applicable to a later theory of dielectrics.

Chapter 4 is on boundary-value problems with perfect conductors. It might have been subtitled "The Field of Unknown Distributions" to emphasize the connection of this problem with the summation problem of surface charges which is that of the field of *known* distributions. In a boundary-value problem, the final state of electrical equilibrium corresponds to *some* surface distribution on the conductors, but one that is unknown *a priori*. The various methods that apply to the summation problem must apply here as well, in a suitably modified way. Thus, for a given geometry, there occur the same solutions of Laplace's equation, the same symmetry considerations, the same far field expansions, etc. Among the special

* The phrase "summation problem" is not mine; it is due to Sommerfeld. (A. Sommerfeld, *Electrodynamics*, Academic Press, N.Y., 1952, p. 38.)

topics of this chapter is, of course, the method of images, which I discuss rather more cursorily than in many other books. On the other hand, I have pointed out what is often neglected, that the basic idea of the method (which is really nothing but inspired guesswork) applies to homogeneous as well as inhomogeneous problems. The classical method of superposition of separated solutions of Laplace's equation is, of course, also discussed. There is a section on the use of integral equations which I have found to be pedagogically rewarding: the physics involved in setting up these equations is enlightening, and they are almost the only practical way of solving problems for other than the simplest geometrical shapes. It usually comes as a relief to the student to be able to treat *some* other boundary than the sphere or cylinder. These equations must usually be solved numerically, of course, but they are well adapted to computer solution. At the worst, some of them can be crudely solved by hand and even this is worth the effort (once). In a further effort to escape the tyranny of the sphere and the cylinder, I discuss *composite problems.* These are problems in which the geometry is, so to speak, made up of a sum of separable parts and, typically, they are problems involving more than one sphere, a sphere and a cylinder, a cylinder and a half-plane, etc.

Chapter 5 treats the general theory of boundary-value problems involving Laplace's equation, i.e., the Dirichlet and Neumann problems. This entails, of course, a discussion of Green's theorem which I have deliberately not introduced until this time, although there is logic enough in discussing it earlier. But I have found that although readers follow the mathematical derivation of Green's theorem easily enough, they are not so quick to understand its real nature, and when it is useful. In particular, if it is introduced *before* the discussion of boundary-value problems with conductors, they somehow feel that Green's theorem should be useful in solving them, which it is not.

Chapter 6 treats dielectrics, and here I break with the usual textbook treatments which I think poorly of, in general. Where they are not vague, they tend to be unsound. One exception to this is in the book by Purcell (B); anyone who is familiar with it will see resemblances here to his discussion and may infer correctly that I am indebted to it in several ways. Not the least of these ways has been the reassurance that someone else thinks poorly of many of the standardized treatments.

The problem of dielectrics is, in effect, the problem of calculating average internal fields of dipole distributions. It is, in large part, a problem in statistical mechanics but there is no satisfactory general solution that derives the macroscopic properties from space and time averages of the microscopic ones. Any theory must then be essentially postulatory. The postulate I make involves the field $\boldsymbol{E}_m$ of the equivalent surface and volume charges that correctly yield the *external* field of dielectric matter; it relates $\boldsymbol{E}_m$ to the mean polarization and applied field. Although the postulate is labeled explicitly as such, I have tried to make clear the rationale for it and have adduced experimental evidence about dielectric-filled capacitors to support it. Finally, and perhaps most importantly, I show that the discussion of dielectrics that is based on this hypothesis is precisely equivalent to the usual formalism involving the vector $\boldsymbol{D}$.

This work on dielectrics temporarily closes out electrostatics. The next natural subjects are perhaps force and energy in the electrostatic field. I have deferred these, however, until magnetostatics has been discussed, and have then discussed force and energy side by side for both electric and magnetic fields. In this way, hopefully, one learns both from the similarities and from the differences.

Chapters 7, 8, and 9 comprise a discussion of the magnetostatic field along lines which parallel, insofar as possible, those for the electrostatic field. Thus, in Chapter 7, Ampère's law for the force between currents is stated, the split is made into a field $\boldsymbol{B}$ produced by one current which then acts on the other, and from the definition of $\boldsymbol{B}$ so obtained, its divergence and curl are calculated. I have found it sounder pedagogically to proceed in this way, i.e., to calculate the properties of the field *directly* from its definition rather than to do this by first introducing auxiliary potentials.

Chapter 8 comprises the second summation problem, that of calculating the field of an arbitrary stationary current distribution. Although many books use the vector potential as the major tool for this problem, the magnetic scalar one is almost always superior in practice, and I have treated it in some detail. The vector potential comes into its own only for time-varying fields. There are few cases in magnetostatics where it is easier, or even as easy, to calculate the three components of the vector potential than it is to do the single integration that yields the scalar potential.

After the summation problem for currents, the parallel with the electrostatic field breaks down. There is no magnetic analog of the electrostatic boundary-value problems with conductors since there are, of course, no conductors of magnetism. But there are magnetic materials which are the analogs of dielectrics, and with analogous problems, and these are the subject of Chapter 9. Once again, one must cope with violently fluctuating internal fields and here I have followed the natural course of patterning the discussion on that of dielectrics. An average internal macroscopic field $\boldsymbol{B}_m$ is defined by equivalent surface currents and a postulate is made that relates the mean magnetization $\boldsymbol{M}$ to $\boldsymbol{B}_m$ and to the applied field. It is then shown that the postulational approach is equivalent to the usual theory involving the vector $\boldsymbol{H}$.

The last chapter in statics is entitled "Force and Energy in Static Fields." I have chosen to treat the electrostatic and magnetostatic cases in this one chapter, to learn from their similarities *and* dissimilarities. One disadvantage is that one must, in setting up the expression for magnetic field energy, simply quote some results that will be derived later from Faraday's induction law. This seems, however, a small price to pay for the economy of treatment that results by treating the electric and magnetic case together.

The second part of this book, starting with Chapter 11, treats the time-varying electromagnetic field. The pattern of the discussion follows those for the static electric and static magnetic fields. First, the differential equations of the fields—in this case the Maxwell equations—are presented. The fields are then related to the retarded potentials, which is to say, integral expressions over the currents. The summation problem for these potentials is discussed and only then are boundary-value problems considered. Finally, the most difficult subject, that of fields inside matter, is treated in general, and dielectrics are discussed in particular.

Chapter by chapter, then, Chapter 11 states Maxwell's equations, following the introduction of the displacement current as the postulate that it really is. The displacement current is sometimes derived by requiring that the continuity equation be satisfied between the plates of a condenser, but this derivation merely camouflages the essentially postulatory nature of the current. The chapter is quite conventional in the main, except that I have spent more than the usual time on the question of conservation laws.

Chapter 12 treats the relation of the special theory of relativity to electrodynamics. In an effort to stay within reasonable bounds and yet keep everything of interest, I have foregone a common approach which starts from the historical experiments (Michelson–Morley, etc.), and have preferred to spend the time and space gained to discuss concepts at the base of the theory, such as absolute time, inertial frames, the nature of a vector, etc., which are perhaps not examined in enough detail in general.

Chapter 13 on time-harmonic currents is the third summation problem in the book and, as such, it can lean to a degree on the previous two, although it is, of course, much more complicated in detail.

Chapter 14 on the fields of point charges in motion is, in effect, also a summation problem. I had originally thought of calling this chapter the "Summation Problem for Point Currents" to emphasize this fact. But the point nature of the current makes this kind of summation problem different enough from the previous ones that such a title is perhaps somewhat labored. I have, however, tried to emphasize the similarity of the approximations necessary to evaluate the fields of point currents to those for the time-harmonic case. The explicit form of the Lienard–Wiechert fields is perhaps somewhat deceiving. They do not constitute the answer to a problem but are, in effect, the problem itself since, in any particular case, approximations must be derived for the retarded time in terms of the present time. These approximations, low velocity, multipole, etc., are essentially the same as those made for the time-harmonic case.

Chapter 15 is on time-harmonic boundary-value problems with perfect conductors. I have chosen to study this idealized case first before discussing the physics of imperfect conductors. There are enough new concepts even in the idealized case —modes, guide wavelengths, cutoff frequencies, etc.—that still other concepts of skin depth and field penetration are best deferred. Important time-harmonic boundary problems are those of diffraction. I have tried to emphasize the approximate nature of the usual diffraction theory of physical optics and at least outline one rigorous solution of a diffraction problem; this is the problem of diffraction by a perfectly conducting half-plane first solved by Sommerfeld but presented here in a version due to Clemmow. Too frequently, the Kirchhoff theory, i.e., the application of Green's theorem to the Helmholtz equation, leads to formulae that are represented as solutions of the diffraction problem or, at least, whose approximate nature is not enough emphasized. The solution of diffraction problems, then, appears to be an exercise in the evaluation of Fresnel or Fraunhofer integrals. This is, of course, not the case.

Chapter 16 on fields in matter was difficult to write. The question of the behavior of time-varying fields inside matter has, of course, all the difficulties of the two static cases and some of its own as well. It is usually presented for dielectrics in terms of the unsatisfactory formalism involving $\boldsymbol{D}$ and the time-dependent polarization vector, $\boldsymbol{P}$. Having shown how these are superfluous for electrostatics, I have tried to do the same for time-dependent fields in matter. I have, moreover, tried to treat all kinds of matter on the same conceptual basis. The common denominator has been to consider that matter of whatever kind is an ensemble of currents: damped currents for conductors; localized polarization currents for dielectrics, etc. The one postulate that is common to almost all classical models of matter is then a generalized Ohm's law: at a point the current is proportional to the field with a proportionality constant Γ which may be complex. The consequences of this postulate are reasonable for conductors. But in dielectrics, the currents are localized at atomic sites whereas the field is ubiquitous so the postulated relation can hold only in some average sense. One is led to basic difficulties involving the difference between the average field in a volume and the effective field acting on a dipole. These are, of course, the same problems that face one in the theory of the static dielectric constant. I have tried to raise these problems to the surface and, without solving them, to give, by means of various one-dimensional and other models, some idea of the physics that is involved in them.

With the properties of matter elucidated, the last chapter discusses boundary value and other problems associated with fields in matter, including reflection at a dielectric interface and surface waves. I began to write a section on Cerenkov radiation for this chapter. Most of its characteristic features can be derived by using the previously shown fact that time-harmonic waves propagate in matter with a modified wave number. But actually to calculate the intensity of Cerenkov radiation, one must also have an expression for the screening effect of the medium. This effect is sometimes expressed by saying that the medium is like free space except that the velocity c is replaced by c/n, where n is the dielectric constant, and a charge c is replaced by the effective value e/n. But the only way known to me of calculating this last result is by using the formal Maxwell equations in terms of $\boldsymbol{D}$ and $\boldsymbol{H}$ and the formal constitutive relations. As the last paragraph suggests, I am not convinced of the accuracy of this standard formal theory and *am* convinced of its ambiguity. Moreover, having done without $\boldsymbol{D}$ and $\boldsymbol{H}$ throughout the rest of the book, I thought it just as well to do without them altogether, at the cost of having to ask the reader to look up the Cerenkov effect elsewhere.

A word about the references. They are meant to supplement the text, but also to be randomly stimulating. I have tried to include some recent journal articles even in the earlier chapters, not with the intent of reviewing the literature completely, but rather of showing by a few essentially arbitrary references that even the older parts of electromagnetic theory are far from closed books to research.

I am indebted to many people for help of one kind or another in the making of this book. I am especially grateful to Dr. John Jasperse, who taught a course based on preliminary notes for the book, and whose advice helped shape large outlines and clarify small details. I have also profited from the comments of Dr. Ronald Newburgh on Chapter 12, and from the errors brought to my attention by Mr. Carl Holmstrom, who read through the whole manuscript. Finally, I owe a considerable debt to Mrs. Connie Friedman and Mrs. Donna Dickinson, whose technical typing skill and ability at deciphering were invaluable in producing the typescript of the bulk of the book.

Despite this help, errors and obscurities undoubtedly remain. I would be made grateful, if not happy, by anyone who brings them to my attention.

L.E.

CONTENTS

PREFATORY NOTES

In this book, references are cited in three different ways. First, the bibliography at the end of the book lists references which are of interest for more than one chapter; these are referred to in the text by citing the author's name, followed by a B (for bibliography) in parentheses. For example: Whittaker (B). Second, the references at the end of a chapter, that are primarily of interest in the chapter itself, are cited by the author's name plus an R in parentheses: Kennedy (R). Finally, there are ordinary footnotes indicated by standard footnote symbols.

The units used in this book are cgs (Gaussian).

THE CLASSICAL ELECTROMAGNETIC FIELD

1
CONCEPTS OF A FIELD THEORY

This is a book on the *electromagnetic* field, one of the many examples of fields that are important in physics and natural science. In the words of Morse and Feshbach (B), "Practically all of modern physics deals with fields: potential fields, probability fields, electromagnetic fields, tensor fields, and spinor fields." Since the concept of field is applied so widely, it is perhaps natural that it has taken on somewhat different contexts with, of course, an underlying common denominator. In this first brief chapter, we shall try to analyze the concept and its variants and so highlight the essential aspects of the electromagnetic field.

One mathematical common denominator is easy to isolate. Mathematically, a field is a function, or a set of functions considered as an entity, of the coordinates of a point in space (and possibly of time). For example, if the temperature is defined at every point in some volume, we say there is a *scalar temperature field* throughout that volume. If, in a moving fluid, the three vector components of velocity are known as a function of position in the fluid, they constitute as a whole a *vector velocity field*. In the theory of elasticity, the relative vector *displacements* of points of an elastic solid from their unstrained positions are described in terms of a double vectorial or, as it is more usually called, a *tensor field*. In modern physics, the Schrödinger probability amplitude ψ or the generalized Dirac spinor amplitudes are examples of fields.

It is worth noting that in this above list, there are really two kinds of fields. The first kind, exemplified by the temperature or velocity fields, is an idealization that is really defined only in a certain approximation of coarseness or fineness. For example, the velocity field of hydrodynamics is meaningful only in an average or continuum approximation in which the atomic and grainy structure of the fluid is not considered. This point is discussed in some detail in Morse and Feshbach (B) and we shall not elaborate on it here. By contrast, the Schrödinger or Dirac fields are not approximations to an underlying discontinuous physical model but must be assumed to exist no matter how finely space is divided.

To describe the evolution of the electromagnetic field concept, we recall some history that begins with Newton (1642–1726). One of the great laws of physics is Newton's universal law of gravitation. This law embodies the concept of *action at a distance*, according to which gravitational masses exert the forces they do on each other by virtue of their positions in space, the *intervening* space playing no active role. This is meant to contrast with forces which work via *contiguous action* whereby two masses at a distance exert forces which are transmitted by the intervening medium. For example, if several billiard balls are in contact in a row on a table and the first one is struck, the last one will move. Thus, the one billiard ball exerts a force on the other distant one, but by a mechanism which involves successive actions of the intermediate balls, the one moving the next, moving the next, etc. The concept of contiguous action is then quite different from that of action at a distance where no intermediate mechanism or medium is considered.

The work of Newton is relevant in a second way. His laws of the motion of point particles and rigid bodies paved the way for the development of the continuum mechanics of fluids and later of elastic bodies. Some tentative beginnings on the subject of fluid flow were made by Newton himself, but the real groundwork was later laid by John Bernoulli (1667–1748) and Euler (1707–1783). They bypassed the problem of the actual microscopic structure of fluids by adopting a *continuum* model and then applied Newton's laws of point mechanics to small elements of the continuum. The same idea was later applied to elastic solids, and the vibrations of these solids was discussed by applying Newton's laws to a small element of the solid, assuming that the forces acting on it were the stresses due to the rest of the solid, plus any external forces. Hydrodynamics was therefore formulated in terms of the velocity and acceleration of the moving fluid at every point, i.e., in terms of *velocity and acceleration vector fields.* The theory of elastic solids was similarly formulated in terms of *stress and strain tensorial fields.*

So much for mechanics; we turn now to electromagnetism. A basic law of electrostatics is Coulomb's law for the force between two charged particles. Except for the fact that the electric force can be either attractive or repulsive, whereas the gravitational force is always attractive, this law is obviously similar to Newton's law of gravitation. It was then considered from the time of its discovery as an example of *action at a distance,* in which two charges act on each other in a way that has nothing to do with the intervening medium. But this view began to be questioned, at least in the mind of Faraday (1791–1867), by his work on dielectric polarization. This phenomenon led him to attribute more and more importance to the intervening medium. We cannot go into the details of Faraday's results and reasoning but, for illustration, shall concentrate on one of his findings. This has to do with the effect of insulators or *dielectrics* on the capacitance of condensers. Consider a parallel-plate capacitor with air between its plates; it has a certain capacitance. If the air is replaced by a dielectric medium, the capacitance will be increased. Faraday viewed the phenomenon of enhanced capacitance as somehow due to the fact that the electric force generated by the charges on the plates was *weakened* by the dielectric medium. But *if changing the medium* that intervened between the charges *changed the force,* then somehow the forces must depend on, or be *transmitted by,* the medium. As a corollary of this view, Faraday considered that the essential feature of the interaction between charged particles was the lines of force that carried the "stresses" of the medium from one charge to another. These lines of force that extend from charge to charge through the medium were considered primary; the charges merely happen to be the places where the lines of force start and stop. Although these views were mainly derived from experiments on dielectric polarization with strongly dielectric substances, they seemed to be equally valid for those whose dielectric constant was close to unity. By extension, then, Faraday considered them valid for that insulator whose dielectric constant is exactly unity, i.e., free space. In short, in Faraday's view, free space was a substance, qualitatively like all other insulating substances, that contained charges which could be separated and displaced, as charges were separated and displaced in the material insulators he used in his capacitor experiments.

The next great name in the history of the electromagnetic field is that of Maxwell (1831–1879). He took up Faraday's ideas on the nature of the force between charges and the importance of the intervening medium. He succeeded in showing that the forces that charge complexes exert on each other and their energies could be expressed not only in terms of the magnitudes of the charges and their positions but in terms of a *stress energy tensor* that was defined throughout the medium (even if that medium was free space) and that had as components certain functions of the field strengths. For example, the force that one point charge exerts on another could be calculated by either using Coulomb's law, or surrounding the charge by an imaginary surface and integrating over that surface the total electrical stress as given by the stress energy tensor. This concept of stresses in the medium gave no new result, but it did at least show that the Faraday–Maxwell conception of the "state of the medium" was consistent with the results, if not the concept, of action at a distance.

The next step in the development of the field concept was also due to Maxwell. This was, of course, the discovery of the equations that bear his name and, as a corollary, the discovery of electromagnetic waves and their identity with light waves. Two basic guides in this work were Faraday's theory of charge polarization or displacement, and the theory, well developed by Maxwell's time, of the vibrations of an elastic solid. By generalizing Faraday's idea of *displacement* to the time-varying case and by introducing the so-called *displacement current* $\boldsymbol{D}$, Maxwell found that electromagnetic phenomena could be described in terms of four field vectors, $\boldsymbol{E}$, $\boldsymbol{B}$, $\boldsymbol{D}$, $\boldsymbol{H}$. But like the velocity fields of hydrodynamics or the stress or strain fields of elasticity, these fields were not considered to exist by themselves but were somehow considered to be vibrations or displacements of an underlying luminiferous ether whose properties were those of a somewhat special kind of elastic solid. Electromagnetic waves were then, so to speak, secondary: the ether could exist without electromagnetic waves but electromagnetic waves could not exist without the ether.

Maxwell's theory was a great triumph. But as the years after its discovery passed, it was accompanied by an increasing perplexity as to the nature of the hypothetical ether that underlay it. The history of the researches and speculations on the nature of the ether is beyond us here. It is well described in Whittaker (B) and some detail is given in Chapter 12 of this book. Suffice it to say here that a famous experiment by Michelson and Morley showed that the ether did not exist. Nonetheless, electromagnetic waves continued doggedly to be generated and propagated. The Michelson–Morley experiment thus served, so to speak, to emancipate the electromagnetic field. After it, the fields could not be considered as "merely" vibrations of an underlying medium. From the time of the disproof of the existence of the ether, the electromagnetic field had to be looked on as an entity in its own right, as real as matter and everywhere on a par with it.

In fact, the essence of the electromagnetic field theory is that the field *does* have properties that we usually associate with matter. It can possess energy, momentum, and angular momentum. The field is thus a *dynamical concept* and is not merely a mathematical function of the space coordinates and time. We shall highlight this essential aspect of the electromagnetic field by a simple example. In it, we presuppose a small amount of elementary knowledge on the reader's part; namely, that accelerated charged particles radiate and that this radiation is propagated with a finite velocity c.

Suppose two charged particles q_1 and q_2 are separated by a distance d. Imagine that we suddenly move particle 1, say, and then quickly bring it to rest again. Having been accelerated, the particle will emit a pulse of radiation which travels at velocity c and hence would make itself felt on particle 2 at a time $t = d/c$. If we look at this system of two charged particles at some time *after* the first particle is brought to rest but at a time which is *less than* d/c, we would see simply two charged particles, each at rest; they would constitute an *isolated system* with kinetic energy zero. Suddenly, however, at a time $t = d/c$, we would find that the second charged particle began to move. Superficially, the energy (and momentum) of this isolated system would appear to change even though there were no external forces acting on it. How can we reconcile this with the conservation laws of energy and momentum? The field theory gives one possible answer to this question. According to it, we have simply overlooked the fact that there are forms of energy other than kinetic; there is in fact another physical entity that we have not mentioned, the *electromagnetic field*, and this entity "contains" energy and momentum. The energy and momentum that begin to be transmitted to particle 2 at time $t = d/c$ were, in fact, contained in the electromagnetic field for earlier times.

We shall spell out these conservation laws involving the electromagnetic field in Chapter 11. For the moment, however, we make another point: the conceptual difficulties with energy and momentum conservation come about primarily because of the *finite* velocity c of electromagnetic propagation. For if c were infinite, we could not set up, even in thought, the above experiment and the concomitant difficulties would not arise. For *static* fields, then, in which propagation velocities are not involved, we shall see that the field concept does not so inevitably impose itself. Many of the results of electrostatics and magnetostatics can be formulated in terms *either* of action at a distance *or* of a field.

In summary, the basic idea of electromagnetism as a field theory is that charges and currents produce at each point of space a field that has a reality of its own, that can contain and propagate energy, momentum, and angular momentum, and that acts on other charges. The field is *produced by*, and *acts on*, charges. Correspondingly, there are two sets of equations. These are, first, *Maxwell's equations* which describe the field produced by a given set of charges and currents. Second, there is the *Lorentz force equation* which shows how a given field acts on charges.

We should not leave the impression that classical field theory itself is without its difficulties. It has these too. There is a conceptual difficulty in describing the self-action of the field on the charge that produces it, which is not soluble in classical field theory. This is outlined in Section 14.9, and an extensive discussion is given in Rohrlich (B). For practical computational purposes, these difficulties are resolved in part by *quantum electrodynamics*, which is the extension of the classical field theory of this book to incorporate quantum concepts; there remain however difficulties in principle connected with the occurrence of infinite quantities. There have been attempts over the years to bypass the difficulties with field theory, by reviving a sophisticated version of the action-at-a-distance theory. Examples are the theories of Wheeler and Feynman (R) and a recent paper by Kennedy (R).

REFERENCES

Hesse, Mary B., *Forces and Fields*, Philosophical Library, New York, 1962.

This book is subtitled, "The concept of action at a distance in the history of physics." The development of this concept is studied from primitive times to the present.

Kennedy, Frederick James, "Instantaneous Action at a Distance Formulation of Classical Electrodynamics," *J. Math. Phys.*, **10,** 1349 (1969).

A recent attempt to provide an alternative to the field description of the interaction of charges. There is a useful set of references.

Wheeler, J. A., and R. P. Feynman, "Interaction with the Absorber as the Mechanism of Radiation," *Rev. Mod. Phys.*, **17,** 157 (1945); "Classical Electrodynamics in Terms of Direct Interparticle Action," *Rev. Mod. Phys.*, **21,** 425 (1949).

An interesting attempt to alter the standard framework of electromagnetic theory.

Williams, L. Pearce, *The Origins of Field Theory*, Random House, New York, 1966.

A nonmathematical, but interesting and detailed account of the history of the field concept.

2
THE ELECTROSTATIC FIELD

It has been known for a long time that there are materials which, having been rubbed, will attract or repel other small pieces of matter. In particular, the Greek philosophers knew that amber possessed this property; it is from the Greek word for amber, $\eta\lambda\varepsilon\kappa\tau\rho o\nu$, that the word "electricity" is derived. There appears to have been nothing extensive made of this knowledge, however, and the beginning of electrostatics as a science must be dated, along with many other beginnings, from the Renaissance. The detailed history of this is complicated and fascinating, but we can mention only some highlights. Gilbert (1540–1605) found that materials could be roughly divided into those that could be electrified by friction (electrics) and those that could not (nonelectrics). Today we call such materials *insulators* and *conductors*, respectively. DuFay (1698–1739) discovered that the "electrical fluid" was of two types, "vitreous" and "resinous"; this was the origin of the concept of positive and negative electrical charges. Moreover, it was found that vitreous and resinous electricity could not be produced from nothing; a change in the amount of one was accompanied by a like change in the amount of the other. This observation was the primitive basis of the idea of the conservation of charge; in effect, it justified our treating amounts of positive and negative electricity according to the algebraic laws for combining positive and negative quantities.

All the early experiments were done with frictional electricity, i.e., with electrical forces generated by rubbing bodies together. It is amusing to note that even today we still do not have a clear understanding of the detailed mechanism of this phenomenon, as Harper (R) discusses.

Be that as it may, the developments over the years and centuries have been distilled into a system of concepts and nomenclature that we have inherited. In brief, a body which is in a state to exhibit the kind of electrical phenomena described here is said to be *electrically charged* or to possess *charge.* Somewhat tautologically, then, charge is a name for that "quantity" which a body possesses when it exhibits electrical forces. This charge is of two kinds, negative and positive, and the amounts of these can, when appropriate, be added algebraically. Charges of like sign tend to repel each other and of unlike to attract. From modern experiments we know even more exactly that charge is invariable in the sense that the total charge of an isolated body cannot be changed. This most important fact is the Law of Charge Conservation. We know moreover that charge is *quantized*, i.e., the total charge on a body is an *integral multiple* of a certain unit of charge.

This unit of charge is that which appears on the electron, which elementary particle is, in general, the mobile particle whose mass motion constitutes the "subtile electrical fluid" of the earlier days.

All this is qualitative and descriptive, however. The science of electrostatics begins with a quantitative version of the above facts. This is embodied in Coulomb's law which is described below. But first we discuss the question of how charge may be distributed in space, and how its distribution is described mathematically.

2.1 THE DISTRIBUTION OF CHARGE

Charge can manifest itself in either *discrete* or *continuous* distributions; here we discuss the relation between these two concepts. Historically the basic charge element was considered to be the *point charge*, with vanishingly small dimensions. Starting from this, one conceives of a *continuous distribution* of charge as one in which very many point charges are distributed in close proximity so that on some spatial or temporal average they are distributed continuously. In much the same way, one often describes a gas or liquid by a smooth mass–density function even though it is known that on an ultimate small scale it is grainy and molecular. For many purposes the graininess is not important, however, and equally, a complex of point charges can be considered for many purposes as a continuous smooth charge density. This may be so, for example, if the observation is made at distances which are large compared to the scale of graininess.

Quantum mechanics on the other hand, yields an opposite point of view; this discipline teaches that it is more natural to think of charge as distributed continuously. Thus in an atom, molecule, or solid, the charge is conceived as being spread out in a continuous smooth cloud, whose density is governed by quantum-mechanical laws. The idea of a continuous charge distribution is then basic. This does not, however, destroy the usefulness of the concept of a point charge. For we shall see later that any charge distribution of one sign whose linear dimensions are all small compared with the distance at which its properties (field and potential) are measured will act *like* a point charge. The precise meaning of this statement will emerge when we discuss *multipoles*. For the moment, we simply say that the properties of such a distribution are given by an expansion whose first term is the point charge term, and whose succeeding terms diminish as powers of the reciprocal of the distance to the point of observation. But before doing that, we should emphasize that a point charge is not one which is small in any absolute sense; there is no such absolute sense! A point charge is one that is *relatively* small, and the scale is set by the distance to the point of observation. Thus the sun, if charged, would be to a good approximation a point charge from the earth as observation point, whereas the proton is not necessarily a point charge if it is probed at distances of the order 10^{-13} cm or less.

We shall have to deal with some charge distributions that are relatively diffuse, and with others that are very sharply localized and concentrated. The various laws and formulas of electrostatics frequently apply equally well to either type of distribution. It is very convenient to have a single mathematical description that is applicable to either case. We shall accomplish this by introducing a *density function* $\rho(x, y, z)$, where x, y, z are space coordinates (not necessarily rectangular ones). We shall often write this more briefly as $\rho(\boldsymbol{r})$, where $\boldsymbol{r}$ is the position vector associated with the point x, y, z. Then for the *continuous* case, the quantity of charge in an infinitesimal volume element dv is defined as

$$\text{Quantity of charge in } dv = \rho(\boldsymbol{r})\, dv. \tag{2.1}$$

Since charge in the Gaussian system is measured in electrostatic units (esu or statcoulombs), ρ has dimensions esu $\times$ (cm)$^{-3}$. A point charge is included in this general description by letting $\rho(\boldsymbol{r})$ be a *δ-function.* This kind of "function" is discussed in Appendix B; briefly, it can be considered as the limit of a sequence of functions which are increasingly singular, but which always have unit integral. In terms of the δ-function, a *point charge* q at $\boldsymbol{r}_1$ is described by

$$\rho(\boldsymbol{r}) = q\delta(\boldsymbol{r} - \boldsymbol{r}_1),$$

and more than one point charge, with charge q_i at $\boldsymbol{r}_i$, by

$$\rho(\boldsymbol{r}) = \sum_i q_i\delta(\boldsymbol{r} - \boldsymbol{r}_i). \tag{2.2}$$

2.2 COULOMB'S LAW AND SUPERPOSITION

There are two fundamental laws of electrostatics. The first is Coulomb's law, which describes the force between *two* point charges; the second is the law of superposition, which extends this to more than two charges.

For Coulomb's law, consider two point charges of magnitudes q and q', a distance r apart. Then the law states that the strength F of the force between them is

$$F = C\frac{qq'}{r^2},$$

where C is a constant of proportionality that depends on the system of units. The direction of the force is along the line joining the two charges and is a repulsion if the charges are like in sign, an attraction if unlike. Both aspects of this law can be stated in vector notation. We refer to Fig. 2.1 in which $\boldsymbol{r}$ is the vector (magnitude r) from the charge q' to the charge q and $\hat{\boldsymbol{r}}$ is the unit vector* in that direction. In cgs units C is unity, and the vector force on q due to q' can be written

$$\boldsymbol{F} = \frac{qq'}{r^2}\hat{\boldsymbol{r}}. \tag{2.3}$$

---

* We shall generally use the caret (ˆ) over a given position vector to denote the corresponding *unit* vector.

Fig. 2.1 Illustration for Coulomb's law.

The second law of electrostatics is that of *superposition*; it has to do with the force exerted on a given charge by a distribution of charges. The law is this: the force on the given charge due to the distribution is the *vector sum of the forces that each charge of the distribution would produce, if it were alone.* In other words, Coulomb's law between any two point charges is not affected by the presence of other point charges. This may sound banal to the point of being trivial, but it is far from that; without superposition, electrostatics would be a quite different (and more difficult) subject.

2.3 THE ELECTRIC FIELD

Although Coulomb's law and superposition are the theoretical bases of electrostatics, it is useful for both practical and conceptual reasons to introduce two secondary ideas, those of electric field and potential. As we indicated in Chapter 1, the concept of field is not essential in electrostatics, but it almost inevitably enters later in the discussion of time-varying phenomena; it then makes for a nicer continuity to deal with it from the beginning. The concept of potential is likewise not so basic as Coulomb's law, but it has many virtues, not the least of which is that it enables one to reduce vectorial problems to scalar ones.

We begin with electric field. From Fig. 2.1, the electric field $\boldsymbol{E}$ at the position of the charge q is defined as the vectorial force exerted on that charge, divided by the magnitude q,

$$\boldsymbol{E} = \frac{\boldsymbol{F}}{q}. \tag{2.4}$$

Comparing Eqs. (2.3) and (2.4), we see that for this case $\boldsymbol{E}$ is also defined by

$$\boldsymbol{E} = \frac{q'}{r^2}\hat{\boldsymbol{r}}. \tag{2.5}$$

The split of Coulomb's law, (2.3), into the two equations (2.4) and (2.5) is, as we have said, somewhat arbitrary. Nonetheless, these equations do point up an essential aspect of the field concept: the field is *produced* (Eq. 2.5) by the charge q' and then *acts* to exert a force (Eq. 2.4) on the charge q. It follows from the law of superposition for forces that the field $\boldsymbol{E}$ obeys the superposition law as well. For an arbitrary distribution of charge, the field is then given by vectorially summing the field due to all the charge elements. For example, for point charges $q'_1, \ldots, q'_N$, the field at a point is, with $\boldsymbol{r}_i$ the vector from q'_i to the field point,

$$\boldsymbol{E} = \sum_{i=1}^{N} \frac{q'_i \hat{\boldsymbol{r}}_i}{r_i^2}, \tag{2.6}$$

Point
charge

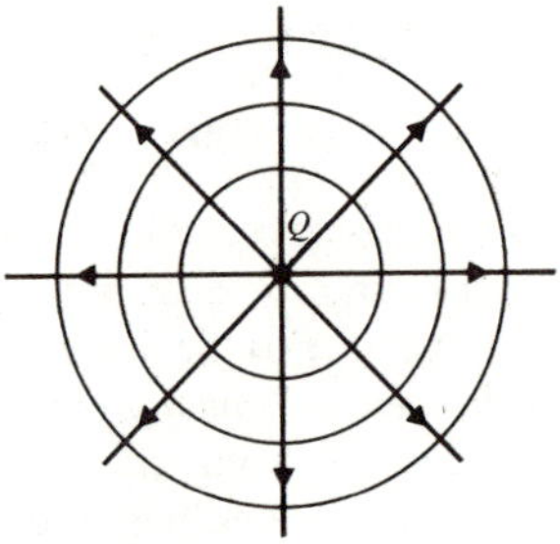

Two point
charges
Q and $-Q$

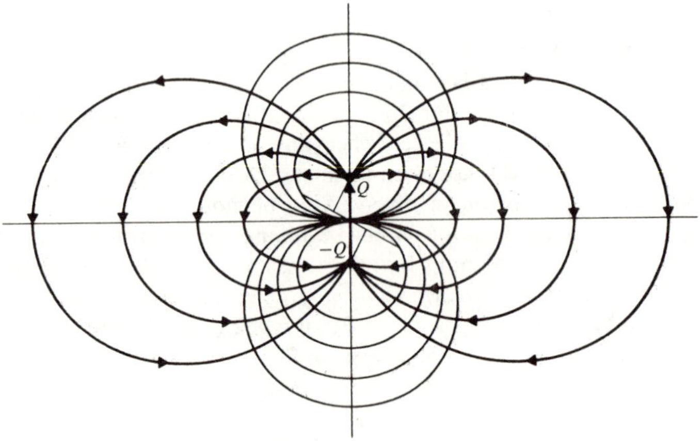

Uniformly charged
circular disk,
in plane containing
disk axis

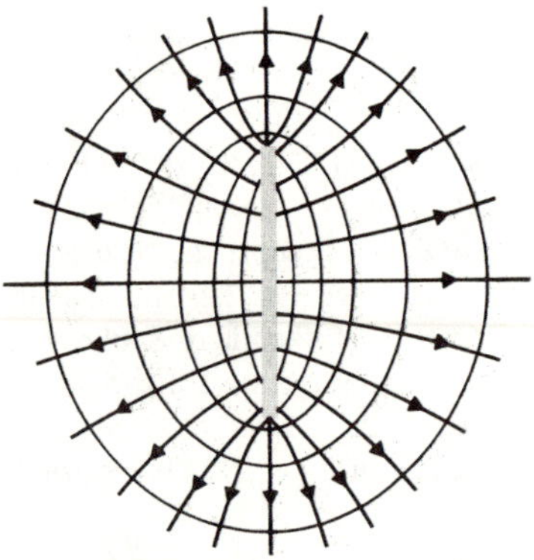

Fig. 2.2 Lines of force and equipotential surfaces for some charge distributions.

Two point charges Q and Q

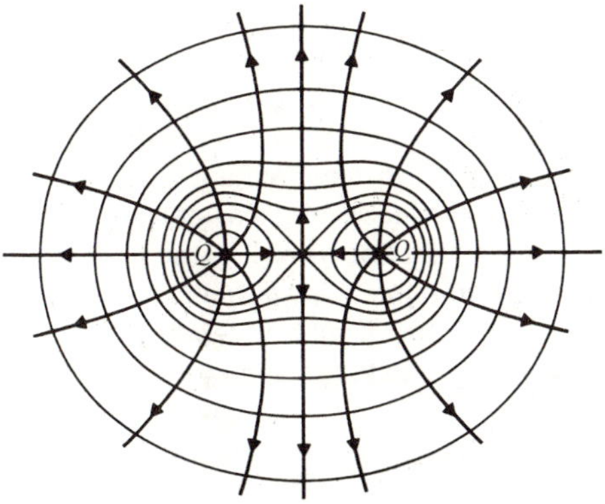

Point dipole

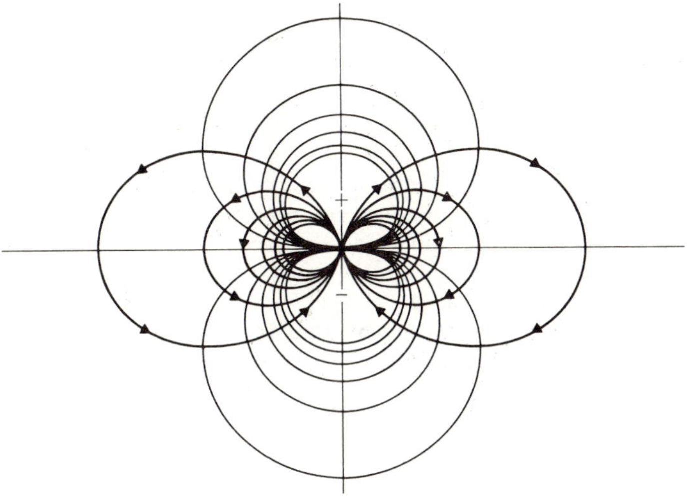

Line charge

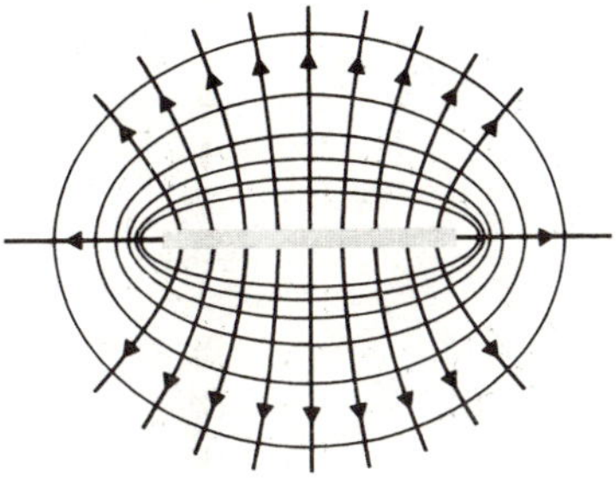

Fig. 2.2 *continued*

and if the charge distribution is continuous, there is an analogous integral formula. From (2.4) we see that the electric field is alternatively defined as the *force on a unit positive test charge* at a point. It is assumed, of course, that the presence of the test charge does not perturb the charge distribution whose field it measures.

The electric field is frequently characterized pictorially by its *lines of force*. These are closed lines whose directed tangent at any point of space gives the direction of the field at that point. Since the field can be defined as the force on a unit positive charge, it points *away from* the immediate vicinity of *positive* charges and *toward* the immediate vicinity of *negative* ones. One says that the field lines start on positive charges and stop on negative ones. Figure 2.2 gives examples of the lines of force produced by various charge distributions.

The electric field of any charge distribution is a vector field defined at all points in space. What are its general properties? Now it is known from vector analysis that a vector field is uniquely determined if its divergence and curl are known throughout space; this is sometimes called *Helmholtz' Theorem* and is discussed in Appendix C. We might then begin to answer this question by investigating the divergence and curl of $\boldsymbol{E}$. If we do this, we ask about the *differential properties* of the field. Alternatively, we might look at *integral properties*, i.e., properties the field has when it is integrated in some specified way. The differential and integral properties are, of course, not independent; the one can be derived from the other, and we can then start with whichever is most convenient.

We shall begin with an integral property that is known as *Gauss' Theorem*. To derive it, consider a point charge q enclosed in some volume bounded by a surface as shown in Fig. 2.3. We want to consider the total *flux* of $\boldsymbol{E}$ through this surface. Recall that the flux of $\boldsymbol{E}$ through any small surface element is defined as the product of the normal component of $\boldsymbol{E}$ times the area of the element. If we let $d\boldsymbol{s}$ be an outward pointing vector element of surface, the flux of $\boldsymbol{E}$ through the whole surface S is $\int_S \boldsymbol{E} \cdot d\boldsymbol{s}$. Using the expression (2.5) for $\boldsymbol{E}$ and recognizing that the element of solid angle $d\Omega$ is $\hat{\boldsymbol{r}} \cdot d\boldsymbol{s}/r^2$, we have

$$\int_S \boldsymbol{E} \cdot d\boldsymbol{s} = q \int_S \frac{\hat{\boldsymbol{r}} \cdot d\boldsymbol{s}}{r^2} = q \int_S d\Omega = 4\pi q.$$

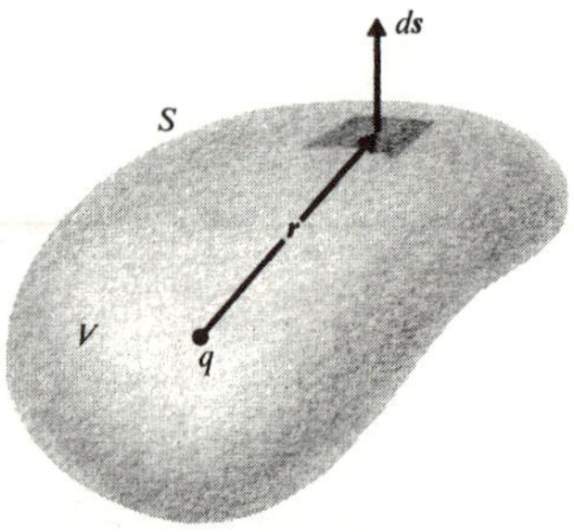

Fig. 2.3 Gauss' theorem.

This result for a single point charge in the volume is obviously independent of where in the volume the point charge is. It follows then from the law of superposition that for a *distribution* of charges in the volume a similar result would hold except that instead of q on the right-hand side there would appear the *total* charge, call it Q, in the volume. We can then write Gauss' Theorem as

$$\int_S \boldsymbol{E} \cdot d\boldsymbol{s} = 4\pi Q. \tag{2.7}$$

If one does a calculation analogous to the above for the case that the point charge is *outside* the surface, then for the simple surface shown, a line of force will enter the surface at one point and exit on the opposite side. Remembering that $d\boldsymbol{s}$ is the *outwardly* drawn vector surface element, we readily see that the net entering flux of a narrow cone of lines of force will just cancel the corresponding exiting flux on the opposite surface. Explicitly we may say what is implied by Eq. (2.7): the net flux through a closed surface that does not enclose any charge is zero.

To get the differential counterpart of Eq. (2.7), we can proceed here in either of two ways, and it is instructive to mention both of them. The first relies on the divergence theorem which is discussed in texts on vector analysis and stated in Appendix C. For any vector field $\boldsymbol{A}(\boldsymbol{r})$ which is defined through a volume V, that is bounded by a simply-connected surface S, as in Fig. 2.3, the divergence theorem is the following relation between integrals:

$$\int_S \boldsymbol{A} \cdot d\boldsymbol{s} = \int_V (\nabla \cdot \boldsymbol{A})\, dv. \tag{2.8}$$

Now imagine a continuous distribution of charge with a density $\rho(\boldsymbol{r})$. Apply Gauss' Theorem to an *arbitrary volume* V_0 in it. Then the Q that appears on the right-hand side of (2.7) can be written as $Q = \int_{V_0} \rho\, dv$, and the left-hand side can be rewritten via the divergence theorem to yield

$$\int_{V_0} (\nabla \cdot \boldsymbol{E} - 4\pi\rho)\, dv = 0.$$

This equation can hold for an *arbitrary* volume only if the integrand vanishes, and so we have

$$\nabla \cdot \boldsymbol{E} = 4\pi\rho. \tag{2.9}$$

This is the differential expression of Gauss' Law. Another way of getting the same result is to apply Eq. (2.7) to an infinitesimal volume ΔV with bounding surface ΔS. Then the right-hand side of Eq. (2.7) can be approximated by $4\pi\bar{\rho}\,\Delta V$, where $\bar{\rho}$ is the average value of ρ through the volume. Equation (2.7) then reads

$$\frac{1}{\Delta V}\int_{\Delta S} \boldsymbol{E} \cdot d\boldsymbol{s} = 4\pi\bar{\rho}.$$

In the limit that the volume ΔV shrinks to a point, the mean value $\bar{\rho}$ simply becomes ρ, the value at that point. Moreover, the left-hand side of the last equation then *defines* $\nabla \cdot \boldsymbol{E}$ at that point, and so we again get Eq. (2.9).

So much for $\nabla \cdot \boldsymbol{E}$. To finish characterizing $\boldsymbol{E}$ as a vector field we now calculate $\nabla \times \boldsymbol{E}$. Here it will be easier to begin with the curl considered explicitly as a differential operator. Although we want to calculate $\nabla \times \boldsymbol{E}$ for the field of an *arbitrary* charge distribution, we shall begin by calculating for a single point charge and later invoking superposition. Without loss of generality we can imagine that this point charge q is at the origin of a system of coordinates and we call its field $\boldsymbol{E}_q$. Then Eq. (2.5) gives for the components E_{qx}, etc.,

$$E_{qx} = \frac{qx}{r^3}, \qquad E_{qy} = \frac{qy}{r^3}, \qquad E_{qz} = \frac{qz}{r^3}.$$

In rectangular coordinates, the x-component of $\nabla \times \boldsymbol{E}_q$ is

$$(\nabla \times \boldsymbol{E}_q)_x = \frac{3qyz}{r^5} - \frac{3qzy}{r^5} = 0.$$

With a similar result for the other components we have for this example

$$\nabla \times \boldsymbol{E}_q = 0.$$

This result is rather more general than might appear at first sight. For, given an *arbitrary* charge distribution, then at any point where the charge density is zero, the field can be considered as the superposition of the fields due to more or less distant charge elements. From the above result, each of these elements produces a field whose curl is zero; the equation $\nabla \times \boldsymbol{E} = 0$ is thus valid at any point "outside" an arbitrary charge distribution.

For a point inside the distribution, where ρ is finite, we can use an argument whose general form we shall have occasion to repeat several times later in this book in other contexts. At such a point we imagine a small sphere, centered at the point. The field there is then the superposition of the fields due to those charges inside the sphere *and* to those charges outside. These latter, as we have just seen, produce a field whose curl is zero. We then consider the field at some point in the sphere *due to the charge in the sphere itself*, and for this we must be more explicit about its size. If the function $\rho(\boldsymbol{r})$ is continuous, which is assumed, we can always take the radius of the sphere small enough that the charge density inside is essentially constant to whatever accuracy we may wish to assign. Call that radius r_0, and the density ρ_0. We then have the problem of calculating the field inside a uniformly charged sphere. This field will be radial, by symmetry, and it can be calculated by Gauss' Theorem. Thus the flux of $\boldsymbol{E}$ through some spherical surface of radius $r < r_0$ is just $4\pi r^2 E$ and the total charge within the surface is $\rho_0 \cdot \frac{4}{3}\pi r^3$. From Eq. (2.7)

$$\boldsymbol{E} = \frac{4\pi\rho_0 \boldsymbol{r}}{3}.$$

The curl of this field is zero everywhere for $r < r_0$ and in particular at $r = 0$. Thus the charges outside the sphere *and* those inside produce fields whose curl is zero. We can therefore state that everywhere *inside or outside of nonsingular charge distributions,*

$$\nabla \times \boldsymbol{E} = 0.$$

An important corollary of the above discussion is that the field is finite inside or outside any continuous charge distribution for which ρ is finite everywhere.

From this differential characterization of $\nabla \times \boldsymbol{E}$ we get the integral counterpart by invoking Stokes' Theorem. This theorem is discussed in Appendix C. In brief, it states that given a closed curve C (element of length $d\boldsymbol{l}$) spanned by a surface S (element of area $d\boldsymbol{s}$), then for a vector field $\boldsymbol{A}(\boldsymbol{r})$ one has, with a sign convention discussed in Appendix C:

$$\int_S (\nabla \times \boldsymbol{A}) \cdot d\boldsymbol{s} = \oint_C \boldsymbol{A} \cdot d\boldsymbol{l}.$$

If we apply this result to the vector field $\boldsymbol{E}$, we have for any *closed path*, signified by $\oint$,

$$\oint \boldsymbol{E} \cdot d\boldsymbol{l} = 0. \tag{2.11}$$

2.4 THE POTENTIAL

The fact that the curl of $\boldsymbol{E}$ is zero enables a simplification. For we know from vector analysis* that this implies that $\boldsymbol{E}$ can be calculated as the gradient of a single scalar, the *potential function*, which we call Φ:

$$\boldsymbol{E} = -\nabla\Phi. \tag{2.12}$$

One virtue of Φ is that it reduces a vectorial problem to a scalar one. Instead of calculating each component of $\boldsymbol{E}$ separately, we can calculate the single function Φ and then use (2.12). Also, Φ has a useful physical interpretation: the work done in taking a unit charge between two points can be simply expressed in terms of it. We will discuss this now.

Note first that the work the field $\boldsymbol{E}$ does in moving a *unit* positive charge from one point to another along some infinitesimal length $d\boldsymbol{l}$ is

$$dW = \boldsymbol{E} \cdot d\boldsymbol{l}.$$

The work done in moving the particle a distance $d\boldsymbol{l}$ *against* the force $\boldsymbol{E}$ is just the negative of this. The total work W_{12} required to move a particle a finite distance between two points 1 and 2 in a field $\boldsymbol{E}$ is then

$$W_{12} = -\int_1^2 \boldsymbol{E} \cdot d\boldsymbol{l}. \tag{2.13}$$

---

* Vector fields whose curl is zero everywhere are called conservative fields.

Using Eq. (2.12), we have

$$W_{12} = \int_1^2 \nabla\Phi \cdot d\boldsymbol{l} = \Phi_2 - \Phi_1. \tag{2.14}$$

Thus the difference of potential between two points is the work that must be done to move a unit charge between those points. Equations (2.13) and (2.14) also show that the line integral of the tangential component of $\boldsymbol{E}$ along any path joining two points is independent of the path and depends only on the endpoints. This is an important property of conservative fields. The equation, $\Phi(\boldsymbol{r}) =$ constant, defines a surface called an *equipotential surface*; such surfaces play an important role in electrostatics. Some equipotential surfaces for simple charge distributions are sketched in Fig. 2.2.

In the spirit of the discussion of $\boldsymbol{E}$, consider the properties of Φ. The differential property is obtained by combining (2.9) and (2.12), and is

$$\nabla \cdot \nabla\Phi \equiv \nabla^2\Phi = -4\pi\rho. \tag{2.15}$$

This is *Poisson's equation.* In a region where ρ is zero it becomes *Laplace's equation*

$$\nabla^2\Phi = 0. \tag{2.16}$$

We can get the integral counterpart of Poisson's equation from the potential for a point charge and the principle of superposition. To do this we first observe that the potential which a single point charge produces at a given distance is the magnitude of the charge divided by that distance. The potential Φ_q of a point charge of magnitude q' at the origin should then be $\Phi_q = q'/r$. This is easily confirmed. Taking the negative gradient of Φ_q, we are led to the expression (2.5) for $\boldsymbol{E}$. Now suppose we have a charge distribution $\rho(\boldsymbol{r}')$ as shown in Fig. 2.4. (It is described in terms of a primed variable to save $\boldsymbol{r}$ for the vector to the field point.) The small element of charge $\rho(\boldsymbol{r}')\,dv'$ produces at $\boldsymbol{r}$ a contribution to the potential which equals the magnitude of this charge element divided by its distance from the

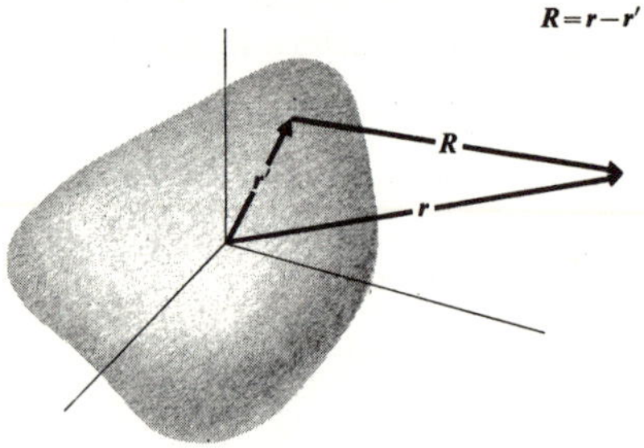

Fig. 2.4 Coordinate system for calculating Φ at the field point $\boldsymbol{r}$ due to the charge distribution $\rho(\boldsymbol{r}')$.

field point. This contribution is $\rho\, dv'/R = \rho\, dv'/|\boldsymbol{r} - \boldsymbol{r}'|$. The total potential at $\boldsymbol{r}$ is, by superposition,

$$\Phi(\boldsymbol{r}) = \int \frac{\rho(\boldsymbol{r}')\, dv'}{|\boldsymbol{r} - \boldsymbol{r}'|}. \tag{2.17}$$

This is a very important equation, since it permits us to calculate Φ, and hence $\boldsymbol{E}$, for an arbitrary charge distribution, continuous or discrete. The function $1/|\boldsymbol{r} - \boldsymbol{r}'|$ that appears in (2.17) is sometimes called a *unit source* or *Green function.** It is called a unit source function because it gives the effect at $\boldsymbol{r}$ due to a unit source at $\boldsymbol{r}'$, and is called a "Green function" for historical reasons that we shall see later. It is one of the simplest examples of a large class of Green functions that we shall be concerned with.

In summary, the differential and integral properties of $\boldsymbol{E}$ and Φ that have been found are:

Differential	Integral
	$\boldsymbol{E}$
$\nabla \cdot \boldsymbol{E} = 4\pi\rho$	$\int \boldsymbol{E} \cdot d\boldsymbol{s} = 4\pi Q$
$\nabla \times \boldsymbol{E} = 0$	$\oint \boldsymbol{E} \cdot d\boldsymbol{l} = 0$
	Φ
$\nabla^2\Phi = -4\pi\rho$	$\Phi = \int \frac{\rho(\boldsymbol{r}')\, dv'}{|\boldsymbol{r} - \boldsymbol{r}'|}$

PROBLEMS

1. Use Gauss' Theorem to calculate the field inside and outside an infinitely long cylinder of circular cross section which contains a uniform density of charge. Can you calculate the field similarly for a cylinder of finite length?
2. Find the distribution of charge giving rise to an electric field whose potential is

$$\Phi(x, y) = 2\left(\tan^{-1}\frac{1 + x}{y} + \tan^{-1}\frac{1 - x}{y}\right),$$

where x and y are Cartesian coordinates. Such a distribution is called a *two-dimensional* one since it does not depend on the third coordinate z.

* There are various definitions of Green functions that differ by constant factors. Sometimes this Green function is taken to be $-1/4\pi|\boldsymbol{r} - \boldsymbol{r}'|$.

3. The potential of a certain charge distribution is

$$\Phi = q\frac{e^{-\alpha r}}{r}\left(1 + \frac{\alpha r}{2}\right),$$

where r is a radial coordinate and q and α are parameters. Find the distribution of charge, continuous *and* discrete, that would produce such a potential.

4. Suppose the force F between two like particles is governed by a "modified" Coulomb's law

$$\boldsymbol{F} = K\frac{e^{-r/a}}{r}\left(\frac{1}{r} + \frac{1}{a}\right)\hat{\boldsymbol{r}}.$$

If a potential exists for the force field $F(\boldsymbol{r})$, find it. Investigate $\nabla \cdot \boldsymbol{F}$ and $\nabla \times \boldsymbol{F}$.

5. The potential of a certain idealized charge distribution in spherical coordinates r, θ, is

$$\Phi = m\cos\theta/r^2.$$

Find the components E_r and E_θ of the field, and sketch the lines of force.

REFERENCES

Canby, Edward Tatnall, *A History of Electricity*, Hawthorn Books, Inc., New York, 1963.
Semipopular, elementary, but accurate, and the illustrations are delightful.

Harper, W. R., "How Do Solid Surfaces Become Charged?" *Static Electrification: 1967 Conference Proceedings*, The Institute of Physics, London, 1967.

Roller, Duane and Roller, Duane H. D., *The Development of the Concept of the Electric Charge*, Harvard University Press, Cambridge, Mass., 1954.
Scholarly, and devoted to a rather specialized topic, but one which has many ramifications.

3
THE SUMMATION PROBLEM FOR CHARGES

In this chapter we turn to the problem of evaluating the expression (2.17) for $\Phi(\boldsymbol{r})$ assuming $\rho(\boldsymbol{r}')$ is given. Following Sommerfeld, we call this the *summation problem.* By this is meant the problem of explicitly working out, in a convenient and practical way, the integral or sum in (2.17). This kind of problem is to be contrasted with the so-called *boundary value problem* that we shall treat later; this latter type is much more difficult, essentially because $\rho(\boldsymbol{r}')$ is not known in advance.

To keep perspective here, we remark that although (2.17) represents a general solution to the problem of calculating the potential, and hence the field, of any distribution, there may be other, more convenient, methods of solution. For example, it may sometimes be easier to use directly Poisson's equation, the differential counterpart of (2.17), than to try to evaluate the integral. Or, with charge configurations of considerable symmetry, Gauss' Law may give a quicker and easier result. And for certain idealized distributions that extend to infinity (two-dimensional or one-dimensional distributions), there may be convergence problems with the integral. But with these minor qualifications, Eq. (2.17) stands as a most useful and general expression.

We shall discuss several kinds of distribution, without being able to treat all the interesting distributions that occur in the various branches of physics. As a small preview, we mention some examples. A charge may be paired with an equal and opposite one at a small distance, forming in the limit of zero distance a so-called *point dipole.* A point dipole may then be paired with another of the opposite sense to produce a *point quadrupole,* which may be paired to form *point octupoles*; the complete generalization of such concepts is embodied in an important general theory of the *multipole expansion.* Charge may be distributed on or throughout long slender cylinders in *two-dimensional distribution,* where quantities depend on only *two* space coordinates, or in *one-dimensional distributions* which depend only on a *single* coordinate. Charge may be distributed in thin *surface layers.* As for point dipoles, two closely spaced surface layers of opposite sign may be paired to form a *double* or *dipole layer.* The charge distribution may have special *symmetries in space* as it does for atoms and molecules, or it may be arranged in *periodic lattices* as it is in crystals. Point dipoles may be arranged periodically to form a *dipolar lattice* important in the theory of dielectrics. And there are many other possibilities which we shall not have space to discuss.

3.1 POINT CHARGE ON z-AXIS

The next simplest distribution after a single point charge at the origin is a point charge not at the origin. We shall consider this in some detail since this seemingly simple problem contains many features applicable to the more general charge distributions that may figure in Eq. (2.17). We shall move in two steps: first, considering a charge on the z-axis, and then considering one at some general point of space.

Assume that a charge q is at $z = a$ on the z-axis. We want to find the potential at some field point with spherical coordinates r, θ as shown in Fig. 3.1.

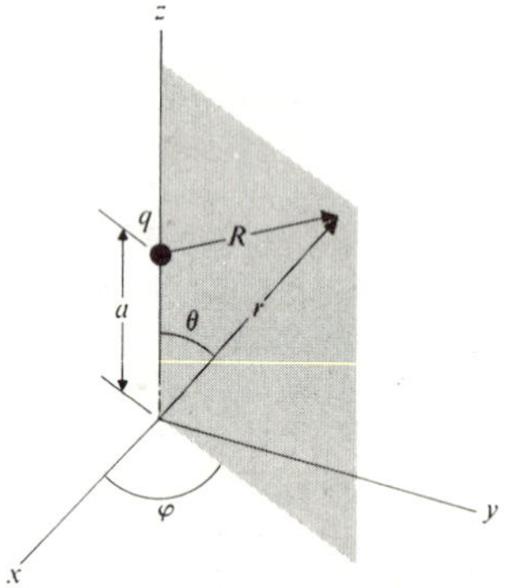

Fig. 3.1 Potential of a charge on the z-axis.

The potential at r, θ is

$$\Phi(r, \theta) = \frac{q}{R} = \frac{q}{(r^2 + a^2 - 2ar\cos\theta)^{1/2}}. \tag{3.1}$$

The dependence on r and θ in (3.1) can be put into a form more convenient for many purposes by expanding this potential in a certain set of functions $P_l(\cos\theta)$, $l = 0, 1, 2, \ldots$ called *Legendre polynomials.* This is a set which arises frequently in other contexts in mathematics and physics. Its properties are set out in some detail in Appendix D, where its orthogonality and completeness are discussed. Here we do not need these properties and shall be brief.

First consider $1/\sqrt{r^2 + a^2 - 2ar\cos\theta}$ for $r > a$. Writing it as

$$1/(r\sqrt{1 + \varepsilon}),$$

where

$$\varepsilon = \left(\frac{a^2}{r^2} - \frac{2a}{r}\cos\theta\right)$$

is less than unity, we can expand $1/\sqrt{1 + \varepsilon}$ by the binomial theorem to get for the first few terms

$$\frac{1}{(r^2 + a^2 - 2ar\cos\theta)^{1/2}} = 1 - \frac{1}{2}\left(\frac{a^2}{r^2} - \frac{2a}{r}\cos\theta\right) + \frac{3}{8}\left(\frac{a^2}{r^2} - \frac{2a}{r}\cos\theta\right)^2 + \cdots.$$

Rearranging the right-hand side in ascending powers of a/r we get initially

$$1 + \left(\frac{a}{r}\right)\cos\theta + \left(\frac{a}{r}\right)^2\left(\frac{3\cos^2\theta - 1}{2}\right) + \cdots. \tag{3.2}$$

The coefficient of $(a/r)^l$ in such an expansion is defined to be the Legendre polynomial P_l. Thus

$$P_0(\cos\theta) = 1,\ P_1(\cos\theta) = \cos\theta,\ P_2(\cos\theta) = \tfrac{1}{2}(3\cos^2\theta - 1). \tag{3.3}$$

Expressions for the polynomials of higher order are given in Appendix D. With these defined, we can write

$$\frac{1}{(r^2 + a^2 - 2ar\cos\theta)^{1/2}} = \frac{1}{r}\sum_{l=0}^{\infty}\left(\frac{a}{r}\right)^l P_l(\cos\theta), \qquad r > a. \tag{3.4}$$

Then the expression (3.1) for the potential Φ can be written compactly as

$$\Phi(r, \theta) = q/r \sum_{l=0}^{\infty} (a/r)^l P_l(\cos\theta), \tag{3.5}$$

and this is the general expansion we want.

Two points about this equation: First, it is an expansion in powers of a/r so that if $a/r \ll 1$, only the first term of the expansion is important, and the field of the point charge away from the origin is, not surprisingly, approximately the same as that of a point charge *at* the origin. For r of the order of a, however, Φ is a quite complicated function, for which the series (3.5) converges slowly, and the angular dependence it yields is not simple. Second, although in (3.5) we have expanded in powers of a/r, the original formula (3.1) is symmetric to interchange of a and r. If we had assumed $r/a < 1$, we could equally well have expanded in powers of r/a and obtained a formula like (3.5) but with r and a interchanged. Again for r/a small, the field would be relatively simple with an uncomplicated angular dependence that would become increasingly complicated as r/a approached unity.

3.2 AXIAL MULTIPOLE EXPANSION

With the above result, we can treat a more general, if not the most general, charge distribution. Consider in particular a so-called *axial distribution* of charge, that is, one in which charge is concentrated along the z-axis. We can think of this as a special case of the general distribution $\rho(\boldsymbol{r}')$ for which

$$\rho(\boldsymbol{r}') = \delta(x')\,\delta(y')\,\sigma(z').$$

Of course, $\sigma(z')$ may represent either a continuous, or by means of δ-functions, a discrete distribution. Since the quantity of charge in dz' is $\sigma(z')\,dz'$, the potential $\Phi(r, \theta)$ is, by superposition,

$$\Phi(r, \theta) = \int \frac{\sigma(z')\,dz'}{\sqrt{r^2 + z'^2 - 2rz'\cos\theta}}. \tag{3.6}$$

Now if $\sigma(z')$ represents a distribution of finite length, i.e., if $\sigma(z') = 0$ for $|z'| > z_0$, we can expand the square root in the denominator of Eq. (3.6) using Eq. (3.4) to get, for $r > z_0$,

$$\Phi(r, \theta) = \int \frac{\sigma(z')}{r} \sum_{l=0}^{\infty} \left(\frac{z'}{r}\right)^l P_l(\cos\theta)\,dz' = \sum_{l=0}^{\infty} \frac{M_l P_l(\cos\theta)}{r^{l+1}}, \tag{3.7}$$

where the coefficients M_l are

$$M_l = \int \sigma(z') z'^l\,dz'. \tag{3.8}$$

These quantities are called the *axial multipole moments* of the system, and the potential for $r > z_0$ is determined by them. We note that M_0, the *monopole* moment, is the total charge in the distribution; M_1 is called the *dipole* moment, M_2 the *quadrupole* moment, etc. Any of these moments may be zero, according to the nature of the distribution $\sigma(z')$, and the potential at large distances then drops off with a power of $1/r$ determined by the first moment that does not vanish.

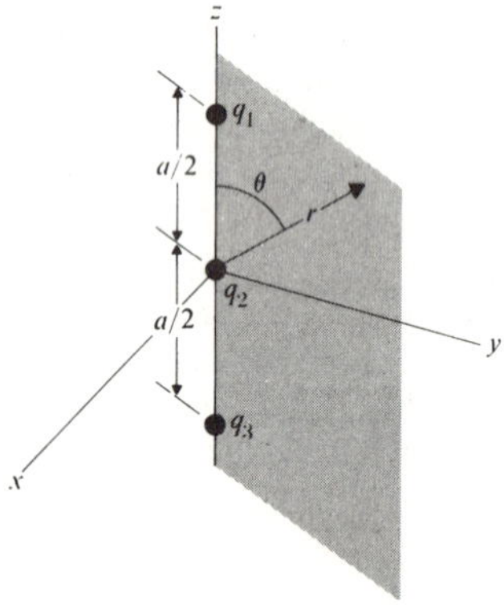

Fig. 3.2 Axial distribution of point charges.

We illustrate the above with two simple examples—one for discrete and one for continuous charge. Suppose first that there are charges as shown in Fig. 3.2. Then $\sigma(z')$ is

$$\sigma(z') = q_1\delta\left(z' - \frac{a}{2}\right) + q_2\delta(z') + q_3\delta\left(z' + \frac{a}{2}\right),$$

and with this we easily find

$$M_0 = q_1 + q_2 + q_3,$$

$$M_l = \begin{cases} \left(\dfrac{a}{2}\right)^l (q_1 - q_3), & l = 1, 3, 5, \dots \\[2ex] \left(\dfrac{a}{2}\right)^l (q_1 + q_3), & l = 2, 4, 6, \dots . \end{cases}$$

Consider some special cases of (3.9). First note that if M_0 is nonzero, the potential falls off at large distances essentially as $1/r$, with correction terms in higher powers of $1/r$. Now suppose that M_0 is zero. For example, take the case $q_1 = -q_3 = q, q_2 = 0$, which corresponds to a plus and a minus charge a distance a apart. Then the even-numbered moments vanish and only the terms with l odd survive in (3.7), which in fact becomes

$$\Phi(r, \theta) = qa \frac{P_1(\cos \theta)}{r^2} + \frac{qa^3}{4} \frac{P_3(\cos \theta)}{r^4} + \cdots . \tag{3.10}$$

Now the first term in (3.10), the *dipole term*, will dominate the expansion for $r \gg a$ but in principle all the other higher terms are always present to some degree. We can, however, define a useful idealized distribution, the *point dipole*, whose potential is given *everywhere* by the first term only in (3.10). To do this we let $a \to 0$ and $q \to \infty$ in such a way that qa remains finite and approaches the limit p:

$$qa \Big|_{\substack{q \to \infty \\ a \to 0}} = p. \tag{3.11}$$

Then all the terms in (3.10), except the first, will vanish. For example, M_3 becomes

$$\frac{qa^3}{4} \equiv \frac{qa \cdot a^2}{4} \to \frac{pa^2}{4} \to 0,$$

and similarly for higher terms. In this limit then we have generated a point dipole whose potential at *any* point in space is

$$\Phi_{\text{dipole}} = \frac{p \cos \theta}{r^2}. \tag{3.12}$$

Calculating the field $\boldsymbol{E} = -\nabla\Phi_{\text{dipole}}$, we find that the only two components E_r and E_θ are given by

$$E_r = \frac{2p \cos \theta}{r^3}, \qquad E_\theta = \frac{p \sin \theta}{r^3}. \tag{3.13}$$

The lines of force of the point dipole have already been sketched in Fig. 2.2.

As we have mentioned, it is quite possible that both the monopole *and* the dipole moments of a system vanish, in which case the behavior of the potential at large distances is determined by the order of the lowest nonvanishing moment. As an example, in Eq. (3.9) take $q_1 = q_3 = q$; $q_2 = -2q$. Then M_0 and all M_l with l odd vanish, and the potential is given by

$$\Phi = \frac{a^2q}{8r^5} P_2(\cos\theta) + \frac{a^4q}{8r^5} P_4(\cos\theta) + \cdots.$$

The first term is now the *quadrupole term*, and it dominates for large distances, subject to the corrections given by the further terms in the series. We can, as for the dipole, form an idealized or *point quadrupole* by imagining that $a \to 0$ and $q \to \infty$ in such a way that a^2q remains finite. Then a^4q and all the other higher multipole moments vanish and, in the limit, we get an entity whose potential is given everywhere by

$$\Phi_{\text{quad}} \propto P_2(\cos\theta)/r^3. \tag{3.14}$$

It goes without saying that we can in much the same way form point multipoles of higher order.

Now consider an example of a continuous distribution of charge. As a particularly simple one, take a finite length of uniform linear charge, for which

$$\sigma(z') = \begin{cases} q/L, & -L/2 < z' < L/2 \\ 0, & \text{otherwise.} \end{cases}$$

Then

$$M_l = \frac{q}{L}\int_{-L/2}^{L/2} z'^l\,dz' = \frac{qL^l(1 - (-)^{l+1})}{2^{l+1}(l+1)},$$

and only the even multipole moments are nonvanishing.

We can get some insight into the physical meaning of axial multipoles by treating the summation problem in rectangular coordinates. In (2.17) we designate the field point $\boldsymbol{r}$ by its coordinates x, y, z, and so have

$$\Phi(x, y, z) = \int \frac{\sigma(z')\,dz'}{\sqrt{x^2 + y^2 + (z - z')^2}} \tag{3.16}$$

Now the square root in the denominator is never zero since it is assumed that the field point is farther from the origin than any part of the charge distribution. We can then expand the function $1/(x^2 + y^2 + (z - z')^2)^{1/2}$ in a Taylor series in z', forgetting for a moment its dependence on x, y, z. We therefore define $f(z')$ by

$$f(z') = \frac{1}{\sqrt{x^2 + y^2 + (z - z')^2}}$$

with the Taylor series expansion

$$f(z') = f(0) + z' \frac{\partial f}{\partial z'}\bigg|_{z'=0} + \frac{z'^2}{2!}\frac{\partial^2 f}{\partial z'^2}\bigg|_{z'=0} + \cdots.$$

It is easy to verify that

$$\frac{\partial^n f}{\partial z'^n}\bigg|_{z'=0} = (-)^n \frac{\partial^n}{\partial z^n}\left(\frac{1}{r}\right).$$

Remembering the definition (3.7) for the multipole moments M_l, we find that the potential can be written as

$$\Phi = M_0\left(\frac{1}{r}\right) - M_1 \frac{\partial}{\partial z}\left(\frac{1}{r}\right) + \cdots + M_n \frac{(-)^n}{n!}\frac{\partial^n}{\partial z^n}\left(\frac{1}{r}\right) + \cdots. \tag{3.17}$$

The first term is, of course, just the monopole term we have found previously. Consider the second term, and recall that the definition of the derivative of a function $g(z)$ is the limit as Δ approaches zero of $(g(z + \Delta) - g(z)/\Delta$. Then

$$\frac{\partial}{\partial z}\left(\frac{1}{r}\right) = \lim_{\Delta\to 0}\left[\frac{1}{\Delta}\left(\frac{1}{(x^2 + y^2 + (z + \Delta)^2)^{1/2}} - \frac{1}{(x^2 + y^2 + z^2)^{1/2}}\right)\right].$$

The first term in the expression in square brackets can be interpreted as the potential of a charge of magnitude $1/\Delta$, at the point Δ on the z-axis; the second term, as the potential of an equal and opposite charge at the origin. The dipole term $(\partial/\partial z)(1/r)$ is the limit of this expression as Δ goes to zero. The higher order terms in (3.17) can be interpreted in a similar way. For example, in the quadrupole term

$$\frac{M_2}{2}\frac{\partial^2}{\partial z^2}\left(\frac{1}{r}\right) = \frac{M_2}{2}\frac{\partial}{\partial z}\left\{\frac{\partial}{\partial z}\left(\frac{1}{r}\right)\right\},$$

we can interpret the term in braces as essentially the dipole potential. The operation of differentiating this term corresponds to taking the difference between two opposite dipole terms a small distance Δ apart and going to the limit $\Delta \to 0$. In exactly the same way, multipoles of any order can be considered to be made up by taking two multipoles of one order lower of opposite sign and carrying out the limiting process analogous to the one above.

3.3 GENERAL MULTIPOLE EXPANSION

We now want to extend the theory of axial multipoles to one that applies for an arbitrary distribution of charge. To do this we shall, following the axial model, first find a convenient expansion for a *single* point charge arbitrarily situated, and then incorporate this expansion into the general theory. Suppose there is a point charge q at $\mathbf{r}'$ (coordinates r', θ', φ') for which we want to calculate the potential conveniently at the field point $\boldsymbol{r}$ (coordinates r, θ, φ). This potential is

$$\Phi(r, \theta, \varphi) = \frac{q}{|\boldsymbol{r} - \boldsymbol{r}'|} = \frac{q}{(r^2 + r'^2 - 2rr' \cos \gamma)^{1/2}}, \tag{3.18}$$

where γ is the angle between $\boldsymbol{r}$ and $\boldsymbol{r}'$. We can expand the square root according to Eq. (3.5) to get, for $r > r'$,

$$\Phi = \frac{q}{r} \sum_{l=0}^{\infty} \left(\frac{r'}{r}\right)^l P_l(\cos \gamma). \tag{3.19}$$

This is a more perspicuous form than (3.18), but the dependence of Φ on the angles θ, φ, θ', φ' is still buried in $P_l(\cos \gamma)$ since γ, and hence $P_l(\cos \gamma)$, is, of course, a function of these variables. For example, for $P_1(\cos \gamma) = \cos \gamma$ one can verify from trigonometry that

$$\cos \gamma = \cos \theta \cdot \cos \theta' + \sin \theta \cdot \sin \theta' \cos (\varphi - \varphi'). \tag{3.20}$$

In principle, we could, since $P_l(\cos \gamma)$ is simply a polynomial in $\cos \gamma$, get the result we want from Eq. (3.20) in terms of linear combinations of powers of $\cos \gamma$. This would be tedious and fortunately is unnecessary since the expansion we seek has been worked out systematically in terms of certain standardized functions that arise in many contexts. These are called *spherical harmonics*, and are discussed in Appendix D. Here we shall record only the properties important to the present purpose.

Spherical harmonics are defined in terms of functions called *associated Legendre polynomials* and written $P_l^m(x)$, for which l can be $0, 1, 2, \ldots$. For a given l, there are $(2l + 1)$ such functions corresponding to the different m values $-l$, $-l + 1, 0, \ldots, l - 1, l$. For m zero or positive, the $P_l^m(x)$ are defined in terms of the ordinary Legendre polynomials $P_l(x)$ by

$$P_l^m(x) = (-)^m (1 - x^2)^{m/2} \frac{d^m}{dx^m} P_l(x). \tag{3.21}$$

The functions for negative m are given in terms of those for positive m by

$$P_l^{-m}(x) = (-)^m \frac{(l - m)!}{(l + m)!} P_l^m(x).$$

With these, the spherical harmonics $Y_{lm}(\theta, \varphi)$ are defined for positive m as

$$Y_{lm}(\theta, \varphi) = \left(\frac{(2l + 1)}{4\pi} \frac{(l - m)!}{(l + m)!}\right)^{1/2} P_l^m(\cos \theta) e^{im\varphi}, \tag{3.22}$$

and for negative m by

$$Y_{l,-m} = (-)^m Y_{lm}^*. \tag{3.23}$$

Now we quote the answer to the original problem of expressing $P_l(\cos\gamma)$ in terms of $\theta', \varphi', \theta, \varphi$. When expressed in terms of spherical harmonics, it is called the *Addition Theorem* for spherical harmonics, and is an important formula. It reads:

$$P_l(\cos\gamma) = \frac{4\pi}{2l+1}\sum_{m=-l}^{l} Y_{lm}^*(\theta', \varphi')Y_{lm}(\theta, \varphi). \tag{3.24}$$

For $l = 0$, and $l = 1$, this is, of course, just a complicated way of writing respectively the result $P_0 = 1$ and the result given by (3.20) for $P_1(\cos\gamma)$.

With these spherical harmonic formulas, we turn to the problem of generating a multipole expansion for a general charge distribution. Such a distribution, with density function $\rho(\boldsymbol{r}')$, is sketched in Fig. 2.4; its potential $\Phi(\boldsymbol{r})$ is given by Eq. (2.17) as

$$\Phi(\boldsymbol{r}) = \int\frac{\rho(\boldsymbol{r}')\,dv'}{|\boldsymbol{r}-\boldsymbol{r}'|} = \int\frac{\rho(\boldsymbol{r}')\,dv'}{(r^2+r'^2-2rr'\cos\gamma)^{1/2}}. \tag{3.25}$$

Consider values of r larger than the maximum value that r' takes on, i.e., field points that are outside the smallest sphere which just circumscribes the distribution. Then the denominator of the integrand in Eq. (3.25) can be expanded much as was Eq. (3.18) to yield

$$\Phi(\boldsymbol{r}) = \int\rho(\boldsymbol{r}')\left(\frac{1}{r}\sum_{l=0}^{\infty}\left(\frac{r'}{r}\right)^l P_l(\cos\gamma)\right)dv'. \tag{3.26}$$

With the expression (3.24) for $P_l(\cos\gamma)$ we get†

$$\Phi(r, \theta, \varphi) = 4\pi\sum_{l,m}\frac{1}{2l+1}\left[\int\rho(\boldsymbol{r}')r'^l Y_{lm}^*(\theta', \varphi')\,dv'\right]\frac{Y_{lm}(\theta, \varphi)}{r^{l+1}}. \tag{3.27}$$

Denoting the integral in this expression by $M_{lm}/\sqrt{4\pi}$

$$M_{lm} = \sqrt{4\pi}\int\rho(\boldsymbol{r}')r'^l Y_{lm}^*(\theta', \varphi')\,dv' \tag{3.28}$$

we have the convenient expression for the potential:

$$\Phi(r, \theta, \varphi) = \sqrt{4\pi}\sum_{l,m}\frac{M_{lm}}{2l+1}\frac{Y_{lm}(\theta, \varphi)}{r^{l+1}} = \sum_{l=0}^{\infty}\Phi_l(r, \theta, \varphi), \tag{3.29}$$

where

$$\Phi_l(r, \theta, \varphi) = \frac{\sqrt{4\pi}}{2l+1}\sum_{m=-l}^{l}\frac{M_{lm}Y_{lm}(\theta, \varphi)}{r^{l+1}}. \tag{3.30}$$

† In summing over spherical harmonics, we shall frequently write for brevity $\sum_{l,m}$ instead of $\sum_{l=0}^{\infty}\sum_{m=-l}^{l}$.

The partial potentials Φ_l are the successive *multipole potentials* and are named as for the axial case: Φ_0 is the monopole potential, Φ_1 the dipole, etc. The quantities M_{lm} are then generalizations of the axial multipole moments we considered for axial distributions.

To enlarge on this, we write out the first few M_{lm} explicitly. We need write only those with positive m, since by (3.23) those with negative m are found from

$$M_{l,-m} = (-)^m M_{lm}^*. \tag{3.31}$$

We start with $l = 0$ which implies $m = 0$; and from (3.28), we have

$$M_{00} = \int \rho(\boldsymbol{r}')\,dv.$$

Thus M_{00} represents the total charge in the system, and if this is nonzero, Eq. (3.30) shows that the dominant term in the expansion of the potential at large distances is just M_{00}/r. From far away, all charge distributions with a net charge look like point charges.

The three terms with $l = 1$ are best displayed by expressing the spherical harmonics in the integrand of (3.29) in rectangular coordinates. We find

$$\begin{gathered} M_{11} = -\sqrt{\tfrac{3}{2}}\int (x' - iy')\rho(\boldsymbol{r}')\,dv', \qquad M_{10} = \sqrt{3}\int z'\rho(\boldsymbol{r}')\,dv', \\ M_{1,-1} = \sqrt{\tfrac{3}{2}}\int (x' + iy')\rho(\boldsymbol{r}')\,dv'. \end{gathered} \tag{3.32}$$

Apart from numerical factors, there are three quantities in M_{11}, M_{10}, $M_{1,-1}$ that are characteristic of the distribution; these are denoted by p_x, p_y, p_z and defined by

$$p_x = \int x'\rho(\boldsymbol{r}')\,dv', \qquad p_y = \int y'\rho(\boldsymbol{r}')\,dv', \qquad p_z = \int z'\rho(\boldsymbol{r}')\,dv'. \tag{3.33}$$

These are the obvious generalizations to three dimensions of the axial dipole moments that were discussed previously. By considering p_x, p_y, p_z as components of a *vector dipole moment* $\boldsymbol{p}$, Eq. (3.33) can be written succinctly as

$$\boldsymbol{p} = \int \boldsymbol{r}'\rho(\boldsymbol{r}')\,dv'. \tag{3.34}$$

The potential Φ_1 is simply expressed in terms of $\boldsymbol{p}$. It turns out to be

$$\Phi_1 = \boldsymbol{p}\cdot\boldsymbol{r}/r^3. \tag{3.35}$$

If $\boldsymbol{p}$ is along the z-axis, this becomes the expression $p\cos\theta/r^2$ that was derived previously for the axial dipole potential. There is another way of expressing the dipole potential that will prove useful. The gradient of $(1/r)$ is $\nabla(1/r) = -\boldsymbol{r}/r^3$, which shows that (3.35) can be written

$$\Phi_1 = -\boldsymbol{p}\cdot\nabla(1/r). \tag{3.36}$$

For $l = 2$, the quantities M_{lm} are again best expressed by using the representations of the spherical harmonics in Cartesian coordinates. Working them out, we find, much as for the dipole case, that they are linear combinations of six basic quantities, $Q_{xx}, Q_{xy}, \ldots, Q_{zz}$. These quantities are components* of the *quadrupole tensor*; they are defined by

$$Q_{xx} = \int (3x'^2 - r'^2)\rho(\mathbf{r}')\, dv',$$

$$Q_{xy} = 3\int x'y'\rho(\mathbf{r}')\, dv',$$

with analogous definitions for the other components. In terms of them, the potential Φ_2 can be written

$$\Phi_2 = \frac{1}{2r^5}(Q_{xx}x^2 + Q_{yy}y^2 + Q_{zz}z^2 + 2Q_{xy}xy + 2Q_{xz}xz + 2Q_{yz}yz). \quad (3.37)$$

There is one point worth mentioning here. There are *five* quantities M_{lm} for $l = 2$ corresponding to the five allowed m-values. On the other hand, we have defined, as their equivalent, *six* components of the quadrupole tensor. This equivalence is possible because the six components are, in fact, not independent, for it is obvious that

$$Q_{xx} + Q_{yy} + Q_{zz} = 0.$$

The multipole expansion can, as for the axial case, also be developed in rectangular coordinates. For if in (3.25) both $\mathbf{r}$ and $\mathbf{r}'$ are expressed in rectangular coordinates, we can again expand

$$1/|\mathbf{r} - \mathbf{r}'| = 1/((x - x')^2 + (y - y')^2 + (z - z')^2)^{1/2}$$

in a Taylor series in x', y', z'. We find for $|\mathbf{r}| > |\mathbf{r}'|$,

$$\frac{1}{|\mathbf{r} - \mathbf{r}'|} = \sum_{\substack{jkl \\ j+k+l=n}} \frac{(-)^n x'^j y'^k z'^l}{j!k!l!} \frac{\partial^n}{\partial x^j\, \partial y^k\, \partial z^l}\left(\frac{1}{r}\right).$$

With this in the integral (3.25), we get for the potential,

$$\Phi(x, y, z,) = \sum_{\substack{jkl \\ j+k+l=n}} R_{jkl} \frac{\partial^n}{\partial x^j\, \partial y^k\, \partial z^l}\left(\frac{1}{r}\right),$$

where the *rectangular moment* R_{jkl} is

$$R_{jkl} = \frac{(-)^n}{j!k!l!}\int x'^j y'^k z'^l \rho(\mathbf{r}')\, dv'.$$

---

* The definitions of these components are not standardized. Some definitions differ by a factor of three from the above, and others do not subtract the r'^2 in the integrand of Q_{xx}, Q_{yy}, Q_{zz}.

This rectangular expansion has no advantage in general over the spherical harmonics expansion, so we shall deal briefly with it. The term involving R_{000} is, of course, the monopole term. The three terms corresponding to $n = 1$ define the dipole term Φ_1,

$$\Phi_1 = -\left(p_x\frac{\partial}{\partial x}\left(\frac{1}{r}\right) + p_y\frac{\partial}{\partial y}\left(\frac{1}{r}\right) + p_z\frac{\partial}{\partial z}\left(\frac{1}{r}\right)\right) = -\mathbf{p}\cdot\nabla\left(\frac{1}{r}\right),$$

so that this gives another derivation of (3.36). Similarly, the terms involving $n = 2$ and the second derivatives of $(1/r)$ can be related to the quadrupole terms that were found previously, but we shall not work these out.

There are two generalizations worth mentioning of this multipole expansion. First, we can define, much as for the axial case, limiting distributions whose potentials are *exactly* Φ_1 or Φ_2, etc. These are then three-dimensional *point dipole, point quadrupole* and in general *point multipole* distributions. Second, we can extend the multipole expansion, which holds at points *exterior* to a given charge distribution, to the opposite case of the potential in the *interior* of a charge distribution which is hollow, i.e., such that there is no charge closer to the origin than a distance r_0. Then Eq. (3.25) can still be used and expanded in terms of $P_l(\cos\gamma)$ except this time in powers of r/r',

$$\Phi(r) = \int \frac{\rho(\mathbf{r}')}{r'} \sum_{l=0}^{\infty} \left(\frac{r}{r'}\right)^l P_l(\cos\gamma)\,dv'. \tag{3.38}$$

If we expand P_l as before, using the Addition Theorem, the interior potential is characterized by what might be called its interior multipole moments, I_{lm},

$$I_{lm} = \int \frac{\rho(\mathbf{r}')}{r'^{l+1}} Y^*_{lm}(\theta', \varphi')\,dv', \tag{3.39}$$

and the general solution for the potential is then

$$\Phi(r, \theta, \varphi) = \sum_{l,m} \frac{4\pi}{2l+1} I_{lm} r^l Y_{lm}(\theta, \varphi). \tag{3.40}$$

3.4 TWO-DIMENSIONAL DISTRIBUTIONS

We have been considering a distribution function $\rho(\mathbf{r})$ which depended on *three* space coordinates. Frequently one has to deal with distributions which are functions of only *two* space coordinates. Think, for example, of a very long and slender cylindrical distribution, parallel to the z-axis. In the limit in which it can be considered as infinitely long, and if the density of charge is independent of z, such a distribution is called *two-dimensional*, since it is a function of only two space dimensions. It follows, of course, that the field components and potential are also independent of z, and depend again on only two coordinates. A two-dimensional distribution can then be defined by a density function $\lambda(\boldsymbol{\rho}')$, where, as in Fig. 3.3,

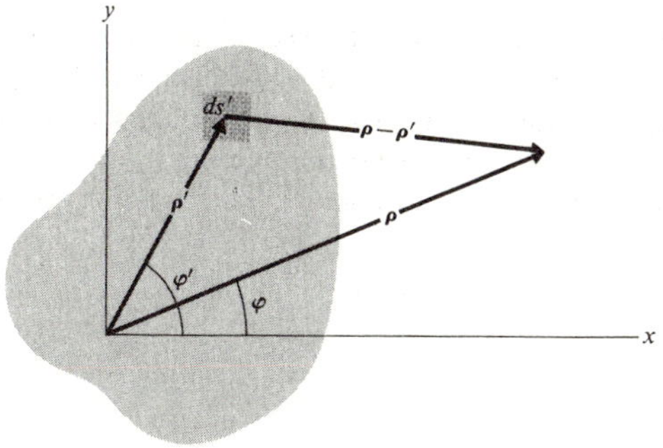

Fig. 3.3 Cross section of a two-dimensional charge distribution.

$\boldsymbol{\rho}'$ is a vector in the xy-plane. If ds' is a small element of area at $\boldsymbol{\rho}'$, then $\lambda(\boldsymbol{\rho}')$ is so defined that $\lambda(\boldsymbol{\rho}')\,ds'$ is the *quantity of charge in a cylinder at* $\boldsymbol{\rho}'$ *of unit height and with base* ds'.

Once again our aim will be to find convenient ways of calculating the potential, this time at the field point $\boldsymbol{\rho}$ shown in Fig. 3.3. And once again we invoke superposition. For if we first find the potential at $\boldsymbol{\rho}$ due to the charge in the infinitely long cylinder at $\boldsymbol{\rho}'$ with infinitesimal cross section, we can by integration find the total potential at $\boldsymbol{\rho}$. Such a long, vanishingly thin charge distribution is called a *line charge.* As the point charge is the "element" out of which all three-dimensional distributions can be synthesized, so the line charge is the two-dimensional element, and we shall begin by calculating the potential it produces.

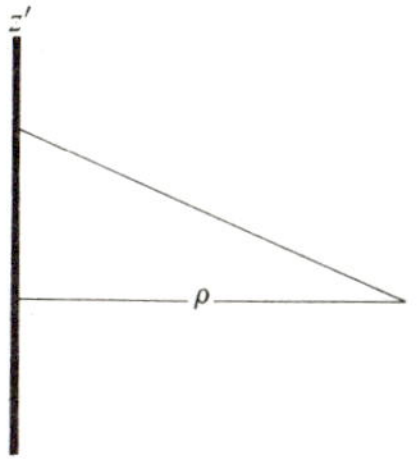

Fig. 3.4 Potential of a line charge.

Figure 3.4 shows such a line charge with the charge λ_0 per unit length; we want to calculate the potential at a distance ρ from the axis. Since $\lambda_0\,dz'$ is the charge in dz', this potential is

$$\Phi(\rho) = \lambda_0 \int_{-\infty}^{\infty} \frac{dz'}{\sqrt{\rho^2 + z'^2}}. \tag{3.41}$$

Now as we have already warned, this integral is infinite: it diverges logarithmically. We shall shortly discuss the meaning of this divergence, and its remedy, but for the moment we adopt another tack to avoid this difficulty; we apply Gauss' Theorem. Imagine a Gaussian surface, consisting of a cylinder of radius ρ and unit height, coaxial with the line charge. Obviously $\boldsymbol{E}$ is directed radially outward, and is a function of ρ only, so we drop the vector notation and simply write it as $E(\rho)$. The only flux of $\boldsymbol{E}$ is through the sides of the cylinder, and this flux is

$$\int E(\rho)\, ds = 2\pi\rho E(\rho).$$

The total charge enclosed by the cylinder is λ_0, and so from Gauss' Theorem we have

$$E(\rho) = 2\lambda_0/\rho. \tag{3.42}$$

To get Φ from this we integrate $E = -d\Phi/d\rho$ from, say, ρ_0 to ρ, to find

$$\Phi(\rho) = -2\lambda_0 \ln \rho + 2\lambda_0 \ln \rho_0. \tag{3.43}$$

The second term in (3.43) is just the constant of integration which, however, we cannot make zero by letting ρ_0 approach infinity. This constant plays no real role since one always wants either differences of potential between two points or, in calculating $\boldsymbol{E}$, the derivative of the potential at a point; in either case, the constant drops out. Frequently one simply drops it from the beginning and writes

$$\Phi(\rho) = -2\lambda_0 \ln \rho. \tag{3.44}$$

as the *two-dimensional logarithmic potential.*

Now we return to the integral expression (3.41) for Φ, which we shall reconcile with (3.44). We consider, as in Fig. 3.4, the potential due to a line charge, but we let its length be finite although very long as compared with ρ. If the charge extends from $-L$ to L, Φ at $z = 0$ is given by the integral in (3.41) but with finite limits L and $-L$,

$$\Phi = \lambda_0 \int_{-L}^{L} \frac{dz'}{\sqrt{\rho^2 + z'^2}} = \lambda_0 \ln \frac{(L^2 + \rho^2)^{1/2} + L}{(L^2 + \rho^2)^{1/2} - L}.$$

If $\rho \ll L$, this potential can be expanded:

$$\ln \frac{(L^2 + \rho^2)^{1/2} + L}{(L^2 + \rho^2)^{1/2} - L} = \ln \frac{(1 + \rho^2/L^2)^{1/2} + 1}{(1 + \rho^2/L^2)^{1/2} - 1} \approx \ln \frac{2}{\rho^2/2L^2} = -2 \ln \rho + \ln 4L^2.$$

Thus

$$\Phi = -2\lambda_0 \ln \rho + \lambda_0 \ln 4L^2. \tag{3.45}$$

Except for the trivial difference in the form of the constants they contain, Eq. (3.45) agrees with Eq. (3.43). And since, for the reasons given above, we can drop the constants anyway, the two expressions agree precisely, provided $\rho \ll L$. Thus the field of a line charge of *finite* length at distances from it that are small compared to its length, is effectively given by the formula for an "infinite" line charge.

With the result (3.44), consider again the potential of the arbitrary two-dimensional distribution of charge, illustrated in Fig. 3.3. With ρ of Eq. (3.44) replaced by $|\boldsymbol{\rho} - \boldsymbol{\rho}'|$, we see that the line-charge element at $\boldsymbol{\rho}'$ contributes an amount $-2\lambda(\boldsymbol{\rho}')\ln|\boldsymbol{\rho} - \boldsymbol{\rho}'|\,ds'$ to the potential at $\boldsymbol{\rho}$; by superposition the total potential $\Phi(\boldsymbol{\rho})$ is then

$$\Phi(\boldsymbol{\rho}) = -2\int \lambda(\boldsymbol{\rho}')\ln(|\boldsymbol{\rho} - \boldsymbol{\rho}'|)\,ds'. \tag{3.46}$$

The function $-2\ln|\boldsymbol{\rho} - \boldsymbol{\rho}'|$ is called a two-dimensional *unit source* or *Green function*; it is, of course, analogous to the three-dimensional function $1/|\boldsymbol{r} - \boldsymbol{r}'|$.

The vectors $\boldsymbol{\rho}$ and $\boldsymbol{\rho}'$ are, so far, quite general and not associated with any special coordinates. Two obvious choices are to express them in rectangular coordinates x, y, or cylindrical ones ρ, φ. We begin with the latter:

$$|\boldsymbol{\rho} - \boldsymbol{\rho}'| = (\rho^2 + \rho'^2 - 2\rho\rho'\cos(\varphi - \varphi'))^{1/2};$$

$\lambda(\boldsymbol{\rho}')$ can be written explicitly as $\lambda(\rho', \varphi')$, and $ds' = \rho'\,d\rho'\,d\varphi'$. Thus Eq. (3.46) can be written

$$\Phi(\rho, \varphi) = -2\int_0^{2\pi} d\varphi' \int_0^{\infty} \lambda(\rho', \varphi')\ln(\rho^2 + \rho'^2 - 2\rho\rho'\cos(\varphi - \varphi'))^{1/2}\rho'\,d\rho'. \tag{3.47}$$

This equation is now in a convenient form for a multipole expansion analogous to the three-dimensional one, and with analogous virtues. For if the distribution is bounded in cross section, i.e., if ρ' takes on some maximum value ρ_0, we can, for $\rho > \rho_0$, expand the logarithm in (3.47), writing

$$\ln(\rho^2 + \rho'^2 - 2\rho\rho'\cos(\varphi - \varphi'))^{1/2} = \ln\rho + \tfrac{1}{2}\ln\left[\left(1 - \frac{\rho'}{\rho}e^{i(\varphi-\varphi')}\right)\left(1 - \frac{\rho'}{\rho}e^{-i(\varphi-\varphi')}\right)\right].$$

Using the expansion in powers of ε of the function $\ln(1 - \varepsilon)$, with $\varepsilon = \rho'/\rho\; e^{\pm i(\varphi-\varphi')}$, we find after some algebra

$$\ln(\rho^2 + \rho'^2 - 2\rho\rho'\cos(\varphi - \varphi'))^{1/2} = \ln\rho - \sum_{n=1}^{\infty}\frac{1}{n}\left(\frac{\rho'}{\rho}\right)^n(\cos n\varphi\cos n\varphi' + \sin n\varphi\sin n\varphi'). \tag{3.48}$$

The potential can then be written

$$\Phi = -2\int_0^{2\pi} d\varphi' \int_0^{\infty} \lambda(\rho', \varphi')\left\{\ln\rho - \sum_{n=1}^{\infty}\frac{1}{n}\left(\frac{\rho'}{\rho}\right)^n(\cos n\varphi\cos n\varphi' + \sin n\varphi\sin n\varphi')\right\}\rho'\,d\rho'. \tag{3.49}$$

Denoting the integrals in (3.49) by A, C_n and S_n,

$$A = -2\int_0^{2\pi} d\varphi' \int_0^{\infty} \lambda(\rho', \varphi')\rho'\, d\rho',$$

$$C_n = \frac{2}{n}\int_0^{2\pi} d\varphi' \int_0^{\infty} \lambda(\rho', \varphi')\rho'^n \cos n\varphi'\, \rho'\, d\rho',$$

$$S_n = \frac{2}{n}\int_0^{2\pi} d\varphi' \int_0^{\infty} \lambda(\rho', \varphi')\rho'^n \sin n\varphi' \rho'\, d\rho',$$

we see that the potential can be written

$$\Phi(\rho, \varphi) = A \ln \rho + \sum_{n=1}^{\infty} \frac{C_n \cos n\varphi + S_n \sin n\varphi}{\rho^n}. \tag{3.50}$$

This expansion in inverse powers of ρ is the two-dimensional multipole expansion, and almost all the general remarks about its three-dimensional analog apply to it as well.

It hardly needs to be mentioned that for a hollow-charge distribution, that is, one for which there is no charge closer to the origin than some distance ρ_0, an interior expansion can be made as in three dimensions. One simply interchanges ρ and ρ' in the expansion (3.48); the functions of ρ and φ that enter the solution are then just $\rho^n \cos n\varphi$ and $\rho^n \sin n\varphi$, and their coefficients are given as integrals over the charge distribution.

The functions $\ln \rho$, $\cos n\varphi/\rho^n$, and $\sin n\varphi/\rho^n$, that crop up in the exterior solution and the analogous functions $\rho^n \cos n\varphi$, and $\rho^n \sin n\varphi$ for the interior solution have a property worth mentioning. Namely, we have derived these functions from an *integral* expression for the potentials, but it is clear that they have a *differential* aspect as well. For since they represent a potential Φ in charge-free space, and this potential satisfies Laplace's equation, it must be that these functions are also solutions of Laplace's equation. And, in fact, in Appendix D, we find just these solutions (among others) emerging from the solutions of Laplace's equation by the method of separation of variables.

3.5 ONE-DIMENSIONAL DISTRIBUTIONS

Idealized charge distributions that depend on only one space coordinate are called *one-dimensional distributions.* As an approximation to one, think, for example, of a space-charge distribution between a flat anode and cathode whose separation is small compared to their dimensions. Such a distribution, along the x-axis, say, is characterized by a density function $\gamma(x)$ such that $\gamma(x)dx$ *is the amount of charge in a cylinder with axis in the x-direction and of unit cross section and thickness dx.* As the point and line charges are the charge elements for three- and two-dimensional distributions, so is the infinitesimally thin charge sheet the one-dimensional element, for we can consider the potential of any distribution as the integral of potentials of such charge sheets. From this integral, we can, much as for two- and three-dimensional distributions, develop a theory of multipoles. In the one-dimensional case, however, the multipole expansion becomes somewhat pedantic, since Poisson's equation is so simple, namely, $d^2\Phi/dx^2 = -4\pi\gamma(x)$, that a single integration of it gives the electric field. It is useful, however, to go far enough to define the one-dimensional *dipole*, or *double layer*, and we shall do this.

We begin by investigating the potential and field produced by the one-dimensional element itself—the infinite charged sheet, with uniform surface density. It is clear that the superposition integral for its potential will diverge, as it did for the two-dimensional case. Here, as there, we shall use Gauss' Theorem to calculate the field, and to investigate the divergence in the integral. If we assume the sheet coincides with the plane $x = 0$, then from symmetry it is clear that the field lines are in the x-direction and point away from the sheet on either side of it. Since there is only an x-component of the field, we can drop the vector notation and write the field as $E(x)$. Applying Gauss' Theorem to a pillbox of *unit* cross section and faces at x and $-x$, we have for the right-hand face

$$\int E(x)\,ds = E(x),$$

with a similar result for the left-hand face, so the total flux through the pillbox is $2E(x)$. If σ is the surface charge density, then by Gauss' Theorem $2E(x) = 4\pi\sigma$; remembering that $\boldsymbol{E}$ is in opposite directions to the right and left of the origin, we can write

$$E(x) = \begin{cases} 2\pi\sigma, & x > 0 \\ -2\pi\sigma, & x < 0. \end{cases} \tag{3.51}$$

The field on either side is independent of x; this result is, of course, a consequence of the fact that we have assumed a sheet of infinite dimensions. Indeed, the result (3.51) applies as an *approximation* to the field of a *finite sheet*, at a distance from it which is small compared with its dimensions. To clarify this, consider a simple case. Let us calculate the potential Φ at a point x on the axis of a uniformly charged disk of radius R and surface charge density σ. By superposition, this potential is

$$\Phi(x) = 2\pi\sigma \int_0^R \frac{\rho\,d\rho}{\sqrt{\rho^2 + x^2}} = 2\pi\sigma(\sqrt{R^2 + x^2} - \sqrt{x^2}).$$

For $x \ll R$, the potential is approximately, on writing $\sqrt{x^2}$ more simply as $|x|$,

$$\Phi = -2\pi\sigma|x| + 2\pi\sigma R. \tag{3.52}$$

We see here the same difficulty as for the infinite line charge; the potential contains an infinite constant in the limit $R \to \infty$. The simplest thing to do is to neglect this, or, if one likes, change the reference point of the potential so it is zero at $x = 0$. With this convention, we can write the potential of an infinite charge sheet as

$$\Phi = -2\pi\sigma|x|. \tag{3.53}$$

This form emphasizes the fact that the potential is *continuous* through the origin, whereas its derivative, the electric field as given by Eq. (3.51), is not. These results for $\boldsymbol{E}$ and Φ due to a charged sheet are sketched schematically in Fig. 3.5.

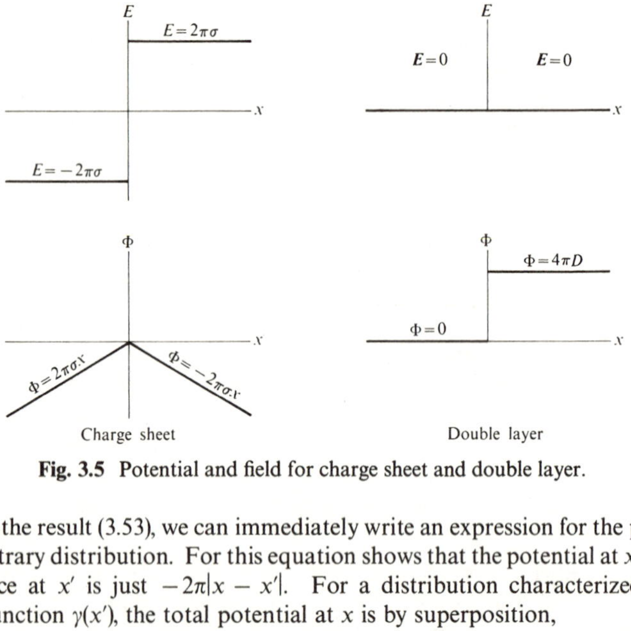

Fig. 3.5 Potential and field for charge sheet and double layer.

With the result (3.53), we can immediately write an expression for the potential of an arbitrary distribution. For this equation shows that the potential at x due to a *unit* source at x' is just $-2\pi|x - x'|$. For a distribution characterized by the density function $\gamma(x')$, the total potential at x is by superposition,

$$\Phi(x) = -2\pi \int \gamma(x')|x - x'|\, dx'. \tag{3.54}$$

The function $-2\pi|x - x'|$ in (3.54) is called the *one-dimensional unit source* or *Green function.*

Beyond the charge sheet, we can form a one-dimensional *dipole* or *double layer* by superposing two charge sheets of opposite sign. Consider such a double layer formed by one charge sheet with density $-\sigma$ at the origin and one with density $+\sigma$ at d. It is easy to see that there is no field outside the charge sheets, i.e., for $x < 0$ or $x > d$. On the other hand, between the sheets there is a field of magnitude $4\pi\sigma$, pointing to the left; the potential difference between $x = 0$ and $x = d$ is then $4\pi\sigma d$. If we let σ become infinite and d go to zero in such a way that σd approaches a finite value D, we generate an *ideal double layer* of strength D. This has properties complementary to those of the charged sheet in that the field is continuous ($\boldsymbol{E} = 0$ for $x > 0$ and $x < 0$), but the potential is discontinuous across it, with a discontinuity of $\Phi_+ - \Phi_- = 4\pi D$.

3.6 SURFACE CHARGES AND DOUBLE LAYERS

Two other important kinds of distributions, the *surface charge* and the *surface double layer*, are generalizations of the plane charge sheet and double layer that we have just discussed.

As the name suggests, a surface distribution is one in which charge is spread on some surface, in a layer idealized to be of vanishing thickness, with a surface density σ that may vary from point to point. A surface *double layer* is obtained by adjoining to such a surface an identical one but with charge of opposite sign and with the two surfaces everywhere displaced by a distance d, in the direction normal to them. An ideal surface double layer is one for which $\sigma d \to D$ as $\sigma \to \infty$ and $d \to 0$.

As we did for their plane counterparts, we want to discuss the continuity properties of field and potential in crossing surface charge sheets and double layers. First, for the surface layer, choose a point on it at which the continuity properties are wanted. Assume the surface is smooth near the point, i.e., that it has no kinks or edges, and also that the density distribution function σ is continuous near the point. Then imagine a small circle drawn on the surface with the point as center and with the property that the small disk it defines is approximately plane and has an essentially constant charge density. By the above regularity and continuity assumptions, it is clear that this will always be possible if we make the circle small enough. Now we want to compare the field at a point infinitesimally close to one side of the disk with the field at a point infinitesimally close to the other side. Thus, however small the radius of the disk, it is very large compared to the distance of these points of observation of the field. Now, the field at the central point of the disk is due, first, to the charge on the disk itself, and, second, to all the other more distant charges. To see the effects of the latter, imagine the charged disk removed, leaving a hole in the surface. It is clear that the more distant charges will produce field lines that are *continuous* through the hole, and therefore across the disk, if we imagine it replaced. The field due to the disk itself will have only a component normal to the disk, and, as we have found, this component is discontinuous by the amount $2\pi\sigma - (-2\pi\sigma) = 4\pi\sigma$. But this means that the *total normal component* of field has this discontinuity, since the sum of a continuous function and one with a jump has itself the jump of the discontinuous part. We can therefore write in for the transition of the normal component of the field

$$(\boldsymbol{E}_2 - \boldsymbol{E}_1) \cdot \boldsymbol{n} = 4\pi\sigma, \tag{3.55}$$

where the subscripts label the opposite sides of the surface, and the unit normal $\boldsymbol{n}$

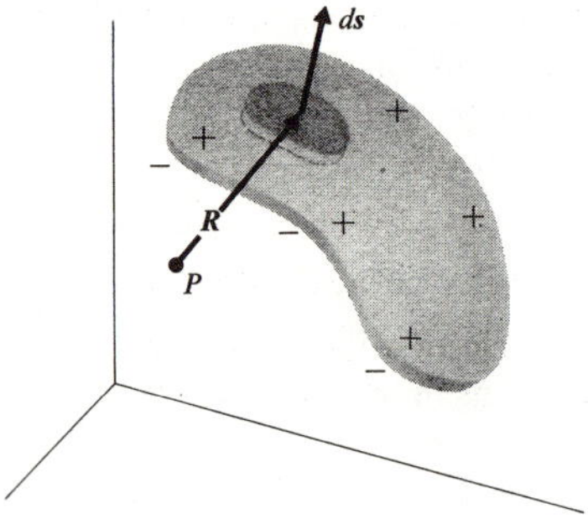

Fig. 3.6 Potential of a double layer at the field point P.

points from the surface into region 2. Incidentally, as a corollary to these arguments, we see that the tangential component of $\boldsymbol{E}$ is continuous across a charged surface, as is the potential.

The result (3.55) is identical with the infinite plane-sheet result for reasons that are obvious in retrospect. It is then clear that the surface double layer can be treated in much the same way with analogous results. We choose a point on the surface as before, imagine a small disk with properties similar to those we have just discussed, divide the field sources into near and distant parts, and generally reproduce the argument of the last paragraph. We find then that the local properties of transition of field and potential across such a double layer are identical with those for an infinite plane one, namely $\boldsymbol{E}$ is continuous, and

$$\Phi_+ - \Phi_- = 4\pi D, \tag{3.56}$$

where the subscript + refers to the positive side of the layer.

For later use we derive here a useful expression for the potential produced by a surface double layer, as shown in Fig. 3.6. Each vector element of surface $\boldsymbol{ds}$ is, effectively, a dipole of moment $D\,\boldsymbol{ds}$, where $\boldsymbol{ds}$ is taken to be positive on going from the negative to the positive side of the layer. Then, using Eq. (3.35) for the dipole potential, the total potential at the field point $\boldsymbol{r}$ is

$$\Phi = -\int \frac{D\boldsymbol{R}\cdot\boldsymbol{ds}}{R^3}. \tag{3.57}$$

The minus sign enters here because the vector $\boldsymbol{R}$ is directed from the field point to the dipolar element; the opposite holds in (3.35). In this last integral $\boldsymbol{R}\cdot\boldsymbol{ds}/R^3$ is just the element of solid angle $d\Omega$ subtended at the field point, so that Eq. (3.57) can be written

$$\Phi = -\int D\,d\Omega. \tag{3.58}$$

For a *constant* surface density D_0 this becomes

$$\Phi = -D_0\,\Omega. \tag{3.59}$$

The last equation, and elementary properties of the solid angle, can be used to derive the result (3.56) for the transition of the potential across a double layer. This is left as a problem for the reader.

3.7 DIPOLE DISTRIBUTIONS—EXTERIOR FIELDS

The problem of calculating the field due to a distribution of dipoles is an important special case of the summation problem. Its importance arises in the theory of dielectrics. In an external electric field a dielectric becomes *polarized*, i.e., it acquires a net dipole moment. This moment arises because each molecule in the dielectric acquires a dipole moment. We must then have some understanding of the potentials and fields due to a collection or *ensemble* of dipoles. We can think of such an ensemble as either arranged randomly, as in a gas or liquid, or in a periodic array, as in a solid. To have a convenient name that applies generally to such ensembles we shall call them *dipolar matter*.

The problem of calculating fields *inside* dipolar matter is a difficult one; here the field is rapidly varying, and depends critically on the detailed structure of the dipolar matter. The problem of calculating the field *outside* is much easier. Here the fields and potentials are smooth and slowly varying and, in fact, independent of the detailed structure. The problem then reduces to a summation problem for certain equivalent *polarization charge* distributions. We treat the easier outside problem first.

It is important to be precise about "outside." To this end, suppose the dipoles are separated from each other by some typical distance a; this might be the mean interatomic distance for a gas or the lattice constant for a solid. Then we say a point is *outside* the distribution if its distance d to the nearest point of the distribution is much greater than a, $d \gg a$. Thus we do not mean outside in a topological sense. A point in the *interior* of some hollow distribution is outside that distribution if it is many interdipole distances away from the nearest dipole.

As a first step in introducing the equivalent polarization charges for calculating the outside field we shall show that we can replace the discrete distribution of dipoles by a smooth continuous distribution. The idea of the equivalence of discrete and continuous distributions for some kinds of approximate calculations will be nothing new. In the multipole expansion, for example, we found that the distant field of point charges in a volume depended only on the net charge and not on how they were arranged. Thus, *discrete* charges $q_1, \ldots, q_n$ produced the same distant field as a *continuous* distribution throughout the volume with total charge $\sum q_i$. It is essentially the same idea that we shall use for dipoles.

To spell this out we shall take a specific model for clarity, but it will be evident that the argument is easily generalized to other models. We shall consider a lattice of point dipoles, one dipole to a unit cell, with cell volume of the order of a^3. In one of the cells we set up a coordinate system with origin at the dipole; its potential at a point is then just given by Eq. (3.36). Imagine now that the single dipole in the cell is divided into two dipoles of the same orientation, but each with half the moment of the original, and suppose further that these two dipoles are arbitrarily positioned somewhere in the cell. We can calculate the potential of this system of two dipoles, at some distance $d \gg a$, by the multipole expansion. What are the multipole moments? It is easy to see that the dipole moment of the system will be the *same* as that of the original dipole, but in addition the system will have a quadrupole (and higher) moments. For calculating the potential at large distances the quadrupole potential will, however, be of the order a/d smaller than the dipole, and the higher multipole potentials will be smaller still. Thus for $d \gg a$ the system of two dipoles will produce the *same potential* as the original.

We can indefinitely continue the process of dividing the dipoles. The two can be made into four, the four into eight, etc., always of course maintaining the original dipole moment of the cell. We generate then a dense cloud of infinitesimal dipoles, which we can think of as ultimately forming a *continuous distribution in which the dipole moment per unit volume* $\boldsymbol{P}$ *is a continuous function of a position vector* $\boldsymbol{r}'$, that is, $\boldsymbol{P} = \boldsymbol{P}(\boldsymbol{r}')$. Moreover, the potential which this distribution produces at some point will differ from the potential of the original discrete lattice by a term of the order of the ratio of the cell size to the mean distance to the observation point. For any point a macroscopic distance from the surface of the sample, the approximation will be very good indeed; for a point near the surface or in the interior, it will break down completely.

The distribution function $\boldsymbol{P}(\boldsymbol{r}')$ is so defined that the dipole moment of the volume dv' at $\boldsymbol{r}'$ is $\boldsymbol{P}(\boldsymbol{r}')\,dv'$. With it the sum of individual dipole potentials is replaced by an integral, and this integral can then be transformed to introduce the equivalent polarization charges we have mentioned. To spell this out we begin by referring to Fig. 3.7. This shows a dipole element of moment $\boldsymbol{P}\,dv'$ at $\boldsymbol{r}'$. By Eq. (3.36) the potential $d\Phi$ of this dipole at the field point $\boldsymbol{r}$, with $\boldsymbol{R} = \boldsymbol{r} - \boldsymbol{r}'$, is

$$d\Phi = -\boldsymbol{P}\cdot\nabla\left(\frac{1}{R}\right)dv' = \boldsymbol{P}\cdot\nabla'\left(\frac{1}{R}\right)dv',$$

where ∇ means gradient with respect to $\boldsymbol{r}$ and ∇' means gradient with respect to $\boldsymbol{r}'$.

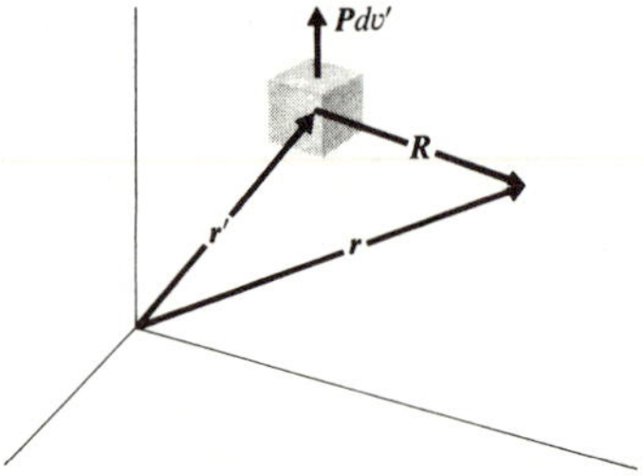

Fig. 3.7 Potential and field at $\boldsymbol{r}$ of dipole $\boldsymbol{P}\,dv'$ at $\boldsymbol{r}'$.

By superposition the total potential at $\boldsymbol{r}$ of all the dipoles in the volume V is

$$\Phi(\boldsymbol{r}) = \int_V \boldsymbol{P}(\boldsymbol{r}') \cdot \nabla'\left(\frac{1}{R}\right) dv'.$$

We use the relation from vector analysis,

$$\nabla' \cdot \left(\frac{\boldsymbol{P}}{R}\right) = \frac{1}{R}\nabla' \cdot \boldsymbol{P} + \boldsymbol{P} \cdot \nabla'\left(\frac{1}{R}\right)$$

to get

$$\Phi = \int_V \nabla' \cdot \left(\frac{\boldsymbol{P}}{R}\right) dv' - \int_V \frac{\nabla' \cdot \boldsymbol{P}}{R} dv'.$$

The divergence theorem applied to the first term gives

$$\Phi = \int_S \frac{P_n\, ds}{R} - \int_V \frac{\nabla' \cdot \boldsymbol{P}}{R} dv', \tag{3.60}$$

where $P_n = \boldsymbol{P} \cdot \boldsymbol{n}$ is the component of $\boldsymbol{P}$ normal to the surface; $\boldsymbol{n}$ is the outwardly drawn unit normal. This formula is our final result, and it is readily interpreted as a potential due to equivalent polarization charges. Thus in the first integral, the factor $1/R$ is characteristic of the Coulomb potential, and it is weighted by the factor $P_n\, ds$. It is then as *if* $P_n\, ds$ were an element of charge, or equivalently as if P_n represented a *surface polarization charge* of surface density (charge unit area) σ_p given by

$$\sigma_p = P_n. \tag{3.61}$$

Similarly, the second integral is the potential due to a fictitious *volume polarization charge* of density (charge/unit volume) ρ_p given by

$$\rho_p = -\nabla \cdot \boldsymbol{P}. \tag{3.62}$$

With these results the problem of summing dipole potentials is reduced to the much easier problem of calculating the potential of an equivalent charge distribution by integration.

To illustrate these ideas we apply them to some simple chunks of dipolar matter. To be specific we take such matter which is made up of dipoles that point in the z-direction at each point of a simple cubic lattice of lattice constant a. If there is a dipole of moment p in each cell of volume a^3, the dipole moment per unit volume P_0 is p/a^3. Consider now a sphere of radius r_0 made of such matter. From (3.61), the surface density is $\sigma_p = P_0 \cos\theta'$, and the volume density ρ_p is, of course, zero. We can then use the basic (3.27) for calculating the potential. If we recognize that $\cos\theta' = \sqrt{4\pi/3}\, Y_{10}(\theta', \varphi')$ so that $\rho(\boldsymbol{r}')$ in (3.27) can be written

$$\rho(\boldsymbol{r}') = P_0 \sqrt{4\pi/3}\, Y_{10}(\theta', \varphi')\delta(r' - r_0),$$

the orthogonality of the Y_{lm} immediately yields

$$\Phi = P_0 \frac{4\pi r_0^3}{3} \frac{\cos\theta}{r^2}. \tag{3.63}$$

This is not a surprising result: the sphere acts like a large dipole with a net moment given by the moment per unit volume times the volume of the sphere.

As another example, we consider the potential inside cavities of various shapes that have been scooped out of dipolar matter. This is a problem that is interesting both in its own right, and also historically, since in the older literature the fields inside matter were defined in terms of cavities of certain standard shapes. These so-called *cavity definitions of the field* are outmoded now, but still yield interesting examples for this section.

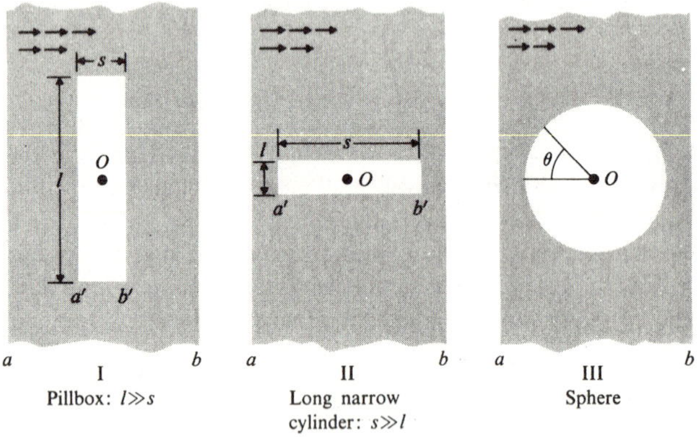

Fig. 3.8 Cavities in a slab of dipolar matter with plane faces a and b.

We consider then a plane slab of dielectric that contains three types of cavities excised as shown in Fig. 3.8. From the equivalent surface charge distributions for each of these geometries one can easily read off the results for the field. For case I the field at O due to the infinite charge sheets a and b is just $4\pi\sigma_p = 4\pi P_0$. The field due to the disk-shaped sheets a' and b' can be obtained from the results of Section 3.6, which show that this field is close to $-4\pi P_0$ when $l \gg s$. The net field at O due to the primed and unprimed sheets is just *zero*. For case II the field due to sheets a and b is again $4\pi P_0$, but now the small disk-shaped sheets a' and b' act like point charges so that the field they produce at the central point O is approximately $2P_0\pi l^2/s^2$. For a long narrow cylinder ($s \gg l$) this field vanishes, and the total field at O is just $4\pi P_0$. For case III we have to calculate the *interior* field produced by a surface charge density $-P_0\cos\theta$. The calculation is much like that which yielded Eq. (3.63) for the exterior field of a sphere. The result is

$$-\tfrac{4}{3}\pi P_0 \tag{3.64}$$

so that the total field inside the cavity is just

$$4\pi P_0 - \tfrac{4}{3}\pi P_0. \tag{3.65}$$

3.8 DIPOLE DISTRIBUTIONS—INTERIOR FIELDS

We now turn to the more difficult problem of calculating internal fields in dipolar matter. Our interest in this is mainly pedagogical. Although it is true that the original motivation for studying dipolar matter was its importance in dielectric theory, it will turn out that there is a macroscopic or formal theory that bypasses the problem of calculating the internal field in detail. This is not to say that it is never of interest to calculate this field. For illustration, we cite one example from solid state theory: There are substances called *ferroelectrics* that, like ferromagnets from which they derive their name, can exhibit a net macroscopic dipole moment even in the absence of an external applied field. Crudely, the molecules that make up a ferroelectric tend to be polarized and aligned by the fields due to the other polarized molecules. The strength of a given induced dipole and its degree of alignment can then be calculated only if we can calculate the field that acts on it; we cannot bypass the problem of calculating in microscopic detail the field at a point in the lattice due to all the other dipoles.

In this section, however, we shall not discuss ferroelectrics or any of the other examples for which one must calculate the internal field. Our aim is more limited and pedagogical. It is, first, to get a better understanding of the nature of the internal field as a help in formulating a formal theory of dielectrics; and second, to show by an example how irregular and rapidly fluctuating such fields are. Since our goal is pedagogical we shall take the simplest model possible: point dipoles arranged in simple cubic lattices. This is doubly artificial in that point dipoles are mathematical abstractions, and simple cubic lattices do not correspond to real substances. But the model does exhibit the essential features of more realistic configurations.

We want to calculate the potential at any point in the unit cell of such a lattice. We mention briefly some techniques that have been used, but will spell out only one in any detail. First, knowing the potential a single dipole of the lattice produces at a point, one might simply set up the triple summation that yields the potential due to all the dipoles. Unfortunately, such sums converge poorly, and although this brute force technique can be made to work with computers, it is not very transparent or illuminating. Another possibility is to begin with the triple summation but transform the poorly convergent series by Fourier analysis into a rapidly converging one; this is the principle of a well-known method due to Ewald (R). Or the summation problem can be considered as a boundary value problem with periodicity, as suggested by Jaynes (R).

The method we shall discuss in some detail is still another one. It is due to Lorentz and it nicely overcomes the convergence difficulties that arise with the straightforward triple summation. These difficulties come from the distant dipoles. We have found, however, that the summation over distant dipole potentials can be replaced by an integration over *equivalent polarization charges.* For a given field point it is then natural to divide the lattice into a local and distant part, to calculate the contribution of the distant part by means of its equivalent polarization charge, and to calculate the contribution of the local dipoles by a straightforward but now limited summation.

Lorentz' classical calculation applies to a cubic lattice of dipoles, each of moment p, that point in the z-direction. It gives the field at a dipole site due to all the dipoles in the lattice except the one at the site. To calculate this field we must assume a definite shape of the sample, since the interior field will depend on the shape of the boundaries. We shall consider a sample in the form of a slab, which is finite in the z-direction, as sketched in Fig. 3.9. To calculate the field at point O, a dipole site, we imagine a sphere centered at O, and of radius r_0 that is large compared with the lattice constant a. Then the field at O is the sum of the field due to the dipoles inside the sphere and of the field due to the dipoles outside the sphere. For any macroscopic sample, the radius r_0 can be taken to be much greater than the lattice constant a. This being so, we can calculate to excellent approximation the field due to the dipoles outside the sphere by their equivalent surface charges. But we have already done this problem, and found in Eq. (3.65) that, with $P_0 = p/a^3$, the field at O is

$$E_z = 4\pi P_0 - 4\pi/3 P_0. \tag{3.66}$$

There remains the contribution of the field due to the dipoles inside the sphere. This is readily seen to be zero, by symmetry; the calculation is left to the reader as a problem. Thus Eq. (3.66) is in fact the final expression for the field at O.

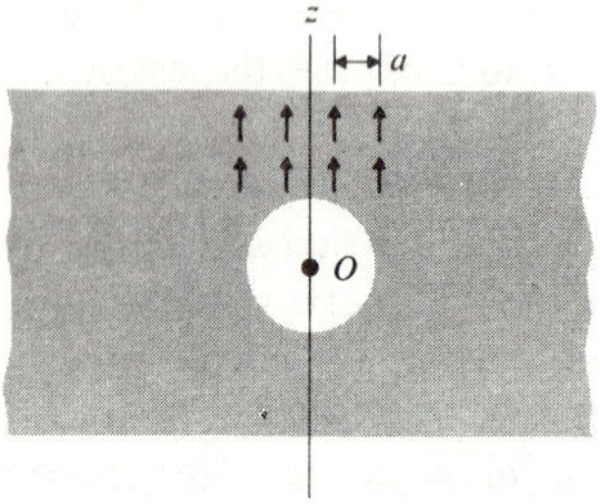

Fig. 3.9 Field at a lattice site of a cubic lattice.

This classical calculation of Lorentz yields the field at only one point in the unit cell. To find the field at other points the method has been extended by McKeehan (R). As before, the dipoles are divided into local and distant ones by an imagined sphere, and the distant ones are taken into account by means of their equivalent surface polarization charges. The fields due to the local dipoles are then summed directly. By doing the calculation for successively larger radii r_o the convergence can be controlled. We shall not go into more detail but simply present the results in Fig. 3.10. The field is written as

$$E_z = 4\pi P_0 - \frac{4\pi P_0}{3} + sP_0,$$

so the various values of s characterize the deviation of the field from the Lorentz value at the origin. Even the few values given in Fig. 3.10 indicate how sharply the field varies over the unit cell.

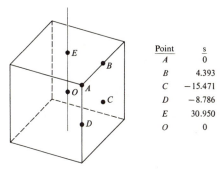

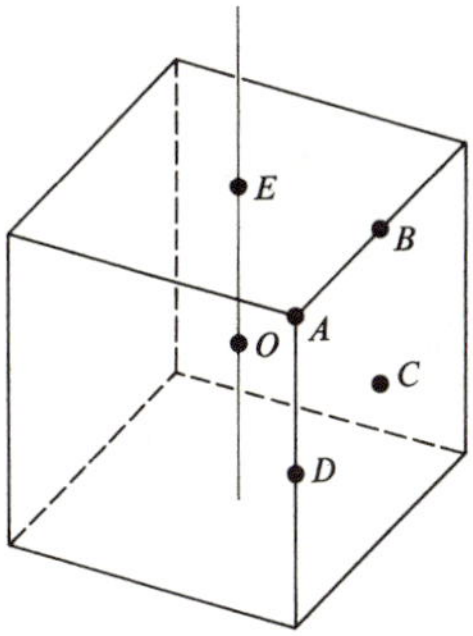

Point	s
A	0
B	4.393
C	−15.471
D	−8.786
E	30.950
O	0

Fig. 3.10 Electric field at various points in the unit cell of a slab-shaped sample of a simple cubic dipole lattice: $E_z = 4\pi P_0 - 4\pi P_0/3 + sP_0$.

PROBLEMS

1. If an axial density distribution $\sigma(z)$ is even about the origin, $\sigma(z) = \sigma(-z)$, show that the expansion of its potential contains only even Legendre polynomials.
2. Point charges q, $-2q$, $2q$, $-q$ are on the z-axis at $2a$, a, $-a$, $-2a$, respectively. Find the first nonvanishing term in the multipole expansion of the potential for $r > 2a$.
3. Devise an axial distribution of point charges for which the potential falls off as $P_6(\cos\theta)/r^7$ for large r.
4. A uniform line charge (length a, total charge q) is between a and $2a$ on the z-axis. A point charge q is at $z = -a$. Find the expansion of the potential in Legendre polynomials for $r < a$.
5. A cube has a uniform distribution of charge throughout it, with total charge Q in the distribution. There is also a point charge $-Q$ at the center. With what power of $1/r$ do the field components fall off at large distances?
6. Two parallel square surfaces (area L^2) are separated by a distance L. The one has a uniform charge density of σ and the other of $-\sigma$. Find the first few terms of the spherical harmonic expansion of the field about an origin at the center of the cube defined by the surfaces.

7. Charge is distributed uniformly throughout an "ice cream cone," i.e., through the portion of space defined in spherical coordinates by $r < R_0$, $\theta < \theta_0$. Find the potential $\Phi(r, \theta)$ everywhere (inside and outside the distribution) as a spherical harmonics expansion.

8. Find the potential and field produced by a slim circular ring of charge of uniform density.

9. Four charges of magnitude $-q$ and one of magnitude $4q$ are placed (Fig. 3.11) in the $z = 0$ plane. Find the components of the quadrupole moment of the distribution.

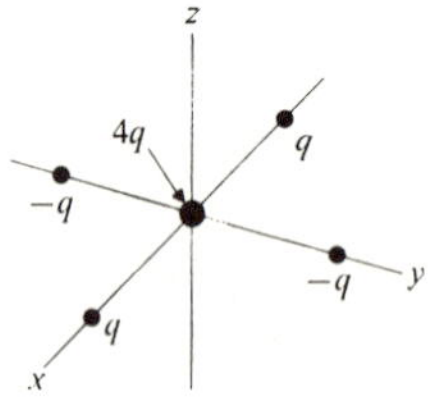

Figure 3.11

10. Two charged rings, radii a and b, where $a > b$, are coaxial and coplanar; the first has total charge Q and the second $-Q$. Find the components of the first nonvanishing multipole moment.

11. Prove the expansion of Eq. (3.48),

$$\ln(\rho^2 + \rho'^2 - 2\rho\rho'\cos(\varphi - \varphi')) = \ln \rho - \sum_{n=1}^{\infty} \frac{1}{n}\left(\frac{\rho'}{\rho}\right)^n (\cos n\varphi \cos n\varphi' + \sin n\varphi \sin n\varphi').$$

12. An infinitely long nonconducting cylinder whose cross section is a right isosceles triangle has uniform charge density on its surfaces. Find a power-series expansion of the potential in x and y (Fig. 3.12) accurate to 1% for $x^2 + y^2 < a^2/5$. Express the potential in cylindrical ρ, φ coordinates.

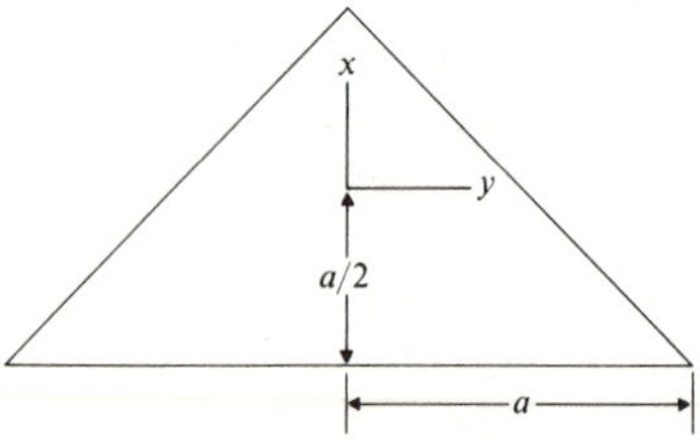

Figure 3.12

13. A cylindrical chunk of dipolar matter, radius a, height $2b$, has dipole moment P per unit volume, parallel to the axis. Find the field for points on the axis outside the chunk.

14. Use Eq. (3.59) to derive the result (3.56) for the discontinuity in potential on traversing a double layer.

REFERENCES

Durand, E., *Electrostatique et Magnetostatique*, Masson, Paris (1953).

Several hundred pages of fine print are devoted to electrostatics and there is a large selection of summation problems.

Ewald, P. P., "Die Berechnung optischer und elektrostatischen Gitterpotentiale," *Ann. Phys.* **64,** 253 (1921).

Jaynes, E. T., *Ferroelectricity*, Princeton University Press, 1953.

McKeehan, L. W., "Magnetic Dipole Fields in Unstrained Cubic Crystals," *Phys. Rev.* **43,** 913, (1933).

Perhaps the simplest possible straightforward summation technique for calculating the internal field in a dipole lattice. The problem is the same for electric dipoles as for magnetic ones.

Weber, E., *Electromagnetic Fields, Theory and Applications*, Vol. 1, Wiley, New York, 1950.

This book contains many worked examples of summation problems.

There have been many methods devised for calculating the field of a lattice of dipoles (or other multipoles). Some recent references chosen almost at random are:

Johnson, F. A., "The Electrostatics of Crystals," *J. Phys. C.* (*Proc. Phys. Soc.*) **2,** 1384 (1969).

De Wette, F. W., and G. E. Schacher, "Internal Fields in General Dipole Lattices," *Phys. Rev.* **137A**, 78 (1965).

4
BOUNDARY VALUE PROBLEMS WITH CONDUCTORS

4.1 THE PHYSICS OF CONDUCTORS IN FIELDS

Substances are broadly divided into those that conduct and those that do not: *conductors* and *dielectrics*. Of course, this division is somewhat arbitrary; in a sense, a dielectric is just a poor conductor and a conductor merely a poor dielectric. Nonetheless, there is an enormous difference between the best examples of each species: the conductivity of a good conductor, such as silver, is about 10^{24} times that of a good dielectric, such as fused quartz. There is also, of course, a whole spectrum of examples of intermediate conductivity; for example, saline water is more or less conductive according to the degree of its salinity. Although modern solid-state theory teaches that there are many detailed mechanisms for conductivity, we shall always assume that we are dealing with a metal. Crudely, in a metal the transport of charge is due to the mobile electrons which move in a background of fixed positive charge. As the electrons move, they experience collisions of various kinds that constitute a resistance to their motion. The degree of this resistance differs among substances; it is characterized quantitatively by the *resistivity* or its reciprocal, *conductivity*. In electrostatics we shall deal almost always with *perfect conductors*. These are (idealized) conductors in which the electrons are *perfectly* free to move. i.e., in which they experience no resistance at all to their motion. This idealization is a sound one, as we shall see later, in that the physics of perfect conductors is often a very good approximation to that of good conductors.

In this chapter we consider conductors in the state of *electrostatic equilibrium* that obtains when all the charges are at rest. It follows from the concept of a perfect conductor that the *total electric field must vanish in its interior*. For any such field, however small, would cause the charges in the interior to move, contrary to the hypothesis of static equilibrium. This fact of a vanishing electric field has important consequences. First, since $\boldsymbol{E}$ is zero everywhere inside the surface, it must be that the *tangential component* of $\boldsymbol{E}$ just inside the surface is *zero*. No work is then needed to carry a charge from one point to another on the surface of the metal. We conclude that *the surface of a perfect conductor in static equilibrium is an equipotential*. The second conclusion is that any charge density can be nonzero only on the surface. For everywhere in the interior $\boldsymbol{E} = 0$; hence from $\nabla \cdot \boldsymbol{E} = 4\pi\rho$, we have $\rho = 0$. This does not mean, of course, that there are neither electrons nor positive charges in the interior but only that they mutually balance everywhere there.

Since the surface of a conductor plays a special role in electrostatics, let us consider it in more detail. Although at the surface the tangential component of $\boldsymbol{E}$ is zero, we must look more carefully at its normal component E_n, for the conclusion that $\boldsymbol{E} = 0$ in the interior was based on the fact that the electrons are perfectly free to move; and so they are everywhere, *except at the surface* and in a direction normal to it. The electrons are, in general, bound to the metal by forces of some kind; whatever the nature of these forces, charges are not free to leave the surface of the metal. Thus the normal component of an electric field at the surface of a perfect conductor plays a special role. To amplify this, we must be more precise as to the meaning of "at the surface." At a small distance just inside the surface it is still true that $\boldsymbol{E}$, and hence E_n, is zero. On the other hand, just outside the surface there *is* a field $\boldsymbol{E}$, and there is no reason to suppose that E_n is zero. We conclude that the *normal component E_n is discontinuous.* This is no surprise. We have previously found in Eq. (3.55) that E_n is discontinuous across a surface charge layer, and have shown above that in a conductor the charge is distributed in such a layer. We can then apply Eq. (3.55) to the present case to get an important relationship. If in that equation, 2 refers to the *exterior* of the metal and 1 the *interior*, we have $\boldsymbol{E}_1 \cdot \boldsymbol{n} = 0$, whence

$$\boldsymbol{E}_2 \cdot \boldsymbol{n} = 4\pi\sigma. \tag{4.1}$$

The surface charge density can thus be found from the normal component of $\boldsymbol{E}$ just outside the surface. Equivalently

$$-\partial\Phi/\partial n = 4\pi\sigma, \tag{4.2}$$

where Φ is the potential just outside the surface.

4.2 HOMOGENEOUS AND INHOMOGENEOUS PROBLEMS

With this understanding of the physics of perfect conductors, we turn to a physical and qualitative discussion of the two main kinds of problem that we shall discuss. First, there are *homogeneous* problems, in which there is no external or applied field, i.e., in which the only sources of the field are charges on the conductors themselves. Second, there are *inhomogeneous* problems, in which there are sources external to the conductors, in addition of course to the charges on the conductors. Concretely, a typical homogeneous problem might be generated when excess charge is introduced to a conductor, for example, by charging it from some electrostatic machine. This charge, under its own mutual repulsion, quickly finds its way around the surface, where it distributes itself in such a way that the surface is an equipotential. In the final state of electrostatic equilibrium there is then a surface with a fixed amount of charge, or equivalently, at a fixed potential but with an *a priori unknown distribution.* Similarly a typical *inhomogeneous* problem might be that of a point charge Q external to a grounded (zero-potential) conducting body. For vividness imagine what would happen as Q approaches the body from infinity. As Q gets nearer to the body, charge will begin to be drawn up from ground attracted by Q; one says that charge is "induced" on the sphere. In the final equilibrium there will be a nonuniform distribution of charge on the surface that is, of course, unknown *a priori*, but is such that the surface of the body is an equipotential. In homogeneous and inhomogeneous problems the physics is basically the same: An electrostatic potential is produced by *all* the charges, the surface charges on the conductor *and*, if there are any, the external charges. The surface charges distribute themselves in such a way that this potential is constant on the surface of the body. If the distribution was known we could calculate the potential everywhere as a summation problem. Of course, in general, the distribution is not known in advance; therein lies the problem.

From one point of view, the main problem in the electrostatics of perfect conductors is: What is the distribution of surface charge that makes the conductor an equipotential? But this viewpoint is too limited. Although the surface charge distribution determines the potential everywhere, it does not follow that we must begin by considering it as the unknown. Alternatively, we can find the surface charge from Eq. (4.2) if the potential is known near the surface, so we might begin with the potential as the unknown. These two points of view have their mathematical counterparts. Thinking of the surface charge distribution as unknown, we are led to an *integral equation* for it; considering the potential as unknown, we have a *differential equation* for it. We discuss these two possibilities in turn.

Consider first the integral equation approach as applied to a homogeneous problem. Figure 4.1 shows a conducting body bounded by a closed surface S on which there is a charge distribution that is a function of $\boldsymbol{r}_s'$, where $\boldsymbol{r}_s'$ is a vector to the surface. The potential at the field point $\boldsymbol{r}$ in space is

$$\Phi(\boldsymbol{r}) = \int_S \frac{\sigma(\boldsymbol{r}_s')\, ds'}{|\boldsymbol{r} - \boldsymbol{r}_s'|}. \tag{4.3}$$

If now $\boldsymbol{r} \to \boldsymbol{r}_s$, where $\boldsymbol{r}_s$ is a point *on the surface*, the potential becomes some constant, say Φ_0:

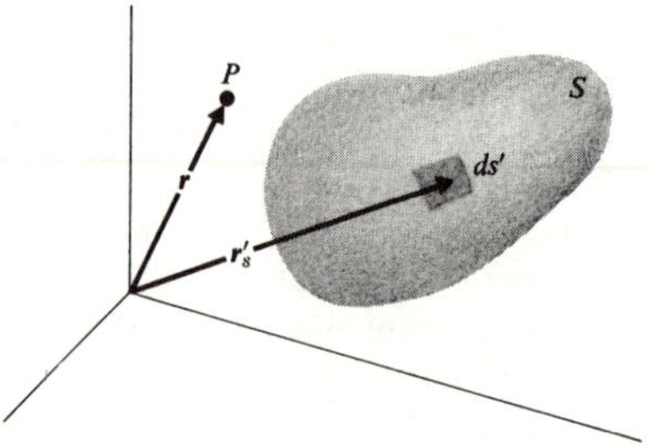

Fig. 4.1 Conductor with surface charge distribution $\sigma(\boldsymbol{r}_s')$.

$$\Phi_0 = \int_S \frac{\sigma(\mathbf{r}'_s)\, ds'}{|\mathbf{r}_s - \mathbf{r}'_s|}. \tag{4.4}$$

This is now an *integral equation* for the determination of the unknown σ; examples of its solution are found in Section 4.7. The singularity in the integrand of Eq. (4.4) is an integrable one and causes no basic difficulty; essentially this point was discussed in Section 2.4, where it was proved that the field is finite in the interior of continuous charge distributions. In this method, σ is treated as the unknown in Eq. (4.4); once it is determined, Φ can be calculated at any exterior point from Eq. (4.3).

There are small variants to this formulation. For example, we might alternatively have asked: A conductor has total charge Q. What is the potential at any point in space? Now the alternative conditions of fixing Φ_0 *or* of fixing Q differ only trivially from each other. Either serves only as a *normalization condition*. The important problem is to find the *relative distribution* of surface charge that makes the surface an equipotential. This relative distribution can then be adjusted, i.e., can be scaled up or down to fit either normalization condition. In fact, any other normalization condition will suffice as, for example, specifying the charge density at a given point. For these reasons it is permissible and frequently convenient in homogeneous problems to take the potential Φ_0 to be unity.

Consider next inhomogeneous problems. Here, the total potential at any point is that due to the *external* or *applied* source, say $\Phi_a(\mathbf{r})$, plus that due to the surface charge distribution $\sigma(\mathbf{r}'_s)$. Thus

$$\Phi(\mathbf{r}) = \Phi_a(\mathbf{r}) + \int_S \frac{\sigma(\mathbf{r}'_s)\, ds'}{|\mathbf{r} - \mathbf{r}'_s|}. \tag{4.5}$$

If we specify that the conductor is at potential Φ_0 and set $\mathbf{r} = \mathbf{r}_s$ in this equation, it becomes an integral equation in the variable $\mathbf{r}_s$ for the unknown function $\sigma(\mathbf{r}_s)$, namely,

$$\Phi_0 = \Phi_a(\mathbf{r}_s) + \int_S \frac{\sigma(\mathbf{r}'_s)\, ds'}{|\mathbf{r}_s - \mathbf{r}'_s|}. \tag{4.6}$$

As for the homogeneous problem there are variants possible; for example the total induced charge might be prescribed. These integral equations will be discussed in more detail in Section 4.7.

We return to the differential viewpoint that concentrates on the potential Φ as the unknown. For the homogeneous problem, the potential must satisfy Laplace's equation outside the bounding surface of the conductors. Thus we are looking for an exterior solution of Laplace's equation for Φ such that on the boundary Φ becomes a constant. This then is *a* solution to the problem; in the next section, we shall prove a uniqueness theorem which shows it is *the* solution. The inhomogeneous problem is quite similar, except that Laplace's equation must be replaced by Poisson's in those exterior regions where the external source density is nonzero.

There are various methods for solving the appropriate Laplace or Poisson equation. One method consists of essentially guessing the correct solution; it is dignified by the name *method of images.* In two dimensions one can exploit the fact that the real and imaginary parts of an analytic function of a *complex variable* satisfy Laplace's equation. Finally, and perhaps most importantly, one can use the *orthogonal functions*, for example, spherical harmonics, that were encountered in treating the summation problem in various geometries. It is not surprising that these functions will be useful. Although they appeared naturally, with no mention of Laplace's equation, as a result of expanding the superposition integral, it is clear that since the functions represent the potential of a charge distribution, they are solutions of that equation. As these functions appeared with *known* coefficients to represent the potentials of *known* distributions, general sums with *unknown* coefficients can be written to represent the potential of *unknown* distributions.

4.3 A UNIQUENESS THEOREM

We now prove the theorem on the uniqueness of solution of boundary value problems with conductors which was mentioned above. The proof begins with the divergence theorem for a vector $\boldsymbol{A}$,

$$\int_V \nabla \cdot \boldsymbol{A}\, dv = \int_S \boldsymbol{A} \cdot d\boldsymbol{s}, \tag{4.7}$$

in which $\boldsymbol{A}$ is the following combination of two *scalar* point functions $F(\boldsymbol{r})$ and $G(\boldsymbol{r})$:

$$\boldsymbol{A} = F\nabla G. \tag{4.8}$$

Putting this into Eq. (4.7), we use

$$\nabla \cdot (F\nabla G) = F\nabla^2 G + \nabla F \cdot \nabla G,$$

for the volume integral and rewrite $F\nabla G \cdot d\boldsymbol{s}$ in the surface integral as $F(\partial G/\partial n)\, ds$, where ds is the scalar element of area and $\partial G/\partial n$ is the derivative normal to the surface. Equation (4.7) becomes what is often called *Green's first identity*:

$$\int_V (F\nabla^2 G + \nabla F \cdot \nabla G)\, dv = \int_S F \frac{\partial G}{\partial n}\, ds. \tag{4.9}$$

Suppose there is a conductor with a closed surface S_0 and for which the exterior volume is V_0. In Eq. (4.9) we identify* V with V_0 and S with S_0. Suppose further

* Somewhat more rigorously, we should let V_0 be bounded by S_0 and by a surface at infinity, but it will be easy to see that this surface contributes nothing, so we neglect it from the beginning.

that the homogeneous problem has been solved and that there is not one but two solutions, Φ_1 and Φ_2, both of which satisfy Laplace's equation in V_0 and reduce to the same constant on S_0. Consider the function

$$\psi = \Phi_1 - \Phi_2,$$

and in Green's identity let $F = G = \psi$ to get

$$\int_{V_0} (\psi \nabla^2 \psi + |\nabla \psi|^2)\, dv = \int_{S_0} \psi \frac{\partial \psi}{\partial n}\, ds. \tag{4.10}$$

Since Φ_1 and Φ_2 both satisfy Laplace's equation, so does their difference and $\nabla^2 \psi = 0$. Moreover, since Φ_1 and Φ_2 reduce to the same value on the boundary, ψ is zero in the integral over S_0 in (4.10). The equation becomes

$$\int_{V_0} |\nabla \psi|^2\, dv = 0.$$

But since $|\nabla \psi|^2$ in the integrand is a positive quantity, the only way this integral can vanish is for $|\nabla \psi|^2$ itself to vanish at every point in the volume. This implies in turn $\nabla \psi = 0$ everywhere in the volume, or

$$\Phi_1 - \Phi_2 = \text{constant}.$$

But the constant in this last equation must be zero since the two solutions agree at the boundary. *The two solutions, Φ_1 and Φ_2, are identical.*

The uniqueness theorem for the inhomogeneous problem can be derived with a minor modification of the above argument. If there are two purportedly different solutions Φ_1 and Φ_2, then each must satisfy not Laplace's equation but Poisson's equation,

$$\nabla^2 \Phi_1 = -4\pi \rho_{\text{ext}}, \qquad \nabla^2 \Phi_2 = -4\pi \rho_{\text{ext}},$$

where ρ_{ext} is the external source density. But again we have $\nabla^2 \psi = 0$, and the remainder of the proof is unchanged from the homogeneous case.

An interesting consequence of the uniqueness theorem is the proof that the field inside any closed hollow conductor (for example, a spherical conducting shell) is zero. If such a shell is under the influence of charge outside it, then both the inner and outer surfaces must be equipotentials. Now it is trivial to modify the above proof to show that the uniqueness theorem holds for this case, in which the surface S_0 is taken to be the *internal* metal boundary and the volume V_0 to be the hollow *interior*. But a solution of this interior boundary value problem is Φ = constant, since this certainly is a constant over the boundary and moreover satisfies Laplace's equation in the interior. By the uniqueness theorem, it is the only solution. But this constant potential means that the electric field is zero inside the volume. It also follows that a homogeneous or inhomogeneous boundary-value problem for a *solid* conductor has the same solution as the same problem for a hollow conductor with the same external shape; in either case the field is zero inside.

4.4 METHOD OF IMAGES

The method of solving boundary value problems with conductors, discussed in this section, is called the "method of images"* when applied to the inhomogeneous problem. It applies equally well to homogeneous problems, however, and we shall discuss it for these as well.

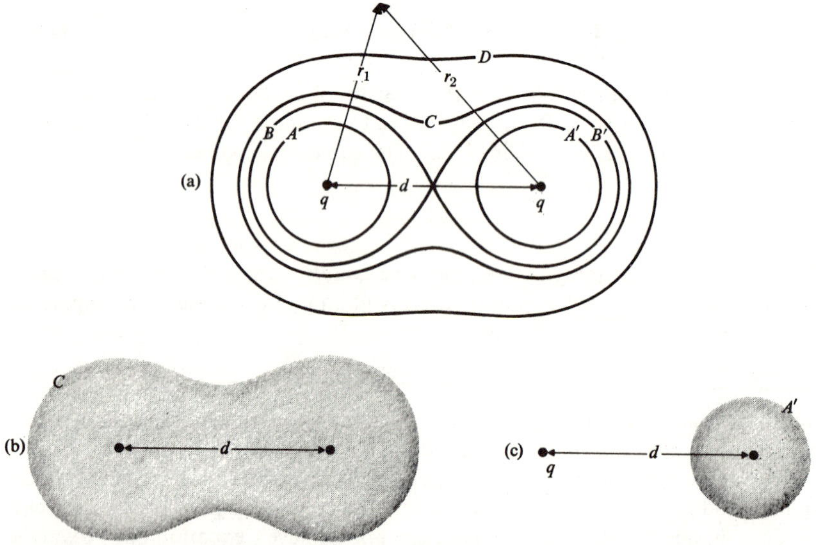

Fig. 4.2 (a) Equipotential surfaces associated with two positive charges a distance d apart. (b) and (c) Two boundary-value problems that can be solved in terms of these surfaces.

To explain the method we begin by supposing that there is a general distribution of charge, continuous or discrete, in some region of space. Associated with this distribution there is a set of equipotential surfaces. Thus, if the charge distribution is simply the two positive charges shown in Fig. 4.2, the equipotential surfaces are the surfaces generated by rotating the curves A, A', B . . . around an axis joining the charges. Consider some *equipotential surface* which encloses the *whole* charge distribution, as for example, the one labeled C in Fig 4.2. Outside this surface the potential Φ_d generated by the distribution is a solution of Laplace's equation; by definition, this potential is constant over the surface. Now consider the homogeneous problem for a *charged conductor with the same shape as the equipotential surface.* To solve it one must find a solution Φ_c which satisfies Laplace's equation outside the body and which becomes constant over the surface of the body. But

* The possible origin of the name is discussed in connection with the problem sketched in Fig. 4.4.

the potential Φ_d defined above has just these properties. It is then *one* solution of the homogeneous boundary value problem; the uniqueness theorem tells us it is the *only* one. In short, we identify the potential Φ_d with Φ_c outside the surface and can relate the normal derivative of this potential to the charge distribution according to Eq. (4.2). There is, of course, no correspondence between Φ_d and Φ_c inside the surface. There the potential Φ_d is what it is according to the charge distribution; Φ_c is constant, in accordance with the discussion above of the physics of a perfect conductor. In the example of Fig. 4.2 the potential is

$$\Phi_d = q(1/r_1 + 1/r_2). \tag{4.11}$$

We can solve the homogeneous problem for conductors whose surfaces are given by the equation $(1/r_1 + 1/r_2) =$ constant. If we assign different values to the constant, a considerable variety of shapes can be generated, as Fig. 4.2. indicates.

These ideas are easily extended to inhomogeneous problems. Consider as before the equipotential surfaces of a charge distribution, but now select a surface which, unlike that in the homogeneous case, has some part of the distribution outside it. In Fig. 4.2, for example, the equipotential surface A' has one of the charges inside and the other outside. For the space outside A' the potential Φ_d in Eq. (4.11) satisfies Poisson's equation for a source density which is that of the point charge q on the left, and this potential becomes constant over the surface A'. We see that Φ_d satisfies the differential equation *and* boundary condition corresponding to a conducting surface of shape A', which is acted on by a point charge q. By the uniqueness theorem this represents the solution to that problem. In the general case of an arbitrarily complicated charge distribution, some of which is inside and some outside a given equipotential surface, the potential Φ_d represents the solution to the problem of a perfect conductor in the shape of an equipotential surface, which is acted upon by whatever charges are exterior to it.

This is fine; the above ideas generate the solution of an infinite number of problems involving conducting shapes and external fields as complicated and bizarre as we might wish. The joker is, of course, that we must take whatever shapes the charge distributions generate, and these may or may not be of interest. Usually we are faced with the converse problem: for a conductor of a *given* shape, we want to solve the homogeneous problem or some inhomogeneous one. The solution then depends on *guessing* the appropriate distribution of charges which will lead to an equipotential of the desired shape. The only way one can really discuss this method is by giving examples which perhaps will suggest means of solving still other problems.

We start with some homogeneous problems. The simplest example is the trivial one with a point charge q. The equipotential surfaces of its potential $\Phi_d = q/r$ are spheres. By adjusting q we find that the solution of the boundary value problem, for which a sphere of radius a is at potential V_0, is

$$\Phi_c = V_0(a/r).$$

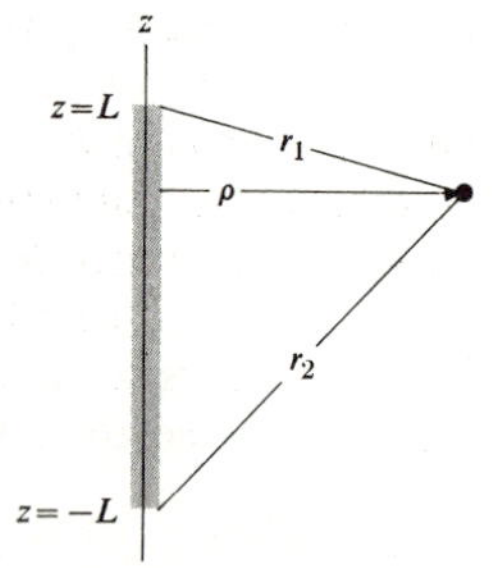

Fig. 4.3 Equipotentials of a line charge.

The next to simplest example perhaps is that for two equal point charges, for which the equipotentials are already sketched in Fig. 4.2. As a third example, we take the potential due to a line charge that extends between $-L$ and L on the z-axis. We discussed the field at $z = 0$ in Section 3.4. In much the same way we find that if λ is the linear density of the charge, from Fig. 4.3 the potential at ρ, z is

$$\Phi = \lambda \ln \frac{L - z + \sqrt{(L - z)^2 + \rho^2}}{-L - z + \sqrt{(L + z)^2 + \rho^2}}. \tag{4.12}$$

We introduce the *elliptic coordinates* u and v, defined in terms of the distances r_1 and r_2; these are

$$u = \tfrac{1}{2}(r_1 + r_2),$$
$$v = \tfrac{1}{2}(r_1 - r_2). \tag{4.13}$$

When r_1 and r_2 remain in a fixed plane, the equation, $u =$ constant, is that of an ellipse; in the geometry of Fig. 4.3 the equation, $u =$ constant, describes an ellipsoid. From the geometry of that figure we have

$$r_1^2 = \rho^2 + (L - z)^2, \qquad r_2^2 = \rho^2 + (L + z)^2, \tag{4.14}$$

so that Φ can be written

$$\Phi = \lambda \ln \frac{L - z + r_1}{-L - z + r_2} = \lambda \ln \frac{L - z + u + v}{-L - z + u - v}. \tag{4.15}$$

Some algebra, and the relation $uv = -zL$ derivable from (4.13) and (4.14), now shows that Eq. (4.15) can be written as

$$\Phi = \lambda \ln \frac{u + L}{u - L}. \tag{4.16}$$

This equation shows that the potential is constant when u is constant, and we have just seen that this implies that the equipotential surfaces are ellipsoids. We have thus "solved" the problem of the potential outside a charged conducting ellipsoid.

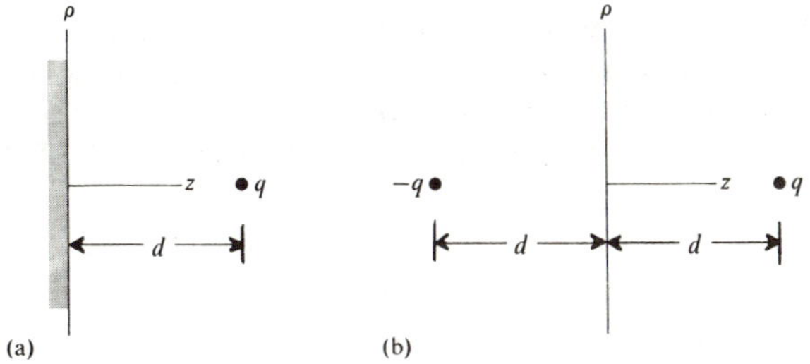

Fig. 4.4 (a) Point charge q a distance d from plane face of a conductor. (b) Equivalent image.

Incidentally this is a quite difficult problem to treat directly, since it involves solving Laplace's equation in the somewhat complicated ellipsoidal coordinates.

We now consider some inhomogeneous problems. The simplest is one which perhaps shows the origin of the name "method of images." It is the problem of a perfect conductor that fills the half-plane $z < 0$, and is at potential zero under the influence of a point charge at $z = d$, as shown in Fig. 4.4. This problem is equivalent to that of a conducting *slab* of finite thickness in the z-direction, whose right-hand face is the plane $z = 0$, under the influence of the point charge. (Why?) If we imagine an "image" charge $-q$ at $z = -d$, the potential of the charge and its image is

$$\Phi = \frac{q}{\sqrt{\rho^2 + (z - d)^2}} - \frac{q}{\sqrt{\rho^2 + (z + d)^2}}. \tag{4.17}$$

This potential correctly reduces to zero at $z = 0$, and for $z > 0$ has the correct singularity to describe the point charge q. We can therefore take it to be the solution of the original problem and with it can calculate quantities of interest. For example, the induced charge density σ on the surface is

$$\sigma = -\frac{1}{4\pi}\frac{\partial \Phi}{\partial z}\bigg|_{z=0} = -\frac{q}{2\pi(\rho^2 + d^2)^{3/2}}. \tag{4.18}$$

The density σ is negative, as we might have suspected, since it represents charge drawn up from infinity, so to speak, by the attraction of the positive charge q. For the same reason it is not surprising that σ is largest for $\rho = 0$, that is, at the point closest to the inducing charge. We can also calculate the force between the conductor and the charge q. It is simplest to do this by recognizing that the force on q is just the force between q and its image. To see this, note that the force on q is due to the field at q produced by all other charges *except* q: the charge cannot act on itself. This field is the gradient of Φ after the part of the potential produced by q has been excluded from Φ. But from (4.17) the part of the potential due to charges other than q is that produced by the image charge; the corresponding force is an attraction of magnitude $q^2/4d^2$. In the physical problem, then, this is the force between the charge and the conducting plane. In a more general case, with not one applied charge q but with a complex of charges (for example, a point multipole), it is clear that the problem can be solved equally easily by imaging each of the charges in the complex.

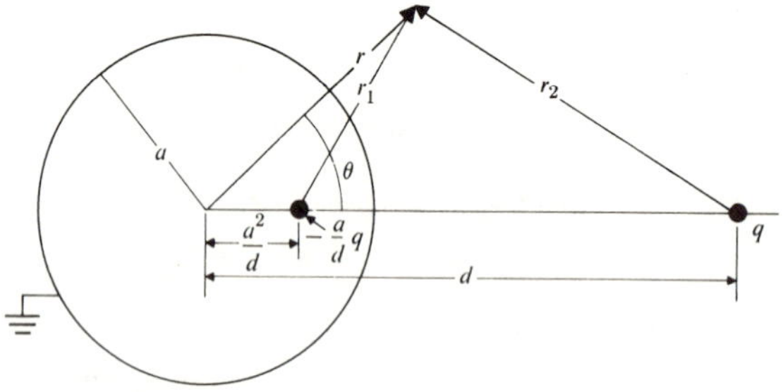

Fig. 4.5 Image for a point charge q a distance d from the center of a grounded sphere of radius a.

As another example, consider a sphere acted on by a point charge q which is a distance d from its center as shown in Fig. 4.5. We shall assume the sphere is at potential zero, or "grounded." Variants on this assumption in which the potential is taken to have some finite value, or in which the total charge or charge density at a point is prescribed, will turn out to be quite simple to handle. We begin by verifying that the charge q and an *image charge* of magnitude $-(a/d)q$, a distance a^2/d from the center, as shown in Fig. 4.5, have as equipotential a sphere of radius a at potential zero. The potential due to these charges is

$$\Phi = q\left(\frac{1}{r_2} - \frac{a}{dr_1}\right) = q\left(\frac{1}{\sqrt{r^2 + d^2 - 2r\,d\cos\theta}} - \frac{1}{\sqrt{(r^2d^2/a^2) + a^2 - 2r\,d\cos\theta}}\right), \tag{4.19}$$

and from this expression it can be verified that Φ does indeed vanish at $r = a$. Equation (4.19) is then the solution to the problem, and with it we can calculate various quantities of interest, as we did for the half-plane problem. For example, the induced charge turns out to be

$$\sigma = -\frac{1}{4\pi}\frac{\partial\Phi}{\partial r}\bigg|_{r=a} = -\frac{q(d^2 - a^2)}{4\pi a(d^2 + a^2 - 2a\,d\cos\theta)^{3/2}}. \tag{4.20}$$

For the same reasons as in the half-plane problem, the force between the charge and the sphere is the same as the force between the charge and its image. A variant of this problem with the sphere not at zero but at some finite potential Φ_0 can readily be solved by putting another image charge of magnitude $\Phi_0 a$ at the center. By adjusting Φ_0, one can then solve the problem of a sphere with a fixed amount of charge.

These examples will illustrate the principles, insofar as there are any, of the method of images. Their successful application demands some intuition based on experience with problems that have been previously solved by the method. Figure 4.6 illustrates some of the problems that can be solved by images.

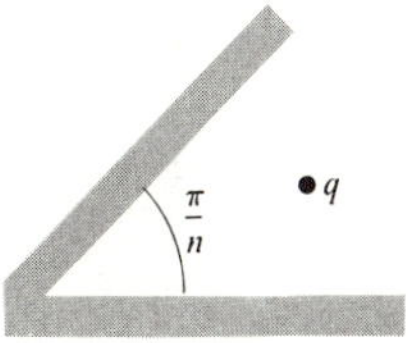

Point charge between two conducting planes that intersect at an angle π/n, with n integral.

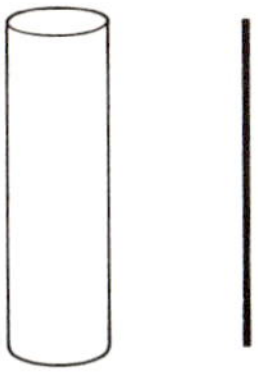

Line charge parallel to axis of infinite circular cylinder.

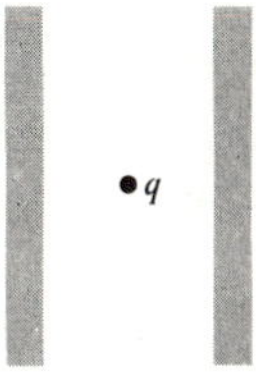

Point charge between two parallel conducting planes; an infinite set of images is required.

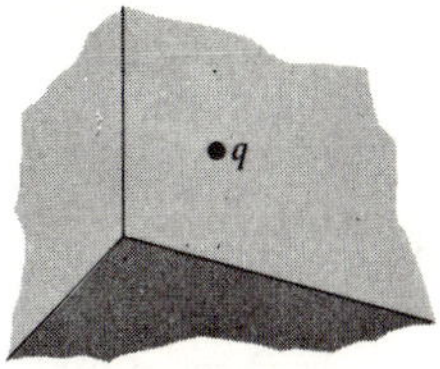

Point charge in region bounded by three mutually perpendicular conducting planes.

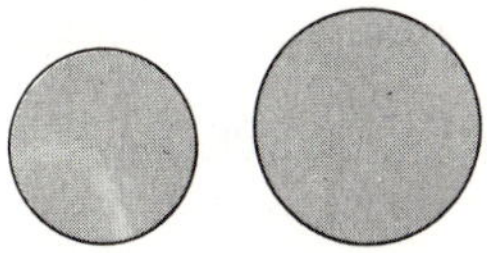

Two charged spheres; an infinite set of images is required. A special case is a sphere and an infinite conducting plane.

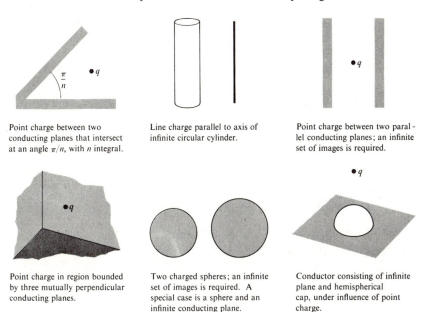

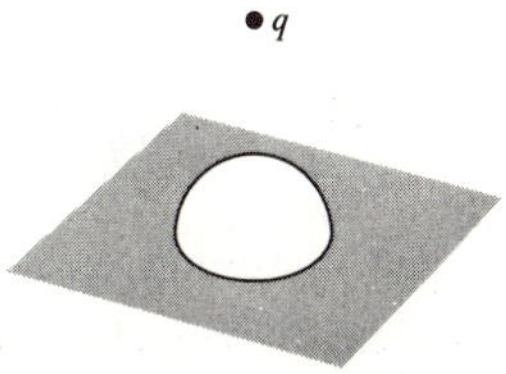

Conductor consisting of infinite plane and hemispherical cap, under influence of point charge.

Fig. 4.6 Examples of problems solvable by images. Problems involving applied point or line charge sources can also be solved for more general sources by imaging each charge element in the source.

4.5 LAPLACE'S EQUATION AND SUPERPOSITION OF ELEMENTARY SOLUTIONS

In the summation problem we encountered various functions, e.g., spherical and cylindrical harmonics, whose superposition gave a general solution of Laplace's equation. Moreover, it will not be a surprise to see in Appendix D that these same functions crop up in the solution of Laplace's equation by the method of separation of variables. In this section, we shall discuss how to use these solutions in solving boundary value problems with conductors.

The two main coordinate systems of interest in the summation problem were the cylindrical and spherical; many of the examples will perforce deal with cylinders and spheres. In addition, in Appendix D, we discuss the solution of Laplace's equation in rectangular coordinates. But even with solutions at hand for rectangular, cylindrical, and spherical coordinates, there is a limited number of physical problems that can be solved. The reason is that for any of these coordinate sets the only problems that can be solved easily are those for which the physical boundary corresponds to one of the variables having a constant value for all values of the other two; for example, the sphere is defined in spherical coordinates by setting $r =$ constant for all θ and φ. But this means that almost the only two boundaries so defined, that are of physical interest, are the sphere and the cylinder. The homogeneous problem for these two cases is solvable but is also trivial, for it is clear by symmetry that any free-charge distribution on a sphere or cylinder will have uniform surface density. Hence we turn to the inhomogeneous problem.

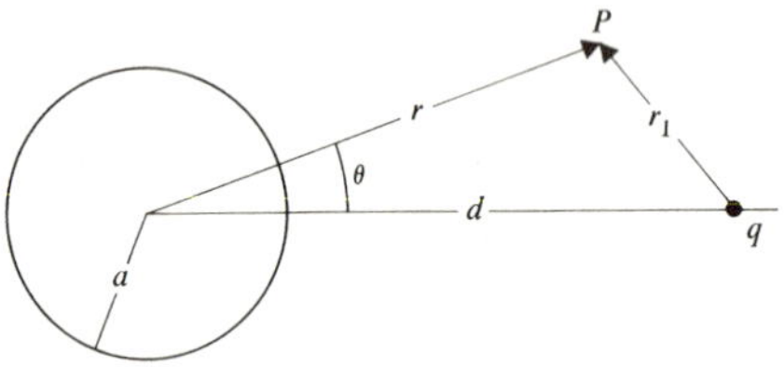

Fig. 4.7 Conducting sphere under influence of point charge.

As a first example, consider a problem that was solved by the method of images; the sphere of radius a under the influence of an external point charge q. As boundary condition we require that the sphere be at potential Φ_0; later we discuss other boundary conditions. From Fig. 4.7, it is clear that the charge density which the point charge *induces* on the sphere depends only on θ. This *induced* charge distribution by itself would produce a potential which we call Φ_i. From the results of the summation problem or from the solutions in Appendix D of Laplace's equation, the appropriate form of Φ_i *outside* the sphere is

$$\Phi_i = \sum_{l=0}^{\infty} \frac{B_l P_l(\cos\theta)}{r^{l+1}}. \tag{4.21}$$

Similarly, the point charge generates an *applied* field and a corresponding potential Φ_a which is

$$\Phi_a = \frac{q}{r_1} = \frac{q}{\sqrt{r^2 + d^2 - 2rd\cos\theta}}. \tag{4.22}$$

The total potential Φ is the sum of these two potentials,

$$\Phi = \Phi_a + \Phi_i, \tag{4.23}$$

and it is Φ which satisfies the boundary condition: $\Phi = \Phi_0$ at $r = a$. Applying this condition we can use the Legendre polynomial expansion of Φ_a in powers of r/d, since this is appropriate at $r = a$, a being less than d. Then Eq. (4.23) is

$$\Phi = \frac{q}{d}\sum_{l=0}^{\infty}\left(\frac{r}{d}\right)^l P_l(\cos\theta) + \sum_{l=0}^{\infty}\frac{B_l P_l(\cos\theta)}{r^{l+1}}, \qquad r < d. \tag{4.24}$$

The boundary condition gives

$$\Phi_0 = \sum_{l=0}^{\infty}\left(\frac{qa^l}{d^{l+1}} + \frac{B_l}{a^{l+1}}\right)P_l(\cos\theta). \tag{4.25}$$

Since the Legendre polynomials are an orthogonal set, we can equate the coefficients of $P_l(\cos\theta)$ on the two sides of Eq. (4.25). The only P_l on the left-hand side is $P_0 = 1$; we find

$$\Phi_0 = \frac{q}{d} + \frac{B_0}{a}, \qquad 0 = \frac{qa^l}{d^{l+1}} + \frac{B_l}{a^{l+1}}, \qquad l \neq 0. \tag{4.26}$$

This effectively completes the solution of the problem, for on combining (4.21), (4.22), (4.23), and (4.26), we find

$$\Phi = \frac{q}{\sqrt{r^2 + d^2 - 2rd\cos\theta}} + \frac{a(\Phi_0 - q/d)}{r} - q\sum_{l=1}^{\infty}\frac{a^{2l+1}P_l(\cos\theta)}{d^{l+1}r^{l+1}}.$$

In *using* this solution, the square root in the first term can now be expanded in powers of r/d or d/r, as appropriate. It is not difficult to verify that this solution represents an expansion in Legendre polynomials of the image solution that was found previously.

As a small variant, let us take another boundary condition: let the total charge on the sphere be Q. From Eq. (4.2) for the surface charge density σ, we have

$$Q = \int \sigma\, ds = -\frac{1}{4\pi}\int \left.\frac{\partial\Phi}{\partial r}\right|_{r=a} ds.$$

By forming $\partial\Phi/\partial r$ from Eq. (4.24) and integrating over the surface of the sphere, we find that all terms except that involving B_0 integrate to zero; the result is

$$Q = \frac{B_0}{4\pi a^2}\int ds = B_0.$$

The other coefficients B_l for nonzero l are unchanged.

As another example, consider a charged ring of radius a, with total charge Q, inside a spherical cavity excised from an infinite conductor, as shown in Fig. 4.8. The potential Φ at any point inside the sphere will be Φ_s, due to the charge induced on the sphere, plus Φ_r, the potential due to the ring:

$$\Phi = \Phi_s + \Phi_r.$$

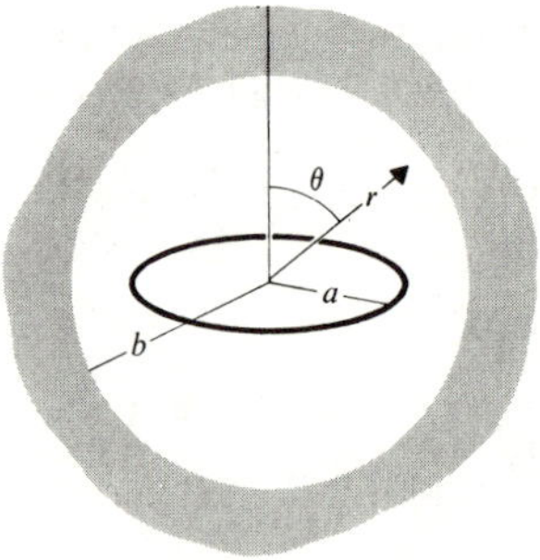

Fig. 4.8 Charged ring inside a sphere excised from an infinite conductor.

The problem of finding the potential due to the ring is a straightforward summation problem. For $r > a$, the answer is

$$\Phi_r = Q \sum_{l=0}^{\infty} (-)^l \frac{(2l-1)!!}{(2l)!!} \frac{a^{2l}}{r^{2l+1}} P_{2l}(\cos\theta).$$

Here $(2l)!! = 2, 4, 6, \ldots, 2l$ and $(2l-1)!! = 1, 3, 5, \ldots, (2l-1)$. Since Φ_r contains only even Legendre polynomials, it is obvious that the same holds true for Φ_s. Anticipating this, and remembering that the solutions must be finite at the origin, we write

$$\Phi_s = \sum_{l=0}^{\infty} C_{2l} r^{2l} P_{2l}(\cos\theta).$$

The condition that Φ be zero at $r = b$ yields

$$C_{2l} = -Q(-)^l \frac{(2l-1)!!}{(2l)!!} \frac{1}{a}\left(\frac{a}{b}\right)^{2l+1}\left(\frac{1}{b}\right)^{2l},$$

and this effectively solves the problem. The solution for $a < r < b$ is then

$$\Phi = \frac{Q}{a} \sum_{l=0}^{\infty} (-)^l \frac{(2l-1)!!}{(2l)!!}\left[\left(\frac{a}{r}\right)^{2l+1} - \left(\frac{a}{b}\right)^{2l+1}\left(\frac{r}{b}\right)^{2l}\right] P_{2l}(\cos\theta), \qquad a < r.$$

For $r < a$ we must use the expansion for the ring potential in powers of r/a; Φ_s remains unchanged, and we find

$$\Phi = \frac{Q}{a} \sum_{l=0}^{\infty} (-)^l \frac{(2l-1)!!}{(2l)!!}\left[\left(\frac{r}{a}\right)^{2l} - \left(\frac{a}{b}\right)^{2l+1}\left(\frac{r}{b}\right)^{2l}\right] P_{2l}(\cos\theta), \qquad r < a.$$

In the above discussions, we have used a physical point of view that might be summarized as follows: Charge and charge alone can produce a potential. If in some problem we pinpoint *all* the different sources of charge and write an expression for the potential produced by each source, the sum of these partial potentials is the total potential that satisfies the boundary conditions. To illustrate this again and to make one further point, consider a problem which really is a special case of the one on page 60: a grounded sphere in a uniform external field of strength E_0 in the z-direction. This *is* a special case, since the uniform external field can be thought of as produced by an increasingly strong point charge which is placed ever farther from the sphere, and whose field then becomes more and more uniform over the sphere. From the above physical point of view, this charge at infinity is the source of the applied potential Φ_a which, in spherical coordinates r, θ, is

$$\Phi_a = -E_0 z = -E_0 r \cos\theta = -E_0 r P_1(\cos\theta). \tag{4.27}$$

To this we must add the potential Φ_i due to the induced charges. As before, this has the form (4.21), so the total potential Φ is

$$\Phi = -E_0 r P_1(\cos\theta) + \sum_{l=0}^{\infty} \frac{B_l}{r^{l+1}} P_l(\cos\theta). \tag{4.28}$$

The condition that $\Phi = 0$ when $r = a$ then determines the coefficients to be

$$B_1 = E_0 a^3, \qquad B_l = 0, \qquad l \neq 1,$$

and the solution is

$$\Phi = E_0\left(\frac{a^3}{r^2} - r\right) P_1(\cos\theta). \tag{4.29}$$

Now we look at this problem from a slightly different and more abstract viewpoint, which leads of course to the same answer but is sometimes illuminating. Instead of pinpointing the sources of potential, we begin with the general solution of Laplace's equation that is appropriate to azimuthal symmetry:

$$\Phi = \sum_{l=0}^{\infty}\left(A_l r^l + \frac{B_l}{r^{l+1}}\right) P_l(\cos\theta). \tag{4.30}$$

The new viewpoint is that we look on the behavior of the solution at infinity as one of the *boundary conditions*, although there is, of course, no physical boundary at infinity. We might better call this behavior a *limiting condition* or a *boundary condition at infinity*. But whatever its name, it implies that for large r the solution must behave like $-E_0 r P_1(\cos\theta)$, and since for very large r the terms in $1/r^{l+1}$ in (4.30) vanish, this means that

$$A_1 = -E_0, \qquad A_l = 0, \qquad l \neq 1.$$

Using the second *physical* boundary condition $\Phi = 0$ at $r = a$, we retrieve the previous result (4.28) but from this slightly different point of view.

As a final problem, we shall consider a grounded sphere for which the external field is not due to a point charge but to a dipole. We refer to Fig. 4.9, which shows a

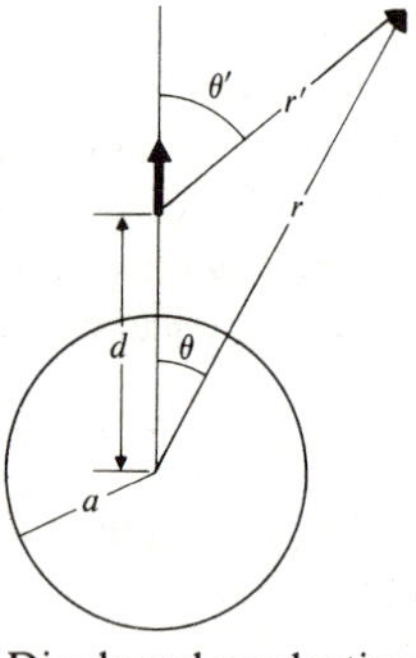

Fig. 4.9 Dipole and conducting sphere.

dipole of moment m on the z-axis and in the z-direction, a distance d from the center of the sphere.

It is convenient initially to work in the two coordinate systems shown in the figure. In the primed system, the applied dipole potential Φ_d is

$$\Phi_d = \frac{m \cos \theta'}{r'^2}; \tag{4.31}$$

and in the unprimed system, the induced potential Φ_i due to the charge on the sphere is

$$\Phi_i = \sum_{l=0}^{\infty} \frac{B_l P_l(\cos \theta)}{r^{l+1}}. \tag{4.32}$$

The total potential Φ is, of course, $\Phi_d + \Phi_i$, and it is Φ that must be zero on the surface of the sphere. To use this condition fruitfully we must express the dipole potential $\cos \theta'/r'^2$ in a Legendre polynomial expansion in $P_l(\cos \theta)$. We can get such a formula if we first recall how in Section 3.2 dipole potentials were generated by differentiating the potential of a point charge. Consider then $1/r'$,

$$\frac{1}{r'} = \frac{1}{\sqrt{x^2 + y^2 + (z - d)^2}}.$$

It can be considered to be the potential of a unit charge a distance d up the z-axis. If we differentiate this with respect to d, we find

$$\frac{\partial}{\partial d}\left(\frac{1}{r'}\right) = \frac{(z - d)}{[x^2 + y^2 + (z - d)^2]^{3/2}} = \frac{P_1(\cos \theta')}{r'^2}. \tag{4.33}$$

But for $r < d$, we also have

$$\frac{\partial}{\partial d}\left(\frac{1}{r'}\right) = \frac{\partial}{\partial d} \sum_{s=0}^{\infty} \frac{r^s}{d^{s+1}} P_s(\cos \theta) = -\sum_{s=0}^{\infty} \frac{(s + 1)}{d^{s+2}} r^s P_s(\cos \theta). \tag{4.34}$$

Equating (4.33) and (4.34) we have

$$\frac{P_1(\cos\theta')}{r'^2} = -\sum_{s=0}^{\infty} \frac{(s+1)r^s}{d^{s+2}} P_s(\cos\theta), \qquad r < d. \tag{4.35}$$

With this formula we can now apply the boundary condition that $\Phi = 0$ at $r = a$ to find, in a standard way,

$$B_l = \frac{m(l+1)a^{2l+1}}{d^{l+2}},$$

and this effectively completes the solution of the problem.

The problem of higher-order multipoles external to a sphere can be solved in much the same way. We need an analogous formula to (4.35), which is quoted here for reference:

$$\frac{P_l(\cos\theta')}{r'^{l+1}} = \frac{(-)^l}{d^{l+1}} \sum_{s=0}^{\infty} \frac{(l+s)!}{l!s!}\left(\frac{r}{d}\right)^s P_s(\cos\theta), \qquad r < d. \tag{4.36}$$

For $r > d$ the corresponding equation is

$$\frac{P_l(\cos\theta')}{r'^{l+1}} = \frac{1}{d^{l+1}} \sum_{s=0}^{\infty} \frac{(l+s)!}{l!s!}\left(\frac{d}{r}\right)^{s+l+1} P_{s+l}(\cos\theta), \qquad r > d. \tag{4.37}$$

We have confined the examples above to spherical geometries, but it is clear that the same physics applies in the two-dimensional cylindrical case. The same is true for the more exotic coordinate systems, but the mathematical details become more complicated.

4.6 COMPOSITE PROBLEMS

Thus far we have discussed boundary value problems for which the conducting surfaces have been mainly spheres, cylinders, or half-planes, since these are the easiest to handle. Now we shall extend the physical discussion to problems in which there is more than one of these kinds of surface; we call such problems *composite*. Examples might include problems involving two or more spheres, perhaps of different radii, or a sphere and cylinder, cylinder and half-plane, and so forth.

As we have mentioned, some problems of this type can be solved by separating Laplace's equations in special coordinate systems. For example, the problem of two cylinders can be solved in bipolar coordinates, as discussed by Wendt (R). But when there are more than two elements, or when the elements are of different kinds, e.g., a sphere and a plane, these special coordinate systems are not applicable. Thus the present method is rather more general; moreover it can be extended to problems involving time-dependent electromagnetic fields, i.e., scattering problems, and to even more general problems. These are discussed by Eyges (R).

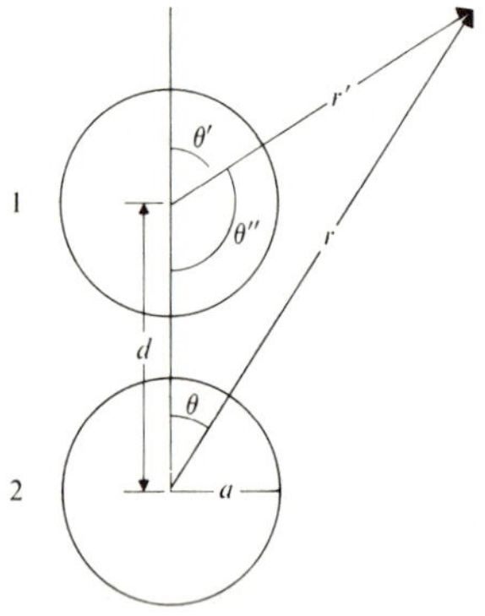

Fig. 4.10 Two conducting spheres, each with charge Q.

We shall present the method by means of an example. Two ungrounded spheres of radius a are a distance d apart; each has total charge Q. How is the charge distributed over the surface of each sphere? To begin the discussion we set up two coordinate systems as shown in Fig. 4.10. From the physical viewpoint discussed above there are two sources of field and potential, the charge on sphere 1 and the charge on sphere 2:

$$\Phi = \Phi_1 + \Phi_2.$$

The potential Φ_1 produced by the charge on 1 is

$$\Phi_1 = \sum_{l=0}^{\infty} \frac{A_l P_l(\cos\theta')}{r'^{l+1}},$$

and Φ_2 is

$$\Phi_2 = \sum_{l=0}^{\infty} \frac{B_l P_l(\cos\theta)}{r^{l+1}},$$

where A_l and B_l are coefficients to be determined. It is clear that the set A_l and the set B_l are not completely independent; from the symmetry of the problem there must be some relation between them. To see this, write Φ_1 in terms of r' and the *interior* angle $\theta'' = \pi - \theta'$. Then

$$P_l(\cos\theta') = (-)^l P_l(\cos(\pi - \theta')) = (-)^l P_l(\cos\theta'').$$

The charge on the first sphere must be the same function of θ'' that the charge on the second is of θ, whence it follows that $A_l = (-)^l B_l$. Finally, it will be convenient to rewrite the coefficients according to

$$B_l = a^{l+1} C_l.$$

The potential Φ then becomes

$$\Phi = \sum_{l=0}^{\infty} \frac{(-)^l a^{l+1} C_l P_l(\cos\theta')}{r'^{l+1}} + \sum_{s=0}^{\infty} \frac{a^{s+1} C_s P_s(\cos\theta)}{r^{s+1}}. \tag{4.38}$$

If in (4.38) we replace $P_l(\cos\theta')/r'^{l+1}$ with the expansion (4.36), we find the expression for Φ valid near $r = a$:

$$\Phi = \sum_{s=0}^{\infty} P_s(\cos\theta)\left\{ C_s\left(\frac{a}{r}\right)^{s+1} + \left(\frac{r}{d}\right)^s \sum_{l=0}^{\infty} C_l\left(\frac{a}{d}\right)^{l+1} \frac{(l+s)!}{l!s!}\right\}. \tag{4.39}$$

The potential is a constant, Φ_0, at $r = a$. (We shall later enforce the requirement that the total charge on a sphere be Q). Using this condition of constant potential we find from (4.39) that

$$\Phi_0 = C_0 + \sum_{l=0}^{\infty} C_l\left(\frac{a}{d}\right)^{l+1}, \tag{4.40}$$

$$0 = C_s + \sum_{l=0}^{\infty} C_l\left(\frac{a}{d}\right)^{s+l+1} \frac{(l+s)!}{l!s!}, \qquad s = 1, 2, \ldots. \tag{4.41}$$

This is an infinite set of equations in the coefficients C_l, and it is not easy to see how to solve them exactly. It is not difficult, however, to solve them approximately by means of a "perturbation expansion" in the parameter a/d, where $a/d < \frac{1}{2}$ always. We know the solution when a/d approaches zero, i.e., when the spheres are infinitely far apart. In this case, it is obvious physically that the potential on each will be spherically symmetric; mathematically, this emerges from Eqs. (4.40) and (4.41) in that only C_0 is different from zero. Beyond this these equations can be solved to whatever power of a/d we like by retaining terms up to that power in the set. Before doing this we shall stipulate that the total charge on a sphere must be Q. Forming the charge density $\sigma = -(1/4\pi)(\partial\Phi/\partial r)$ from Eq. (4.39) and requiring that this density integrate to Q over sphere 2, we find, on using the orthogonality of the $P_s(\cos\theta)$, that $C_0 = Q/a$. This equation replaces (4.40), and (4.41) now enables us to calculate the higher C_s in terms of C_0 to whatever power of a/d we wish. For example, to the third power we get

$$C_1 = -\frac{Q}{a}\frac{(a/d)^2}{(1 + 2(a/d)^3)}, \qquad C_2 = -\frac{Q}{a}\left(\frac{a}{d}\right)^3.$$

4.7 INTEGRAL EQUATIONS

In this section we discuss the use of the integral equations (4.4) and (4.6) in solving boundary value problems. These equations for the unknown charge distributions have the advantage of being valid for any surface, not only those for which separated solutions of Laplace's equation exist. Moreover, they reduce by one the dimensionality of the unknown function, since a three-dimensional problem involves an

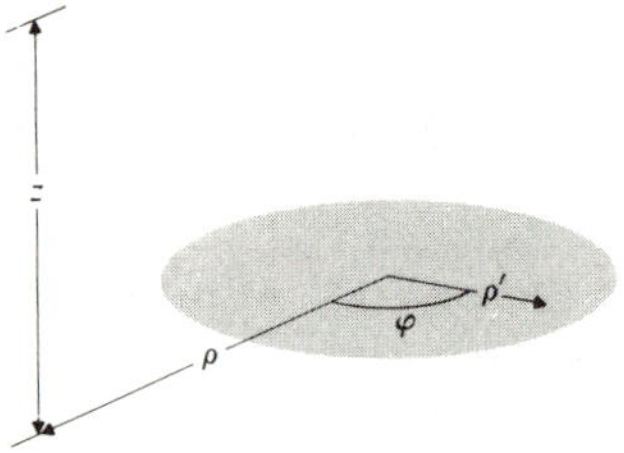

Fig. 4.11 Coordinates for the charged disc.

unknown charge distribution on a surface, that is, a function of two variables, and a two-dimensional problem reduces to an unknown function of one variable. The disadvantage of such equations is that they are not easily solved analytically, and those special examples that can be solved involve rather sophisticated mathematical techniques, in particular, the theory of functions of a complex variable. On the other hand, they can frequently be solved numerically to adequate accuracy on modern computers. We shall illustrate these remarks by presenting an analytic example and some examples of numerical solution.

For the analytic example, we take the homogeneous problem of the potential due to charge on an infinitely thin disk. This problem was first discussed by Green in 1838, and then solved by Weber in 1873 using so-called *dual integral equations*, whose solution relied on properties of discontinuous integrals involving Bessel functions. Another solution was given in 1884 by Lord Kelvin, who used the known separated solutions of the Laplace equation in ellipsoidal coordinates. These coordinates become those of an oblate (watch-shaped) spheroid if two of the ellipsoidal axes are taken equal, and become essentially those of a disk if the third axis becomes very small. We mention this chronology to emphasize that the relatively simple solution we are about to discuss, which is due to Copson (R), is far from obvious. The solution involves a relatively small excursion into the complex plane, for which we ask the indulgence of the reader who may not be familiar there.

Consider then a charged disk of radius a in the plane $z = 0$ of the coordinate system shown in Fig. 4.11. The field point has the cylindrical coordinates ρ and z. The surface density σ, which is taken to be the *combined* density on both sides of the disk, can be a function only of ρ', $\sigma = \sigma(\rho')$. By superposition the potential Φ is

$$\Phi(\rho, z) = \int_0^a \sigma(\rho')\rho'\,d\rho' \int_0^{2\pi} \frac{d\varphi}{\sqrt{\rho^2 + \rho'^2 - 2\rho\rho'\cos\varphi + z^2}}. \tag{4.42}$$

On the disk, i.e., for $z = 0$ and $0 < \rho < a$, this potential is a constant which we can take to be unity; with this condition, Eq. (4.42) becomes an integral equation for the unknown $\sigma(\rho')$:

$$1 = \int_0^a \sigma(\rho')\rho'\,d\rho' \int_0^{2\pi} \frac{d\varphi}{\sqrt{\rho^2 + \rho'^2 - 2\rho\rho'\cos\varphi}}, \qquad 0 < \rho < a. \tag{4.43}$$

This equation is, of course, a particular example of Eq. (4.4).

Copson's solution begins with a formula for transforming the φ integral in Eq. (4.43). We state the formula and then prove it. It reads

$$\int_0^{2\pi} \frac{d\varphi}{\sqrt{\rho^2 + \rho'^2 - 2\rho\rho' \cos\varphi}} = 4\int_0^{\mathrm{Min}(\rho,\rho')} \frac{dx}{\sqrt{(\rho^2 - x^2)(\rho'^2 - x^2)}}, \tag{4.44}$$

where $\mathrm{Min}(\rho, \rho')$ is the lesser of ρ and ρ'. To prove this formula consider first the following integral, in which we assume $b < 1$:

$$\int_0^{2\pi} \frac{d\varphi}{\sqrt{1 + b^2 - 2b\cos\varphi}}.$$

Transformed to the complex variable $z = e^{i\varphi}$, the integral becomes one over the unit circle C in the complex plane:

$$\int_0^{2\pi} \frac{d\varphi}{\sqrt{1 + b^2 - 2b\cos\varphi}} = \frac{1}{i}\int_C \frac{dz}{\sqrt{z(z-b)(1-bz)}};$$

the branch of the integrand is taken to be that which has the value $1 - b$ when $z = b$. The branch cuts extend from zero to b and from $1/b$ to infinity. For $b < 1$ the integral is regular over C, and can be shrunk down to a loop that just surrounds that portion of the cut which is between zero and b. The top and bottom parts of this flat loop contribute equally, so we have

$$\begin{aligned}\int_0^{2\pi} \frac{d\varphi}{\sqrt{1 + b^2 - 2b\cos\varphi}} &= 2\int_0^b \frac{dx}{\sqrt{x(b-x)(1-bx)}} \\ &= 4\int_0^b \frac{dy}{\sqrt{(b^2 - y^2)(1 - y^2)}}\end{aligned} \tag{4.45}$$

The third integral in this equation is derived from the second by the substitution $y^2 = bx$.

Now we return to the proof of Eq. (4.44). For definiteness, assume $\rho'/\rho < 1$ so that $\mathrm{Min}(\rho', \rho)$ is ρ'. Take a factor ρ from the square root of the integral on the left-hand side of Eq. (4.44), and thereby write it in a form to which (4.45) can be applied, with $b = \rho'/\rho$. This gives

$$\frac{1}{\rho}\int_0^{2\pi} \frac{d\varphi}{\sqrt{1 + (\rho'/\rho)^2 - 2(\rho'/\rho)\cos\varphi}} = \frac{4}{\rho}\int_0^{\rho'/\rho} \frac{dy}{\sqrt{((\rho'/\rho)^2 - y^2)(1 - y^2)}}.$$

The substitution $y = x/\rho$ yields the formula which, with its counterpart for $\rho/\rho' < 1$, is comprised in Eq. (4.44).

We return to the integral equation (4.43) and break the ρ' integration into one from zero to ρ and one from ρ to a. If we use the appropriate version of (4.44) in each of these two parts, it becomes

$$\frac{1}{4} = \int_0^\rho \rho'\sigma(\rho') \int_0^{\rho'} \frac{dx}{\sqrt{(\rho'^2 - x^2)(\rho^2 - x^2)}} d\rho' + \int_\rho^a \rho'\sigma(\rho') \int_0^{\rho} \frac{dx}{\sqrt{(\rho'^2 - x^2)(\rho^2 - x^2)}} d\rho'.$$

By inverting the orders of integration, this can be condensed to

$$\frac{1}{4} = \int_0^\rho \left(\int_x^a \frac{\sigma(\rho')\rho'\, d\rho'}{\sqrt{\rho'^2 - x^2}} \right) \frac{dx}{\sqrt{\rho^2 - x^2}}. \tag{4.46}$$

The virtue of the transformations above is that they have reduced the original equation (4.43), which involved the unknown $\sigma(\rho')$ in a *double integral*, to two integral equations, each involving only a *single integral*. For on defining $S(x)$ by

$$S(x) = \int_x^a \frac{\sigma(\rho')\rho'\, d\rho'}{\sqrt{\rho'^2 - x^2}}, \tag{4.47}$$

we can write Eq. (4.46) as

$$1 = 4 \int_0^\rho \frac{S(x)\, dx}{\sqrt{\rho^2 - x^2}}. \tag{4.48}$$

We may first solve Eq. (4.48) for $S(x)$ and use the result to solve Eq. (4.47) for $\sigma(\rho')$. Copson's solution of these equations depended on the observation that they could be transformed to the *Abel integral equation*, whose solution is well known. Here we shall simply confirm his result. The solution of Eq. (4.48) for $S(x)$ is

$$S(x) = (1/2\pi);$$

this is immediately verified by the substitution $x = \rho \sin\theta$ in (4.48). Equation (4.47) then becomes

$$\frac{1}{2\pi} = \int_x^a \frac{\sigma(\rho')\rho'\, d\rho'}{\sqrt{\rho'^2 - x^2}},$$

and by standard integration formulae it can be verified that the solution is

$$\sigma(\rho') = \frac{1}{\pi^2\sqrt{a^2 - \rho'^2}}. \tag{4.49}$$

The charge density is singular at the edge; a singularity is not uncommon in problems involving geometries with edges. The inhomogeneous problem of the grounded disk in a uniform external field has also been solved by Copson, by an extension of the method above.

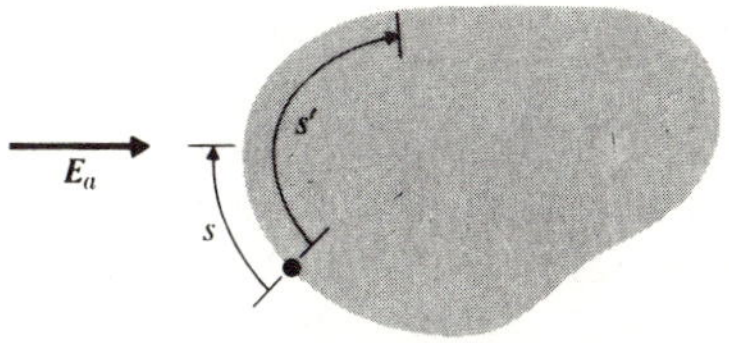

Fig. 4.12 Cylinder of arbitrary cross section in an applied field $\boldsymbol{E}_a$.

We now consider *numerical* solutions of the integral equations, and begin with the inhomogeneous equation corresponding to the two-dimensional problem of an infinitely long cylinder in a field perpendicular to its axis. The specific example we treat will be that of a cylinder of rectangular cross section in a uniform field, but we shall begin the development by considering a cylinder of arbitrary cross section in an arbitrary applied field $\boldsymbol{E}_a$ whose potential is Φ_a. Figure 4.12 shows such a cylinder. Let a variable s measure distance around the perimeter of the cylinder. The external potential Φ_a is, in general, a function of variables in space but on the cylinder it can be considered a function of s, $\Phi_a(s)$. The unknown surface charge density per unit length, called λ, is a function $\lambda(s')$ of a similar variable s'. The potential produced at s by the charge in the strip of width ds' is $-2\lambda(s')\ln D(s, s')ds'$, where $D(s, s')$ is the (straight-line) distance between the point at s and that at s'. The equation expressing the fact that the total potential is constant as a function of s is then

$$\Phi_a(s) - 2\int \lambda(s') \ln [D(s, s')]\, ds' = \text{constant}. \tag{4.50}$$

We shall indicate only one of the simpler methods of solving equations of this type; refinements are discussed by Harrington (R).

We begin by breaking the full range of s into N equal parts, and letting s_i be the perimetric distance to the midpoint of the ith part. We can evaluate Eq. (4.50) at these values of s_i to find, still exactly,

$$\Phi_a(s_i) - 2\int \lambda(s') \ln [D(s_i, s')]\, ds' = \text{constant}, \qquad i = 1, \ldots, N. \tag{4.51}$$

A simple approximation to this set comes about if we replace the integral by its approximate expression as a sum. Taking the width of each of the N parts to be Δ, we have

$$\Phi_a(s_i) - 2\Delta \sum_{j=1}^{N} \lambda(s_j) \ln D(s_i, s_j) = \text{constant}, \qquad i = 1, 2, \ldots, N. \tag{4.52}$$

This is now a set of N inhomogeneous equations in the unknowns $\lambda(s_j)$, which can be solved on a computer. Various practical details must be decided in a numerical solution of these equations. The number N must be large enough to assure that the discrete sum is a good approximation to the continuous integral; symmetries may exist that effectively reduce the number of unknowns. More importantly, when $i = j$ in the sum in Eq. (4.52), there is a logarithmic singularity. The source of this singular term is the potential at s_j due to the charge in the strip centered at s_j. Obviously, the infinite expression $-2\Delta \ln D(s_j, s_j)$ does not constitute a valid approximation to this potential. The remedy is to replace this term by an accurate expression for the potential at the center of the strip of width Δ due to charge uniformly distributed over it. With this modification Eq. (4.52) becomes perfectly regular.

So much for the general case. The specific results for the rectangular cylinder in a homogeneous field are taken from the work of Mei and Van Bladel (R). The geometry is shown in Fig. 4.13 along with the results for the relative charge density on either side of the cylinder. In these calculations the sides were typically divided into 20 intervals. The results shown in Fig. 4.13 are, of course, smoothed-out versions of the computer results.

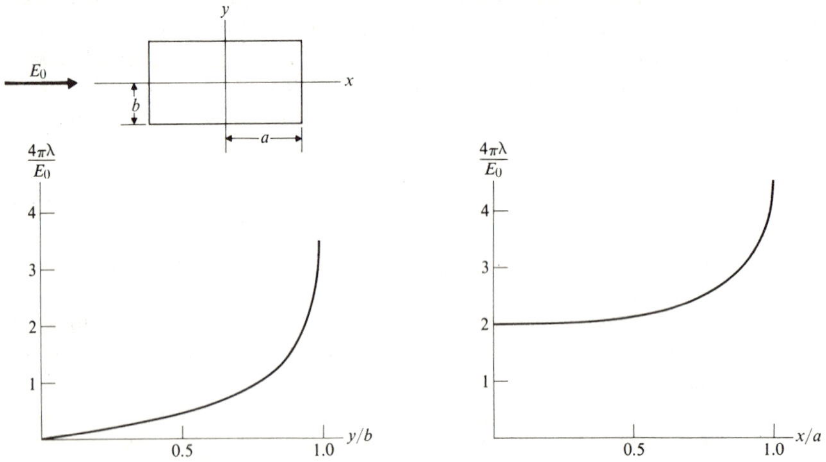

Fig. 4.13 Charge distribution on the sides of a rectangular cylinder in a uniform external field E_0.

Another interesting problem that has been solved numerically is that of the capacitance of a charged cube. If charge Q is introduced onto a conducting cube, the charge will distribute itself in a manner which will make the surface of the cube an equipotential at some potential Φ_0 that is unknown *a priori*. The capacitance C is then the ratio Q/Φ_0. Once the distribution of charge has been found numerically, Φ_0, and hence the capacitance can be determined. Details are found in Reitan (R). Prior to the numerical calculation carried out by Reitan, the capacitance had been bounded by various analytic approximation methods, and the best result had been $0.62211a < C < 0.71055a$, with a being the cube side. The numerical calculation yielded the result $C = 0.655a$.

PROBLEMS

1. A point charge q is a distance b from the center of two concentric grounded spherical conducting surfaces of radii a and c $(a < b < c)$. Show that for $a < r < b$, the potential is

$$\Phi = q \sum_{n=0}^{\infty} \frac{b^{2n+1} - c^{2n+1}}{b^{n+1}(a^{2n+1} - c^{2n+1})}\left(r^n - \frac{a^{2n+1}}{r^{n+1}}\right) P_n(\cos\theta),$$

where the z-axis is the diameter through the charge.

2. A point dipole of moment $\boldsymbol{p}$ is a distance d from the closer face of a perfectly conducting plane slab of thickness t, and the vector $\boldsymbol{p}$ is directed toward that face. Find the charge distributions on both faces of the slab.

3. A charge q is placed between two infinite perpendicular conducting planes, a perpendicular distance d from each plane. Find the force on the charge.

4. Find the image, or images, that solves the problem of a line charge with linear density λ which is parallel to an infinitely long conducting circular cylinder.

5. The center of a conducting sphere of radius a that carries total charge Q is a distance d from an infinite plane conducting boundary. Find the force on the sphere in an expansion in powers of a/d, including terms up to $(a/d)^3$.

6. A spherical conducting shell occupies the region $a < r < b$. Find the field everywhere, assuming that (a) a point charge q is in the interior $(r < a)$ of the shell at a distance d from the center; (b) the point charge is outside the shell.

7. A point charge *outside* a hollow conducting sphere will produce no field in the hollow interior; the metal "shields" the interior. The solution to Problem 6 shows that a point charge *inside* the hollow interior will produce a field outside; the same metal ceases to act as a "shield." Explain.

8. Show that the equipotential surfaces of a uniformly charged two-dimensional (infinitely long) strip are elliptic cylinders.

9. A point charge q is a distance d from the center of a charged conducting sphere of radius a. What is the minimum *total* charge on the sphere so that the charge *density* on it has the same sign everywhere?

10. A charge q is a perpendicular distance d from each of two parallel, infinite, conducting planes. Find an expression for the induced surface charge density on the planes.

11. A conducting sphere has total charge Q on it, and is in the field of two external point charges q and $-q$ that are each a distance d from the center of the sphere and are colinear with the center. Find the charge distribution on the sphere.

12. A dipole of moment $\boldsymbol{p}$ is a distance d from the center of a grounded conducting sphere and points to the center of it. What is the charge distribution on the sphere?

13. An infinite plane conducting slab has thickness a. On an axis perpendicular to the slab is a charge q_1 to the left of it, a distance d_1 from the left-hand face, and a charge q_2 to the right, a distance d_2 from the right-hand face. Find the field everywhere.
14. A dipole with fixed moment p_0 is at the center of a conducting spherical shell with inner and outer radii a and b. A uniform applied field $\boldsymbol{E}_0$ is perpendicular to the dipole direction. Find the potential everywhere.
15. Solve numerically the appropriate integral equation for the charge distribution on a two-dimensional (infinitely long) conducting strip.
16. Find the images that solve the problem of a point charge and a plane with a hemispherical cap (Fig. 4.6).

REFERENCES

Copson, E. T., "On the Problem of the Electrified Disc," *Proc. Edin. Math. Soc.* **8**, 14 (1947).

Eyges, Leonard, "Some Nonseparable Boundary Value Problems and the Many Body Problem," *Ann. Phys.* **2,** 101 (1957).

The idea of treating certain nonseparable boundary problems as composite ones is fruitful beyond electrostatics. It can be applied to multi-center potential problems in quantum mechanics and to the quantum mechanical three and many body problems. When the central features of the method, namely the decomposition of amplitudes and simultaneous use of different coordinate systems, is applied to the three-body scattering problem, it leads to the so-called *Fadeev equations.*

Harrington, R. F., *Field Computation by Moment Methods,* Macmillan, New York (1968).

A discussion of the numerical solution, by digital computer, of electromagnetic field problems, frequently by formulating them as an integral equation and then solving the equation as an approximate set of linear equations. Chapter 2 is on electrostatics. There is an extensive set of references.

Hess, J. L., and A. M. O. Smith, "Calculation of Potential Flow about Arbitrary Bodies," *Prog. Aero. Sciences* **8,** Pergamon Press, London (1967).

The problem discussed in this paper is not an electrostatic one but is closely related in that it involves the solution of Laplace's equation, with certain boundary conditions. The technique involves formulating the problem as an integral equation which is then solved by breaking it up into an approximate set of linear equations. What is of present interest is the considerable discussion of the techniques of efficiently solving these equations on a computer.

Mei, K., and J. van Bladel, "Low Frequency Scattering by Rectangular Cylinders," *IEEE Trans. on Antennas and Prop.*, **AP-11,** 52–56; January 1963.

Reitan, D. K., and T. J. Higgins, "Calculation of the Electrical Capacitance of a Cube," *J. Appl. Phys.* **22,** 223 (1951).

Wendt, G., Statische Felder und Stationäre Ströme, in *Handbuch der Physik,* Band XVII, Springer, Berlin (1958).

5
GENERAL BOUNDARY VALUE PROBLEMS

5.1 DIRICHLET AND NEUMANN PROBLEMS

In this chapter, we generalize boundary value problems with conductors to a wider class. For a conductor of given shape, the mathematical problem of the last chapter was to find a solution of Laplace's equation which was constant over a surface of that shape. By specifying the potential as a constant at every point on a closed surface, we determined a unique solution of Laplace's equation. The generalization of this problem will center on two questions. First, beyond conductors and the boundary condition of constant potential, what are the *general* boundary conditions that specify a unique solution of Laplace's equation? For example, is it enough (or perhaps too much) to prescribe the potential not as a constant but as a function defined at every point on a closed surface? The answer to this question is somewhat complicated and relies on the theory of partial differential equations, but we shall take the answer given by theory and try to make it plausible. With the first question answered, we then ask: Given the appropriate boundary conditions, how does one actually find solutions of Laplace's equation that satisfy them?

The first question will be clarified by considering the analogous problem with an ordinary differential equation. We shall take it to be of the second order, since such equations are familiar from mechanics:

$$\frac{d^2y}{dx^2} = f(x)\frac{dy}{dx} + g(x)y(x) + h(x). \tag{5.1}$$

To solve (5.1) initial conditions must be specified; we may assume that these consist of specifying the function and its derivative at some point, i.e., of specifying $y(x_0)$ and $y^{(1)}(x_0)$.* In mechanics, for example, where y may represent a displacement and x time, this simply corresponds to the familiar specification of an initial position and velocity. But why are just $y(x_0)$ and $y^{(1)}(x_0)$ needed, no more and no less? The answer becomes clear if we imagine the solution of (5.1) expanded in a

* We use the notation

$$\left.\frac{d^ny}{dx^n}\right|_{x=x_0} = y^{(n)}(x_0).$$

Taylor series about the point x_0; for this pedagogical purpose, we assume that the functions $f(x)$, $g(x)$, $h(x)$ are regular and differentiable. A Taylor series solution has the form

$$y(x) = y(x_0) + y^{(1)}(x_0)x + y^{(2)}(x_0)\frac{x^2}{2!} + \cdots$$

and to obtain it we must know not only $y(x_0)$ and $y^{(1)}(x_0)$ but also all the higher derivatives, $y^{(2)}, y^{(3)}, \ldots$. But if we assume that $y(x_0)$ and $y^{(1)}(x_0)$ are given, we may use them to calculate the higher derivatives. Thus $y^{(2)}(x_0)$ is given directly as a function of these two quantities by evaluating (5.1) at $x = x_0$. Similarly, $y^{(3)}(x_0)$ is expressible in terms of $y^{(2)}$ and $y^{(1)}$ by differentiating (5.1) once, and so on for $y^{(4)}$, $y^{(5)}$, etc. Thus the Taylor series solution *can* be constructed from the initial conditions. More conditions than these are excessive; for example, $y^{(2)}(x_0)$ cannot be prescribed *independently*. Fewer conditions do not suffice to give an unambiguous solution.

Consider now the analogous problem for Laplace's equation which we shall for simplicity assume to be two-dimensional:

$$\frac{\partial^2 \Phi}{\partial x^2} + \frac{\partial^2 \Phi}{\partial y^2} = 0. \tag{5.2}$$

Can we by analogy fix Φ and some derivatives, for example, $\partial\Phi/\partial x$ and $\partial\Phi/\partial y$, at some point x_0, y_0 and so construct a (two-dimensional) Taylor series in the manner above? It is easy to see that this is not possible. To construct the Taylor series all possible mixed partial derivatives are needed; but a little reflection shows that if a finite number of derivatives is assigned initially, it is not possible to construct all the others from the differential equation itself. This is not too surprising in retrospect, since we found for perfect conductors that initial conditions were not prescribed at a point, but over a closed curve. We adopt another mode of reasoning, based on this remark. The *one-dimensional function* $y(x)$ that satisfies the second-order equation (5.1) is fixed by the value of the function and its derivative at a *point*. One might then expect that the *two-dimensional function* satisfying Laplace's second-order partial differential equation is fixed by prescribing Φ and its first derivatives over a *line or curve*. We investigate this idea. First, it is clear that the derivatives need not be the obvious choices $\partial\Phi/\partial x$ and $\partial\Phi/\partial y$; $\partial\Phi/\partial s$ and $\partial\Phi/\partial n$, the derivatives *along* the curve and *normal* to it, will serve as well. But clearly, if $\Phi(s)$ is known, $\partial\Phi/\partial s$ is determined; Φ and $\partial\Phi/\partial s$ are not independent. We are thus led to ask about specifying Φ and $\partial\Phi/\partial n$. But there is an example which shows that, in general, both of these cannot be specified. For in the special case of the constant potential boundary condition, we have seen that $\partial\Phi/\partial n$, which is essentially the charge density, cannot be assigned *a priori* but must be calculated from the solution. It would appear from these very loose arguments that either Φ *or* $\partial\Phi/\partial n$ can be prescribed over a curve. But what about the nature of the curve?

Can it be open or must it be closed?* In electrostatics, of course, all curves were closed; it turns out that the rigorous theory of initial conditions, discussed, for example, in Morse and Feshbach (R), shows that in the present more general case they must also be closed.

The problem of determining a regular, finite Φ everywhere in the interior (exterior) of a closed boundary curve when Φ is specified *on* the curve is the two-dimensional interior (exterior) *Dirichlet problem.* The analogous problem when $\partial\Phi/\partial n$ is specified on the boundary is the *Neumann problem.* Three-dimensional problems are similarly named and classified, except that *closed curve* is replaced by *closed surface.*

5.2 SUPERPOSITION OF ELEMENTARY SOLUTIONS

In this section we present, by means of examples, a straightforward method of attacking the Dirichlet and Neumann problems. The examples are chosen primarily for their pedagogical value, rather than for their physical interest. The method uses the separated solutions of Laplace's equations that are discussed in Appendix D. Aside from this Appendix, however, many of these solutions have already appeared in the work on the summation problem.

For the first example consider the exterior Dirichlet problem for a sphere: The potential distribution on the surface of a sphere of radius a is given as a certain function $F(\theta, \varphi)$. What is the potential Φ for $r > a$? A general expression for the potential, i.e., a general solution of Laplace's equation, is the familiar

$$\Phi(r, \theta, \varphi) = \sum_{l,m} \left(A_{lm} r^l + \frac{B_{lm}}{r^{l+1}} \right) Y_{lm}(\theta, \varphi). \tag{5.3}$$

This expression must be fitted to the prescribed boundary values. The condition that the potential be finite at infinity excludes the solutions involving r^l. For $r > a$ then we have

$$\Phi(r, \theta, \varphi) = \sum_{l,m} \frac{B_{lm} Y_{lm}(\theta, \varphi)}{r^{l+1}}.$$

The boundary condition reads

$$\sum_{l,m} \frac{B_{lm}}{a^{l+1}} Y_{lm}(\theta, \varphi) = F(\theta, \varphi),$$

whence

$$B_{lm} = a^{l+1} \int Y_{lm}^*(\theta, \varphi) F(\theta, \varphi)\, d\Omega$$

* A closed curve is one for which a point that moves unidirectionally along the curve will eventually return to its starting point.

and this constitutes the solution to the problem. Similarly the *interior Dirichlet problem* with the same boundary conditions allows us to set B_{lm} in (5.3) equal to zero and to determine the A_{lm} as

$$A_{lm} = \frac{1}{a^l} \int Y^*_{lm}(\theta, \varphi) F(\theta, \varphi)\, d\Omega.$$

For the second example consider the *Neumann problem* for the plane, wherein the surface over which $\partial\Phi/\partial n$ is prescribed is the plane $z = 0$, and the volume in which the solution is sought is the upper half-space $z > 0$. We assume that $\partial\Phi/\partial z$ at $z = 0$ is a known function $H(x, y)$,

$$\left.\frac{\partial \Phi}{\partial z}\right|_{z=0} = H(x, y), \tag{5.4}$$

and want to find a potential Φ that satisfies Laplace's equation for $z > 0$ and the boundary condition (5.4). We begin with the elementary separated solutions $\Phi_{\alpha\beta}$ of Laplace's equation derived in Appendix D and write

$$\Phi_{\alpha\beta} = e^{i(\alpha x + \beta y)} e^{-\sqrt{\alpha^2 + \beta^2} z}.$$

Aside from the Appendix, it is trivial to verify directly that this is a solution of Laplace's equation for any value of α and β. A sum (or integral) over α and β remains a solution, as does such an integral when weighted with some function $h(\alpha, \beta)$. We take such a weighted integral as an expression for Φ which can be fitted to the boundary values,

$$\Phi(x, y, z) = \frac{1}{2\pi} \int_{-\infty}^{\infty} \int_{-\infty}^{\infty} h(\alpha, \beta) e^{i(\alpha x + \beta y)} e^{-\sqrt{\alpha^2 + \beta^2} z}\, d\alpha\, d\beta. \tag{5.5}$$

Assuming that this solution can be differentiated under the integral sign, we have from (5.4)

$$H(x, y) = -\frac{1}{2\pi} \int_{-\infty}^{\infty} \int_{-\infty}^{\infty} \sqrt{\alpha^2 + \beta^2}\, h(\alpha, \beta) e^{i(\alpha x + \beta y)}\, d\alpha\, d\beta.$$

This equation is now in a form to which the Fourier integral theorem is applicable. That is, it can be inverted to yield

$$h(\alpha, \beta) = -\frac{1}{2\pi\sqrt{\alpha^2 + \beta^2}} \iint H(x, y) e^{-i(\alpha x + \beta y)}\, dx\, dy. \tag{5.6}$$

Equation (5.6) solves the problem in principle. For a given $H(x, y)$ the integral defines $h(\alpha, \beta)$, and the potential is then given by putting this function in (5.5). Whether or not the integration can be carried out for any practical case is, of course, another question.

Our third example involves a summation rather than an integration over elementary solutions, and illustrates the *quantization of separation parameters.*

Consider the two-dimensional interior Dirichlet problem for the rectangular boundary shown in Fig. 5.1. The boundary conditions we shall adopt are that Φ be zero on the three sides of the rectangle shown, and that it reduce to a known function $f(x)$ on the lower side, i.e., for $y = 0$ $(0 < x < a)$.

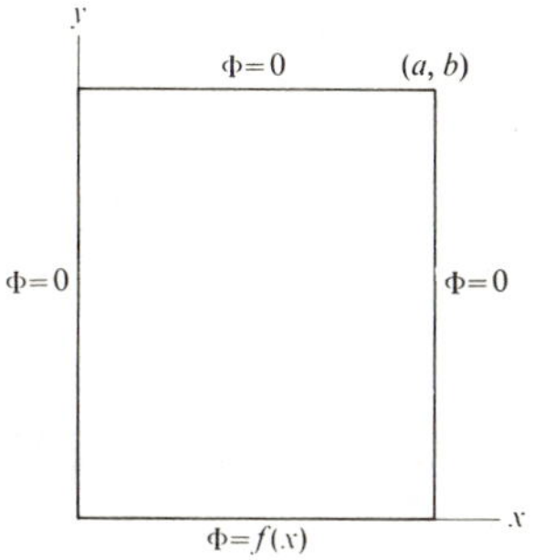

Fig. 5.1 Interior Dirichlet problem for a rectangular boundary.

First, with a little foresight, we choose a solution from the separated solutions of the Laplace equation in Appendix D which involves trigonometric functions of x and hyperbolic functions of y:

$$(A \sin \lambda x + B \cos \lambda x)(C \sinh \lambda y + D \cosh \lambda y). \tag{5.7}$$

Apart from Appendix D it is trivial to show directly that (5.7) is a solution for any value of λ. The constants in Eq. (5.7) must now be chosen to fit the boundary conditions. The condition $\Phi = 0$ when $x = 0$ can be met by eliminating $\cos \lambda x$, that is, by setting $B = 0$. The requirement that $\Phi = 0$ when $y = b$ is satisfied by choosing C and D such that

$$C \sinh \lambda b + D \cosh \lambda b = 0.$$

The condition that $\Phi = 0$ when $x = a$ can be met by letting λ take on only the values $n\pi/a$, with n integral, so the values of λ we permit are

$$\lambda_n = n\pi/a.$$

By combining constants and using the addition formula for the difference of hyperbolic functions, we can see that the elementary solution (5.7) that satisfies all the boundary conditions except that at $y = 0$ has become the following:

$$\sin \frac{n\pi x}{a} \sinh \frac{n\pi}{a}(b - y). \tag{5.8}$$

Since this is a solution for any value of n, we may multiply it by an arbitrary constant F_n and sum over n, to obtain a general solution. Explicitly we take for the superposed solution:

$$\Phi(x, y) = \sum_{n=1}^{\infty} F_n \sin \frac{n\pi x}{a} \sinh \frac{n\pi}{a}(b - y). \tag{5.9}$$

Now it is possible to satisfy the boundary condition at $y = 0$. The expression (5.9) for $\Phi(x, y)$ is essentially a Fourier sine series in x, and the boundary condition determines the coefficients F_n in a standard way:

$$\Phi(x, 0) = \sum_{n=1}^{\infty} F_n \sinh \frac{n\pi b}{a} \sin \frac{n\pi x}{a},$$

and by the orthogonality of the set $\sin n\pi x/a$,

$$F_n \sinh \frac{n\pi b}{a} = \frac{2}{a} \int_0^a \Phi(x, 0) \sin \frac{n\pi x}{a}\, dx.$$

Finally,

$$\Phi(x, y) = \frac{2}{a} \sum_{n=1}^{\infty} \frac{\int_0^a \Phi(x, 0) \sin (n\pi x/a)\, dx}{\sinh (n\pi b/a)} \sin \frac{n\pi x}{a} \sinh \frac{n\pi}{a}(b - y). \tag{5.10}$$

It is worth interpolating here that not all Dirichlet and Neumann problems can be solved as simply as we have done and are about to do. The chosen boundary conditions are more pedagogical than practical, and facilitate the determination of the coefficients in the various expansions. An arbitrary problem will generally not yield so readily as these have.

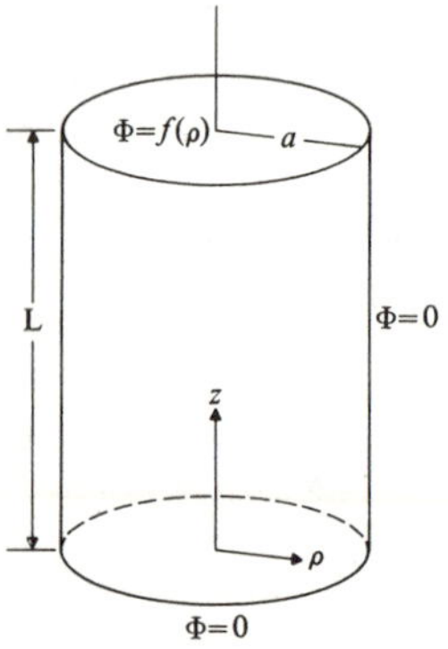

Fig. 5.2 Interior Dirichlet problem for cylinder.

As an example in cylindrical coordinates, consider the interior Dirichlet problem for a finite cylinder of radius a and height L (see Fig. 5.2). For boundary conditions we choose a set quite similar to that for the rectangular cylinder, since this gives an instructive comparison between the two cases. Specifically, we require $\Phi = 0$ on the sides and, in this case, on the bottom of the cylinder, and $\Phi = f(\rho)$ on the top, i.e., for $z = L$. We also assume $f(a) = 0$ in order that the potential be continuous in going from the top to the side. An elementary separated solution of Laplace's equation in the cylindrical coordinates ρ, φ, z is, from Appendix D,

$$(AJ_n(\gamma\rho) + BN_n(\gamma\rho))(C \sinh \gamma z + D \cosh \gamma z)e^{in\varphi}. \tag{5.11}$$

For a cylindrically symmetric potential to correspond to the cylindrically symmetric boundary condition, we must choose $n = 0$. The term in $N_0(\gamma\rho)$ can then be thrown out since it becomes infinite at $\rho = 0$. The boundary condition at $z = 0$ is satisfied by striking out the term in $\cosh \gamma z$. The elementary solution is effectively reduced to

$$J_0(\gamma\rho) \sinh \gamma z. \tag{5.12}$$

This is a solution for any γ, and a sum or integral over a spectrum of values of γ will also be one. We can try to satisfy the condition at $z = L$ first by choosing the γ's properly, and by appropriately choosing the coefficients in such a sum, similar to the rectangular case. Now, however, the sum will not be a Fourier expansion but its analog in Bessel functions; part of the point of this example is to introduce such an expansion. We digress to quote the appropriate theorem.

It can be shown that an infinite series of Bessel functions can represent an arbitrary function over a finite interval, say from 0 to a, much as Legendre polynomials can represent an arbitrary function over the interval from minus one to one. If $f(\rho)$ is such a function, this *Fourier-Bessel expansion theorem** reads

$$f(\rho) = \sum_{n=1}^{\infty} A_{pn} J_p\left(x_{pn}\frac{\rho}{a}\right), \tag{5.13}$$

where x_{pn} is the nth root of $J_p(x)$,

$$J_p(x_{pn}) = 0.$$

The coefficients in such an expansion are determined from the orthogonality relation*

$$\int_0^a \rho J_p\left(x_{pn'}\frac{\rho}{a}\right) J_p\left(x_{pn}\frac{\rho}{a}\right) d\rho = (a^2/2)[J_{p+1}(x_{pn})]^2\delta_{n'n},$$

for on multiplying (5.13) by $\rho J_p(x_{pn'}\rho/a)$ and integrating we find

$$A_{pn} = \frac{2}{a^2 J_{p+1}^2(x_{pn})} \int_0^a \rho f(\rho) J_p\left(x_{pn}\frac{\rho}{a}\right) d\rho. \tag{5.14}$$

---

* See, for example, E. Butkov, *Mathematical Physics*, Addison-Wesley (1968).

With this result at hand, let us reconsider the problem of the cylinder. We want to superpose solutions like (5.12) over a spectrum of values of γ, and begin by choosing a sequence $\gamma_1, \gamma_2, \ldots$ whose members are related to the roots x_{0n} of J_0 by

$$\gamma_n = x_{0n}/a.$$

We then form a sum with arbitrary coefficients C_n of the elementary solutions (5.12) with these values of γ_n, and take this to represent Φ; thus

$$\Phi(\rho, z) = \sum_{n=1}^{\infty} C_n J_0\left(x_{0n}\frac{\rho}{a}\right)\sinh\frac{x_{0n}z}{a}. \tag{5.15}$$

This solution, of course, vanishes at $\rho = a$. Now the final boundary condition

$$\Phi(\rho, L) = f(\rho), \qquad 0 < \rho < a \tag{5.16}$$

can be satisfied, for we have

$$f(\rho) = \sum_{n=1}^{\infty} C_n \sinh\left(\frac{x_{0n}L}{a}\right) J_0\left(x_{0n}\frac{\rho}{a}\right).$$

The inversion formula for the Bessel expansion then yields

$$C_n \sinh\left(\frac{x_{0n}L}{a}\right) = \frac{2}{a^2 J_1^2(x_{0n})}\int_0^a \rho f(\rho) J_0\left(x_{0n}\frac{\rho}{a}\right) d\rho, \tag{5.17}$$

and this effectively solves the problem.

5.3 GREEN'S THEOREM AND AN INTEGRAL IDENTITY FOR THE POTENTIAL

In the general theory of boundary value problems, an important role is played by a mathematical theorem called *Green's Theorem* (or sometimes *Green's second identity*) and by certain integral expressions for the potential that are derived from it. We discuss these subjects now, first proving the mathematical theorem itself.

Consider a volume V bounded by a closed surface S; let $\boldsymbol{n}$ be the outwardly drawn normal to the surface. Let $\boldsymbol{r}'$ be a position vector in a coordinate system with arbitrary origin (not necessarily in V). Now take two arbitrary functions $\psi(\boldsymbol{r}')$ and $\chi(\boldsymbol{r}')$ that are appropriately continuous in V and form the function

$$\boldsymbol{A}(\boldsymbol{r}') = \psi(\boldsymbol{r}')\nabla\chi(\boldsymbol{r}').$$

Then

$$\nabla\cdot\boldsymbol{A} = \nabla\psi\cdot\nabla\chi + \psi\nabla^2\chi.$$

The divergence theorem

$$\int_V \nabla\cdot\boldsymbol{A}\, dv' = \int_S \boldsymbol{A}\cdot d\boldsymbol{s}'$$

yields

$$\int_V (\nabla\psi \cdot \nabla\chi + \psi\nabla^2\chi)\, dv' = \int_S (\psi\nabla\chi) \cdot ds'.$$

Write the similar equation with the roles of ψ and χ interchanged, subtract the two equations, and use $(\psi\nabla\chi) \cdot ds' = \psi(\partial\chi/\partial n)\, ds'$ to obtain

$$\int_V (\psi\nabla^2\chi - \chi\nabla^2\psi)\, dv' = \int_S \left(\psi \frac{\partial\chi}{\partial n} - \chi \frac{\partial\psi}{\partial n}\right) ds'. \tag{5.18}$$

This relation between surface and volume integrals is Green's Theorem.*

Now we begin to apply this result to physics. Consider some charge distribution which is described in the coordinate system $\boldsymbol{r}'$ by a density function $\rho(\boldsymbol{r}')$. This charge is completely arbitrary; it may be distributed in a way which bears no relation to the volume in that it may be partially or totally inside or outside of V. However that may be, it produces a potential† $\Phi(\boldsymbol{r}')$ which is defined in V, and which we choose to identify with the function χ in Eq. (5.18):

$$\chi(\boldsymbol{r}') \rightarrow \Phi(\boldsymbol{r}'). \tag{5.19}$$

For the other function $\psi(\boldsymbol{r}')$ we choose

$$\psi(\boldsymbol{r}') \rightarrow \frac{1}{|\boldsymbol{r} - \boldsymbol{r}'|} \equiv \frac{1}{R}, \tag{5.20}$$

where $\boldsymbol{r}$ is for the present purposes simply a vectorial constant. We now put these functions back into Green's theorem, for which purpose we must form $\nabla^2\Phi$ and $\nabla^2\psi$. For $\nabla^2\Phi$ we have Poisson's equation

$$\nabla^2\Phi(\boldsymbol{r}') = -4\pi\rho(\boldsymbol{r}'). \tag{5.21}$$

The calculation of $\nabla^2\psi$ is more delicate, since ψ is singular for $\boldsymbol{r} = \boldsymbol{r}'$. This is discussed in Appendix B, where it is concluded that

$$\nabla^2\left(\frac{1}{|\boldsymbol{r} - \boldsymbol{r}'|}\right) = -4\pi\delta(\boldsymbol{r} - \boldsymbol{r}'). \tag{5.22}$$

With Eqs. (5.19) through (5.22) in Green's Theorem, we get the *first integral identity, also* sometimes known as Green's Theorem:

$$\Phi(r) = \int_V \frac{\rho(\boldsymbol{r}')\, dv'}{R} + \frac{1}{4\pi}\int_S \left[\frac{1}{R}\left(\frac{\partial\Phi}{\partial n}\right) - \Phi\frac{\partial}{\partial n}\left(\frac{1}{R}\right)\right] ds'. \tag{5.23}$$

---

* There are, in fact, several theorems that may go by the name of Green's Theorem. In addition to (5.18), there is its two-dimensional analog, and Eq. (5.23) below.

† The reader should be alert to the fact that the present convenient notation differs from the familiar notation of the summation problem, where $\boldsymbol{r}'$ always referred to a source point and $\boldsymbol{r}$ to a field point.

The nature of this identity deserves some comment. The first term in it is certainly familiar: It appears to be simply the superposition volume integral. Why then are there two surface integrals in addition? The answer lies in the fact that the volume integral includes *only that part of the charge that is within V*. If some charge is outside V, its effect on the potential inside is *simulated* by the two surface integrals. The volume integral in (5.23) is then not the superposition integral, but only part of it. It follows that if the volume happens to contain all the charge, the surface integrals must vanish,* and we recover the simple superposition result. When the surface integrals do not vanish they have an interesting physical interpretation. The first surface integral can be thought of as the potential due to a surface charge layer of density $(1/4\pi)(\partial\Phi/\partial n)$ and the second as the potential due to a surface double layer of strength $\Phi/4\pi$.

We now consider the relation of Green's identity to the Dirichlet and Neumann problems. For this suppose that the volume V contains *none* of the charge so that Φ satisfies the Laplace, not the Poisson equation, inside V. Then Eq. (5.23) reads

$$\Phi(\boldsymbol{r}) = \frac{1}{4\pi}\int_S \left[\frac{1}{|\boldsymbol{r} - \boldsymbol{r}'|}\left(\frac{\partial\Phi}{\partial n}\right) - \Phi\,\frac{\partial}{\partial n}\left(\frac{1}{|\boldsymbol{r} - \boldsymbol{r}'|}\right)\right] ds'. \tag{5.24}$$

Now there are two things that Eq. (5.24) is *not*. First, it is not a solution of the Dirichlet or Neumann problems, since if Φ is known on the right-hand side, $\partial\Phi/\partial n$ is not known, and conversely. Second, although (5.24) is an equation involving integrals, it is not an integral equation with the usual understanding that an integral equation can be solved for the unknown function. The reason for this is that neither of the functions under the integral sign, namely Φ and $\partial\Phi/\partial n$, is the same function Φ that stands on the left-hand side: Φ on the left is a function of a space variable $\boldsymbol{r}$ and the quantities on the right are functions of a variable $\boldsymbol{r}'$ on the surface. Equation (5.24) can, however, be converted to a *singular integral equation* by letting $\boldsymbol{r}$ become a point on the surface ($\boldsymbol{r} \to \boldsymbol{r}_s$), in which case it becomes an equation for $\Phi(\boldsymbol{r}_s)$ if $\partial\Phi/\partial n$ is prescribed, and conversely.

What Eq. (5.24) *is*, then, is an identity which defines a function over an extended domain (the volume V) in terms of some of its properties on a more limited domain (the surface S). In that sense, it is not basically different from a Taylor series, say, which defines a function over an extended domain in terms of its properties (the derivatives) over a more limited domain, namely, a point. As another analog, Eq. (5.24) is closely akin to Cauchy's integral theorem which expresses an analytic function in the interior of a closed curve in terms of its value on the curve.

---

* There are two cases for which this can be demonstrated explicitly. First, if the volume V extends to infinity, thereby necessarily including all the charge, the surface integrals are also at infinity and can be shown to vanish individually. Second, if the charge distribution is bounded in space and the volume V is finite but includes all the charge, the multipole expansion for the potential can be used to show that the two surface integrals cancel each other.

5.4 A GENERALIZED INTEGRAL IDENTITY AND GREEN FUNCTIONS

Although (5.23) is not especially useful as it stands for the solution of the Dirichlet or Neumann problem, there is a generalization of it that is much more so, and we shall discuss this now. In the above discussion we have chosen $\psi(\boldsymbol{r}')$ to be $1/|\boldsymbol{r} - \boldsymbol{r}'|$ but we can make a more general choice and still carry out a derivation analogous to the one that led to (5.23). The generalization consists of taking $\psi(\boldsymbol{r}')$ to be $1/|\boldsymbol{r} - \boldsymbol{r}'|$ *plus* a solution of Laplace's equation, which solution we call $L(\boldsymbol{r}, \boldsymbol{r}')$. It is also convenient to make a change of notation. We recognize that ψ is a function of $\boldsymbol{r}$ as well as $\boldsymbol{r}'$ and denote this function of two variables by $G(\boldsymbol{r}, \boldsymbol{r}')$, so that

$$\psi \equiv G(\boldsymbol{r}, \boldsymbol{r}').$$

This generalization for ψ, in this new notation, is

$$G(\boldsymbol{r}, \boldsymbol{r}') = \frac{1}{|\boldsymbol{r} - \boldsymbol{r}'|} + L(\boldsymbol{r}, \boldsymbol{r}'). \tag{5.25}$$

With this we see that the input for the derivation of (5.23) is much the same as before. Equations (5.19) and (5.21) are unchanged; Eq. (5.20) is replaced by Eq. (5.25) above; and, most importantly, we still have

$$\nabla^2 G(\boldsymbol{r}, \boldsymbol{r}') = -4\pi\delta(\boldsymbol{r} - \boldsymbol{r}'). \tag{5.26}$$

Thus the only change in (5.23) is to replace $1/|\boldsymbol{r} - \boldsymbol{r}'|$ by $G(\boldsymbol{r}, \boldsymbol{r}')$. The function $G(\boldsymbol{r}, \boldsymbol{r}')$ is called a *Green function*, and with it the *generalized integral identity* is

$$\Phi(\boldsymbol{r}) = \frac{1}{4\pi}\int_V G(\boldsymbol{r}, \boldsymbol{r}')\rho(\boldsymbol{r}')\,dv' + \frac{1}{4\pi}\int_S \left(G(\boldsymbol{r}, \boldsymbol{r}')\frac{\partial\Phi}{\partial n} - \Phi\frac{\partial}{\partial n}G(\boldsymbol{r}, \boldsymbol{r}')\right) ds'. \tag{5.27}$$

When using this to help solve the Dirichlet or Neumann problems, we shall, for simplicity, consider the case where there is no charge in the volume so that Laplace's equation is satisfied. In this case, (5.27) becomes

$$\Phi(\boldsymbol{r}) = \frac{1}{4\pi}\int_S \left(G(\boldsymbol{r}, \boldsymbol{r}')\frac{\partial\Phi}{\partial n} - \Phi\frac{\partial}{\partial n}G(\boldsymbol{r}, \boldsymbol{r}')\right) ds'. \tag{5.28}$$

The essential point of this new method will be to exploit the considerable arbitrariness that remains in the function $L(\boldsymbol{r}, \boldsymbol{r}')$, which has been required to be only *some* solution of Laplace's equation within the region V. Specifically, consider Dirichlet conditions. Suppose we could choose $L(\boldsymbol{r}, \boldsymbol{r}')$ so that $G(\boldsymbol{r}, \boldsymbol{r}')$ *was zero whenever* $\boldsymbol{r}'$ *was on the boundary*. On labeling this special choice of $G(\boldsymbol{r}, \boldsymbol{r}')$ by the subscript D, we find that Eq. (5.28) becomes

$$\Phi(\boldsymbol{r}) = -\frac{1}{4\pi}\int_S \Phi(\boldsymbol{r}')\frac{\partial G_D(\boldsymbol{r}, \boldsymbol{r}')}{\partial n}\,ds' \tag{5.29}$$

This equation expresses Φ in V in terms of its value on S, that is, it solves the Dirichlet problem.

It might appear difficult to choose $L(\boldsymbol{r}, \boldsymbol{r}')$ as we have described, but it turns out that it can be so chosen, and the Green function $G_D(\boldsymbol{r}, \boldsymbol{r}')$ does indeed exist. In fact, as a bit of serendipity, the boundary value problems we have already solved for perfect conductors in the field of a point charge have essentially produced such functions. To see what is meant by this, consider two problems in conjunction: (a) The interior Dirichlet problem for the volume V of Fig. 5.3. (b) The *auxiliary* problem of a volume of the shape of V excised from an infinite *perfect conductor*, with a unit point charge at $\boldsymbol{r}_0$. For the auxiliary problem (b) we want to find the potential $\tilde{\Phi}$ at some field point which we shall call $\tilde{\boldsymbol{r}}$. This potential will be, by familiar reasoning, the sum of the *applied* potential $1/|\boldsymbol{r}_0 - \tilde{\boldsymbol{r}}|$, and the *induced* potential Φ_i due to the charges on the perfectly conducting surface S. This latter potential is, of course, a function of $\boldsymbol{r}_0$ as well as a function of $\tilde{\boldsymbol{r}}$ and we shall indicate this explicitly by writing $\Phi_i = \Phi_i(\tilde{\boldsymbol{r}}, \boldsymbol{r}_0)$. The potential $\tilde{\Phi}$ is a similar function, so we can write

$$\tilde{\Phi}(\tilde{\boldsymbol{r}}, \boldsymbol{r}_0) = \frac{1}{|\tilde{\boldsymbol{r}} - \boldsymbol{r}_0|} + \Phi_i(\tilde{\boldsymbol{r}}, \boldsymbol{r}_0). \tag{5.30}$$

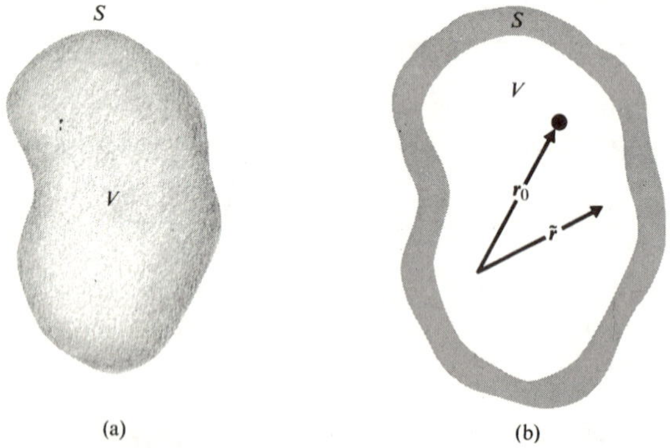

Fig. 5.3 (a) Volume V and surface S for interior Dirichlet problem. (b) Volume V of the same shape excised from conducting matter, with unit point charge at $\boldsymbol{r}_0$.

As a function of $\tilde{\boldsymbol{r}}$ the last expression has two properties of interest here. First, Φ_i is a solution of Laplace's equation; second, $\tilde{\Phi}$ vanishes whenever $\tilde{\boldsymbol{r}}$ is on the surface. But these are just the properties that the Green function of Eq. (5.25) (for Dirichlet conditions) must have as a function of its variable $\boldsymbol{r}'$. Changing the names of the variables in (5.30) to correspond with (5.25) then yields the *Green function for Dirichlet conditions*. That is, with $\tilde{\boldsymbol{r}} \to \boldsymbol{r}$ and $\boldsymbol{r}_0 \to \boldsymbol{r}'$, we have

$$G_D(\boldsymbol{r}, \boldsymbol{r}') = \tilde{\Phi}(\boldsymbol{r}, \boldsymbol{r}').$$

Therefore, we have at hand the solutions to as many Dirichlet problems as there are to the problem of a grounded conductor under the influence of a point charge, and we shall shortly consider some examples.

First, however, we shall discuss the Neumann problem. A natural temptation is to choose $L(\mathbf{r}, \mathbf{r}')$ in (5.25) so that G_N, the Green function for Neumann conditions, satisfies

$$\frac{\partial G_N(\mathbf{r}, \mathbf{r}')}{\partial n} = 0 \tag{5.31}$$

for $\mathbf{r}'$ on the surface. Equation (5.31) is, however, not consistent with (5.26). For if (5.26) is integrated over the volume V, the right-hand side becomes -4π, and the Laplacian on the left, as a divergence of a gradient, can be transformed to a surface integral by the divergence theorem to yield

$$\int \frac{\partial G_N(\mathbf{r}, \mathbf{r}')}{\partial n} ds' = -4\pi.$$

This equation can be satisfied if we take $\partial G_N/\partial n$ to be a constant; if S_0 is the area of the surface, we can write

$$\frac{\partial G_N(\mathbf{r}, \mathbf{r}')}{\partial n} = \frac{-4\pi}{S_0}.$$

Frequently S_0 is infinite, in which case we get back (5.31). This will be so for the exterior Neumann problem where the volume V must be considered to be bounded by some finite inner surface, and by another at infinity. Since we shall deal only with the exterior problem we shall always use (5.31), in which case Eq. (5.28) becomes

$$\Phi(\mathbf{r}) = \frac{1}{4\pi} \int G_N(\mathbf{r}, \mathbf{r}') \frac{\partial \Phi}{\partial n} ds'. \tag{5.32}$$

5.5 APPLICATION OF GREEN FUNCTIONS

We present now some examples of the Green function method of solving the Dirichlet and Neumann problems. These examples will generally exploit the Green functions that have been found, almost incidentally, in solving boundary value problems with conductors. Thus the work of this section is really the trivial part of that involved in using the Green function method; at this point it simply becomes an exercise in integration. The part that is difficult, and that makes the method something less than the royal road to the solution of all Dirichlet and Neumann problems, is, of course, the difficulty in the first place of finding the appropriate Green functions.

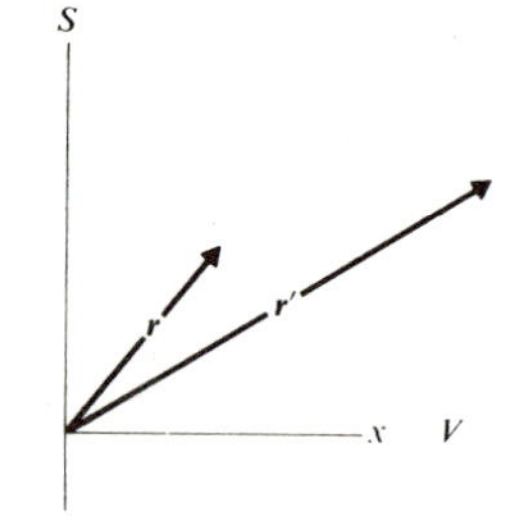

Fig. 5.4 For problems in which the surface S is the plane $x = 0$ and the volume V the half space $x > 0$.

With this *caveat* consider the first example, in which the surface S is the plane $x = 0$ and the volume V is the half-space $x > 0$, as sketched in Fig. 5.4. We assume that $\Phi(0, y', z')$, the potential on the surface, is prescribed. From the image solution for the auxiliary electrostatic problem, the Green function for Dirichlet conditions is

$$G_D(\mathbf{r}, \mathbf{r}') = \frac{1}{\sqrt{(x - x')^2 + (y - y')^2 + (z - z')^2}} - \frac{1}{\sqrt{(x + x')^2 + (y - y')^2 + (z - z')^2}}.$$

To use this we must form $\partial G_D/\partial n$, the normal outward derivative at the surface. Since outward is to the left in this case, we have

$$\frac{\partial G_D}{\partial n} = -\left.\frac{\partial G_D}{\partial x'}\right|_{x'=0} = -\frac{2x}{((x^2 + (y - y')^2 + (z - z')^2)^{3/2}}.$$

With (5.29), the potential is then

$$\Phi(x, y, z) = \frac{x}{2\pi}\iint \frac{\Phi(0, y', z')\, dy'\, dz'}{(x^2 + (y - y')^2 + (z - z')^2)^{3/2}}. \tag{5.33}$$

It is perhaps worth pointing out that although (5.33) represents the closed-form solution, the integral cannot usually be evaluated explicitly. Except for the exponent $\frac{3}{2}$ instead of $\frac{1}{2}$ in the denominator, Eq. (5.33) resembles the integrals in the summation problem which could only be evaluated by expansions and some kind of approximation.

For the Neumann problem with the same geometry it is almost trivially obvious that the Green function is

$$G_N(\mathbf{r}, \mathbf{r}') = \frac{1}{\sqrt{(x - x')^2 + (y - y')^2 + (z - z')^2}} + \frac{1}{\sqrt{(x + x')^2 + (y - y')^2 + (z - z')^2}},$$

from which, using (5.32), we find

$$\Phi(x, y, z) = -\frac{1}{2\pi}\iint \frac{\left.\dfrac{\partial\Phi}{\partial x'}\right|_{x'=0}}{[x^2 + (y - y')^2 + (z - z')^2]^{1/2}}\,dy'\,dz'. \tag{5.34}$$

The same remarks as to the possibility of evaluating the integral apply as for the Dirichlet case. Since the integral is intractable, it is not easy to directly demonstrate the equivalence of this form of solution to the solution by Fourier integrals embodied in Eq. (5.5).

As another example, consider the exterior Dirichlet problem for a sphere (see Fig. 5.5). The auxiliary electrostatic problem yields an expression for the Green function in either of two forms: first, using images, as embodied in Eq. (4.19) and second, involving spherical harmonics. We consider the image form first since it yields an elegant solution of the problem in closed form. To apply the image solution in the present context, we use Eq. (4.19) with $\boldsymbol{r}$ replaced by $\boldsymbol{r}'$ and d by $\boldsymbol{r}$.

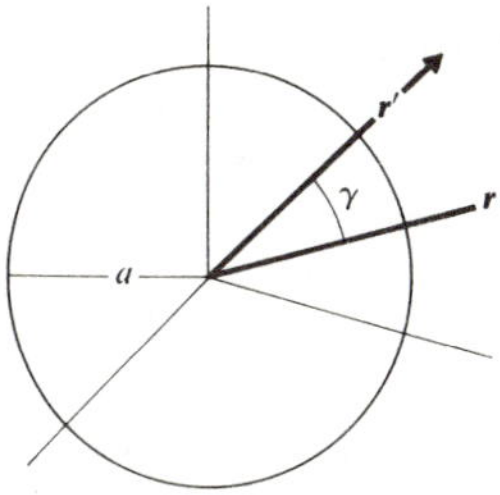

Fig. 5.5 · Dirichlet problem for the sphere.

The potential Φ defined by (4.19) then becomes $G_D(\boldsymbol{r}, \boldsymbol{r}')$:

$$G_D(\boldsymbol{r}, \boldsymbol{r}') = \frac{1}{(r^2 + r'^2 - 2rr'\cos\gamma)^{1/2}} - \frac{1}{(r^2r'^2/a^2 + a^2 - 2rr'\cos\gamma)^{1/2}}.$$

For finding $\partial G/\partial n$ we remember that $\boldsymbol{n}$ and $\boldsymbol{r}'$ are oppositely directed, and so have

$$\frac{\partial G_D}{\partial n} = -\left.\frac{\partial G_D}{\partial r'}\right|_{r'=a} = \frac{(a^2 - r^2)}{a(a^2 + r^2 - 2ar\cos\gamma)^{3/2}}.$$

When we put this into (5.29) the solution to the Dirichlet problem is, with $ds' = a^2 d\Omega'$

$$\Phi(r, \theta, \varphi) = \frac{a(r^2 - a^2)}{4\pi}\int \frac{\Phi(a, \theta', \varphi')\,d\Omega'}{(a^2 + r^2 - 2ar\cos\gamma)^{3/2}}, \tag{5.35}$$

where γ is the angle between $\boldsymbol{r}$ and $\boldsymbol{r}'$. The Green function for the Neumann problem for the sphere is also known, but it is not trivial or obvious to find. It is presented in Problem 5.8.

The solution to the two-dimensional Dirichlet problem of finding the potential $\Phi(\rho, \varphi)$ in the interior or exterior of a circular cylinder of radius a when the potential is prescribed on the surface can be solved much as was the problem for the sphere. The answer to the *interior problem* is the famous *Poisson integral formula*:

$$\Phi(\rho, \varphi) = \frac{a^2 - \rho^2}{2\pi} \int_0^{2\pi} \frac{\Phi(a, \varphi')\, d\varphi'}{a^2 - 2a\rho \cos(\varphi - \varphi') + \rho^2}. \tag{5.36}$$

5.6 GENERAL PROPERTIES OF GREEN FUNCTIONS

In this section we discuss some of the common denominators of the Green functions we have encountered. We note first that we have worked mainly with three-dimensional examples, but these generally have their two- and even one-dimensional counterparts. To keep the present discussion general, we adopt a new notation in which P and P' stand for points in space in any number of dimensions. Then for three, two, and one dimensions, P can be taken to be respectively $\boldsymbol{r}$, $\boldsymbol{\rho}$, x.

Essentially four different kinds of Green functions have been considered. The first encompass those that entered the superposition integral for the potential of a charge distribution. The three-, two-, and one-dimensional examples of this integral are embodied in Eqs. (2.17), (3.46), and (3.54), in which the Green functions are, respectively, $1/|\boldsymbol{r} - \boldsymbol{r}'|$, $-2 \ln |\boldsymbol{\rho} - \boldsymbol{\rho}'|$, and $-2\pi|x - x'|$. We can label these functions with a subscript F, which stands for "free space," since the functions refer to space that contains only a charge distribution and is free of conductors and dielectrics. In the present general notation, the three equations just mentioned can all be written as the integral of a product of a *generalized source function* $\rho(P')$, and a Green function $G_F(P, P')$:

$$\Phi(P) = \int G_F(P, P')\rho(P')\, dP'. \tag{5.37}$$

There is a second kind of problem that we have discussed whose solution can be expressed in terms of a Green function, i.e., in a form analogous to (5.37). It is that of a grounded conductor under the influence of some *arbitrary* applied charge distribution $\rho_a(P')$. For suppose the solution is known to the problem of the potential at a field point P when a grounded conductor is under the influence of a unit point charge at P'. Call this potential $G_C(P, P')$. Then the potential Φ at P, when the conductor is under the influence of an *arbitrary* distribution $\rho_a(P')$, is

$$\Phi(P) = \int G_C(P, P')\rho_a(P')\, dP'. \tag{5.38}$$

The third and fourth examples of the use of Green functions occur in the

Dirichlet and Neumann problems. In the present notation, and with a point on the surface denoted by P', the solution of the Dirichlet problem can be written

$$\Phi(P) = -\frac{1}{4\pi}\int \Phi(P')\frac{\partial G_D(P, P')}{\partial n}\,ds', \tag{5.39}$$

and that of the Neumann problem,

$$\Phi(P) = \frac{1}{4\pi}\int \frac{\partial \Phi}{\partial n} G_N(P, P')\,ds'. \tag{5.40}$$

These four examples have in common the fact that they represent the potential at a point as a superposition of a *source density* $S(P')$ and *unit source* strength function $U(P, P')$ which gives the effect at P of a unit source at P'. That is, all the equations (5.37) through (5.40) can be written in the form

$$\Phi(P) = \int U(P, P')S(P')\,dP'. \tag{5.41}$$

In Eqs. (5.37) and (5.38), the identification of U and S is obvious. In Eq. (5.39), $U(P, P')$ is just $\partial G_D(P, P')/\partial n$ and $S(P')$ is $-\Phi(P')/4\pi$. In Eq. (5.40), $U(P, P')$ is $G_N(P, P')$ and $S(P')$ is $1/4\pi(\partial\Phi/\partial n)$. The second common characteristic is that all four Green functions satisfy Poisson's equation for a source density which is a δ-function, that is,

$$\nabla^2 G(\mathbf{r}, \mathbf{r}') = -4\pi\delta(\mathbf{r} - \mathbf{r}'), \tag{5.42}$$

and in addition satisfy boundary conditions of one or another kind.

There are, however, conceptual differences among the four Green function formulas. The last two are tied to the Poisson equation, since they have been derived by the aid of Green's identity, which is special to that equation. On the other hand, although the first two Green function formulas, (5.37) and (5.38), have been presented in connection with Poisson's equation, they are really special cases of a much more general viewpoint, which we now discuss.

Consider some linear differential *operator* $\mathscr{L}$ that may act in any number of dimensions. For example, it might be the Laplacian in three dimensions or $d^3/dx^3 + 7\,d/dx$ in one dimension, or whatever. Consider also the inhomogeneous equation for an unknown function $\Phi(P)$,

$$\mathscr{L}_P\Phi(P) = F(P), \tag{5.43}$$

where $F(P)$ is assumed known, and we have added a subscript to $\mathscr{L}$ to emphasize that it acts on P. Equation (5.43) can be solved if the similar equation with a δ-function on the right-hand side can be solved. For suppose the solution of

$$\mathscr{L}_P G_0(P) = \delta(P) \tag{5.44}$$

is known. Now form $G_0(P - P')$ where P' is just a constant as far as $\hat{\mathscr{L}}_P$ is concerned. Since $\hat{\mathscr{L}}_P$ is a linear operator, $G_0(P - P')$ satisfies

$$\hat{\mathscr{L}}_P G_0(P - P') = \delta(P - P'). \tag{5.45}$$

In terms of $G(P, P') \equiv G_0(P - P')$ the solution to Eq. (5.43) is*

$$\Phi(P) = \int F(P')G(P, P')\,dP', \tag{5.46}$$

since operating on Eq. (5.46) with $\hat{\mathscr{L}}_P$ reproduces the original equation. We can look at the derivation of the superposition integral (2.17) from the Poisson equation (2.15) as an example of the method of solution we have just outlined, although it is not, in fact, the way we happened to derive that integral. For the Poisson equation the operator $\hat{\mathscr{L}}_P$ is the Laplacian, the function $F(P)$ is the charge density, and the Green function† is what we have previously called $G_F(P, P')$. A similar technique can be used if, in addition to satisfying the differential equation, $\Phi(P)$ is required to satisfy some boundary conditions. We have then to find a Green function which satisfies Eq. (5.45), and, in addition, satisfies the same boundary condition (as a function of P) that $\Phi(P)$ does. We can look at the problem of finding the potential due to a grounded conductor under the influence of an arbitrary applied distribution, *via* the function $G_C(P, P')$, as an example of this kind.

An historical note which may clarify the difference between (5.37), (5.38) and (5.39), (5.40) is this: The first use of a Green function was made in 1838 by George Green, who introduced the Dirichlet and Neumann functions, much as we have done, in connection with Poisson's equation and the Green identity. The generalization of these functions to operators other than the Laplacian, i.e., to the general kind of operator we have described above, was due to Bôcher (R) in 1907. Although the generalization does not rely on Green's identity, the functions Bôcher introduced were, in his opinion, sufficiently analogous to the functions introduced by Green that he chose also to call them Green functions.

Finally we note that the Green functions can be used not only to integrate an inhomogeneous equation, as above, but to convert a homogeneous differential

---

* We shall use the standard notation $G(P, P')$ for the Green function as a function of its two variables, although as the notation $G_0(P - P')$ indicates, as we use it, it is here a function of the difference between P and P', and in fact is usually a function only of the absolute magnitude of that difference. For example, the Green function for the Poisson equation is a function only of $|\boldsymbol{r} - \boldsymbol{r}'|$.

† In the Green function equation (5.42) for the Poisson equation the right hand contains an extra factor—4π, as compared to Eq. (5.44). For the present general development, this factor is just a nuisance. It is much more convenient to use (5.44) as it stands, and consider that the Green function is only defined up to an arbitrary factor which can be absorbed in the integral of formula (5.46). Different authors do choose these factors differently.

equation to an integral equation. This is often a useful technique.* Consider, for example, the differential equation

$$\hat{\mathscr{L}}_P \Phi(P) = v(P)\Phi(P). \tag{5.47}$$

If the Green function that satisfies Eq. (5.45) is known, this last equation can be converted to the *homogeneous integral equation*

$$\Phi(P) = \int G(P, P')v(P')\Phi(P')\, dP'. \tag{5.48}$$

The equivalence of these last two equations can be seen by applying the operator $\hat{\mathscr{L}}_P$ to (5.48), thereby reproducing (5.47). If the differential equation (5.47) has to be solved subject to boundary conditions (e.g., conditions at infinity), a more general integral equation equivalent to it is the *inhomogeneous* one,

$$\Phi(P) = \Phi_a(P) + \int G(P, P')v(P')\Phi(P')\, dP', \tag{5.49}$$

where $\hat{\mathscr{L}}_P \Phi_a(P) = 0$, and where $\Phi_a(P)$ is chosen with an eye to the boundary conditions. When the operator $\hat{\mathscr{L}}_P$ is $(\nabla^2 + k^2)$, these equations are important in quantum mechanics and in the theory of wave propagation in dielectric media. Equation (5.48) is then useful for so-called *bound-state problems* in quantum mechanics. Similarly, Eq. (5.49) is useful for *scattering problems*, either in quantum mechanics or with dielectrics, in which case $\Phi_a(P)$ frequently represents an incident wave.

The discussion above shows that Green functions can be useful in a variety of problems, provided they can be found. We shall consider the problem of calculating them by more general methods than have been presented. Two important general techniques are discussed briefly: expansion in eigenfunctions and direct solution of the differential equation. We begin with the former.

Given the operator $\hat{\mathscr{L}}_P$, we can formally write an eigenvalue equation in which the effect of $\hat{\mathscr{L}}_P$ operating on some function $\Phi(P)$ is simply to multiply the function by a constant λ:

$$\hat{\mathscr{L}}_P \Phi = \lambda \Phi.$$

The question of whether or not this equation has a solution, and the nature of the solution, is a very involved one. Frequently the boundary conditions, or conditions of regularity or finiteness, select only certain allowed values λ_n for which acceptable solutions exist. These allowed values are the *characteristic values*, and the corresponding allowed functions are the *characteristic functions*, both of which we label with an index n. This index may stand for a discrete or continuous set of numbers. The functions then satisfy

$$\hat{\mathscr{L}}_P \Phi_n = \lambda_n \Phi_n. \tag{5.50}$$

* The integral equations of Section 4.6, which were derived by physical arguments, can also be derived by using Green's Theorem and the Green functions of this section.

It is often the case that the set Φ_n is a complete one for expansion of an arbitrary function of its variables. If this is so, we have from Appendix B, Eq. (B.10), the basic orthogonality relation:

$$\sum_n \Phi_n(P)\Phi_n^*(P') = \delta(P - P'). \tag{5.51}$$

We now claim that $G(P, P')$ has the following representation:

$$G(P, P') = \sum_n \frac{\Phi_n(P)\Phi_n^*(P')}{\lambda_n}. \tag{5.52}$$

To prove this we operate on (5.52) with $\hat{\mathscr{L}}_P$; using (5.50) we conclude that the last expression does satisfy the basic defining equation (5.45) for the Green function.

For an example of this type of expansion, take $\hat{\mathscr{L}}_P$ to be the Laplacian in three dimensions. The complete set of functions Φ_n can be taken to be the set $\Phi_{\boldsymbol{k}}$ labeled by the continuous vectorial index $\boldsymbol{k}$:

$$\Phi_{\boldsymbol{k}} = e^{i\boldsymbol{k}\cdot\boldsymbol{r}}/(2\pi)^{3/2}.$$

This set satisfies Eq. (5.50), in which λ_n is now $-k^2$. We know that a Green function for Poisson's equation is $1/4\pi|\boldsymbol{r} - \boldsymbol{r}'|$, remembering that the right-hand side of (5.45) does not contain a factor -4π. From (5.52), with the sum replaced by an integral, we have the well-known relation

$$\frac{1}{4\pi|\boldsymbol{r} - \boldsymbol{r}'|} = \frac{1}{(2\pi)^3}\int \frac{e^{i\boldsymbol{k}\cdot(\boldsymbol{r}-\boldsymbol{r}')}}{k^2}\, d\boldsymbol{k}. \tag{5.53}$$

As an example of an expansion of a Green function that satisfies boundary conditions, consider that for the two-dimensional interior Dirichlet problem for the rectangular region of Fig. 5.1, i.e., the region bounded by the planes $x = 0$, $x = a$ and $y = 0$, $y = b$. As eigenfunctions we take the set

$$\Phi_{lm}(x, y) = \sqrt{\frac{4}{ab}}\sin\frac{l\pi x}{a}\sin\frac{m\pi y}{b}.$$

These functions vanish on the boundaries and satisfy (5.50) with the index n replaced by lm. The eigenvalue λ_n in that equation is similarly replaced by k_{lm} defined by

$$k_{lm} = \pi^2\left(\frac{l^2}{a^2} + \frac{m^2}{b^2}\right).$$

The Green function defined by (5.52) is then

$$G(x, y; x', y') = \frac{4}{ab\pi^2}\sum_{l=1}^{\infty}\sum_{m=1}^{\infty}\frac{\sin\dfrac{l\pi x}{a}\sin\dfrac{l\pi x'}{a}\sin\dfrac{m\pi y}{b}\sin\dfrac{m\pi y'}{b}}{\left(\dfrac{l^2}{a^2} + \dfrac{m^2}{b^2}\right)}. \tag{5.54}$$

A second technique for finding the Green function is simply to solve (5.44) as a differential equation, and to build into the solution the δ-function singularity on the right-hand side. This frequently involves solutions which are of one form in some domain of the variables, and of a different form in another domain. As a simple illustration we shall take the operator $\hat{\mathscr{L}}_P$ to be the one-dimensional Helmholtz operator:

$$\hat{\mathscr{L}}_P = \frac{d^2}{dx^2} + k_0^2.$$

This becomes the one-dimensional Laplace operator for the special case $k_0 = 0$, and we can check the result against the previous result for this equation. To follow convention, we define the Green function in this case with the factor -4π multiplying the δ-function. It is convenient to begin by solving for $G_0(x)$ which satisfies

$$\frac{d^2G_0(x)}{dx^2} + k_0^2G_0(x) = -4\pi\delta(x). \tag{5.55}$$

Solutions of this equation valid for $x \neq 0$ are e^{ik_0x} and e^{-ik_0x}, and it is from these that we must synthesize the combination that satisfies the δ-function condition. This condition is obtained by integrating (5.55) over the small interval from $-\varepsilon$ to ε to get

$$\left.\frac{dG_0}{dx}\right|_{+\varepsilon} - \left.\frac{dG_0}{dx}\right|_{-\varepsilon} + k_0^2\int_{-\varepsilon}^{\varepsilon} G_0(x)\,dx = -4\pi. \tag{5.56}$$

It will be self-consistent to assume that G is continuous across the origin, in which case the integral in (5.56) vanishes in the limit $\varepsilon \to 0$, and the equation becomes a condition on the discontinuity of the left- and right-hand derivatives of $G_0(x)$ at the origin:

$$\left.\frac{dG_0}{dx}\right|_{+} - \left.\frac{dG_0}{dx}\right|_{-} = -4\pi. \tag{5.57}$$

If we are to piece together the solutions $e^{\pm ik_0x}$ to satisfy this, it is clear that we cannot take either e^{ik_0x} or e^{-ik_0x} by itself over the whole range, since these have continuous derivatives across the origin. The only alternative is to choose, say, Ae^{ik_0x} for $x > 0$ and Ae^{-ik_0x} for $x < 0$. We try

$$G_0(x) = \begin{cases} Ae^{ik_0x}, & x > 0 \\ Ae^{-ik_0x}, & x < 0 \end{cases}.$$

The condition (5.57) then yields $A = 2\pi i/k_0$ and we see that $G_0(x)$ can be written in the compact form

$$G_0(x) = \frac{2\pi i e^{ik_0|x|}}{k_0}.$$

From this result we have

$$G(x, x') = \frac{2\pi i e^{ik_0|x-x'|}}{k_0}. \tag{5.59}$$

In the limit $k_0 = 0$, this simply becomes $-2\pi|x - x'|$, the Green function for the one-dimensional Laplace equation.

PROBLEMS

1. Solve the Dirichlet problem for the half-plane with a Fourier integral expansion analogous to that used in the text for the Neumann problem.
2. Show that for a charge distribution bounded in space, the two surface integrals in Eq. (5.23) vanish individually when the volume V becomes infinite.
3. Charge is confined to a finite volume; the two surface integrals in (5.23) are taken over a sphere which completely encloses the volume. Use the multipole expansion to show that the surface integrals mutually cancel.
4. Find the Green function for Neumann conditions when the volume V is the quarter-space $y > 0, z > 0$, that is, for boundary conditions on the planes $y = 0, z > 0$ and $z = 0, y > 0$.
5. The potential on the surface of a sphere varies as $\Phi_0(1 + \cos\theta + \cos^2\theta)$, and there is no charge inside. Find the potential in the interior.
6. Deduce a two-dimensional Green's identity analogous to (5.23), and use it to derive the Poisson integral formula (5.36).
7. Show explicitly that $-2\pi|x - x'|$, the Green function for the one-dimensional Laplace equation, satisfies $d^2G/dx^2 = -4\pi\delta(x - x')$. [*Hint*: The first derivative of G is a step function; to form the second derivative somehow represent the step function as the limit of a sequence of differentiable functions.]
8. Verify that the Green function $G_N(\mathbf{r},\mathbf{r}')$ for the exterior Neumann problem on the sphere is, with γ the angle between $\mathbf{r}$ and $\mathbf{r}'$,

$$\frac{1}{(r^2 + r'^2 - 2rr'\cos\gamma)^{1/2}} + \frac{1}{((r^2r'^2/a^2) + a^2 - 2rr'\cos\gamma)^{1/2}}$$
$$+ \frac{1}{a}\ln\frac{2a^2}{a^2 - 2rr'\cos\gamma + (r^2r'^2 + a^4 - 2rr'a^2\cos\gamma)^{1/2}}.$$

9. By actually evaluating the integral, verify the relation (5.53).

REFERENCES

Bôcher, M., "Green's Functions in Space of One Dimension," *Bull Am. Math. Soc.* **7**, 297 (1901).

Butkov, E., *Mathematical Physics*, Addison-Wesley, Reading, Mass. (1968).
This book contains a concise but readable account of Green functions and related topics.

Günter, N. M., *Potential Theory*, Frederick Ungar, New York (1967).

A more recent text than Kellogg below, and with somewhat different emphasis.

Kellogg, O. D., *Potential Theory*, Frederick Ungar, New York (1929).

The mathematical rigor that is absent from our treatment of Green's theorem, Green functions, etc. can be found in this book, as well as many topics we have not touched on.

Morse, P., and H. Feshbach, *Methods of Mathematical Physics*, McGraw-Hill, New York, (1953).

A complete and varied discussion of Green functions in many contexts.

6
DIELECTRICS

6.1 INTRODUCTION

We have touched lightly on dielectrics before. In Chapter 4 on conductors, dielectrics were partially characterized as substances which did not conduct electricity. And in Chapter 3 on the summation problem, we remarked that the problem of calculating the field of a dipole lattice was a precursor to the theory of dielectrics. These two passing references point up the essential nature of dielectrics. They are nonconductors which can also be the source of electric fields in that they can be polarized, i.e., they can be considered as an assemblage of microscopic charge distributions with dipole moments.

Several questions must then be answered in formulating a theory of dielectrics. First, the nature of the polarization must be discussed. What kinds of dipoles make up the different kinds of dielectrics? How are they produced and how are they distributed in space? The model of point dipoles in a cubic lattice is a simple and specialized one; real dielectrics are more complicated. And what are the specific experimental facts that must be understood? Presumably, all properties of a given dielectric could be explained if the microscopic and violently fluctuating internal fields in it could be calculated in detail. But as we have seen, this is difficult even for the simple model in Chapter 3 and is essentially impossible for more complicated models of real dielectrics. We must then ask whether at least some of the experimental results can be explained without a full knowledge of this detailed field, and clarify what aspects of the experiments depend on what aspects of the fields. We will start with a discussion of the physical mechanisms whereby a state of polarization can be set up in a dielectric.

Consider some charge distribution, with *net* charge zero; for simplicity we shall call it a *molecule*. The potential it produces can be expanded in a multipole expansion. Since the molecule is neutral, the first term in the expansion may be, although not necessarily, a dipole term. In fact, most commonly the dipole moment vanishes; the molecule is then called a *nonpolar* one. But suppose that an external electric field is applied to such a molecule. The field will pull the positive charge in one direction and the negative in the other, subject to whatever atomic restoring forces may exist. It is plausible, and is in fact usually the case, that this distorted charge distribution has a multipole expansion in which the dipole term does not vanish. We then say that the molecule is *polarized*, and if it is one of many similar ones making up a medium, that the *medium is polarized.*

There is another mechanism for polarization of a medium. Suppose a gas or liquid is made up of *polar* molecules, i.e., ones which, in contrast to the case above, each have a *permanent* dipole moment. Even given this, however, there will be no *net* dipole moment in any small volume containing a statistically significant number of such molecules. For the orientation of the molecules will tend to be random and, roughly speaking, for every dipole pointing in one direction, there will be one pointing in the opposite direction. Thus any small but macroscopic volume will not possess a dipole moment. By contrast, if an external field is applied, the dipoles will tend to line up along its direction subject to the disorienting tendency of their thermal motions. Without going into quantitative detail, we argue that it is plausible physically that an applied electric field will also *polarize* a gas or liquid of polar molecules by making them into a medium in which any small macroscopic volume does have a net dipole moment.

All states of matter, gases, liquids and solids, can be dielectrics, and any theory of dielectrics must be generally applicable to all of them. A clue toward such a general theory is given by a general experimental result on the capacitance of condensers. It is found that if the space between the plates of a condenser is filled with a dielectric substance, the capacitance is increased by a factor which (neglecting edge effects) is *independent of the shape of the condenser*, but depends only on the substance. On the basis of this, a constant ε characteristic of the substance can be defined as the ratio of the capacitance of *any* condenser with and without dielectric; ε is called the *dielectric constant.*

Let us see some implications of this experimental fact. Consider a parallel-plate condenser, with charge Q on one plate and $-Q$ on the other. There will be a field $\boldsymbol{E}$ between the plates, and a corresponding potential difference V,

$$V = \int \boldsymbol{E} \cdot d\boldsymbol{l}; \tag{6.1}$$

the capacitance C is defined as the ratio of the charge to the potential difference:

$$C = Q/V. \tag{6.2}$$

If then the region between the plates is filled with dielectric without changing the charge Q on the plates, the capacitance is increased. From (6.2) we see that if Q is unchanged, this can only mean that the potential difference V is smaller, implying that on the average $\boldsymbol{E}$ is smaller. How can $\boldsymbol{E}$ be smaller? Since the charges on the plates are the same, the field they produce remains unchanged. Hence some mechanism must operate to produce a new field which partially compensates the field of the charges on the plates, so that the *resultant* net field is smaller. This mechanism is, of course, *polarization*. In the next sections we shall turn these qualitative ideas into quantitative ones.

6.2 POLARIZATION CHARGE AND THE AVERAGE INTERNAL FIELD

At first sight the problem of calculating the reduction in field produced by the dielectric seems hopeless, for it would appear to lead into the formidable problem of calculating the highly irregular internal field in matter. But if, for the moment, we have only the more modest objective of explaining the basic fact of capacitors expressed above, there is one saving feature. That is, macroscopic measurements on capacitors tell nothing about the details of this fluctuating interior field; they yield information only on the line integral of this field, i.e., the potential difference.

To see why this is relevant, recall that for finding the *exterior* field of a distribution of dipoles, we could, to good approximation, replace the discrete distribution by an imagined continuous one with dipole moment $\boldsymbol{P}(\boldsymbol{r})$. The fields could then be calculated as if there were a *volume* distribution of charge

$$\rho_p = -\nabla \cdot \boldsymbol{P}, \tag{6.3}$$

and a *surface* distribution

$$\sigma_p = P_n. \tag{6.4}$$

In the interior of matter, on the other hand, the conditions that make this equivalence valid are not satisfied, since any point is necessarily close to one or another dipole. But, and this is the important point, we shall now show that the *line integral* of the field through the dielectric can be calculated just as if the interior field were produced by the *same* surface and volume charges, i.e., by those given by (6.3) and (6.4). We refer to Fig. 6.1, which shows a dielectric and its equivalent surface and volume charges. The closed paths with arrows represent possible paths of integration for the line integral of the field. We divide these paths into parts: for the dielectric, a part e exterior to it and i interior to it, and similarly e' and i' for the model of equivalent surface and volume charges. The line integral V of the field is defined by (6.1), in general, with whatever subscripts e, e', etc., that are necessary to denote a particular path.

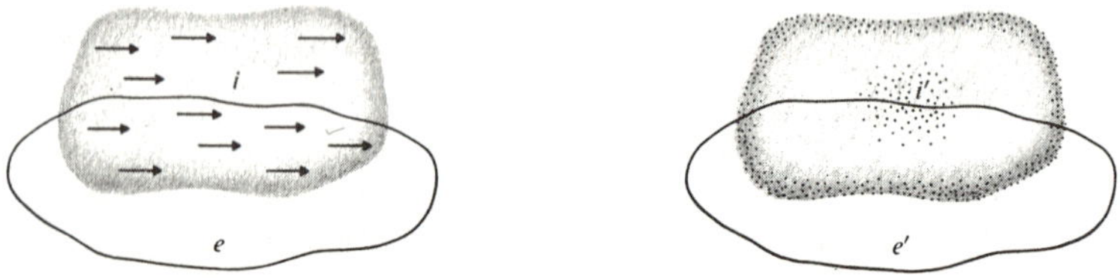

Fig. 6.1 A dielectric sample and its equivalent surface and volume charges.

The dipoles that make up the dielectric create an electric field which is, of course, conservative, so that its line integral around any closed path is zero. Thus

$$V_e + V_i = 0.$$

Similarly for the model of equivalent volume and surface charges, we have

$$V_{e'} + V_{i'} = 0.$$

But the integrals over the *exterior* paths are, as we have explained, the same,

$$V_e = V_{e'},$$

from which we conclude that the integrals* over the interior paths must be the same:

$$\int_i \boldsymbol{E} \cdot d\boldsymbol{l} = \int_{i'} \boldsymbol{E} \cdot d\boldsymbol{l}. \tag{6.5}$$

Since the average value of the field along a path can be defined as the line integral of $\boldsymbol{E}$ along the path, divided by the path length, Eq. (6.5) states that the equivalent surface and volume distributions can be used to calculate the *average value* of the interior field correctly. This does *not* mean, of course, that the local microscopic field is also given correctly. It is clear that this local field will depend on the exact nature of the dipolar distribution that is contained in the cell. On the other hand, the average field depends only on the equivalent polarization charges, and these are fixed by the *net* dipole moment of the cell. The same net dipole moment can be constituted in an infinite number of different ways, e.g., by a single point dipole in the cell, by two or more weaker dipoles at different points of the cell, by a continuous charge distribution containing both positive and negative charges, etc. The local field inside the cell will be quite different for these models, all of which will have the same line integral from cell face to cell face.

It is useful to have a name for the field that is calculated in the interior of a dielectric using the equivalent polarization charges of Eqs. (6.3) and (6.4). We shall call this field $\boldsymbol{E}_m$, since in some literature it is called the *macroscopic* field. However, we might with equal enlightenment consider the subscript m to stand for "model" since the field does not really exist in nature, but is a model of the real local fluctuating field in the usual sense that a model of a physical system is a construct which has some but not all of the properties of the real physical system. In the present case, the model field has the property of giving the correct line integral but not of giving the local fluctuations.

6.3 THE RELATION OF POLARIZATION TO THE APPLIED FIELD

We have been discussing the field in the interior of a dipolar distribution that was assumed given. Nothing was said about how the given distribution is produced; we turn to this now. We shall assume that the polarization of the medium is maintained by some applied electric field; the main point of this remark is to exclude from consideration *electrets*, which are bodies that maintain a polarized state in the absence of any external or applied field.

We shall then be considering a sample of dipolar matter in an external field. Consider first a single one of its molecules isolated in space. If a field $\boldsymbol{E}$ is applied,

* In the integral on the left $\boldsymbol{E}$ is, of course, the actual field; on the right it is the equivalent field.

a dipole moment $\boldsymbol{p}$ will be induced in it. Frequently in practice $\boldsymbol{p}$ will be proportional to $\boldsymbol{E}$,*

$$\boldsymbol{p} = \alpha \boldsymbol{E}. \tag{6.6}$$

The constant α is called the *polarizability* or *molecular polarizability*. The field $\boldsymbol{E}$ in this equation is, loosely, the field acting on the element but it merits two comments. First, we assume that this field is essentially constant throughout the (small) region of space occupied by the element; otherwise, of course, there would be ambiguity as to where to evaluate the $\boldsymbol{E}$ that enters (6.6). Second, the field $\boldsymbol{E}$ in (6.6) is the *external* or *applied* field, i.e., it is the field due to all other sources *except* the element itself; the field which the element itself produces does not act to polarize it.

With this in mind we return to the sample of dielectric matter in an external field $\boldsymbol{E}_a$. The molecules of the sample will be arranged with some regularity in a lattice for solid matter, and more or less randomly in a gas or liquid. Whatever the case, we shall assume only that the molecules are small enough compared with their separation to be considered point dipoles. The external field $\boldsymbol{E}_a$ is that produced by all the sources of charge that are, roughly speaking, external† to the dielectric. Consider the polarization, in some arbitrary ordering, of the ith molecule. A moment $\boldsymbol{p}_i$ proportional to the field $\boldsymbol{E}$ will be induced in it according to Eq. (6.6). The field $\boldsymbol{E}$ that enters this equation is the *total* field at the molecule exclusive of the field produced by the molecule itself. In the present case, the total field is the sum of the *applied field* $\boldsymbol{E}_a$ and the field $\boldsymbol{E}_d$ produced by all the *other dipoles* in the dielectric. For $\boldsymbol{p}_i$, we have

$$\boldsymbol{p}_i = \alpha(\boldsymbol{E}_a + \boldsymbol{E}_d). \tag{6.7}$$

The index i runs over the number of dipoles in the lattice, so there is an equation like (6.7) for every lattice site, and their totality constitutes, in general, an intractable self-consistency problem. It is even more difficult than the difficult problem of finding the field due to a *given* distribution which we have already discussed. Here we do not even know what the $\boldsymbol{p}_i$ are in advance.

Since Eq. (6.7) cannot be solved as it stands it must somehow be modified or replaced. In principle, there is the possibility of averaging (6.7) over an ensemble, using statistical mechanics, but even that is too difficult to be practical in general. What we shall do here is average (6.7) in a heuristic and unrigorous way to get a smoothed-out version of it, and take this version as a postulate. For averaging the left-hand side of Eq. (6.7), we consider a volume that is small compared to the size of the sample and to the distance over which the external field varies, but large

---

* In general, it is not uncommon in anisotropic media for each of the components of $\boldsymbol{p}$ to be linear functions of all the components of $\boldsymbol{E}$, so that $\boldsymbol{p}$ and $\boldsymbol{E}$ are related by means of a polarizability *tensor*.

† This is not meant to exclude charges that are placed on the surface of the dielectric.

enough to contain a statistically significant number of dipoles. This might be a volume whose typical linear dimension was at least several interdipole distances. The position of this volume is characterized with small ambiguity by a position vector $\boldsymbol{r}$. Imagine the dipoles in this volume smeared out to form a continuous dipolar distribution, as they were for calculating the external field and, as before, define $\boldsymbol{P}(\boldsymbol{r})$ as the dipole moment density. Then for the smoothed-out counterpart of Eq. (6.7), we consider an equation in which $\boldsymbol{P}(\boldsymbol{r})$ is on the left-hand side.

Now we consider the two terms on the right-hand side of (6.7). Since the applied field $\boldsymbol{E}_a$ is already a smooth function which by hypothesis is constant over the small volume we consider, we need not average it but simply retain it. The essential part of the hypothesis is the averaging of the microscopic field $\boldsymbol{E}_d$. It is natural to replace this field by its model counterpart $\boldsymbol{E}_m$ that correctly reproduces the line integral of $\boldsymbol{E}_d$, and we are tempted to take, as preliminary hypothesis, the equation

$$\boldsymbol{P} = \lambda(\boldsymbol{E}_a + \boldsymbol{E}_m), \tag{6.8}$$

where λ is some constant of proportionality.

This equation must, however, be qualified in an important way. We have remarked that for a *single* polarized element the polarizing field has its source *outside* the element; the element is not polarized by its own field. Since there are no sources of the polarizing field at the element, the *divergence* of the polarizing field must vanish there. In Eq. (6.8) we have not a single element but a polarization density $\boldsymbol{P}$; it is, however, proportional to a polarizing field $(\boldsymbol{E}_a + \boldsymbol{E}_m)$ on the right-hand side, and we are led to investigate the divergence of this equation. To begin, we note explicitly that $\boldsymbol{E}_m$ is composed of a surface term $\boldsymbol{E}_s$ due to σ_p and a volume term $\boldsymbol{E}_v$ due to ρ_p:

$$\boldsymbol{E}_m = \boldsymbol{E}_s + \boldsymbol{E}_v. \tag{6.9}$$

We put this into (6.8) and form $\nabla \cdot \boldsymbol{P}$, assuming a uniform dielectric for which λ is a constant. Inside the dielectric volume, $\nabla \cdot \boldsymbol{E}_a = \nabla \cdot \boldsymbol{E}_s = 0$ since there are no sources of $\boldsymbol{E}_a$ or $\boldsymbol{E}_s$ in the volume. Then

$$\nabla \cdot \boldsymbol{P} = \lambda \nabla \cdot \boldsymbol{E}_v. \tag{6.10}$$

The field $\boldsymbol{E}_v$ can also be calculated directly. It is the gradient of a potential Φ_v which can be found from the polarization charge density ρ_p by means of a superposition integral. That is,

$$\boldsymbol{E}_v = -\nabla \int \frac{\rho_p(\boldsymbol{r}')}{|\boldsymbol{r} - \boldsymbol{r}'|}\, dv', \tag{6.11}$$

where the gradient is, of course, with respect to $\boldsymbol{r}$. On applying the divergence operator to both sides of (6.11) and using the fact that

$$\rho_p = -\nabla \cdot \boldsymbol{P} \qquad \text{and} \qquad \nabla^2(1/|\boldsymbol{r} - \boldsymbol{r}'|) = -4\pi\delta(\boldsymbol{r} - \boldsymbol{r}'),$$

we find

$$\nabla \cdot \boldsymbol{P} = -\frac{1}{4\pi} \nabla \cdot \boldsymbol{E}_v. \tag{6.12}$$

But this is consistent with (6.10) only if $\boldsymbol{E}_v$ and $\nabla \cdot \boldsymbol{P}$ are zero. Thus we conclude *that* $\boldsymbol{E}_v$ *is zero and only the surface term stands on the right-hand side of Eq.* (6.8). Our basic postulate is

$$\boldsymbol{P} = \chi_e(\boldsymbol{E}_a + \boldsymbol{E}_s), \tag{6.13}$$

where the coefficient of proportionality χ_e is called the *electrical susceptibility*. It is convenient to introduce one new definition and call the field on the right-hand side of (6.13) the *average* field $\bar{\boldsymbol{E}}$:

$$\bar{\boldsymbol{E}} = \boldsymbol{E}_a + \boldsymbol{E}_s. \tag{6.14}$$

We can then write the assumed law of polarization as

$$\boldsymbol{P} = \chi_e \bar{\boldsymbol{E}}. \tag{6.15}$$

Since the basic equation (6.15) is a postulate, we may ask whether there is any evidence to support it. There is; it leads to the central empirical fact about condensers that we have mentioned. A condenser which has first a vacuum between its plates and is then filled with some dielectric material will increase its capacitance by an amount which depends *only on the material and is independent of the shape of the plates*. In the next section we show how this fact is explained by the hypothesis of Eq. (6.15).

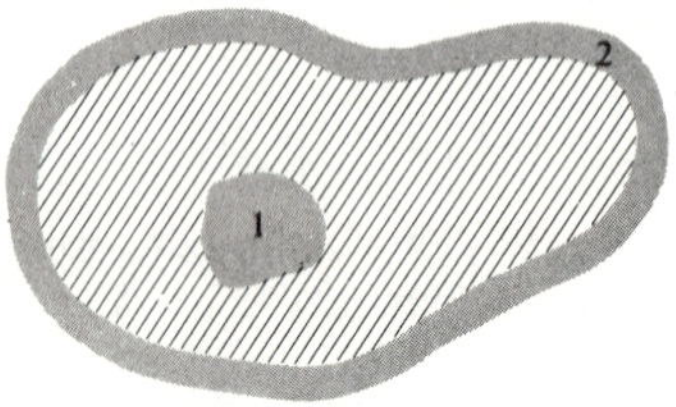

Fig. 6.2 A condenser filled with dielectric.

6.4 CAPACITORS AND THE DIELECTRIC CONSTANT

Consider a general condenser which can be taken to consist of two metallic surfaces with the one separated completely from the other, as shown in Fig. 6.2. For the sake of the illustration, we have drawn them as surfaces of a cylinder long enough that end effects are negligible. The shapes of the surfaces are not necessarily similar, but we suppose, for the purpose of defining capacitance, that there is a charge Q on one of them and a charge $-Q$ on the other. From the previous discussion of boundary value problems with conductors, we know that the charge on the surfaces will so distribute itself that *each* metallic surface becomes an equipotential. An important point to remember is that the *relative* distributions of surface charge are fixed by the shape of the plates and are *independent of the total charge*; that is, with x times as much charge added to each plate, the charge density will be everywhere x times as large, but the relative distribution will remain unchanged.

With this in mind, we return to the condenser which we now suppose to be filled uniformly with a dielectric of susceptibility χ_e. A polarization will be set up in the dielectric, and the average field $\bar{\boldsymbol{E}}$ that it produces can be calculated from an equivalent surface charge. This polarization surface charge σ_p is so close to the true surface charge σ_c on the conductors that we can assume the two charges add: in effect, there is a total surface charge $\sigma_t = \sigma_c + \sigma_p$ on the conductors. Even with the dielectrics in place, the conductors are equipotentials, with the potential now being produced by the total charge σ_t. But since the conductors are still of the same shape, the *relative distribution of the charge σ_t must be the same as before the dielectric was introduced.* This implies that the *relative variation* of the average field point by point through the dielectric is the same as if the dielectric were absent. The effect of the dielectric is simply to change the absolute magnitude of this field, and we can fix the magnitude by calculating the new field at some convenient point. For example, take a point close to the inner surface of the outer conductor just within the outer boundary of the dielectric. The field there will be essentially normal to the conductor. Let E_n be this normal component before the dielectric is introduced. From Eq. (4.1) with the normal taken inward, i.e., from conductor to dielectric, we have

$$E_n = 4\pi\sigma_c.$$

Let $\bar{E}_n$ be the normal component of the average field *after* the dielectric has been introduced. Then

$$\bar{E}_n = 4\pi(\sigma_c + \sigma_p) = E_n + 4\pi\sigma_p.$$

But from Eqs. (6.4) and (6.15), with a minus sign demanded by the present convention for the direction of the normal, we have

$$\sigma_p = -P_n = -\chi_e \bar{E}_n.$$

Thus

$$\bar{E}_n = E_n - 4\pi\chi_e \bar{E}_n$$

or

$$\bar{E}_n = E_n/\varepsilon, \tag{6.16}$$

where the *dielectric constant* ε is

$$\varepsilon = 1 + 4\pi\chi_e. \tag{6.17}$$

This shows that the average field at the surface is decreased by a factor $1/\varepsilon$, and from the remarks above this means that the *average field* is everywhere decreased by the same amount. So is the *potential difference* between the two plates, and Eq. (6.2) for capacitance shows that for a condenser of any shape which is completely filled with dielectric, the *capacitance is increased by a factor* ε.

6.5 BOUNDARY VALUE PROBLEMS WITH DIELECTRICS

In the real world, dielectrics do not occur only sandwiched between condenser plates. We now discuss how to find the average field $\bar{\boldsymbol{E}}$ when a uniform dielectric is polarized by an applied field $\boldsymbol{E}_a$. The discussion above leads us to replace, as in Fig. 6.3, the dielectric body by a *model volume* of the same shape with a surface polarization charge that is *a priori* unknown but with no charge in the interior. The field $\bar{\boldsymbol{E}}$ can then be derived from a potential,

$$\bar{\boldsymbol{E}} = -\nabla\bar{\Phi}, \tag{6.18}$$

where $\bar{\Phi}$ satisfies Laplace's equation inside the model volume and the Poisson equation appropriate to the sources of $\boldsymbol{E}_a$ outside the volume. The problem is now much like electrostatic problems with conductors, in which case there are regions of space where Laplace's or Poisson's equation is satisfied, and surface charges that are *a priori* unknown. For these problems, the general solutions of these equations are made specific by the boundary conditions, and the unknown surface charge can be determined from the normal derivative of the potential. Here we shall follow much the same course and begin by investigating the boundary conditions.

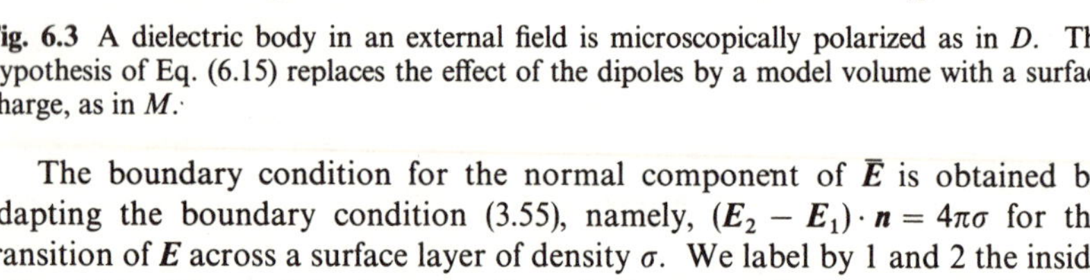

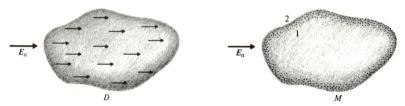

Fig. 6.3 A dielectric body in an external field is microscopically polarized as in *D*. The hypothesis of Eq. (6.15) replaces the effect of the dipoles by a model volume with a surface charge, as in *M*.

The boundary condition for the normal component of $\bar{\boldsymbol{E}}$ is obtained by adapting the boundary condition (3.55), namely, $(\boldsymbol{E}_2 - \boldsymbol{E}_1)\cdot\boldsymbol{n} = 4\pi\sigma$ for the transition of $\boldsymbol{E}$ across a surface layer of density σ. We label by 1 and 2 the inside and outside of the model volumes, so that $\boldsymbol{n}$ is the outwardly drawn normal. Since we know that for the present case

$$\sigma = \boldsymbol{P}\cdot\boldsymbol{n}, \tag{6.19}$$

this condition (3.55) becomes

$$(\bar{\boldsymbol{E}}_2 - \bar{\boldsymbol{E}}_1) \cdot \boldsymbol{n} = 4\pi \boldsymbol{P} \cdot \boldsymbol{n}. \tag{6.20}$$

From Eq. (6.15),

$$\boldsymbol{P} = \chi_e \bar{\boldsymbol{E}}_1 \tag{6.21}$$

and with this we can write (6.20) in terms of the dielectric constant ε defined by (6.17), as*

$$(\bar{\boldsymbol{E}}_2 - \varepsilon \bar{\boldsymbol{E}}_1) \cdot \boldsymbol{n} = 0. \tag{6.22}$$

It was also found previously that the tangential component of $\bar{\boldsymbol{E}}$, or equivalently the potential $\bar{\Phi}$, is continuous across a boundary containing surface charge:

$$\bar{\Phi}_2 = \bar{\Phi}_1. \tag{6.23}$$

With the boundary conditions (6.22) and (6.23) the problem is almost as simple as electrostatic ones with conductors. In sum, one solves Laplace's or Poisson's equation for $\bar{\Phi}$ in the interior and exterior of a *model volume* of the *same shape* as the dielectric. The boundary conditions then fix the unknown parameters in the solution, and the mean field $\bar{\boldsymbol{E}}$ is given by Eq. (6.18).

As an illustration consider a dielectric sphere in the field of a point charge q, as shown in Fig. 6.4. The exterior potential $\bar{\Phi}_2$ is that due to the point charge plus that due to the (unknown) surface polarization charge. It can then be written

$$\bar{\Phi}_2 = \frac{q}{r_1} + \sum_{l=0}^{\infty} \frac{B_l}{r^{l+1}} P_l(\cos\theta).$$

For matching boundary conditions at $r = a$, we expand $1/r_1$ in the form appropriate for $r < d$ to get, using $P_l(\cos(\pi - \theta)) = (-)^l P_l(\cos\theta)$,

$$\bar{\Phi}_2 = \frac{q}{d} \sum_{l=0}^{\infty} (-)^l \left(\frac{r}{d}\right)^l P_l(\cos\theta) + \sum_{l=0}^{\infty} \frac{B_l}{r^{l+1}} P_l(\cos\theta).$$

The interior potential $\bar{\Phi}_2$ is again that due to the point charge *plus* that due to the unknown surface charge, but both of these may be lumped into the expression

$$\bar{\Phi}_1 = \sum_{l=0}^{\infty} A_l r^l P_l(\cos\theta).$$

---

* For the general case of a transition between two media of dielectric constants ε_1 and ε_2, at the interface of which charge has been *applied* with a surface density σ_a, Eq. (6.22) can be easily generalized to read

$$(\varepsilon_2 \bar{\boldsymbol{E}}_2 - \varepsilon_1 \bar{\boldsymbol{E}}_1) \cdot \boldsymbol{n} = 4\pi\sigma_a.$$

This condition is also sometimes written in terms of the *displacement vector* $\boldsymbol{D}$ defined by $\boldsymbol{D} = \bar{\boldsymbol{E}} + 4\pi\boldsymbol{P}$; it then reads

$$(\boldsymbol{D}_2 - \boldsymbol{D}_1) \cdot \boldsymbol{n} = 4\pi\sigma_a.$$

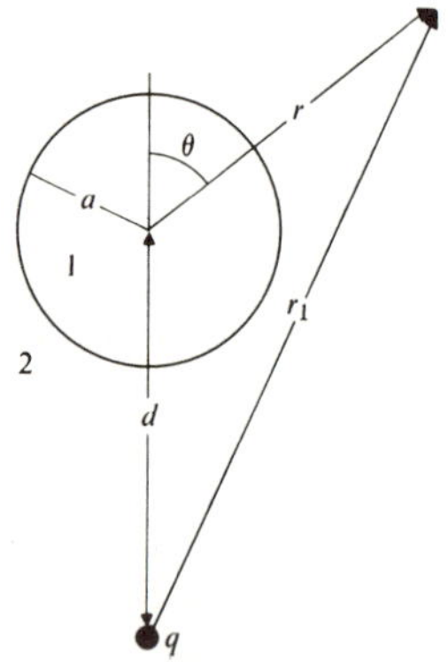

Fig. 6.4 Point charge outside dielectric sphere.

On applying the boundary conditions (6.22) and (6.23) we find, in a straightforward way:

$$A_l = (-)^l \frac{q(2l + 1)}{d^{l+1}(l\varepsilon + l + 1)}$$

$$B_l = (-)^l \frac{qa^{2l+1}l(1 - \varepsilon)}{d^{l+1}(l\varepsilon + l + 1)}.$$

and this determination of the coefficients effectively completes the solution of the problem.

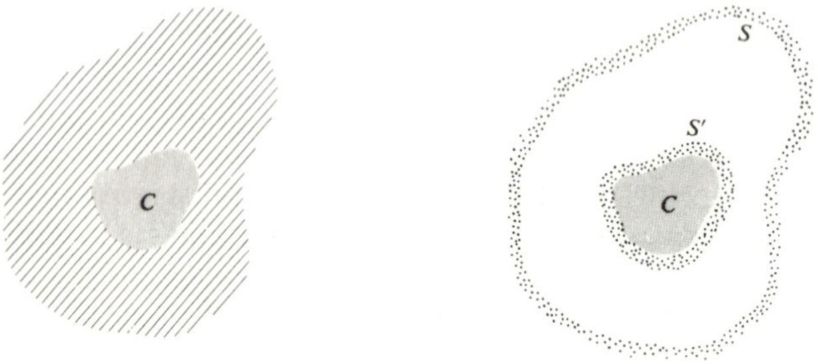

Fig. 6.5 A conductor C embedded in a dielectric and the surface charges S and S' of the equivalent model.

As a second example consider the problem of point charges or conductors *embedded* in dielectric material. Here we shall have to be somewhat careful as to the meaning of point charges, and in fact will consider them as limiting cases of conductors with vanishingly small dimensions. Consider first a conductor embedded in a dielectric as shown in Fig. 6.5. According to the discussion above, for calculating the average field $\bar{E}$ the physical dielectric is replaced by the model, also shown in the figure, in which the two surfaces of the dielectric are replaced by unknown surface charges. For the sake of the illustration we have drawn the inner surface as very close to, but not quite touching, the conductor. Suppose first that the dielectric sample is very large, i.e., that the outer surface has receded to infinity so that the conducting body is embedded in an infinite sea of dielectric. Then if we assume that the surface at infinity plays no further role, it is clear that the field the conductor produces will be $1/\varepsilon$ times what it would produce in vacuum. The argument is essentially identical with the one for the dielectric-filled condenser: The conductor alone in space must have a charge distribution which makes it an equipotential; when surrounded by dielectric it still must be an equipotential, except under the influence of its charge σ_c *plus* the surface polarization charge σ_P; when boundary conditions are applied, the field produced by the net charge $\sigma_t = \sigma_p + \sigma_c$ is seen to be $1/\varepsilon$ times the field in vacuum. In this universe of dielectric, the field of a charged conductor is reduced everywhere by a factor $1/\varepsilon$ with respect to its field in vacuum.

With this result in mind, we discuss the field of a *point* charge in a sample of dielectric of *finite* size, for which the outer surface has not receded to infinity. As a model of a point charge we must take some finite distribution whose dimensions shrink to zero. For example, suppose the conductor in Fig. 6.5 is a small sphere of radius a with a *fixed* total charge; let us investigate what happens when a becomes very small. We must once more apply the boundary conditions for the transition of $\bar{E}$ across the surface layer of polarization charge S' just surrounding the conductor. The sources of the field $\bar{E}$ are three: the charge on the conductor, and the polarization charge on S, and that on S'. However, as the conductor gets smaller the field it produces near S' becomes larger and larger, whereas the field produced by S remains essentially constant. In this limit the field due to the (almost) point charge dominates the boundary conditions at S'. For the combined effect of the field of the charge *cum* inner surface, the outer surface plays no role; it is as if it were absent. In short, the point charge and the polarization charge on S' taken together produce a field of magnitude $q/\varepsilon r^2$ and a potential $q/\varepsilon r$.

In any problem that involves point charges embedded in a dielectric, we need not go through the preceding analysis again, but can assume that the potential the charge produces is obtained by dividing q/r by the dielectric constant *of the medium in which it is embedded*. This answers a question that is sometimes puzzling on first sight. Suppose there are two adjoining media of dielectric constants ε_1 and ε_2, and a point charge q is in medium 1. What is the potential that the point charge q produces in medium 2? Is it $q/\varepsilon_1 r$, or $q/\varepsilon_2 r$, or does some linear combination of ε_1 and ε_2 enter into the denominator of the Coulomb potential? One might tend to think the latter, since the lines of force traverse both medium 1 and medium 2. The discussion above shows, however, that the combined effect of the point charge and its associated polarization charge is to produce the usual potential q/r divided by the dielectric constant of the *medium in which the charge is located*. In the above example, the potential the point charge produces is $q/\varepsilon_1 r$ both in medium 1 and medium 2.

These ideas can be applied in solving some problems involving dielectrics by the method of images. As in our treatment of conductor problems, the method is not systematic but depends on guessing configurations of charges that will satisfy the appropriate Laplace and Poisson equation and the boundary conditions as well. An example involving a point charge embedded in a medium of one dielectric constant that adjoins a medium of a second dielectric constant is given as Problem 6.6.

6.6 MACROSCOPIC AND MICROSCOPIC PROPERTIES

We have presented two points of view for discussing dielectrics. There are the microscopic one defined by Eq. (6.7) and its macroscopic counterpart defined by (6.15). As we have emphasized, Eq. (6.15) is an independent hypothesis. That is, although (6.15) hopefully represents an averaged form of (6.7), it is too difficult mathematically to show that *whatever* the nature of the system (6.7) refers to (gaseous, liquid, or solid) and whatever the nature of the dipoles (point or otherwise), Eq. (6.15) *always* follows from correctly averaging (6.7). But for a particularly simple system we may be able to show that (6.15) follows from (6.7), and in so doing relate the macroscopic parameter χ_e to the microscopic parameter α.

We shall now show that this can be done for a system already discussed, namely that of point dipoles in a cubic lattice of a body of arbitrary shape. We want to calculate the field at a lattice point due to all the other dipoles in the system. As in Chapter 3 we do this by imagining a sphere drawn about one of the lattice points, thereby dividing the dipoles into "distant" ones (outside the sphere) and "near" ones (inside the sphere). The effect of the distant dipoles is evaluated from the polarization charges they produce on the surface (called *s*) of the specimen and on the spherical surface (called *l*, after Lorentz, who first outlined this calculation). The corresponding fields are then called $\boldsymbol{E}_s$ and $\boldsymbol{E}_l$. The field $\boldsymbol{E}$ at a lattice point is the sum of the applied field $\boldsymbol{E}_a$, the fields $\boldsymbol{E}_s$ and $\boldsymbol{E}_l$, and $\boldsymbol{E}_n$, the field due to the *near* dipoles in the interior of the sphere; that is,

$$\boldsymbol{E} = \underbrace{\boldsymbol{E}_a}_{\text{Applied}} + \underbrace{\boldsymbol{E}_s + \boldsymbol{E}_l}_{\text{Distant}} + \underbrace{\boldsymbol{E}_n}_{\text{Near}}. \tag{6.24}$$

It is left to the reader to show that the near field $\boldsymbol{E}_n$ of the cubically symmetric dipole array vanishes by reasons of symmetry. From Eq. (3.64) the Lorentz field $\boldsymbol{E}_l$ is* $4\pi\boldsymbol{P}/3$, where $\boldsymbol{P}$ is the local macroscopic average dipole moment. Of course $\boldsymbol{P}$ varies throughout the medium but we assume that the sphere is small enough so

* It is clear that $\boldsymbol{E}_l$ and $\boldsymbol{P}$ are parallel. The minus sign in Eq. (3.64) arises in that it refers to a case in which $\boldsymbol{P}$ is directed to the left, and hence is negative.

that $\boldsymbol{P}$ is sensibly constant over it. Finally $\boldsymbol{E}_s$, together with $\boldsymbol{E}_a$, makes up the *average* field $\bar{\boldsymbol{E}}$, so that $\bar{\boldsymbol{E}} = \boldsymbol{E}_a + \boldsymbol{E}_s$. We can then write (6.24) as

$$\boldsymbol{E} = \bar{\boldsymbol{E}} + 4\pi\boldsymbol{P}/3. \tag{6.25}$$

From Eq. (6.6), the relation between the polarizability $\boldsymbol{p}$ and the local field is

$$\boldsymbol{p} = \alpha\boldsymbol{E} = \alpha(\bar{\boldsymbol{E}} + 4\pi\boldsymbol{P}/3).$$

Since $N\boldsymbol{p} = \boldsymbol{P}$, we can multiply this by N, the number of dipoles per unit volume, and obtain

$$\boldsymbol{P} = N\alpha(\bar{\boldsymbol{E}} + 4\pi\boldsymbol{P}/3). \tag{6.26}$$

By putting $\boldsymbol{P} = \chi_e\bar{\boldsymbol{E}}$ in (6.26) we can solve for χ_e to get

$$\chi_e = \frac{N\alpha}{1 - (4\pi/3)N\alpha}. \tag{6.27}$$

This equation is called the *Clausius-Mosotti* relation. Although it has been deduced for a lattice of dipoles, it is frequently applied to gases, yielding moderate agreement with experiment, although there is no real reason that it should be valid for this case.

PROBLEMS

1. A spherical shell of material of dielectric constant ε, and inner and outer radii a and b, is centered at the origin. A uniform field $\boldsymbol{E}_0$ is applied in the z-direction. Find the field everywhere.
2. An infinitely long dielectric cylinder is in a uniform field $\boldsymbol{E}_0$ that is perpendicular to its axis. Find $\boldsymbol{E}$ inside and outside the cylinder.
3. A point charge q is embedded in a dielectric sphere of dielectric constant ε, radius a, at a distance d from the center. Find the field everywhere.
4. Show that a dielectric body of dielectric constant ε behaves, in the limit $\varepsilon \to \infty$, like a perfect conductor.
5. A conducting sphere of radius a, surrounded by a dielectric shell of outer radius b, has total charge Q and is under the influence of a point charge q at $z = d$ ($d > b$) on the z-axis. Find the field everywhere.
6. The space $z > 0$ is filled with matter of dielectric constant ε_1, and the space $z < 0$ with matter of dielectric constant ε_2. A charge q is at $z = d$ on the z-axis. Show that for $z > 0$ the field can be calculated as if it were due to a charge of magnitude q/ε_1 at $z = d$ plus an image charge q' at $z = -d$, and for $z < 0$ the field is as if it were due to a charge q'' at $z = d$. Find q' and q''.

REFERENCES

Durand, E., *Electrostatique et Magnetostatique*, Masson, Paris (1953).

This thorough work has a large collection of problems involving dielectrics.

Mahan, G. D., "Local Field Corrections to Coulomb Interactions," *Phys. Rev.* **153,** 983 (1967).

According to the smoothing and averaging assumptions of this chapter, two point charges q and q' a distance r apart in dielectric exert a force on each other of magnitude $qq'/\varepsilon r^2$. On the other hand, the concept of dielectric constant obviously ceases to be meaningful when r is so small that few or no dipoles of the medium intervene between the charges. This paper discusses the problem of the short-range corrections to the concept of dielectric constant.

Theimer, O., and R. Paul, "Effective Polarizing Field in Ionic Crystals," *J. App. Phys.* **36,** 3678 (1965).

A discussion of corrections to the Lorentz result of Eq. (6.125) due to the fact that real dipoles are not point dipoles.

van Bladel, J., *Electromagnetic Fields*, McGraw-Hill, New York (1961).

The problem of a dielectric body in an external field can be reduced to an integral equation, much as for the problem of a conductor in an external field. This reduction is discussed on p. 131 of this book, which has additional references.

7
THE MAGNETOSTATIC FIELD

We begin with the briefest of histories of magnetism. The Greeks were aware that certain natural stones (lodestones) had the property of attracting iron. The word *magnetism* is, in fact, derived from *Magnesia*, a place in Greece, where such stone is found. As early as the twelfth century, one practical aspect of magnetism was known, in that the magnetic compass was used by various peoples to determine direction at sea. In the thirteenth century, Pierre deMaricourt undertook to measure the deflection of a magnetic needle near the surface of a spherical sample of lodestone. He found that the lines, whose tangent at a point gave the direction of the magnetic needle, resembled the meridians of longitude on the earth. Like the meridians, they started at one point on the surface and led to a diametrically opposite point. From the geographical analogy, he called these points the *poles*, a name that has endured until today, and he regarded these poles as the seat of the "magnetic power." These experiments were greatly extended by William Gilbert (1540–1603), who made the most important discovery that the magnetic compass works because the earth itself is a great magnet.

It was not until the nineteenth century, however, that a relation was established between electricity and magnetism, although a connection had been suspected for some time. In 1820, Oersted found that electricity in motion, that is, an electric current, could produce an effect on magnetic needles which appeared identical to that produced by natural lodestones. In modern language, we would say he showed that a current, as well as natural magnetic stone, produces a *magnetic field*. Oersted's results were soon made quantitative by Biot and Savart, who formulated the law governing the production of a magnetic field by a long straight wire. Moreover, they showed that the field could be assumed to be the sum of fields produced by each differential element of the wire according to a formula (Eq. 7.8), which bears their name. Following this, Ampère extended the experiments and discussed the field due to, and forces on, currents in wires of any shape. His work culminated in a formula for the force between two current loops of arbitrary shape, and it is this basic *Ampère law* that we shall take as the keystone of our discussion.

7.1 CURRENT AND RELATED CONCEPTS

Insofar as the details of motion can be defined, the microscopic motion of charged particles in matter is rapid and, in a sense, random. If on the *average*, however, the charge is moving with a regular ordered motion, this motion constitutes an *electric current*. As with current in a liquid, there are *streamlines* whose directed tangent at any point gives the direction of the (average) motion of the charge at that point. In terms of these streamlines, we can define an electric current density $\boldsymbol{J}$. At any point, the *direction* of $\boldsymbol{J}$ is given by the streamline through the point; the magnitude of $\boldsymbol{J}$ is the quantity of charge per unit time and unit area that crosses an infinitesimal surface normal to the streamline through the point. If at the point the charge density is ρ and its velocity is $\boldsymbol{v}$, then

$$\boldsymbol{J} = \rho \boldsymbol{v}.$$

In the present Gaussian units in which charge is measured in statcoulombs, the dimensions of current density are statcoulombs/cm^2 · sec.

Suppose that at some point on a streamline a plane perpendicular to it is passed and a closed curve which encloses the streamline is drawn in that plane. If the streamlines that pass through each point of the curve are drawn, a tubelike shape will be generated; this is a *tube of current*. The current density is not necessarily constant across such a tube but if the cross-sectional area is small enough, the variation in density can be made arbitrarily small. If the density is constant, we shall call the tube a *current filament*. An arbitrary current distribution can be considered to be made up of current filaments, much as an arbitrary charge distribution can be considered to be made up of small constant-density charge elements.

Consider now some time-dependent current density $\boldsymbol{J}(\boldsymbol{r}, t)$. Imagine an infinitesimal surface $d\boldsymbol{s}$ placed in the current and denote the charge crossing the surface per second by dI. Clearly $dI = \boldsymbol{J} \cdot d\boldsymbol{s}$; hence, the total amount of charge passing per second through a finite surface S is

$$I = \int_S \boldsymbol{J} \cdot d\boldsymbol{s}; \tag{7.1}$$

I is called the *current* through the surface, or sometimes the *total current*. Now suppose that the surface S is a *closed* one; i.e., the surface divides space into two parts which are accessible to one another only by crossing the surface. If we assume that charge is not created or destroyed, it is clear that the charge flowing out must result in a decrease of the charge in the volume V enclosed by the surface. If the charge density is $\rho(\boldsymbol{r})$, the total charge in the volume is $\int_V \rho \, dv$, and its time rate of change is $(\partial/\partial t) \int_V \rho \, dv$. Hence, the conservation of charge is expressed by the equation

$$\int_S \boldsymbol{J} \cdot d\boldsymbol{s} + \frac{\partial}{\partial t} \int_V \rho \, dv = 0. \tag{7.1}$$

The left-hand side of this equation can be changed to $\int_V \nabla \cdot \boldsymbol{J} \, dv$ by the divergence theorem. Since the resulting equation holds for an *arbitrary* volume, we can, as in the derivation of Eq. (2.9), equate the integrands to find the important *equation of continuity*

$$\nabla \cdot \boldsymbol{J} + \partial\rho/\partial t = 0. \tag{7.2}$$

If the charge density is a function of $\boldsymbol{r}$ but does not vary with time, Eq. (7.2) becomes

$$\nabla \cdot \boldsymbol{J} = 0. \tag{7.3}$$

Equation (7.3) implies that the streamlines of current flow are continuous and close on themselves, i.e., they form closed loops.

7.2 FORCES BETWEEN CONDUCTORS AND THE MAGNETOSTATIC FIELD *B*

We shall, insofar as we can, model the discussion in magnetostatics on our previous discussion of electrostatics. The analog of Coulomb's law is Ampère's law (mentioned in the introduction to this chapter) for the force between current-carrying loops. Figure 7.1 shows two closed-current loops or circuits, carrying currents I_1 and I_2 respectively. Ampère's experiments showed that the total force $\boldsymbol{F}_1$ on circuit 1, due to its interaction with circuit 2, is proportional to the two currents I_1 and I_2 and to an integral which depends only on geometry. More precisely, Ampère's law is

$$\boldsymbol{F}_1 \propto I_1 I_2 \oint\oint \frac{d\boldsymbol{l}_1 \times (d\boldsymbol{l}_2 \times \boldsymbol{r}_{12})}{r_{12}^3}. \tag{7.4}$$

The constant of proportionality in (7.4) depends on the system of units. Moreover, this constant is fixed once the units of charge, length, and time are fixed. Expression (7.4) is free of the arbitrariness that was present in Coulomb's law in which the right-hand side could be multiplied by an arbitrary factor and that factor compensated for by redefining the unit of charge. Once the units that enter (7.4) are fixed, there is nothing left to redefine, and the constant of proportionality is a purely empirical one. It turns out that in the Gaussian system the constant of proportionality is $1/c^2$, where c is precisely the velocity of light. This is not an extraordinary numerical coincidence; it has a deeper basis in physics. We shall

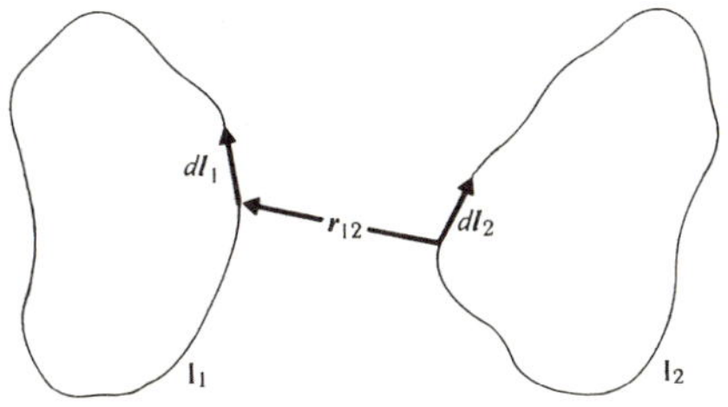

Fig. 7.1 Ampère's law.

see later that the magnetic forces are a consequence of the theory of special relativity as applied to the electric field of moving charged particles. Since the velocity c plays a central role in relativity and, in particular, in the *Lorentz transformation*, it is not surprising that this velocity enters into the expression for forces between moving charged particles, which is just what Ampère's law is about. From the present, more limited point of view, this value of the constant of proportionality is a fact to be accepted. Using it, we can write the basic force law as

$$\boldsymbol{F}_1 = \frac{I_1 I_2}{c^2} \oint\oint \frac{d\boldsymbol{l}_1 \times (d\boldsymbol{l}_2 \times \boldsymbol{r}_{12})}{r_{12}^3}. \tag{7.5}$$

One further point. Equation (7.5) gives the force on current loop 1 due to current loop 2. Hence the force $\boldsymbol{F}_2$ on current loop 2 should be given by changing the roles of 1 and 2; i.e., it should be

$$\boldsymbol{F}_2 = \frac{I_1 I_2}{c^2} \oint\oint \frac{d\boldsymbol{l}_2 \times (d\boldsymbol{l}_1 \times \boldsymbol{r}_{21})}{r_{12}^3}.$$

From Newton's third law on action and reaction, we expect $\boldsymbol{F}_1 + \boldsymbol{F}_2 = 0$. If, however, we form $\boldsymbol{F}_1 + \boldsymbol{F}_2$, we find, using the formula

$$\boldsymbol{A} \times (\boldsymbol{B} \times \boldsymbol{C}) = \boldsymbol{B}(\boldsymbol{A} \cdot \boldsymbol{C}) - \boldsymbol{C}(\boldsymbol{A} \cdot \boldsymbol{B})$$

on each of the integrands, that

$$\boldsymbol{F}_1 + \boldsymbol{F}_2 = \frac{I_1 I_2}{c^2} \oint\oint \left(\frac{d\boldsymbol{l}_2 (d\boldsymbol{l}_1 \cdot \boldsymbol{r}_{12})}{r_{12}^3} + \frac{d\boldsymbol{l}_1 (d\boldsymbol{l}_2 \cdot \boldsymbol{r}_{21})}{r_{12}^3} \right).$$

The integrals on the right-hand side are not obviously zero, but they can be shown to vanish. In the first integral, for example, since $\boldsymbol{r}_{12}/r_{12}^3 = -\nabla(1/r_{12})$, the integration over $d\boldsymbol{l}_1$ is $-\oint \nabla(1/r_{12}) \cdot d\boldsymbol{l}_1$, and the integral of a gradient around a closed loop vanishes. Newton's third law is satisfied, not because there is equality of action and reaction between the current elements $I_1\, d\boldsymbol{l}_1$ and $I_2\, d\boldsymbol{l}_2$ but because of a more global property.

It is now convenient, as it was in electrostatics, to split Eq. (7.5) into two parts; that is, we imagine that one of the currents produces a *field* which then exerts a force on the other current. As we noted for electrostatics, this split is not imperative until time-varying phenomena are considered; in that case, the concept of a field is necessary if we wish to retain the laws of conservation of energy and momentum. But with an eye to the future, we shall follow the electrostatic model and arbitrarily single out the current I_2 as the one that produces the field and define the *magnetic induction field** B that I_2 produces by

$$\boldsymbol{B} = \frac{I_2}{c} \oint \frac{d\boldsymbol{l}_2 \times \boldsymbol{r}_{12}}{r_{12}^3}. \tag{7.6}$$

---

* Somewhat loosely, but in keeping with custom, we shall frequently refer to $\boldsymbol{B}$ as simply the magnetic field. In the present cgs units, $\boldsymbol{B}$ is measured in gauss.

We then postulate that this field $\boldsymbol{B}$ exerts a force $\boldsymbol{F}_1$ on current I_1 given by

$$\boldsymbol{F}_1 = \frac{I_1}{c}\oint d\boldsymbol{l}_1 \times \boldsymbol{B}. \tag{7.7}$$

Then the two definitions (7.6) and (7.7) for the production of the $\boldsymbol{B}$ field and the force exerted by this field are equivalent to the single force law (7.5). Finally, we note that when static magnetic forces between currents are measured, the currents necessarily flow in closed loops. Strictly, perhaps, one should always use only the integral definition (7.6) for $\boldsymbol{B}$, but it is convenient and consistent to extend this definition somewhat by assuming that each element $d\boldsymbol{l}$ of a current I produces a field $d\boldsymbol{B}$ given by

$$d\boldsymbol{B} = \frac{I}{c}\frac{d\boldsymbol{l} \times \boldsymbol{r}}{r^3}, \tag{7.8}$$

where $\boldsymbol{r}$ is the vector from the element to the field point. This expression is sometimes known as the *law of Biot and Savart*, although this name is also used for a formula derived later, for the field of an infinitely long wire. This split of Ampère's law into the sum of fields produced by current elements implicitly uses the principle of *superposition* of magnetic forces, a principle as important for magnetostatics as it was for electrostatics.

7.3 DIFFERENTIAL AND INTEGRAL PROPERTIES OF *B*

We now investigate the properties of the vector field $\boldsymbol{B}$. As in the case of the field $\boldsymbol{E}$, this means a calculation of its divergence and curl. For this calculation, we shall work directly with Eq. (7.8) that defines $\boldsymbol{B}$. The general technique is similar to the one used for $\boldsymbol{E}$ in that the *sources* of the field will be divided into local and distant parts, and the divergence and curl of each of these parts will be investigated separately.

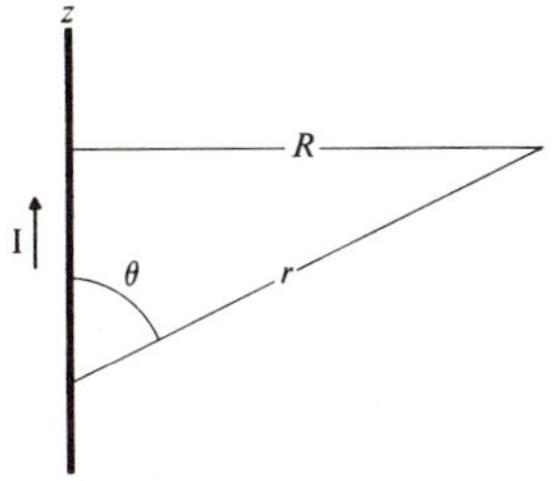

Fig. 7.2 Field of an infinitely long current filament.

As a preliminary step, we derive an expression for the field of an infinitely long, straight, thin current filament as sketched in Fig. 7.2. We apply Eq. (7.8) and, with

$\hat{\varphi}$ the unit vector in the φ-direction, i.e., into the plane of the paper, and $r = \sqrt{R^2 + z^2}$, we have

$$d\boldsymbol{l} \times \boldsymbol{r} = (r \sin\theta\, dz)\hat{\boldsymbol{\varphi}} = (R\, dz)\hat{\boldsymbol{\varphi}}.$$

Then, $\boldsymbol{B}$ has only a φ-component, B_φ, given by*

$$B_\varphi = \frac{IR}{c}\int_{-\infty}^{\infty} \frac{dz}{(z^2 + R^2)^{3/2}} = \frac{2I}{cR}. \tag{7.9}$$

To understand what is meant by *infinitely long*, suppose that the integration in (7.9) was extended not from $-\infty$ to ∞ but from $-L/2$ to $L/2$. For this case,

$$B_\varphi = \frac{2I}{cR\sqrt{1 + 4R^2/L^2}} \approx \frac{2I}{cR}\left(1 - \frac{2R^2}{L^2} + \cdots\right),$$

so that for $R \ll L$, we obtain the formula for the infinitely long wire. As expected, *infinitely long* means "long compared to the distance to the point of observation."

With this result, consider the calculation of $\nabla \cdot \boldsymbol{B}$ and $\nabla \times \boldsymbol{B}$ for the general current distribution $\boldsymbol{J}(\boldsymbol{r})$ sketched in Fig. 7.3. We first choose a point P at which we want to calculate the divergence and curl, and consider a current filament FF' whose cross section is a circle centered at P. The field $\boldsymbol{B}$ in the neighborhood of the point will be the sum of $\boldsymbol{B}^i$, the field due to the current *interior* to the filament, and $\boldsymbol{B}^e$, the field due to that *exterior* to it: $\boldsymbol{B} = \boldsymbol{B}^i + \boldsymbol{B}^e$. We can then investigate divergence and curl of $\boldsymbol{B}^e$ and $\boldsymbol{B}^i$ separately, and we begin with $\boldsymbol{B}^e$.

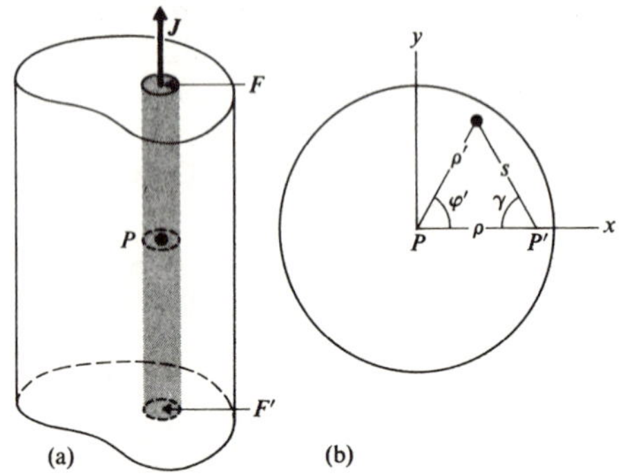

Fig. 7.3 (a) Section of current distribution $\boldsymbol{J}(\boldsymbol{r})$ containing the filament FF' of radius a, and a point P in the filament. (b) Magnified cross section through P.

Consider some other current filament in the region exterior to FF'. A segment of it of length Δl can always be chosen such that Δl is small compared to the smallest distance of any part of the segment to P, and then the radius of the segment

* This equation, as well as Eq. (7.8), is sometimes called the *Biot-Savart law*.

can be taken small compared to Δl. The field the segment produces at P is then given by the Biot-Savart law of Eq. (7.8), and it is easy to verify that both the curl and divergence of this field vanish. Since the whole current distribution exterior to the filament FF' can be considered to be made up of such segments, for each of which the curl and divergence vanish at P, we can conclude that $\nabla \cdot \boldsymbol{B}^e = \nabla \times \boldsymbol{B}^e = 0$.

Now we calculate $\boldsymbol{B}^i$. Let the length of the current filament FF' be L. The points F and F' can always be chosen close enough together so that the filament is essentially straight over the length L. Moreover, the radius a of the filament can be taken small compared to L, so that for field points near P it is effectively infinitely long, and we can employ the Biot-Savart law (7.9) for an infinitely long current. To calculate $\nabla \cdot \boldsymbol{B}^i$ and $\nabla \times \boldsymbol{B}^i$ at P, we must be able to calculate $\boldsymbol{B}^i$ in the *neighborhood* of P, say at some point P'. We then set up the cylindrical coordinate system ρ', φ', as shown in Fig. 7.3(b), with x-axis along the line joining PP'. Consider the field $d\boldsymbol{B}^i$ at P' produced by the current contained in the cylinder of infinitesimal cross section $\rho'\, d\rho'\, d\varphi'$. From the Biot-Savart law, its *magnitude* is

$$|d\boldsymbol{B}^i| = \frac{2J\rho'\, d\rho'\, d\varphi'}{cs},$$

where

$$s = \sqrt{\rho^2 + \rho'^2 - 2\rho\rho' \cos\varphi'}.$$

We want to break $d\boldsymbol{B}^i$ into its components and then integrate over the circle to find the components of the total field $\boldsymbol{B}^i$. First, note that there will be no net component B^i_ρ at P'. When we integrate over the circle, every infinitesimal current element above the axis PP' can be paired with a symmetrically placed one below, and the ρ-components of these two elements will cancel. Consider then B^i_φ which at P' is identical to B^i_y. The field dB^i_φ at P' is

$$dB^i_\varphi = |d\boldsymbol{B}^i| \cos\gamma,$$

where

$$\cos\gamma = \frac{\rho - \rho'\cos\varphi'}{s}.$$

Integrating dB^i_φ over the circle, we have for B^i_φ,

$$B^i_\varphi = \frac{2J}{c} \int_0^a \int_0^{2\pi} \frac{(\rho - \rho'\cos\varphi')\rho'\, d\rho'\, d\varphi'}{\rho^2 + \rho'^2 - 2\rho\rho'\cos\varphi'}. \tag{7.10}$$

The φ' integral in this equation is not an elementary one, but since the integrand is a rational function in $\cos\varphi'$, it can be done by contour integration; the result is given in extended tables of integrals* as

$$\int_0^{2\pi} \frac{b + d\cos x}{b^2 + d^2 + 2bd\cos x}\, dx = \begin{cases} 2\pi/b, & |b| > |d| \\ 0, & |b| < |d|. \end{cases}$$

* For example, Gröbner, W., and N. Hofreiter, Integraltafeln, Part II. *Bestimmte Integrale*, Springer Verlag, Wien (1961).

To apply this to (7.10), we break the ρ' integral into two parts: from zero to ρ and from ρ to a. The integral from ρ to a contributes nothing and that from zero to ρ yields

$$B^i_\varphi = 2\pi J\rho/c. \tag{7.11}$$

Given (7.11), the formula from Appendix C for curl in cylindrical coordinates shows that $\nabla \times \boldsymbol{B}^i$ is in the z-direction and has the value $4\pi J/c$. Since $\nabla \times \boldsymbol{B}^e$ is zero, and since the z-direction is an arbitrary one, this result can be written as the vectorial relation

$$\nabla \times \boldsymbol{B} = 4\pi \boldsymbol{J}/c. \tag{7.12}$$

Applying the expression from Appendix C for divergence in cylindrical coordinates to (7.11), we find that $\nabla \cdot \boldsymbol{B}^i$ is zero, as was $\nabla \cdot \boldsymbol{B}^e$. The second differential relation for B is then

$$\nabla \cdot \boldsymbol{B} = 0. \tag{7.13}$$

As for electrostatics, we can use the divergence theorem and Stoke's theorem to write integral counterparts of these differential laws. Thus, corresponding to (7.13), we have for any *closed* surface S,

$$\int_S \boldsymbol{B} \cdot d\boldsymbol{s} = 0. \tag{7.14}$$

Stoke's theorem applied to (7.12) yields

$$\int_C \boldsymbol{B} \cdot d\boldsymbol{l} = 4\pi I/c, \tag{7.15}$$

where C is a closed loop and I is the current through any surface that spans that loop. Equation (7.15) is known as *Ampère's circuital law.*

7.4 THE MAGNETIC SCALAR POTENTIAL

In electrostatics, the useful scalar potential has two quite different properties. First, its difference at two points is equal to the work needed to move a unit charge between those points; second, it simplifies field calculations in that one can do a single summation problem for the potential, and then differentiate it, instead of having to do a separate summation problem for each field component. In magnetostatics, there is, under certain circumstances, a scalar function with just this second property: its gradient gives the field $\boldsymbol{B}$. On the other hand, it does not have an analogous first property because the scalar function is not single valued. This is of little importance in practice; since there are no free magnetic charges, we rarely wish to calculate the work required to move one from point to point.

This scalar function is the *magnetic scalar potential.* The name "potential" is perhaps unfortunate since, conditioned by electrostatic experience, we tend to be

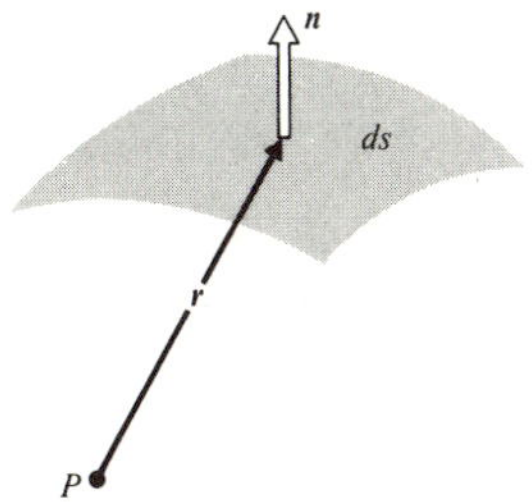

Fig. 7.4 Defining a solid angle.

made uneasy by the fact that it is not single valued. This uneasiness might be lessened if the function were not called a potential, but was called, say, the magnetic scalar *semipotential* to emphasize that it has only one of the two properties of the ordinary potential. We shall, however, defer to common usage.

The magnetic scalar potential depends on a certain solid angle, so we shall begin by reviewing some properties of solid angles. Suppose, as in Fig. 7.4, there is a vectorial element of area $\boldsymbol{ds}$ and a positive* normal $\boldsymbol{n}$ to the element such that $\boldsymbol{ds} = \boldsymbol{n}\,ds$. The element of solid angle $d\Omega$ subtended by $\boldsymbol{ds}$ at P is defined by

$$d\Omega = \frac{\boldsymbol{ds}\cdot\hat{\boldsymbol{r}}}{r^2} \equiv \frac{\boldsymbol{ds}\cdot\boldsymbol{r}}{r^3}, \tag{7.16}$$

where, as usual, $\hat{\boldsymbol{r}}$ is the unit vector along $\boldsymbol{r}$. Consider now a closed curve C spanned by some surface S. The solid angle subtended by S is

$$\Omega = \int_S \frac{\boldsymbol{r}\cdot\boldsymbol{ds}}{r^3}. \tag{7.17}$$

It is easy to prove that Ω is independent of the shape of the surface that spans C, that is, it is essentially defined by C itself. A point inside a spherical surface obviously subtends a solid angle of 4π; it follows that a point in the interior of a *closed* surface of *any* shape also subtends 4π. In particular, if the closed surface has the form of a very thin pillbox, each of the disk-shaped surfaces must subtend an angle 2π from a point in the center. This essentially proves that an infinite plane subtends a solid angle 2π. Consider such a plane with a certain positive normal $\boldsymbol{n}$. With reference to (7.16), $\boldsymbol{r}$ reverses direction as the surface is crossed so that if the solid angle subtended at a point just on one side of the surface is 2π, that on the opposite side is -2π. The solid angle, therefore, *jumps discontinuously by* 4π *on crossing the plane.* The same result holds for crossing any surface since, for points close enough to any small section of a surface, the surface is effectively an infinite plane. This discontinuity is another aspect of the multivaluedness we have referred to above.

* A convention for choosing the positive normal precedes Eq. (7.18).

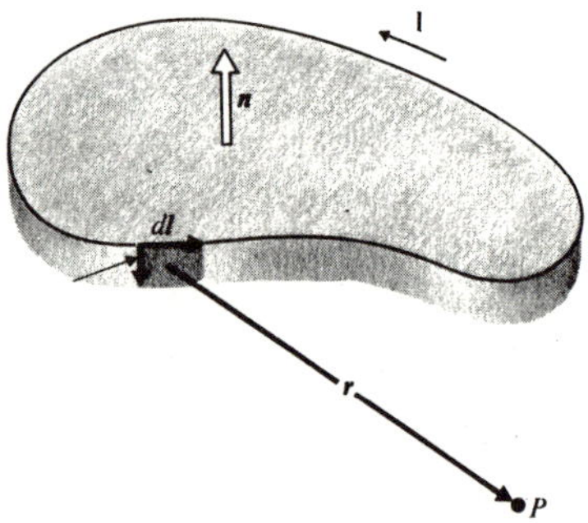

Fig. 7.5 Magnetic scalar potential.

Consider, as in Fig. 7.5, a current-carrying loop which is spanned by some surface. To define the positive normal to the surface (and hence, the sign of the solid angle), we make the direction of the current bear the same relation to the positive normal as the direction of rotation of a right-handed screw does to its direction of advance. The field of this loop at the point P is

$$\boldsymbol{B} = \frac{I}{c}\int \frac{d\boldsymbol{l} \times \boldsymbol{r}}{r^3}. \tag{7.18}$$

Now consider the solid angle Ω that is subtended by the loop at the point P, which point has coordinates x, y, z. Then Ω is a function of x, y, z, and the change $d\Omega$, when P is displaced by an amount $d\boldsymbol{u}$, is

$$d\Omega = \nabla\Omega \cdot d\boldsymbol{u}. \tag{7.19}$$

Moving P by $d\boldsymbol{u}$ is equivalent, as far as change in solid angle is concerned, to moving the *loop* bodily by an amount $-d\boldsymbol{u}$. If the loop is so moved, it generates a ribbon-like surface indicated by the shaded area in Fig. 7.5. The parallelogram formed by $d\boldsymbol{l}$ and $-d\boldsymbol{u}$ of area $-d\boldsymbol{u} \times d\boldsymbol{l}$ is an element of area for the ribbon. The change of solid angle that this parallelogram generates in the displacement $-d\boldsymbol{u}$ is

$$-(d\boldsymbol{u} \times d\boldsymbol{l}) \cdot \left(-\frac{\boldsymbol{r}}{r^3}\right) = d\boldsymbol{u} \cdot \frac{d\boldsymbol{l} \times \boldsymbol{r}}{r^3}.$$

The total change of solid angle $d\Omega$ is

$$d\Omega = d\boldsymbol{u} \cdot \int \frac{d\boldsymbol{l} \times \boldsymbol{r}}{r^3}. \tag{7.20}$$

Comparing this with (7.19), we see that

$$\nabla\Omega = \int \frac{d\boldsymbol{l} \times \boldsymbol{r}}{r^3}.$$

Using (7.18) we can then write

$$B = -\nabla\Phi_M,$$

where the *magnetic scalar potential* Φ_M is

$$\Phi_M = -\frac{I}{c}\Omega, \tag{7.21}$$

with Ω the solid angle subtended by the current loop.

It is now clear why it was important to provide the solid angle with a sign. If two identical loops of current are very close together, but carry currents in opposite directions, it is clear that the fields they produce must mutually cancel. With the above convention for the sign of solid angles, the mathematics reproduces this obvious physical result.

7.5 THE VECTOR POTENTIAL

By its derivation, the magnetic scalar potential does not hold in the interior of a current, and this is a disadvantage. In fact, since $\nabla \times \boldsymbol{B}$ is not zero, there can be *no* true scalar potential from which $\boldsymbol{B}$ is everywhere derivable by differentiation. But a generalized potential does exist: $\boldsymbol{B}$ can be calculated by differentiation of a vector function which we now introduce. Moreover, the result will hold in the interior of currents.

The two basic equations for $\boldsymbol{B}$ are those for its divergence and curl. If we try to express $\boldsymbol{B}$ as the curl of another vector, the so-called *vector potential* $\boldsymbol{A}$,

$$\boldsymbol{B} = \nabla \times \boldsymbol{A}, \tag{7.22}$$

we have automatically $\nabla \cdot \boldsymbol{B} = 0$. The curl equation (7.12) is expressed in terms of $\boldsymbol{A}$ by

$$\nabla \times (\nabla \times \boldsymbol{A}) = 4\pi \boldsymbol{J}/c. \tag{7.23}$$

This is effectively a set of three coupled partial differential equations involving the components of $\boldsymbol{A}$, and we must verify that they can be solved and are not overdetermined or underdetermined. We show that they can be solved by exhibiting the solution. We use the mnemonic, valid in rectangular coordinates,*

$$\nabla \times (\nabla \times \boldsymbol{A}) = \nabla\nabla \cdot \boldsymbol{A} - \nabla^2 \boldsymbol{A}. \tag{7.24}$$

Now Eq. (7.22) specifies only the curl of $\boldsymbol{A}$, and by Helmholtz's theorem, $\boldsymbol{A}$ is not defined until $\nabla \cdot \boldsymbol{A}$ is also specified. It is convenient to take advantage of this arbitrariness in $\nabla \cdot \boldsymbol{A}$ and set $\nabla \cdot \boldsymbol{A} = 0$. This leaves us with

$$\nabla^2 A_x = -\frac{4\pi J_x}{c} \tag{7.25}$$

---

* See Appendix C for a discussion of this equation. It is not a general vector relation.

and similarly for A_y, A_z. Now Eq. (7.25) and each of its two companions are of the same form as the Poisson equation in electrostatics and their solutions can be written by analogy with that equation. Thus

$$A_x(\boldsymbol{r}) = \frac{1}{c}\int \frac{J_x(\boldsymbol{r}')}{|\boldsymbol{r} - \boldsymbol{r}'|}\, dv', \tag{7.26}$$

with similar equations for A_y and A_z. All three equations can be written

$$\boldsymbol{A}(\boldsymbol{r}) = \frac{1}{c}\int \frac{\boldsymbol{J}(\boldsymbol{r}')}{|\boldsymbol{r} - \boldsymbol{r}'|}\, dv', \tag{7.27}$$

where it is again understood that this is not a true vector equation but rather one which is shorthand for the three Cartesian equations similar to Eq. (7.26).

Equation (7.27) is useful beyond the fact that it shows that Eq. (7.23) has a solution. Analogously to the superposition integral of electrostatics, it yields (in principle) the field of an arbitrary current distribution.

We collect here, for convenience, some of the differential and integral properties of $\boldsymbol{B}$, Φ_M, and $\boldsymbol{A}$:

Differential	Integral
$\boldsymbol{B}$	
$\nabla \cdot \boldsymbol{B} = 0$	$\int_S \boldsymbol{B} \cdot d\boldsymbol{s} = 0$
$\nabla \times \boldsymbol{B} = \dfrac{4\pi \boldsymbol{J}}{c}$	$\oint \boldsymbol{B} \cdot d\boldsymbol{l} = \dfrac{4\pi I}{c}$
Φ_M	
$\nabla^2 \Phi_M = 0$	$\Phi_M = -\dfrac{I}{c}\nabla\Omega$
$\boldsymbol{A}$	
$\nabla^2 \boldsymbol{A} = -\dfrac{4\pi \boldsymbol{J}}{c}$	$\boldsymbol{A} = \dfrac{1}{c}\displaystyle\int \frac{\boldsymbol{J}(\boldsymbol{r}')}{\vert\boldsymbol{r} - \boldsymbol{r}'\vert}\, dv'$
(Cartesian coordinates)	

PROBLEMS

1. A current I flows around a circle of radius a, and a current I' flows in a very long straight wire in the same plane. Show that the mutual attraction is

$$\frac{4\pi II'}{c}(\sec \alpha - 1),$$

where 2α is the angle subtended by the circle at the nearest point of the wire.

2. A circuit carrying current I has the form of a regular polygon of n sides inscribed in a circle of radius a. Show that the magnitude of $\boldsymbol{B}$ at the center is

$$\frac{2nI}{ac}\tan\frac{\pi}{n}.$$

3. Two infinitely long parallel wires carrying current I_1 and I_2 are a distance d apart. Show that the force per unit length between them has magnitude $2I_1I_2/c^2d$. Find the sign of the force for the currents in the same direction.

4. A circular loop carries current I. Use the Biot-Savart law to find the field at a point on the axis.

5. A wire carrying current I is bent so that it forms an angle α (see Fig. 7.6). Show that the field at a point d from the bend, in the plane of the wire, has the magnitude

$$B = \frac{2I}{cd}(1 - \cos\frac{\alpha}{2})$$

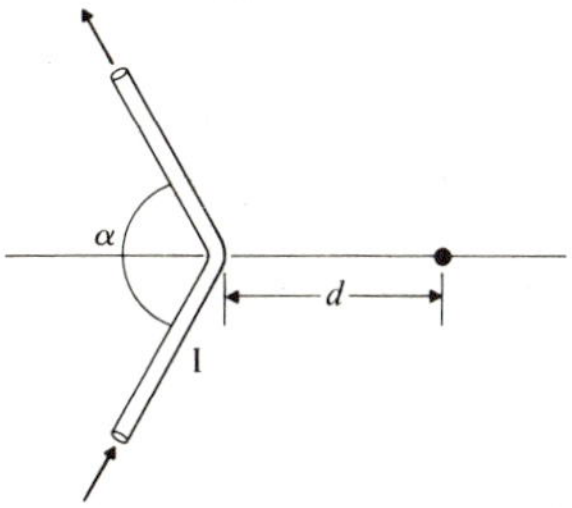

Figure 7.6

6. A disk of radius a carrying a uniform charge density σ spins on its axis with angular velocity ω (Rowland disk). Show that the field B_z at a point z on the axis has the value

$$\frac{2\pi\sigma\omega}{c}\left[\frac{2z^2 + a^2}{\sqrt{z^2 + a^2}} - 2z\right].$$

7. An infinitely long coaxial cable carries a uniformly distributed current I that flows down in the inner conductor $(0 < \rho < a)$ and up in the outer conductor $(b < \rho < c)$; ρ is a cylindrical coordinate. Find $\boldsymbol{B}$ everywhere.

REFERENCE

Mattis, D. C., *The Theory of Magnetism*, Harper & Row, New York (1965).

This is a book on the quantum theory of magnetism, but the first chapter contains an excellent sketch of the history of magnetism and of theories of magnetism.

8 THE SUMMATION PROBLEM FOR STATIONARY CURRENTS

8.1 INTRODUCTION

In this chapter we treat the magnetic *summation problem*, i.e., the problem of calculating the field due to a prescribed current distribution. As in electrostatics, this problem contrasts with *boundary value problems* in which there are not only prescribed currents but induced ones that are unknown *a priori*. These boundary value problems will be discussed later. The strategy of the summation problem for currents is not quite so straightforward as it was for charges. For the simpler cases we can apply the Biot-Savart law (7.8), or, if the symmetry is great enough, Ampère's circuital law. But for a more systematic development, we must choose between two potentials, neither of which is as satisfactory as the universally applicable electrostatic scalar potential. On the one hand is the magnetic scalar potential, which has the disadvantage that it does not hold in the interior of currents, whereas the electrostatic potential did hold in the interior of charge distributions. On the other hand is the universally applicable vector potential which does, in general, require the evaluation of a separate integral for each of its three components. The vector potential is, however, of considerable importance later in the theory of time-varying fields, and partly for pedagogical reasons we include some examples of its use now. For the simple purpose of calculating static magnetic fields it is usually not advantageous.

In the following sections we have tried to indicate by example the situations in which one or the other of the various possible methods is favored. For example, for currents in fine wires the magnetic scalar potential in its integral form is often particularly useful, since it involves only a single integration for calculating a solid angle. On the other hand, if there is a surface current which divides space into an exterior and an interior, it is sometimes convenient to use the potential in differential form. That is, away from the surface and in the regions of no current, $\nabla \cdot \boldsymbol{B} = 0$ and $\nabla \times \boldsymbol{B} = 0$; hence, $\boldsymbol{B}$ can be derived from a scalar potential, Φ_M, which satisfies Laplace's equation inside and outside the boundary. The equation $\nabla \times \boldsymbol{B} = 4\pi \boldsymbol{J}/c$ then becomes a *boundary condition* for matching the exterior solution to the interior one. Some simple examples involving the vector potential are also presented, but perhaps its greatest importance for this chapter is that we may use it to derive a *multipole expansion*, useful in treating the problem of dipolar magnetic matter. We show how the external field of such matter can be calculated by equivalent surface and volume currents (amperian currents), much as the external field of electric dipolar matter was calculable from equivalent surface and volume polarization charges.

8.2 INTEGRAL MAGNETIC SCALAR POTENTIAL

As a first example, we use the magnetic scalar potential to calculate the field of a circular current loop, and begin by considering the field on the axis. Let the loop lie in the plane $z = 0$, as in Fig. 8.1, thereby defining a disk-like surface. The solid angle subtended by the element of area $2\pi\rho\, d\rho$ is $-(2\pi\rho\, d\rho)/(\rho^2 + z^2)\cdot(z/\sqrt{\rho^2 + z^2})$. The minus sign results from the fact that the normal $\boldsymbol{n}$ to the loop and the vector from P to the area element have opposite senses. The whole solid angle subtended is

$$\Omega = -2\pi z \int_0^a \frac{\rho\, d\rho}{(\rho^2 + z^2)^{3/2}} = -2\pi\left[\frac{z}{\sqrt{z^2}} - \frac{z}{\sqrt{a^2 + z^2}}\right]. \tag{8.1}$$

The factor $z/\sqrt{z^2}$ is unity for z positive and minus unity for z negative. It is this factor that gives the required discontinuity of 4π on traversing the disk. From $\boldsymbol{B} = (I/c)\nabla\Omega$, the field on the axis is

$$B_z = \frac{I}{c}\frac{d\Omega}{dz} = \frac{2\pi I a^2}{c(a^2 + z^2)^{3/2}}. \tag{8.2}$$

This result is, of course, also easy to obtain by applying the Biot-Savart law (7.8) directly.

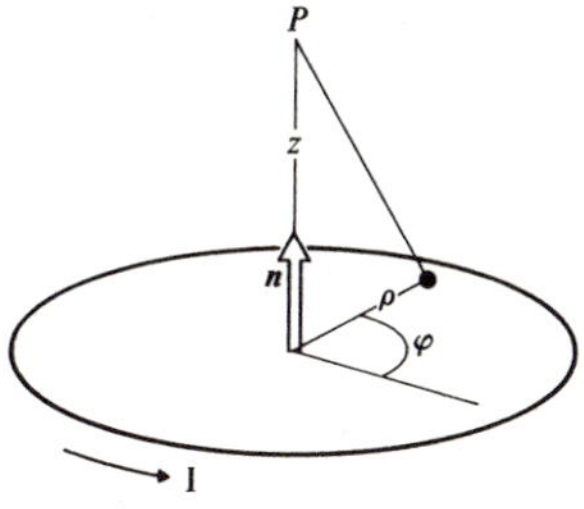

Fig. 8.1 Field of a current loop.

Now consider the field off axis. Note first that for both $z > 0$ and $z < 0$, it can be calculated from $\boldsymbol{B} = (I/c)\nabla\Omega'$ where $\Omega' = -2\pi[1 - (z/\sqrt{a^2 + z^2})]$ since Ω' differs from Ω only by a constant. Since Ω' satisfies $\nabla^2\Omega' = 0$, we can apply the Legendre polynomial expansion of the solution of this equation, and use the device of fixing the coefficients in a Legendre expansion in $P_l(\cos\theta)$ by fitting the expansion to a *known* one on the axis, i.e., for $\theta = 0$. Thus, Ω' can be expanded for $z < a$ and $z > a$ to yield

$$\Omega' = -2\pi\left(1 - \frac{z}{a} + \sum_{n=1}^{\infty}(-)^{n+1}\left(\frac{z}{a}\right)^{2n+1}\frac{(2n-1)!!}{(2n)!!}\right), \qquad z < a \tag{8.3}$$

$$\Omega' = -2\pi \sum_{n=1}^{\infty} (-)^{n+1} \left(\frac{a}{z}\right)^{2n} \frac{(2n-1)!!}{(2n)!!}, \qquad z > a. \tag{8.4}$$

Here $l!!$ is $1 \cdot 3 \cdot 5 \cdots l$ for l odd and is $2 \cdot 4 \cdot 6 \cdots l$ for l even.

Now we turn to the calculation of Ω' off axis, i.e., as a general function of spherical coordinates r and θ. If we assume first for definiteness that $r < a$, Ω' can be expanded in the form

$$\Omega' = \sum_{n=0}^{\infty} A_n r^n P_n(\cos\theta). \tag{8.5}$$

Along the axis, $\theta = 0$, and $r^n P_n(\cos\theta) = z^n$. Then (8.5) becomes $\Omega' = \Sigma_{n=0}^{\infty} A_n z^n$ and by comparison with (8.3) the coefficients can be determined. In this way we find for $r < a$:

$$\Omega' = -2\pi \left[1 - \frac{r}{a} P_1(\cos\theta) + \sum_{n=1}^{\infty} (-)^{n+1} \frac{(2n-1)!!}{(2n)!!} \left(\frac{r}{a}\right)^{2n+1} P_{2n+1}(\cos\theta) \right], \tag{8.6}$$

and for $r > a$:

$$\Omega' = -2\pi \sum_{n=1}^{\infty} (-)^{n+1} \frac{(2n-1)!!}{(2n)!!} \left(\frac{a}{r}\right)^{2n} P_{2n-1}(\cos\theta). \tag{8.7}$$

The field components B_r and B_θ can now be derived from Eqs. (8.6) and (8.7) by differentiation. For example, one finds for $r > a$,

$$B_r = \frac{2\pi I}{ca} \sum_{n=0}^{\infty} \frac{(-)^n (2n+1)!!}{(2n)!!} \left(\frac{a}{r}\right)^{2n+3} P_{2n+1}(\cos\theta), \tag{8.8}$$

and

$$B_\theta = -\frac{2\pi I}{ca} \sum_{n=0}^{\infty} \frac{(-)^n (2n+1)!!}{(2n+2)!!} \left(\frac{a}{r}\right)^{2n+3} P_{2n+1}^1(\cos\theta). \tag{8.9}$$

For large distances $r \gg a$, we can take the leading terms in (8.8) and (8.9) and use $P_1^1(\cos\theta) = -\sin\theta$ and $m = I\pi a^2/c$ to find

$$B_r \approx \frac{2m\cos\theta}{r^3}, \qquad B_\theta \approx \frac{m\sin\theta}{r^3}. \tag{8.10}$$

The quantity m is a special case of a vector magnetic dipole moment defined later by Eq. (8.13) or (8.32). The field $\boldsymbol{B}$ of Eq. (8.10) is similar to the electric field $\boldsymbol{E}$ of a dipole of electric moment p. We shall soon learn that this is a general result in that any current distribution bounded in space has a multipole expansion whose first term corresponds to a magnetic dipole.

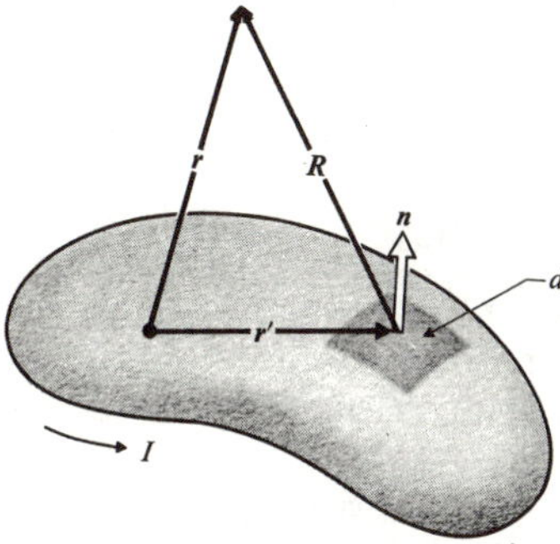

Fig. 8.2 Magnetic scalar potential of a current loop.

As the second example of the use of the magnetic scalar potential, we consider the field due to a current-carrying loop as sketched in Fig. 8.2. We imagine the loop spanned by some surface S. Let $\boldsymbol{r}$ be a vector to the field point from some origin in the vicinity of the loop, and $\boldsymbol{r}'$ a vector to an element $d\boldsymbol{s}' = \boldsymbol{n}\,ds'$ of the surface. The normal $\boldsymbol{n}$ will, of course, be oriented differently at different points of the surface and must therefore be considered a function $\boldsymbol{n}(\boldsymbol{r}')$ of $\boldsymbol{r}'$. We want to calculate $\Omega(\boldsymbol{r})$, the solid angle subtended by the loop at $\boldsymbol{r}$. Now the element of solid angle $d\Omega$ subtended by ds' is, with $\boldsymbol{R} = \boldsymbol{r} - \boldsymbol{r}'$,

$$d\Omega = -\frac{d\boldsymbol{s}' \cdot \boldsymbol{R}}{R^3} = -\frac{\boldsymbol{n}(\boldsymbol{r}') \cdot (\boldsymbol{r} - \boldsymbol{r}')\,ds'}{|\boldsymbol{r} - \boldsymbol{r}'|^3}$$

and if we write out the dot product explicitly, the whole solid angle Ω subtended by the loop is

$$\Omega = -\int \frac{n_x(\boldsymbol{r}')(x - x') + n_y(\boldsymbol{r}')(y - y') + n_z(\boldsymbol{r}')(z - z')}{((x - x')^2 + (y - y')^2 + (z - z')^2)^{3/2}}\,ds'. \tag{8.11}$$

Of course, this integral cannot be evaluated without knowing the shape of the loop. But if the dimensions of the loop are small compared to the distance to the field point, it can be evaluated approximately by a multipole expansion, much as was done with the superposition integral in electrostatics. To accomplish this, we expand the integrand in powers of x'/r, y'/r, z'/r by expanding $(\boldsymbol{r} - \boldsymbol{r}')/|\boldsymbol{r} - \boldsymbol{r}'|^3$ in a

Taylor series in x', y', z'. The expansion is straightforward; one finds up to terms linear in x', y', z';

$$\Omega = -\frac{\boldsymbol{N}\cdot\boldsymbol{r}}{r^3} + \frac{3}{r^5}\left(\begin{array}{l} T_{xx}\left(x^2 - \dfrac{r^2}{3}\right) + T_{yy}\left(y^2 - \dfrac{r^2}{3}\right) + T_{zz}\left(z^2 - \dfrac{r^2}{3}\right) \\ + xy(T_{xy} + T_{yx}) + xz(T_{xz} + T_{zx}) + yz(T_{yz} + T_{zy}) \end{array}\right), \tag{8.12}$$

where $\boldsymbol{N} = \int \boldsymbol{n}\, ds'$.

The quantities T_{xx}, etc., are defined typically by

$$T_{xx} = \int x' n_x(\boldsymbol{r}')\, ds' \qquad \text{and} \qquad T_{yz} = \int y' n_z(\boldsymbol{r}')\, ds'.$$

We may call the terms in (8.12) involving T_{xx}, etc., the *quadrupole terms.* As for the electrostatic case, they are smaller by a factor of the order a/r than the dipole term associated with the vector $\boldsymbol{N}$. We shall discuss only the dipole term in detail and turn to it now.

If we define the *vector dipole moment* $\boldsymbol{m}$ by

$$\boldsymbol{m} = \frac{I}{c}\boldsymbol{N}, \tag{8.13}$$

the field corresponding to the dipole term in (8.12) can be derived from

$$\boldsymbol{B} = -\nabla\left(\frac{\boldsymbol{m}\cdot\boldsymbol{r}}{r^3}\right). \tag{8.14}$$

This dipole field is like that for the electric case, namely,

$$\boldsymbol{B} = \frac{3\hat{\boldsymbol{r}}(\boldsymbol{m}\cdot\hat{\boldsymbol{r}}) - \boldsymbol{m}}{r^3}, \tag{8.15}$$

where $\boldsymbol{r}$ is a vector from the dipole to the field point, and $\hat{\boldsymbol{r}}$ is the unit vector in that direction. The vector $\boldsymbol{m}$ is, of course, the vector generalization of the moment $\mathrm{m} = I\pi a^2/c$ defined for a circular loop. For a *plane* loop lying in the x-y plane, $\boldsymbol{m}$ is a vector in the z direction with magnitude equal to the area of the loop:

$$\boldsymbol{m} = (I/c)(\text{Area})\,\hat{\boldsymbol{z}}. \tag{8.16}$$

8.3 DIFFERENTIAL MAGNETIC SCALAR POTENTIAL

It is sometimes more convenient to use a magnetic scalar potential in differential rather than integral form. Suppose that we want to find the field due to currents on some surface; for example, a metal-plated insulating sphere. We could, if we chose, use the integral form of the magnetic scalar potential to calculate the field of one of the closed current filaments that make up the current distribution and then integrate over all filaments. This would be like calculating the field of a sphere with surface charge in electrostatics by the superposition integral that sums up the contribution of each surface charge element. Such a calculation is correct in theory. In practice it is easier to use a scalar potential, Φ, to recognize that exterior to the sphere, $\Phi = A/r$, and then to find A by the boundary condition, $\partial\Phi/\partial n = -4\pi\sigma$. It is the analogous situation in magnetostatics that we shall discuss now. In regions where the current density $\boldsymbol{J}$ is zero we have

$$\nabla \times \boldsymbol{B} = 0,$$

$$\nabla \cdot \boldsymbol{B} = 0.$$

The first of these equations shows that $\boldsymbol{B}$ can be derived in these regions from a scalar potential, Φ_M:

$$\boldsymbol{B} = -\nabla\Phi_M, \tag{8.17}$$

and the second shows that

$$\nabla^2\Phi_M = 0.$$

Note incidentally that Φ_M need not necessarily be considered as a *differential* form of the integral potential of the last section. Even if that integral formulation did not exist, the equations above would still hold in their own right. The fact that $\nabla \times \boldsymbol{B} = 0$ both inside and outside the region does not mean, of course, that we can neglect the equation $\nabla \times \boldsymbol{B} = 4\pi\boldsymbol{J}/c$. It must still hold, but in the present case $\boldsymbol{J}$ is a singular function and it is best to apply the equation in its integral form, i.e., as *Ampère's circuital law*. In this case it becomes a *boundary condition* relating values of $\boldsymbol{B}$ on either side of the current sheet.

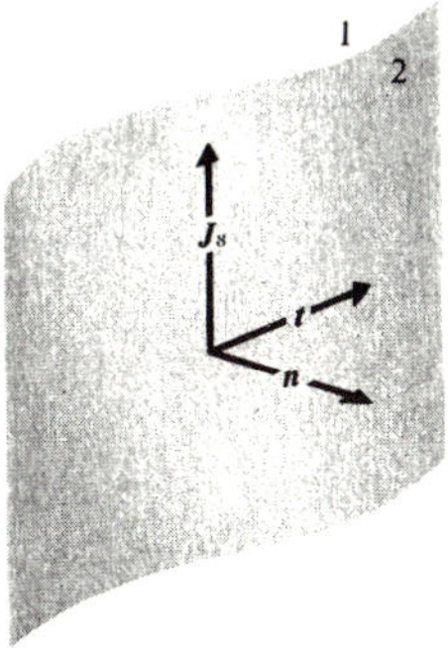

Fig. 8.3 Current sheet, the vector $\boldsymbol{J}_s$, the unit normal $\boldsymbol{n}$ and the unit tangent $\boldsymbol{t}$.

We shall derive this boundary condition and refer to Fig. 8.3 which shows a surface with surface current of density $\boldsymbol{J}_s$ and normal $\boldsymbol{n}$. $\boldsymbol{J}_s$ has the dimensions of charge/(length × time) and is so defined that the amount of charge per second crossing a segment of length dl perpendicular to $\boldsymbol{J}_s$ is $J_s\,dl$. At a point on the surface, we draw a unit vector $\boldsymbol{t}$ tangential to it and perpendicular to $\boldsymbol{J}_s$, and such that $\boldsymbol{n}$, $\boldsymbol{t}$, and $\boldsymbol{J}_s$ form a right-handed system. Then we apply Ampère's circuital law,

$$\oint \boldsymbol{B} \cdot dl = 4\pi I/c,$$

to a circuit consisting of a narrow infinitesimal rectangle which straddles the surface. The longer sides, of length Δl, are parallel to $\boldsymbol{t}$ and the plane of the rectangle is perpendicular to $\boldsymbol{J}_s$. With $\boldsymbol{B}_1$ and $\boldsymbol{B}_2$ the fields on either side of the sheet, $\oint \boldsymbol{B} \cdot d\boldsymbol{l}$ is approximately $(\boldsymbol{B}_2 - \boldsymbol{B}_1) \cdot \boldsymbol{t}\, \Delta l$, and the current I is approximately $J_s\ \Delta l$. Ampère's circuital law becomes the *boundary condition on the tangential component of* $\boldsymbol{B}$:

$$(\boldsymbol{B}_2 - \boldsymbol{B}_1) \cdot \boldsymbol{t} = \frac{4\pi}{c} J_s. \tag{8.18}$$

When we take a rectangle whose normal is in the plane interface, but not necessarily in the direction of $\boldsymbol{J}_s$, the boundary condition can be written in a form frequently seen:

$$(\boldsymbol{B}_1 - \boldsymbol{B}_2) \times \boldsymbol{n} = \frac{4\pi \boldsymbol{J}_s}{c}. \tag{8.19}$$

We get a second boundary condition on applying $\int_s \boldsymbol{B} \cdot d\boldsymbol{s} = 0$ to a small pillbox with faces normal to $\boldsymbol{n}$, and with one face in region 1 and the other in region 2. In the limit that the thickness of the pillbox goes to zero, all the flux of $\boldsymbol{B}$ is through the faces, and for the net flux to be zero we must have

$$(\boldsymbol{B}_2 - \boldsymbol{B}_1) \cdot \boldsymbol{n} = 0. \tag{8.20}$$

To apply the magnetic scalar potential, we must know the solutions of Laplace's equation on either side of the boundary. In practice it is useful mainly for the simple geometries (sphere, cylinder, etc.) that are familiar from electrostatics. As an example, consider surface currents on a sphere. Suppose that the current is in the φ direction (along lines of constant latitude) but that its magnitude may depend on the latitude angle θ. A simple case of physical interest is when the dependence on θ is as $\sin\theta$:

$$\boldsymbol{J}_s = J(\theta)\hat{\boldsymbol{\varphi}} = J_0 \sin\theta\hat{\boldsymbol{\varphi}}. \tag{8.21}$$

This dependence is interesting in that it can be interpreted in terms of three different physical models as in Fig. 8.4. First, suppose a uniformly charged sphere of radius a with surface density σ spins about its polar axis with angular velocity ω. Then at some point this corresponds to a surface current density in the φ direction of magnitude σv, where v is the velocity of the point. This velocity is $\omega a \sin\theta$, thus the current density is $\sigma\omega a \sin\theta$. The same current density could be realized (approximately) by winding the sphere with fine wire and spacing the windings appropriately. Finally, we shall find later that the external field of a solid ferro-

Fig. 8.4 Three physical models for the surface current $J_\varphi(\theta) = J_0 \sin\theta$.

magnetic sphere, which is uniformly magnetized in the z-direction, can be calculated from a so-called equivalent surface current distribution of the form (8.21).

With this physical interpretation, we return to the mathematical problem. Let the regions 1 and 2 refer to the inside and outside of the sphere. There is obviously no φ dependence; hence the interior and exterior solutions, Φ_M^1 and Φ_M^2, can be written

$$\Phi_M^1 = \sum_{l=0}^{\infty} C_l r^l P_l(\cos\theta),$$

$$\Phi_M^2 = \sum_{l=0}^{\infty} \frac{D_l}{r^{l+1}} P_l(\cos\theta).$$

Since $\boldsymbol{B}\cdot\boldsymbol{n} = B_r = -\partial\Phi_M/\partial r$, the condition (8.20) on the equality of the normal components of $\boldsymbol{B}$ is

$$\sum_{l=1}^{\infty} lC_l a^{l-1} P_l(\cos\theta) = -\sum_{l=0}^{\infty} \frac{(l+1)D_l P_l(\cos\theta)}{a^{l+2}}.$$

Then, with $\boldsymbol{B}\cdot\boldsymbol{t} = B_\theta = -(1/r)(\partial\Phi_M/\partial\theta)$, condition (8.18) becomes

$$-\frac{1}{a}\sum_{l=0}^{\infty} \frac{D_l}{a^{l+1}}\frac{dP_l}{d\theta} + \frac{1}{a}\sum_{l=0}^{\infty} C_l a^l \frac{dP_l}{d\theta} = \frac{4\pi}{c} J_0 \sin\theta.$$

It is readily verified that these equations are solved by setting all C_l and D_l equal to zero except C_1 and D_1. They become

$$C_1 + \frac{2D_1}{a^3} = 0,$$

$$-C_1 + \frac{D_1}{a^3} = \frac{4\pi J_0}{c},$$

whence

$$\Phi_M^1 = -\frac{8\pi J_0}{3c} r \cos\theta,$$
$$\Phi_M^2 = \frac{4\pi a^3 J_0}{3c} \frac{\cos\theta}{r^2}. \tag{8.22}$$

From (8.22) we see the field inside is uniform and of magnitude $B_z = (8\pi J_0/3c)$, and the field outside is that of a dipole of moment $(4\pi a^3 J_0/3c)$. For the model of the uniformly magnetized sphere, it is, of course, only the outside field that is meaningful.

8.4 THE VECTOR POTENTIAL AND A MULTIPOLE EXPANSION

We now consider some examples of the use of the vector potential. In two of these we shall suppose that the current flows in a fine wire, and that the current density is constant over the cross section of the wire. In these cases the volume element dv' in Eq. (7.27) is the product $S\,dl$ where S is the cross-sectional area of the wire and dl is an element of length along it. The current $\boldsymbol{I}$ is $\boldsymbol{J}S$ so $\boldsymbol{J}\,dv \to \boldsymbol{I}\,dl = I\,\boldsymbol{dl}$ and

$$\boldsymbol{A} = \frac{I}{c}\int \frac{d\boldsymbol{l}'}{|\boldsymbol{r} - \boldsymbol{r}'|}. \tag{8.23}$$

Consider first a long straight wire coincident with the z-axis. In this case dl' is dz'; for a field point a distance ρ from the wire $|\boldsymbol{r} - \boldsymbol{r}'| = \sqrt{\rho^2 + z'^2}$. Obviously $\boldsymbol{A}$ has only a component A_z, which is

$$A_z(\rho) = \frac{I}{c}\int_{-\infty}^{\infty} \frac{dz'}{\sqrt{\rho^2 + z'^2}}. \tag{8.24}$$

Unfortunately the integral diverges. This can be remedied by the assumption that the wire has a long but finite length $2L$, that is, by integrating only from $-L$ to L, and expanding the result in powers of ρ/L. One finds to first order in ρ/L:

$$A_z = -(2I/c)\ln\rho. \tag{8.25}$$

We have omitted from the right-hand side of (8.25) constants which become infinite when $L \to \infty$. These are irrelevant since they drop out on differentiation. Equation (8.25) is correct in that on calculating $\boldsymbol{B}$ from it, using $\boldsymbol{B} = \nabla \times \boldsymbol{A}$, we get the familiar result for B_φ, the only nonvanishing component:

$$B_\varphi = 2I/c\rho. \tag{8.26}$$

We use the result (8.25) and superposition to solve the second problem: a surface current which flows in the z-direction on a strip of width $2d$ and infinite length. We refer to Fig. 8.5 which shows the cross section of the strip between $-d$ and d on the y-axis.

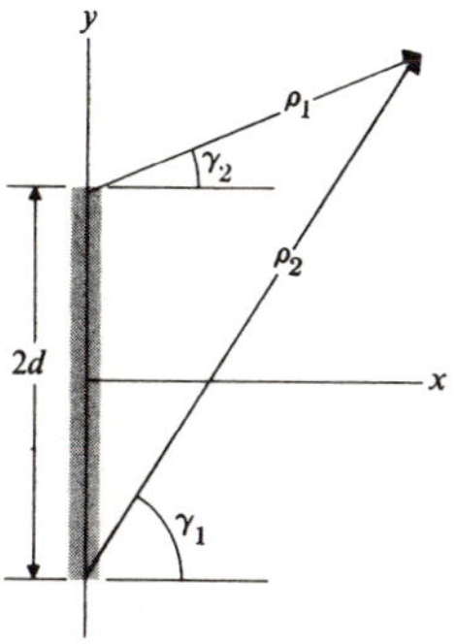

Fig. 8.5 A uniform current flows in the z-direction over a strip of width $2d$.

We assume that the surface current density J_0 in the z-direction, i.e., out of the plane of the paper, is constant as a function of y. Obviously only A_z will be nonvanishing. The contribution that the current in the infinitesimal strip of width dy' makes to A_z is given by Eq. (8.25) with $I = J_0\,dy'$, and $\rho = \sqrt{x^2 + (y - y')^2}$. Thus,

$$A_z = -\frac{2J_0}{c}\int_{-d}^{d} \ln\sqrt{x^2 + (y - y')^2}\,dy'.$$

The integral is elementary and one finds

$$A_z = \frac{2J_0}{c}\{(y - d)\ln\rho_1 - (y + d)\ln\rho_2 + 2d + x(\gamma_1 - \gamma_2)\},$$

where

$$\rho_1 = \sqrt{x^2 + (y - d)^2}, \qquad \rho_2 = \sqrt{x^2 + (y + d)^2},$$

and γ_1 and γ_2 are the angles shown. The field components then are

$$B_x = \frac{\partial A_z}{\partial y} = -\frac{2J_0}{c}\ln\frac{x^2 + (y + d)^2}{x^2 + (y - d)^2},$$

$$B_y = -\frac{\partial A_z}{\partial x} = \frac{2J_0}{c}\left(\tan^{-1}\frac{y + d}{x} - \tan^{-1}\frac{y - d}{x}\right).$$

As a final example of the vector potential, we consider a circular loop of current, which we have already discussed by other methods. Referring to Fig. 8.6 we can, without loss of generality, take the field point $\boldsymbol{r}$ in the plane $\varphi = 0$. Equation (8.23) must be evaluated in rectangular coordinates. Now the components of $d\boldsymbol{l}'$ are:

$$dl'_x = -\sin\varphi'(a\,d\varphi'), \qquad dl'_y = \cos\varphi'(a\,d\varphi').$$

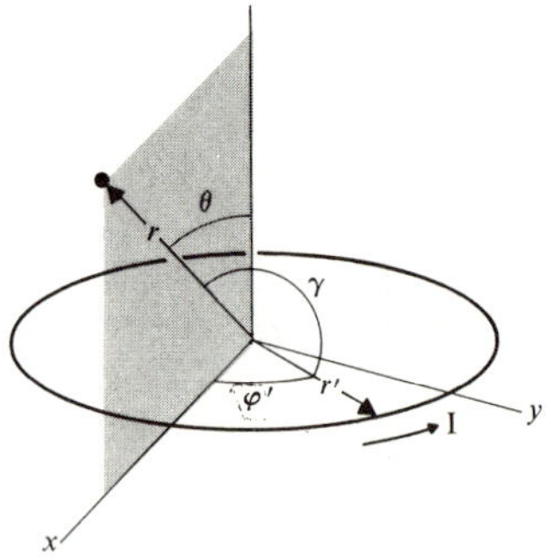

Fig. 8.6 A circular loop carrying current I.

Using these equations in (8.23), it is easy to see that A_x vanishes, since for every element of integration $d\varphi'$ that contributes to A_x there is another element an equal distance away that makes an opposite contribution. We also see from Fig. 8.6 that A_y is identical with the φ-component A_φ, and because it is more convenient to work in spherical coordinates we shall consider that this component is A_φ. The cosine of the angle between $\boldsymbol{r}$ and $\boldsymbol{r}'$ is $\sin\theta\cos\varphi'$, and thus

$$A_y \equiv A_\varphi = \frac{Ia}{c}\int_0^{2\pi}\frac{\cos\varphi'\,d\varphi'}{\sqrt{a^2 + r^2 - 2ar\sin\theta\cos\varphi'}}. \tag{8.27}$$

The integral is not elementary but it can be done in terms of the elliptic integrals* K and E of the first and second kind:

$$A_\varphi(r,\theta) = \frac{4Ia}{c\sqrt{a^2 + r^2 + 2ar\sin\theta}}\left[\frac{(2-\lambda^2)K(\lambda) - 2E(\lambda)}{\lambda^2}\right]$$

with

$$\lambda^2 = \frac{4ar\sin\theta}{a^2 + r^2 + 2ar\sin\theta}.$$

The field components can now be calculated from

$$B_r = \frac{1}{r\sin\theta}\frac{\partial}{\partial\theta}(\sin\theta A_\varphi),$$

$$B_\theta = -\frac{1}{r}\frac{\partial}{\partial r}(rA_\varphi),$$

$$B_\varphi = 0.$$

* See, for example, Abramowitz and Stegun (B).

The results above are correct but not perspicuous. They can be put in a more transparent form by writing the field components in a Legendre polynomial expansion. This can be done by expanding the square root in (8.27) in the standard way, according to whether a/r is smaller or larger than unity, and then applying the addition theorem. The final result for $\boldsymbol{B}$ is, of course, identical with that in Section 8.2, so we shall not undertake this expansion here. It will be convenient, however, to go a limited way and display the magnetic-dipole term which we know must emerge as the first term of such an expansion.

In the integrand of (8.27), for $a/r \ll 1$:

$$\frac{1}{\sqrt{a^2 + r^2 - 2ar\sin\theta\cos\varphi'}} \approx \frac{1}{r}\left(1 - \frac{a^2}{2r^2} + \frac{a}{r}\sin\theta\cos\varphi' + \cdots\right).$$

Applying this equation, we find that the first two terms in (8.27) vanish on integration and, in this approximation,

$$A_\varphi = \frac{I\pi a^2}{cr^2}\sin\theta. \tag{8.28}$$

Using the definition (8.16) for the vector magnetic moment $\boldsymbol{m}$ of a plane current loop, Eq. (8.28) can be written more generally, with subscript d for *dipole*, as

$$\boldsymbol{A}_d = \frac{1}{r^3}(\boldsymbol{m} \times \boldsymbol{r}). \tag{8.29}$$

When we evaluate $\boldsymbol{B} = \nabla \times \boldsymbol{A}_d$, the result of Eq. (8.15) is, of course, recovered.

The vector potential for an *arbitrary* current distribution can be expanded in a multipole expansion, which is qualitatively similar to the electrostatic expansion and to the magnetostatic one that uses the magnetic scalar potential. We sketch here the derivation of the dipole term of that expansion. Consider Eq. (7.27) for the vector potential of an arbitrary current distribution. With an origin in the distribution, we have for large r:

$$\boldsymbol{A}(\boldsymbol{r}) = \frac{1}{cr}\int \boldsymbol{J}(\boldsymbol{r}')\,dv' - \frac{1}{c}\int \boldsymbol{r}' \cdot \nabla\left(\frac{1}{r}\right)\boldsymbol{J}(\boldsymbol{r}')\,dv' + \cdots.$$

For a stationary distribution, the first integral $\int \boldsymbol{J}\,dv'$ vanishes since the current flows in closed loops. For the second integral, which defines the dipole term of the expansion, we suppose that the current distribution is broken into N filaments with current I_i in the ith filament and $\boldsymbol{r}'_i$ a vector to a point in that filament. Since this current is in the direction $d\boldsymbol{r}'_i$, we can write

$$\int \boldsymbol{r}' \cdot \nabla\left(\frac{1}{r}\right)\boldsymbol{J}(\boldsymbol{r}')\,dv' = \sum_{i=1}^{N} I_i \int \boldsymbol{r}'_i \cdot \nabla\left(\frac{1}{r}\right) d\boldsymbol{r}'_i.$$

Now

$$\boldsymbol{r}_i' \cdot \boldsymbol{\nabla}\left(\frac{1}{r}\right) d\boldsymbol{r}_i' = \frac{1}{2}(\boldsymbol{r}_i' \times d\boldsymbol{r}_i') \times \boldsymbol{\nabla}\left(\frac{1}{r}\right) + \frac{1}{2} d\left[\boldsymbol{r}_i' \cdot \boldsymbol{\nabla}\left(\frac{1}{r}\right) \boldsymbol{r}_i'\right],$$

and the total differential vanishes when integrated over the closed filament. The contribution $\boldsymbol{A}_i$ of the ith filament to $\boldsymbol{A}$ is then

$$\boldsymbol{A}_i = -\boldsymbol{m}_i \times \boldsymbol{\nabla}\left(\frac{1}{r}\right),$$

where the vector dipole moment of this filament is

$$\boldsymbol{m}_i = \frac{I_i}{2c}\int \boldsymbol{r}_i' \times d\boldsymbol{r}_i'. \tag{8.30}$$

For a plane loop it is easy to show that Eq. (8.30) defines the same quantity as Eq. (8.13). For a general current distribution, one finds on superposing the effect of the constituent filaments that the dipole term $\boldsymbol{A}_d$ is given by

$$\boldsymbol{A}_d = -\boldsymbol{m} \times \boldsymbol{\nabla}\left(\frac{1}{r}\right), \tag{8.31}$$

where

$$\boldsymbol{m} = \frac{1}{2c}\int \boldsymbol{r}' \times \boldsymbol{J}(\boldsymbol{r}')\, dv'. \tag{8.32}$$

8.5 EXTERNAL FIELD OF DIPOLAR MAGNETIC MATTER

In electrostatics we dealt with the summation problem for electric dipolar matter, i.e., the problem of calculating the external field of a large assemblage of electric dipoles. Here we shall discuss the analogous problem for magnetic dipolar matter. The problem is even more important than it was in electrostatics, since practical examples of dipolar matter abound in the form of permanent magnets. As for electrostatics, we shall assume only that magnetic dipolar matter consists of dipoles arranged either regularly or at random, but with some distance a characterizing their mean spacing, and that the field point is a distance away from the nearest dipole which is large compared with a. In the magnetic case, we shall find that there are two methods of calculating this field. We can use certain *equivalent currents* or certain equivalent *magnetic charges*. Each of these methods will be useful in its place, therefore we shall discuss them both.

We begin with the equivalent magnetic charges, since for this case the derivation can rely on the analogous one for electrostatics. Recall first that the field $\boldsymbol{B}$ of a magnetic dipole of moment $\boldsymbol{m}$ has the same form as the field $\boldsymbol{E}$ of an electric dipole of moment $\boldsymbol{p}$. A magnetic dipolar distribution will then produce a $\boldsymbol{B}$-field of the same form as the $\boldsymbol{E}$-field of a similar electric dipolar distribution. But we have found that the external field of such electric dipolar matter can be calculated to a good approximation by integrating over an equivalent continuous distribution $\boldsymbol{P}(\boldsymbol{r}')$. It follows that the $\boldsymbol{B}$-field can be calculated from a similar continuous distribution $\boldsymbol{M}(\boldsymbol{r}')$. That is, for the sake of calculating the *external* field, it is as if there were surface and volume densities σ_m and ρ_m of magnetic charge given by

$$\sigma_m = \boldsymbol{M}\cdot\boldsymbol{n}, \qquad \rho_m = -\nabla\cdot\boldsymbol{M}. \tag{8.33}$$

Then $\boldsymbol{B}$ can be calculated from $\boldsymbol{B} = -\nabla\Phi^*$, where

$$\Phi^* = \int_S \frac{\sigma_m\,ds'}{|\boldsymbol{r}-\boldsymbol{r}'|} + \int_V \frac{\rho_m\,dv'}{|\boldsymbol{r}-\boldsymbol{r}'|}. \tag{8.34}$$

It hardly needs be emphasized that these results do not show that magnetic charge "really exists." They are merely an algorithm for calculating the external field in a certain approximation. Nothing more can be read into them.

Now we shall discuss the method of equivalent currents. We could, in principle, calculate the vector potential, and, hence, the field of a sample of magnetic matter by summing over the vector potentials of the individual dipoles. Such a sum is, in general, impractical, but it can be replaced by an integral over a continuous distribution $\boldsymbol{M}(\boldsymbol{r}')$ under the same conditions as for the electrostatic case. We shall therefore not elaborate on this, but rather assume that we have constructed the distribution $\boldsymbol{M}(\boldsymbol{r}')$ that approximates a given dipolar distribution. Using (8.29) with $\boldsymbol{m}$ replaced by $\boldsymbol{M}(\boldsymbol{r}')\,dv'$, integrating over the distribution, and with $\boldsymbol{R} = \boldsymbol{r} - \boldsymbol{r}'$, we have for the vector potential $\boldsymbol{A}$ at the field point $\boldsymbol{r}$

$$\boldsymbol{A}(\boldsymbol{r}) = \int_V \frac{\boldsymbol{M}(\boldsymbol{r}')\times\boldsymbol{R}}{R^3}\,dv' = \int_V \boldsymbol{M}\times\nabla'\left(\frac{1}{R}\right)dv'. \tag{8.35}$$

From vector analysis we have for Φ, a scalar, and $\boldsymbol{F}$, a vector function of $\boldsymbol{r}'$,

$$\nabla'\times(\Phi\boldsymbol{F}) = \Phi\nabla'\times\boldsymbol{F} - \boldsymbol{F}\times\nabla'\Phi.$$

We apply this with $\Phi = 1/R$ and $\boldsymbol{F} = \boldsymbol{M}$. Therefore,

$$\boldsymbol{M}\times\nabla'\left(\frac{1}{R}\right) = \frac{1}{R}\nabla'\times\boldsymbol{M} - \nabla'\times\left(\frac{\boldsymbol{M}}{R}\right)$$

and

$$\boldsymbol{A} = \int_V \frac{(\nabla'\times\boldsymbol{M})\,dv'}{R} - \int_V \nabla'\times\left(\frac{\boldsymbol{M}}{R}\right)dv'. \tag{8.36}$$

We can convert the second term in (8.36) to a surface integral, by using the theorem from vector analysis:

$$\int_V \nabla\times\boldsymbol{G}\,dv = \int_S \boldsymbol{n}\times\boldsymbol{G}\,ds,$$

where V is a volume bounded by a surface S that has normal $\boldsymbol{n}$. Then

$$\boldsymbol{A} = \int_V \frac{\nabla \times \boldsymbol{M}}{R} dv' - \int_S \frac{\boldsymbol{n} \times \boldsymbol{M}}{R} ds'. \tag{8.37}$$

Comparing this result with that in (7.27), we see that the volume distribution of magnetic moment is equivalent to a *volume current density* $\boldsymbol{J}_v$ given by

$$\boldsymbol{J}_v = c(\nabla \times \boldsymbol{M}), \tag{8.38}$$

and a surface current density $\boldsymbol{J}_s$ given by

$$\boldsymbol{J}_s = c(\boldsymbol{M} \times \boldsymbol{n}). \tag{8.39}$$

For a medium with uniform magnetization, only the surface currents exist; these are called *amperian currents.*

PROBLEMS

1. The planes of two circular current loops of radius R are parallel, their centers are at $z = R/2$ and $z = -R/2$ on the z-axis, and each carries current I in the same direction. Such an arrangement is an example of a *Helmholtz coil.* Show that the field components in cylindrical coordinates, B_z and B_ρ, are approximately for z and ρ small,

$$B_z \approx \frac{32\pi I}{5\sqrt{5}Rc}\left(1 - \frac{144}{125}\frac{z^4}{R^4} + \frac{432z^2\rho^2}{125R^4} + \cdots\right), \qquad B_\rho \approx \frac{32\pi I}{5\sqrt{5}Rc}\frac{72z\rho(4z^2 - 3\rho^2 + \cdots)}{125R^4}.$$

2. Suppose the currents in the two loops of Problem 1 were in opposite directions. Find $\boldsymbol{B}$ at large distances ($r \gg R$).
3. Devise a distribution of currents for which the field components fall off as $1/r^5$ at large distances, i.e., for which both the dipole and quadrupole moments vanish.
4. A finite circular cylinder of dipolar matter is uniformly magnetized along the direction of the cylinder axis. Use both the equivalent surface charges and equivalent currents to calculate the external field.
5. Use the surface charge and surface currents of Problem 4 to formally calculate a "field" inside the magnetized cylinder. Is the field calculated by means of the charges similar to that calculated by means of the equivalent currents?
6. Two infinitely long parallel straight wires separated by a distance d carry current in opposite directions; they form a so-called transmission line. Find B_x and B_y in a coordinate system where the x-axis is along the line that joins the wires with origin halfway between them.
7. An infinite cylindrical conductor of radius a contains an infinite cylindrical hole of radius b whose center is offset a distance c from the center of the cylinder. The conductor carries a current I uniformly distributed over its cross section. Find the magnetic field $\boldsymbol{B}$ in the hole.
8. Derive the results (8.8) and (8.9) for the fields B_r and B_θ produced by a circular loop from the vector potential A_φ of Eq. (8.27).
9. For a current loop of arbitrary shape, show the equivalence of the two definitions (8.13) and (8.30) of magnetic moment, namely $(I/c)\int_S \boldsymbol{n}\, ds$ and $(I/2c)\oint \boldsymbol{r} \times d\boldsymbol{r}$.

REFERENCES

Brick, D. B. and A. W. Snyder, "External D.C. Field of a Long Solenoid," *Am. J. Phys.* **33** 905 (1965).

Lest it be thought that in the year 1965 the magnetic summation problem is closed to research, here is an example to the contrary. The field of a long but finite solenoid can be expressed exactly, but inconveniently for computational purposes, in terms of elliptic integrals. This paper develops convenient series expansions that apply to many cases of practical interest.

Durand, E., *Electrostatique et Magnetostatique*, Masson, Paris (1953).

This book in French is as thorough in its variety of magnetic summation problems as it is in electric ones.

Smythe, W. R., *Static and Dynamic Electricity*, McGraw-Hill, New York (1950).

Several examples are presented here that are not treated in the present text, most of them using the vector potential.

9
MAGNETIC MATERIALS AND BOUNDARY VALUE PROBLEMS

9.1 INTRODUCTION

In this chapter we deal with the magnetic analogs of the problems discussed for dielectrics. These are problems involving *paramagnetic* or *diamagnetic* matter, in which volume magnetic moments are set up and maintained by *applied* or *external* fields. We shall not treat the various kinds of magnetism, as *ferro-*, *antiferro-*, and *ferri-*, in which the *internal* fields that induce or maintain the magnetization are of quantum-mechanical origin.

We shall, therefore, be considering two kinds of elementary dipoles. Those of the first kind are analogous to electric polar molecules; they can be thought of as small, permanent magnets which, like polar molecules, tend to line up in the direction of an applied field. Substances composed of these are called *paramagnetic*. There are, however, microscopic magnetic dipoles that are, speaking loosely, associated with microscopic orbits of charged particles, and they behave differently. In a sense, they tend to *anti*polarize in an applied field. Such dipoles make up *diamagnetic* matter and are the second kind we shall consider. The theory presented applies equally to paramagnetic and diamagnetic matter.

Aside from the complication of the two kinds of magnetic matter, the problems that we shall treat are similar to those with dielectrics. We can take the primary problem to be this: A sample of magnetic matter is placed in a fixed applied field; it becomes polarized, i.e., magnetized, and thereby itself acts as a source of field. What is the *resultant* field and what is the induced polarization? We are then interested, as we were for dielectrics, in the perturbation of an applied field by the presence of magnetic matter. With dielectrics, we were also interested explicitly in the field *inside* matter in connection with the theory of capacitors. This explicit motivation is absent with magnetic materials, since, of course, there are no conductors of magnetism and no magnetic capacitors. But the problem of the internal field cannot be bypassed completely, since it is the internal field that produces the magnetization which is the source, in part, of the external field.

We are then faced with the same difficulties as with dielectric matter: the local field that determines the magnetic moments of the individual magnetic dipoles varies irregularly inside the matter and is difficult to calculate. For dielectrics, this difficulty was bypassed by a postulate, which related the smooth model fields given by the equivalent surface polarization charges to the polarization density. This postulate was in fact equivalent to a traditional theory involving the supplementary vector $\boldsymbol{D}$. We shall follow essentially the same course here except that the comparison with the traditional theory involving a vector $\boldsymbol{H}$ will be somewhat devious, since the historical notation is rather illogical from our viewpoint.

9.2 AN AVERAGE INTERNAL FIELD

For dielectrics, we began by showing that the *average internal field* inside a sample of dipolar matter was correctly given by the equivalent volume and surface charges of the model that gave the external field. If we follow the dielectric pattern, there are *two* analogous choices for the present case: we can begin with the model of equivalent magnetic *charges*, or of equivalent surface and volume *currents*. Let us discuss how we might choose between them.

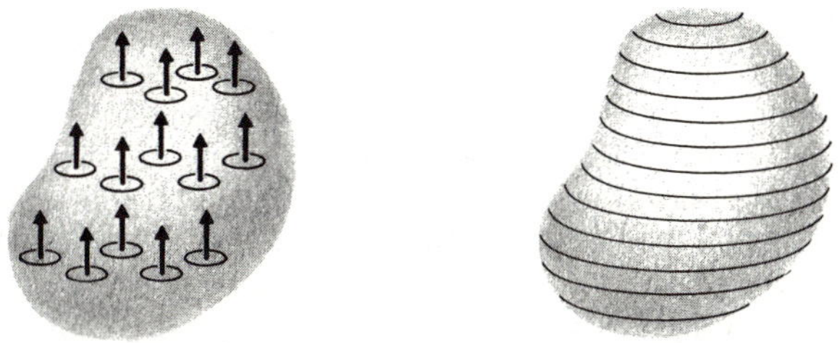

Fig. 9.1 Sample of magnetic dipolar matter and the model of equivalent currents for calculating the external field.

We want to find a smooth, macroscopic field in matter which is in some sense an *average* of the true microscopic field $\boldsymbol{B}$. This latter field, of course, satisfies the relation $\nabla \cdot \boldsymbol{B} = 0$, and it is reasonable to require that the divergence of the average field also be zero. If, however, we work with the model of equivalent magnetic surface charges, the field it predicts will not have zero divergence. The lines of force will stop or start on the surface charges just as electric lines of force stop or start on electric charges. This is sufficient reason to reject this model. We consider then the macroscopic smooth magnetic field that can be calculated from the equivalent surface and volume currents. We shall show that in a certain sense this field yields the same average value in the interior as does the true, local, fluctuating microscopic field. With the exception of one small modification, the argument is so similar to that for the electrostatic case that we can afford to be brief. We refer to Fig. 9.1, which shows a sample of polarized magnetic matter and its equivalent model for calculating the external field. Imagine two similar closed surfaces S and S' which intersect the matter and its model in a geometrically similar way. Divide S and S' into parts e and e' external to the matter and model, and parts i

and i' internal to them. Then since $\nabla \cdot \boldsymbol{B} = 0$ for both the matter and its model, we have for the flux integrals over the closed surfaces S and S':

$$\int_S \boldsymbol{B} \cdot d\boldsymbol{s} = \int_{S'} \boldsymbol{B} \cdot d\boldsymbol{s} = 0.$$

But the flux integral over the exterior parts e and e' is the same for the matter and its model since the model correctly predicts the exterior field:

$$\int_e \boldsymbol{B} \cdot d\boldsymbol{s} = \int_{e'} \boldsymbol{B} \cdot d\boldsymbol{s}.$$

It follows that the flux integrals are the same for the interior parts of the surfaces as well:

$$\int_i \boldsymbol{B} \cdot d\boldsymbol{s} = \int_{i'} \boldsymbol{B} \cdot d\boldsymbol{s}. \tag{9.1}$$

Now this interior integral for the matter, when divided by the surface area, represents a special kind of average that might be called a *surface average* of the irregular, rapidly varying field $\boldsymbol{B}$. Equation (9.1) shows that what we may call the *model* or *macroscopic field* $\boldsymbol{B}_m$ given by the equivalent volume and surface current, correctly reproduces this average of the microscopic field $\boldsymbol{B}$. Thus $\boldsymbol{B}_m$ is analogous to the field $\boldsymbol{E}_m$ that we introduced for dielectrics. It is then natural to develop the theory of magnetic boundary value problems by analogy with dielectric ones, and this is how we shall proceed.

A comparison with the electrostatic case, however, does raise some disconcerting questions that we first discuss. Suppose we calculate for the present sample of magnetic matter and its model the same kind of *line* average as for the electrostatic case. That is, we imagine similar closed integration *loops*, similarly oriented in the matter and the model, and calculate $\oint \boldsymbol{B} \cdot d\boldsymbol{l}$ for both loops. To either loop we can apply Ampère's circuital law, $\oint \boldsymbol{B} \cdot d\boldsymbol{l} = 4\pi I/c$, where I is the current through any surface spanning the loop. But the line integrals of $\boldsymbol{B}$ are *not* necessarily the same for the matter and the model. For example, the integral for the matter will depend sensitively on the detailed shape of the path of integration in the interior, and on whether or not it happens to cut through the microscopic current dipoles that make up the sample. The integral for the model, on the other hand, will depend primarily on where it cuts through the surface, since this will determine the amount of surface current that is enclosed. Thus, although the *surface average* of $\boldsymbol{B}$ is given correctly by the model field $\boldsymbol{B}_m$, the *line average* is not. In electrostatics, the opposite situation obtains. There the line average for the matter was given correctly by the model field $\boldsymbol{E}_m$, but it is easy to see that the analogous surface average is incorrect. For electrostatics, we can apply Gauss' flux theorem, $\int \boldsymbol{E} \cdot d\boldsymbol{s} = 4\pi Q$, where Q is the total charge enclosed. We consider surface integrals over similar surfaces of the kind sketched in Fig. 9.1 for the dielectric matter and model. We cannot conclude that the integrals i and i' over the surfaces interior to the matter and model will be identical, even though the exterior integrals are identical, since the total surface integrals are obviously not identical. In short, for electrostatics the surface average is not given correctly by the model, whereas the line average is, and conversely for magnetostatics. The fact of the incorrect surface average is not too unsettling for electrostatics, since it is the line average, or potential difference, that is of primary concern and is of physical significance in calculating the properties of condensers. Unfortunately, we are on rather shakier ground in the magnetostatic case. The surface average that is given correctly by the model is of no more physical significance than the line average given incorrectly. Nonetheless, for lack of any better prescription, we shall base our discussion of the fields of magnetized matter on the surface average. It is in effect this average that, in a somewhat ambiguous way, plays a central role in the standard theory of magnetized matter involving the vector $\boldsymbol{H}$.

9.3 A POSTULATE RELATING MEAN MAGNETIZATION AND THE APPLIED FIELD

Consider a permeable body in some applied field $\boldsymbol{B}_a$. Microscopically it becomes an array of dipolar elements, in which the induced dipole moment of one of the elements depends on both the external field and the field due to all the other elements. We cannot solve the problem of calculating the strength of the individual moments, as we could not its electrostatic analog. Thus, as we did in electrostatics, we shall introduce a postulate in which *average* magnetic moment is related to *average* field.

The postulate will have much the same *rationale* as its electrostatic analog, so we shall not detail the arguments for it. In any case, the postulate must stand or fall on empirical grounds independently of the *rationale*. Explicitly, the postulate is the obvious counterpart of Eq. (6.8). It relates the mean magnetic moment $\boldsymbol{M}(\boldsymbol{r})$, which is the analog of $\boldsymbol{P}(\boldsymbol{r})$ and is similarly defined, to the applied field $\boldsymbol{B}_a$ by means of

$$\boldsymbol{M} = \lambda(\boldsymbol{B}_a + \boldsymbol{B}_m). \tag{9.2}$$

If we write explicitly that $\boldsymbol{B}_m$ is composed of a surface and a volume term,

$$\boldsymbol{B}_m = \boldsymbol{B}_s + \boldsymbol{B}_v, \tag{9.3}$$

then Eq. (9.2) is

$$\boldsymbol{M} = \lambda(\boldsymbol{B}_a + \boldsymbol{B}_s + \boldsymbol{B}_v). \tag{9.4}$$

In electrostatics we found it necessary to modify the analogous tentative hypothesis since we were able to show that the term $\boldsymbol{E}_v$ was not present for a uniform medium. This came about, roughly speaking, because $\boldsymbol{E}_v$ was proportional to a field whose sources were *external* to the polarized volume element. The divergence of this field was zero at the element; hence the volume polarization charge vanished. Since the analogy to the dielectric case is very close here, we suspect that a similar result may hold for equivalent volume currents. To see that it does, we begin by taking the curl of Eq. (9.4). Now $\nabla \times \boldsymbol{B}_a = \nabla \times \boldsymbol{B}_s = 0$ inside the volume since there are no sources of $\boldsymbol{B}_a$ or $\boldsymbol{B}_s$ there. We conclude that

$$\nabla \times \boldsymbol{M} = \lambda \nabla \times \boldsymbol{B}_v. \tag{9.5}$$

Since the sources of $\boldsymbol{B}_v$ are the volume currents $\boldsymbol{J}_v = c(\nabla \times \boldsymbol{M})$, we can calculate $\boldsymbol{B}_v$ by applying Eq. (7.27). That is,

$$\nabla \times \boldsymbol{B}_v = \frac{1}{c}\nabla \times \left(\nabla \times \int \frac{\boldsymbol{J}_v(\boldsymbol{r}')dv'}{|\boldsymbol{r} - \boldsymbol{r}'|}\right).$$

We expand the right-hand side of this equation by $\nabla \times \nabla \times \boldsymbol{F} = \nabla\nabla \cdot \boldsymbol{F} - \nabla^2 \boldsymbol{F}$, with $\boldsymbol{F} = \boldsymbol{J}_v/|\boldsymbol{r} - \boldsymbol{r}'|$, to find

$$\nabla \times \boldsymbol{B}_v = \frac{1}{c}\nabla\nabla \cdot \int \frac{\boldsymbol{J}_v(\boldsymbol{r}')}{|\boldsymbol{r} - \boldsymbol{r}'|}dv' - \frac{1}{c}\nabla^2 \int \frac{\boldsymbol{J}_v(\boldsymbol{r}')}{|\boldsymbol{r} - \boldsymbol{r}'|}dv'.$$

Now the first integral vanishes, and in the second we use $\boldsymbol{J}_v = c(\nabla \times \boldsymbol{M})$ and $\nabla^2(1/|\boldsymbol{r} - \boldsymbol{r}'|) = -4\pi\delta(\boldsymbol{r} - \boldsymbol{r}')$ to find

$$\nabla \times \boldsymbol{B}_v = 4\pi\nabla \times \boldsymbol{M}. \tag{9.6}$$

But on comparing (9.6) with (9.5), we must conclude that $\boldsymbol{B}_v$ is zero. The basic hypothesis (9.4) becomes

$$\boldsymbol{M} = \lambda(\boldsymbol{B}_a + \boldsymbol{B}_s). \tag{9.7}$$

It is convenient to follow the dielectric pattern in defining an *average field* $\overline{\boldsymbol{B}}$:

$$\overline{\boldsymbol{B}} = \boldsymbol{B}_a + \boldsymbol{B}_s, \tag{9.8}$$

and then Eq. (9.7) can be written

$$\boldsymbol{M} = \lambda\overline{\boldsymbol{B}}. \tag{9.9}$$

9.4 THEORY OF PERMEABLE MATTER IN AN APPLIED FIELD

With the hypothesis (9.9) we turn in detail to the problem of a permeable body in some applied field $\boldsymbol{B}_a$. The problem of calculating the field that results, which is effectively the problem of calculating the density of average magnetization $\boldsymbol{M}$ inside the body, is replaced by a model problem. We consider a model volume which has the same shape as the body. Currents are assumed to flow on the surface of the model, but the interior of the model is considered to be free space. The magnitude and direction of these currents $\boldsymbol{J}_s$ are, of course, *unknown a priori*. But whatever they are, $\overline{\boldsymbol{B}}$ must satisfy

$$\nabla \times \overline{\boldsymbol{B}} = \frac{4\pi\boldsymbol{J}_s}{c}. \tag{9.10}$$

Since the right-hand side of (9.10) is nonzero only on the surface, $\nabla \times \bar{\boldsymbol{B}} = 0$ both in the interior and exterior of the model volume. We can then write $\bar{\boldsymbol{B}}$ in these regions as the gradient of a potential,

$$\bar{\boldsymbol{B}} = -\nabla\bar{\Phi}, \tag{9.11}$$

and, since $\nabla \cdot \bar{\boldsymbol{B}} = 0$, we also have both inside and outside that volume:

$$\nabla^2\bar{\Phi} = 0. \tag{9.12}$$

As we discussed previously, Eq. (9.10) manifests itself conveniently as the boundary condition of Eq. (8.19). Applied to this case, it is

$$(\bar{\boldsymbol{B}}_1 - \bar{\boldsymbol{B}}_2) \times \boldsymbol{n} = \frac{4\pi}{c}\boldsymbol{J}_s, \tag{9.13}$$

where the normal points from region 1 to region 2, and $\bar{\boldsymbol{B}}_1$ and $\bar{\boldsymbol{B}}_2$ are the fields on either side of the boundary. We take region 1 to be the interior, so $\bar{\boldsymbol{B}}_1$ is the interior field and the normal is the outwardly drawn one. According to the model, however,

$$\boldsymbol{J}_s = c(\boldsymbol{M} \times \boldsymbol{n}), \tag{9.14}$$

which gives

$$(\bar{\boldsymbol{B}}_1 - 4\pi\boldsymbol{M}) \times \boldsymbol{n} = \bar{\boldsymbol{B}}_2 \times \boldsymbol{n}. \tag{9.15}$$

With the assumption of Eq. (9.9) relating magnetization and mean field, Eq. (9.15) becomes the boundary condition

$$\bar{\boldsymbol{B}}_1(1 - 4\pi\lambda) \times \boldsymbol{n} = \bar{\boldsymbol{B}}_2 \times \boldsymbol{n}. \tag{9.16}$$

To this must be added the boundary condition on the continuity of the normal component of $\bar{\boldsymbol{B}}$, derivable from $\nabla \cdot \bar{\boldsymbol{B}} = 0$:

$$(\bar{\boldsymbol{B}}_1 - \bar{\boldsymbol{B}}_2)\cdot \boldsymbol{n} = 0 \tag{9.17}$$

With these last two equations we have a well-defined problem, soluble in principle. Laplace's equation must be solved inside and outside the volume and the two boundary conditions (9.16) and (9.17) must be applied to the field. There will presumably be a unique solution, and the techniques for deriving it will be very similar to those for the electrostatic case. The solution yields the magnetization and mean field inside the body, and the external field as well.

9.5 THE VECTOR *H*

The above discussion is not the conventional one. The usual treatment is, in fact, equivalent to the above, but it involves additional (and superfluous) terminology and definitions. These are, however, so well-entrenched in the literature that we shall spell them out and briefly discuss the motivation for introducing them.

Frequently, one begins by observing that $\bar{\boldsymbol{B}}$ and $\boldsymbol{M}$, and hence their tangential components, are *dis*continuous across the boundary between free space and magnetic dipolar matter. We see from (9.15), however, that the tangential component of $\bar{\boldsymbol{B}} - 4\pi\boldsymbol{M}$ is *continuous* across such a boundary; partly for this reason, this linear combination of $\bar{\boldsymbol{B}}$ and $\boldsymbol{M}$ is dignified by giving it the name $\boldsymbol{H}$:

$$\boldsymbol{H} = \bar{\boldsymbol{B}} - 4\pi\boldsymbol{M}. \tag{9.18}$$

In a similar way, the vector $\boldsymbol{D} = \bar{\boldsymbol{E}} + 4\pi\boldsymbol{P}$ is that linear combination of $\boldsymbol{P}$ and $\bar{\boldsymbol{E}}$ whose *normal* component is continuous across a dielectric boundary.

A second common *rationale* for the introduction of $\boldsymbol{H}$ involves the distinction between bound and free currents. *Free currents* are the usual currents produced by batteries and generators, and that flow in wires or other conductors. *Bound currents* are those associated with Eqs. (8.38) and (8.39), that is, they are the equivalent currents due to magnetic matter. The (dubious) argument is now frequently made that in a volume of magnetized matter through which a certain current $\boldsymbol{J}_{\text{free}}$ flows, the mean magnetic field will be produced not only by these currents, but by the bound ones as well. Equation (7.12) must then be taken to be

$$\nabla \times \bar{\boldsymbol{B}} = (4\pi/c)(\boldsymbol{J}_{\text{free}} + \boldsymbol{J}_{\text{bound}}),$$

or, on using $\boldsymbol{J}_{\text{bound}} = c(\nabla \times \boldsymbol{M})$,

$$\nabla \times (\bar{\boldsymbol{B}} - 4\pi\boldsymbol{M}) = \nabla \times \boldsymbol{H} = (4\pi/c)\boldsymbol{J}_{\text{free}}. \tag{9.19}$$

In a sense, then, $\boldsymbol{H}$ keeps track of the free currents and that is one of its merits. There is a good discussion of the usefulness of $\boldsymbol{H}$ in the book by Purcell (B).

The physical assumption is now usually made (for paramagnets and diamagnets) that the magnetization $\boldsymbol{M}$ is proportional to $\bar{\boldsymbol{B}}$, but this is done in a somewhat roundabout way. Namely, one assumes first that

$$\boldsymbol{M} = \chi_m \boldsymbol{H}, \tag{9.20}$$

where χ_m is the *magnetic susceptibility*. From (9.18), this implies that

$$\bar{\boldsymbol{B}} = \boldsymbol{H}(1 + 4\pi\chi_m) = \mu\boldsymbol{H}, \tag{9.21}$$

where the *permeability* μ is defined to be

$$\mu = 1 + 4\pi\chi_m. \tag{9.22}$$

By combining (9.20) and (9.21), we find that

$$\boldsymbol{M} = \frac{\chi_m}{1 + 4\pi\chi_m}\bar{\boldsymbol{B}} = \frac{1}{4\pi}\frac{\mu - 1}{\mu}\bar{\boldsymbol{B}}. \tag{9.23}$$

In essence, this conventional treatment involving $\boldsymbol{H}$ really reduces to a redefinition of the constant λ of Eq. (9.9) in terms either of the magnetic susceptibility χ_m or the permeability μ, according to

$$\lambda = \frac{\chi_m}{1 + 4\pi\chi_m} = \frac{1}{4\pi}\frac{\mu - 1}{\mu}. \tag{9.24}$$

Boundary value problems in this formalism are solved in a way that is completely equivalent to the discussion of the previous section. In a permeable medium in which there is no free current, one concludes from (9.19) that $\boldsymbol{H}$ is derivable from a potential. If the permeability is constant, Eq. (9.21), along with $\nabla \cdot \boldsymbol{B} = 0$, shows that this potential satisfies Laplace's equation. Once again the problem is to find solutions of this equation that satisfy the appropriate boundary conditions.

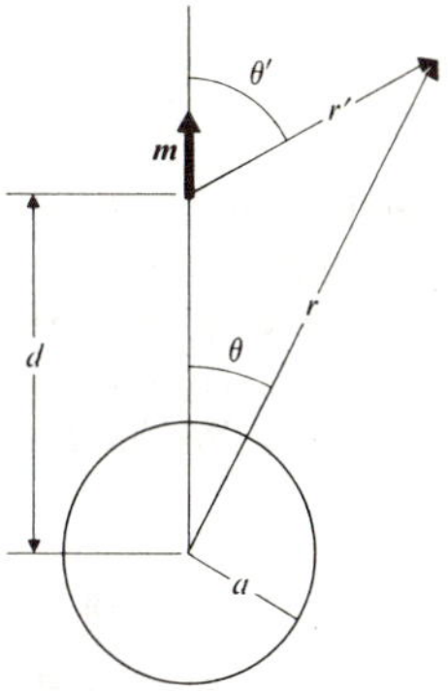

Fig. 9.2 Permeable sphere and external magnetic dipole.

9.6 A BOUNDARY VALUE PROBLEM

To illustrate the treatment of boundary value problems with permeable matter in external fields, we discuss an example. There will be little novelty here, after the experience of similar problems with conductors and dielectrics, since we shall once more be limited to the same simple geometrics and familiar solutions of Laplace's equation. As the example, we take the problem of a small bar magnet, which we shall consider to be a magnetic dipole of moment $\boldsymbol{m}$, external to a spherical sample of permeable matter as shown in Fig. 9.2. The geometry is similar to that in the problem of a point electric dipole outside a conducting sphere (discussed in Section 4.6) so we can borrow from those results. As we have discussed, we must imagine the material body replaced by a spherical model, whose surface carries a current, but which is hollow. As in Eq. (9.11) the interior and exterior fields $\overline{\boldsymbol{B}}_i$ and $\overline{\boldsymbol{B}}_e$ are derivable from potentials. The potential Φ_e exterior to the sphere is that due to the dipole plus that due to the unknown currents, i.e., it is of the form:

$$\Phi_e = \frac{m \cos \theta'}{r'^2} + \sum_{l=0}^{\infty} \frac{A_l}{r^{l+1}} P_l(\cos \theta).$$

The interior potential Φ_i can be taken to be

$$\Phi_i = \sum_{l=0}^{\infty} C_l r^l P_l(\cos \theta).$$

We now apply the boundary conditions. First we must expand the dipole potential, much as in Section 4.6, in a form appropriate for $r < d$, namely,

$$\frac{m\cos\theta'}{r'^2} = -m\sum_{l=0}^{\infty}\frac{(l+1)r^l}{d^{l+2}}P_l(\cos\theta).$$

The condition (9.17) on the normal component of $\bar{B}$ becomes $\partial\Phi_e/\partial r = \partial\Phi_i/\partial r$ whence we have

$$-\frac{m(l+1)la^{l-1}}{d^{l+2}} - \frac{A_l(l+1)}{a^{l+2}} = lC_l a^{l-1}. \tag{9 25}$$

Consider the condition (9.16) on the tangential component, $(1 - 4\pi\lambda)(\bar{B}_i)_\theta = (\bar{B}_e)_\theta$. However, this condition will be satisfied if the potentials satisfy $(1 - 4\pi\lambda)\Phi_i = \Phi_e$ at the boundary. In much the same way, the electrostatic boundary condition—that the tangential components of E be continuous—was equivalent to the condition that the potential be continuous. We have then

$$(1 - 4\pi\lambda)C_l a^l = -m(l+1)a^l/d^{l+2} + A_l/a^{l+1}. \tag{9.26}$$

From Eqs. (9.25) and (9.26) we find, in terms of the permeability $\mu = 1/(1 - 4\pi\lambda)$

$$A_l = \frac{(\mu - 1)ml(l+1)a^{2l+1}}{d^{l+2}(\mu l + l + 1)}$$

$$C_l = -\frac{\mu m(l+1)(2l+1)}{d^{l+2}(\mu l + l + 1)}$$

Beyond this method of solution that has used superposition of elementary solutions of Laplace's equations, some problems can be solved by guesswork, i.e., by the method of images. The details are quite similar to those involved in solving dielectric problems so we shall not discuss this further. We also note that, in the limit of high permeability, the problem of a permeable body in an applied magnetic field resembles the problem of a conducting body in an applied electric field, since the magnetic lines of force tend to be perpendicular to the body. Problems involving high permeability can then be solved approximately in terms of the solution of the analogous electrostatic problem.

PROBLEMS

1. A small bar magnet (magnetic dipole) is at the center of a spherical shell of permeable matter. Find the magnetic field $\bar{B}$ everywhere.
2. A spherical shell of permeability μ is placed in a uniform field $\boldsymbol{B}_0$. Find the magnetic field in the hollow interior and show that in the limit of large permeability the field is of the order $\boldsymbol{B}_0/\mu$. This is an example of magnetic shielding.
3. A line current, I, is parallel to an infinite circular cylinder of permeable matter of radius a, permeability μ. Find the force per unit length on the current.

4. The half-space $z < 0$ is filled with a medium of permeability μ, and the half-space $z > 0$ contains a current distribution $\boldsymbol{J}(\boldsymbol{r})$. Show that this boundary value problem is solved by images as follows: The field for $z > 0$ is as if due to the current $\boldsymbol{J}(\boldsymbol{r})$ and an image current with components

$$\left(\frac{\mu - 1}{\mu + 1}\right) J_x(x, y, -z), \qquad \left(\frac{\mu - 1}{\mu + 1}\right) J_y(x, y, -z), \qquad -\left(\frac{\mu - 1}{\mu + 1}\right) J_z(x, y, -z).$$

The field for $z < 0$ is as if it were produced by an image current $\boldsymbol{J}(2\mu)/(\mu + 1)$.

5. Use the result of the last problem to find the force per unit length on an infinite wire carrying current I that is parallel to the plane face of a semi-infinite permeable medium and a distance d from it.

6. The equator of a sphere of radius a, permeability μ, is wound with an insulated wire carrying current I. Find the fields inside and outside the sphere.

REFERENCES

Josephson, B. D., *Macroscopic Field Equations for Metals in Equilibrium*, *Phys. Rev.* **152,** 21 (1966).

A recent discussion of the meaning and the definition of fields, currents, and magnetization densities in metals.

Hague, B., *Electromagnetic Problems in Electrical Engineering*, Oxford University Press, London (1929).

This book discusses a variety of practical problems involving fields in the presence of permeable matter.

10
FORCE AND ENERGY IN STATIC FIELDS

Until now, this book has been devoted to calculating electric or magnetic fields in one context or another. In the final analysis, these fields are not intrinsically important, except insofar as they exert forces on some body and, hence, are agents for performing work. We have not yet explicitly discussed this problem of force and work (or energy), although we have touched on it here and there, since a complete treatment presupposes the discussion of dielectric and permeable media which we have just completed.

We have considered the fields of *given* distributions of charge and current but have not inquired into the process of forming these distributions. One aspect of this process is the question of how much energy is required, and this is discussed in the first two sections of this chapter. For charge distributions, the energy can be calculated from the laws of electrostatics we have already treated. But for calculating the energy of formation of a current distribution, magnetostatic laws do not suffice. For this case, induced electromotive forces (Faraday's law of induction) are brought into play and this time-dependent phenomenon goes beyond the magnetostatic principles already known. What we have done then is quote in advance some later results so that in this chapter we can treat the energies of charge and current distributions side by side.

Frequently, we are not interested in a charge or current distribution as a whole, but consider that one part of the distribution forms an *external* field for a second part, and ask for the energy of (or force or torque on) that second part in the external field. Sections 10.3 and 10.4 treat this problem for the case of *rigid* charge and current distributions. These are distributions that remain unchanged as they are brought into the external field. The next section discusses the force on a conductor in an external field. This involves a distribution that is *not* rigid; as we have seen, the charge on a conductor redistributes itself as the conductor is brought into a field. Another important nonrigid distribution is that of a dielectric sample which, unpolarized at infinity, is brought into an external field which polarizes it. The force on, and energy of, such a distribution is discussed in Section 10.6, and Section 10.7 treats the analogous problem for magnetic dipolar distributions.

10.1 ENERGY OF ASSEMBLAGE OF CHARGE DISTRIBUTION

Two point charges of the same sign repel each other, and work must be done to make the charges approach each other from infinity. The amount of this work is the potential energy of the system, or the *electrostatic potential energy*. In this section, we want to calculate the analogous energy for an *arbitrary* charge distribution.

Consider two small charge distributions, containing total charges q_1 and q_2, respectively. We do not think of these as point charges since, as we shall find, the concept of point charge entails certain infinities that are best avoided. Instead we think of them as small, in a certain sense. If these distributions are brought to a distance r_{12} of one another, and if the distributions are *small compared to* r_{12}, we know from multipole theory that they interact like point charges, and the work W_2 required to bring them together, i.e., their mutual potential energy, is

$$W_2 = \frac{q_1 q_2}{r_{12}}. \tag{10.1}$$

The work W_N required to assemble N such charges $q_1, \ldots, q_N$ with mutual distances $r_{12} \cdots r_{13} \cdots r_{N-1,N}$ is then the sum of all possible mutual pair energies of the form (10.1). It can be written in either of two different ways:

$$W_N = \sum_{i<j} \sum_{j=2}^{N} \frac{q_i q_j}{r_{ij}} = \frac{1}{2} \sum_{i=1}^{N} \sum_{j=1}^{N} \frac{q_i q_j}{r_{ij}}. \tag{10.2}$$

In the second form, it is understood that terms in the double sum with $i = j$ are omitted. The factor $\frac{1}{2}$ in this form ensures that each pair is counted only once. From this form, we can find an expression for the energy of assemblage W_E of a *continuous distribution* $\rho(\boldsymbol{r})$ by replacing q_i by $\rho(\boldsymbol{r})\,dv$, q_j by $\rho(\boldsymbol{r}')\,dv'$, r_{ij} by $|\boldsymbol{r} - \boldsymbol{r}'|$, and the sum by an integral to get

$$W_E = \frac{1}{2} \int\int \frac{\rho(\boldsymbol{r})\rho(\boldsymbol{r}')\,dv\,dv'}{|\boldsymbol{r} - \boldsymbol{r}'|}. \tag{10.3}$$

If Eq. (10.3) is applied to find the energy of a point charge model—for example, that of a uniformly charged sphere whose radius goes to zero—W_E is infinite. It is for this reason that we avoided the introduction of point charges in the beginning discussion.

A variant of (10.3) is obtained on recognizing that one of the two integrals in it is the potential,

$$\Phi(\boldsymbol{r}) = \int \frac{\rho(\boldsymbol{r}')}{|\boldsymbol{r} - \boldsymbol{r}'|}\,dv'.$$

Then (10.3) can be written as

$$W_E = \frac{1}{2} \int \rho(\boldsymbol{r})\Phi(\boldsymbol{r})\,dv. \tag{10.4}$$

This last form can be used to express the energy in terms of the electric field $\boldsymbol{E}$, rather than in terms of the charge distribution. To see this, replace the charge distribution ρ in (10.4) by $\rho = \nabla \cdot \boldsymbol{E}/4\pi$. Moreover, although the integral in (10.4) is over the charge distribution, it can be formally extended to one over all space; there will still be a contribution to (10.4) only where $\rho(\boldsymbol{r})$ is nonvanishing. Then Eq. (10.4) can be written as

$$W_E = \frac{1}{8\pi}\int_{\text{space}} (\nabla \cdot \boldsymbol{E})\Phi \, dv. \tag{10.5}$$

In (10.5) we use $\Phi(\nabla \cdot \boldsymbol{E}) = \nabla \cdot (\Phi \boldsymbol{E}) - \boldsymbol{E} \cdot \nabla\Phi$, and replace the volume integral involving $\nabla \cdot (\Phi\boldsymbol{E})$ by its equivalent surface integral according to the divergence theorem, to find

$$W_E = \frac{1}{8\pi}\left[\int (\Phi \boldsymbol{E}) \cdot d\boldsymbol{s} - \int_{\text{space}} \boldsymbol{E} \cdot \nabla\Phi \, dv\right].$$

Now the surface integral vanishes since Φ falls off at least as rapidly as $1/\bar{r}$, where $\bar{r}$ is some mean distance from the distribution, and E falls off as $1/\bar{r}^2$, whereas ds increases only as $\bar{r}^2$. The integrand goes as $1/\bar{r}^3$ and, since the original (10.5) is integrated over all space, we must consider that $\bar{r} \to \infty$, whence the surface integral vanishes. In the remaining volume integral, we replace $-\nabla\Phi$ by $\boldsymbol{E}$ to find

$$W_E = \frac{1}{8\pi}\int_{\text{space}} |\boldsymbol{E}|^2 \, dv. \tag{10.6}$$

This formula expresses the energy in terms only of the field $\boldsymbol{E}$. It is as if the energy resides in the field, whereas (10.4) seems to imply it resides in the charges. Of course, both statements are correct since the one formula is equivalent to the other; in fact, this equivalence shows that the one formulation is not more basic than the other. The form of (10.6) suggests that one can attribute an *energy density* $|\boldsymbol{E}|^2/8\pi$ to space in which an electric field $\boldsymbol{E}$ exists, but this is merely suggestive; all one knows for sure is that the integrated formula is correct.

10.2 ENERGY OF FORMATION OF CURRENT DISTRIBUTION

We discuss now the analog for stationary currents of the problem for charges which we have treated: Given some distribution $\boldsymbol{J}(\boldsymbol{r})$, how much energy is required to bring it into being? At first sight, it would seem that this energy could be calculated in much the same way as its electrostatic analog. Namely, we might imagine the current distribution as being composed of current filaments that have been assembled from infinity. We could then calculate the work required to bring together a pair of these filaments (*required* because one filament exerts a force on the other according to Ampère's law), add the work required to bring up a third filament in the field of the first two, then a fourth, etc. Such a procedure would yield a formula, analogous to Eq. (10.3), for the energy of assemblage of the whole current distribution. There is only one flaw in this procedure: it is obviously wrong. There is a basic difference between the formula for the force between two charge elements of the same sign and two (parallel) current filaments of the same sign, i.e., carrying currents in the same direction. The electrostatic force is a repulsion so that an external force is required to form a charge distribution; it is the work done by this external force that we call the energy of assemblage. But the opposite may obtain for a current distribution. Think, for example, of such a distribution in the form of an infinitely long cylinder of arbitrary but finite cross section and consider it broken into infinitely long filaments of infinitesimal cross section. The force per unit length between two such parallel filaments carrying currents I_1 and I_2 in the same direction is *attractive* and of magnitude $I_1 I_2/c^2 d$, where d is the perpendicular distance between them. Energy need not be expended in bringing them in from infinity; in fact, energy is gained, i.e., work can be done *on* some external mechanical system in so doing. By bringing up a third filament, still more work can be done, and by assembling an arbitrarily large current system, arbitrarily large amounts of work can be generated at will.

Obviously, there is more here than meets the eye. What has been omitted is a new law of electromagnetism that modifies the above results. This law is *Faraday's law of induction*, which involves time-varying fields and so has not come into our purview. It is discussed in the next chapter. Pending that discussion, we shall quote the result it yields so that we can discuss here the energy of assemblage of a stationary current distribution and compare it with the analogous formulas for charge distributions.

Qualitatively, Faraday's law states that an electric field is generated by a changing magnetic field. If the current changes in a wire, consequently changing the magnetic field in the neighborhood of the wire, then the electric field thereby generated will do work on the charged particles that constitute the current. Energy is therefore required to alter a current and, in particular, to form a current distribution, i.e., to bring the current carriers from the state in which they are at rest to the state in which they constitute a final stationary current with density $\boldsymbol{J}(\boldsymbol{r})$. In one form, analogous to Eq. (10.4), this energy W_M is *localized*, since it is expressed in terms of the current density and vector potential. This form is derived in Section 11.3, and is quoted here:

$$W_M = \frac{1}{2c} \int \boldsymbol{A} \cdot \boldsymbol{J} \, dv. \tag{10.7}$$

In this form, where the integration goes over only the region of space in which the current density is nonvanishing, the energy appears to reside in the currents, much as in (10.4) the energy is associated with the charge. Equation (10.7) can, however, be transformed into a form where the energy appears to reside in the field, resulting in an expression analogous to the electrostatic Eq. (10.6). We replace $\boldsymbol{J}$ in (10.7) by $(c\nabla \times \boldsymbol{B})/4\pi$ to find

$$W_M = \frac{1}{8\pi} \int \boldsymbol{A} \cdot (\nabla \times \boldsymbol{B}) \, dv.$$

We now put $\boldsymbol{A} \cdot \nabla \times \boldsymbol{B} = \boldsymbol{B} \cdot \nabla \times \boldsymbol{A} - \nabla \cdot (\boldsymbol{A} \times \boldsymbol{B})$ into the last equation and use the divergence theorem to express the second term as a surface integral which can be shown to vanish when the integral is extended over all space, much as for the electrostatic case. On writing $\boldsymbol{B} = \nabla \times \boldsymbol{A}$ for the first term we have

$$W_M = \frac{1}{8\pi} \int_{\text{space}} |\boldsymbol{B}|^2 \, dv. \tag{10.8}$$

Here again the energy is expressed in terms of the field, and it is as if there were a *magnetic energy density* $|\boldsymbol{B}|^2/8\pi$ associated with the field. But as for the electrostatic analog, all we can know with certainty is that the integrated result (10.8) is correct.

10.3 RIGID CHARGE DISTRIBUTION IN APPLIED ELECTRIC FIELD

Frequently we are not interested in the total energies represented by Eqs. (10.6) and (10.7). More often, we are interested in a partial energy of some portion of a charge or current distribution which is then said to be in the *external* or *applied* field of the remainder of the distribution. As an example from electrostatics, consider an electron in an electronic device in which some shaped field is produced by an arrangement of charges on condenser plates or conductors. These charges, along with the electron, constitute a charge distribution whose *mutual* electrostatic energy is given by (10.4), but this is often not of interest. We are usually concerned with the energy of the electron itself and not with the energy required to assemble the charge whose field constitutes the applied field for it.

An *applied* field generates a potential $\Phi_a(\boldsymbol{r})$. The energy of a point charge q in this field is, by definition, $q\Phi_a(\boldsymbol{r})$. If, then, there is a *rigid* distribution of charge of density $\rho(\boldsymbol{r})$, the potential energy W_a of this distribution in the applied field is, by superposition,

$$W_a = \int \rho(\boldsymbol{r})\Phi_a(\boldsymbol{r}) \, dv. \tag{10.9}$$

The concept of rigid distribution is central to (10.9). By this we mean a distribution that is *fixed* in shape and dimensions, as if it were composed of charges attached to a rigid scaffolding. Equation (10.9) applies only to such distributions since it purports to give the work necessary to bring the distribution from infinity in the external field; if the *internal distances*, and hence *internal potential energy*, changes, these changes must be taken into account in a way that is not comprised in Eq. (10.9). Thus, if we had two charges on the end of a spring, the work necessary to bring such a system in from infinity would be the work done on each charge individually, as in Eq. (10.9), *plus* whatever work is done in compressing or extending the spring and thereby also changing the mutual potential energy of the two charges. This case is discussed in Section 10.6 on dielectrics in external fields. Of interest also is the force $\boldsymbol{F}$ on a charge distribution. This is, of course, just the sum of the forces on the individual elements:

$$\boldsymbol{F} = \int \rho(\boldsymbol{r})\boldsymbol{E}_a(\boldsymbol{r})\, dv. \tag{10.10}$$

Not much can be said about Eqs. (10.9) and (10.10) for arbitrary applied fields $\boldsymbol{E}_a$; they are simply integrals that must be evaluated. An important special case, however, arises when these fields vary slowly over the distribution. We can then make a systematic development that involves the familiar multipole coefficients, and evaluate the integral term by term. Consider first the approximate evaluation of Eq. (10.9) for the energy W_a. Assume the origin of coordinates is somewhere in the distribution, and expand $\Phi_a(\boldsymbol{r})$ about this origin:

$$\Phi_a(\boldsymbol{r}) = \Phi_a(\mathbf{0}) + \boldsymbol{r}\cdot\boldsymbol{\nabla}\Phi_a(\mathbf{0}) + \frac{1}{2}\left[x^2\frac{\partial^2\phi_a(\mathbf{0})}{\partial x^2} + y^2\frac{\partial^2\phi_a(\mathbf{0})}{\partial y^2} + z^2\frac{\partial^2\phi_a(\mathbf{0})}{\partial z^2}\right.$$
$$\left. + 2xy\frac{\partial^2\phi_a(\mathbf{0})}{\partial x\partial y} + 2xz\frac{\partial^2\phi_a(\mathbf{0})}{\partial x\partial z} + 2yz\frac{\partial^2\phi_a(\mathbf{0})}{\partial y\partial z}\right] + \cdots$$

Using $\boldsymbol{E}_a(\mathbf{0}) = -\boldsymbol{\nabla}\Phi_a(\mathbf{0})$ and relations like

$$\partial^2\Phi_a/\partial x\partial y = -\partial E_{ax}/\partial y,$$

we find

$$\Phi_a(\boldsymbol{r}) = \Phi_a(\mathbf{0}) - \boldsymbol{r}\cdot\boldsymbol{E}_a(\mathbf{0}) - \frac{1}{2}\left(x^2\frac{\partial E_{ax}(\mathbf{0})}{\partial x} + y^2\frac{\partial E_{ay}(\mathbf{0})}{\partial y} + z^2\frac{\partial E_{az}(\mathbf{0})}{\partial z}\right.$$
$$\left. + 2xy\frac{\partial E_{ax}(\mathbf{0})}{\partial y} + 2yz\frac{\partial E_{ay}(\mathbf{0})}{\partial z} + 2xz\frac{\partial E_{az}(\mathbf{0})}{\partial x}\right) + \cdots$$

We next subtract zero in the form of $r^2/6\boldsymbol{\nabla}\cdot\boldsymbol{E}_a(\mathbf{0})$ from the quadratic terms in this expansion and put the resultant expression into (10.9). Then we can express W_a as an expansion involving the multipole moments of the system: q, the total charge; $\boldsymbol{p}$, the dipole moment; Q_{xx}, Q_{xy}, etc., the quadrupole moments. We find, on writing ϕ_a, $\boldsymbol{E}_a$, for $\phi_a(\mathbf{0})$, $\boldsymbol{E}_a(\mathbf{0})$, etc., that

$$W_a = W_{\text{mono}} + W_{\text{dip}} + W_{\text{quad}} + \cdots,$$

where

$$\begin{aligned} W_{\text{mono}} &= q\phi_a \\ W_{\text{dip}} &= -\boldsymbol{p}\cdot\boldsymbol{E}_a \\ W_{\text{quad}} &= -\frac{1}{6}\left[Q_{xx}\frac{\partial E_{ax}}{\partial x} + Q_{yy}\frac{\partial E_{ay}}{\partial y} + Q_{zz}\frac{\partial E_{az}}{\partial z} + 2Q_{xy}\frac{\partial E_{ax}}{\partial y}\right. \\ &\qquad \left. + 2Q_{yz}\frac{\partial E_{ay}}{\partial z} + 2Q_{xz}\frac{\partial E_{ax}}{\partial z}\right]. \end{aligned} \tag{10.11}$$

Consider next a similar approximation for the evaluation of the integral (10.10) for force. Take one component, say E_{ax}, and expand it about the origin. With the first two terms of such an expansion,

$$F_x = \int \rho(\boldsymbol{r})[E_{ax}(\mathbf{0}) + \boldsymbol{r} \cdot \nabla E_{ax}(\mathbf{0}) + \cdots]\, dv. \tag{10.12}$$

The lowest (monopole) term in this expansion, combined with the similar result for the other two components, leads to the obvious result for $\boldsymbol{F}_{\text{mono}}$, the force due to the *monopole moment*:

$$\boldsymbol{F}_{\text{mono}} = \boldsymbol{E}_a \int \rho(\boldsymbol{r})\, dv.$$

The next (*dipole*) term involves the dipole moment $\boldsymbol{p} = \int \boldsymbol{r}\rho(\boldsymbol{r})\, dv$. The result, which comprises (10.12) and its y and z counterparts, is the vector force $\boldsymbol{F}_{\text{dip}}$ due to the dipole moment of the distribution.

$$\boldsymbol{F}_{\text{dip}} = (\boldsymbol{p} \cdot \nabla)\boldsymbol{E}_a. \tag{10.13}$$

This last expression for the force on a dipole can also be derived from an intuitively simple model. Consider two point charges, q and $-q$, separated by a small distance, Δ. In an *inhomogeneous* field, the two charges will be acted on by forces of different strengths, leaving a net translational force on the system. In the point dipole limit, $q \to \infty$, $\Delta \to 0$ but $q\Delta \to p$, this effect persists; the force that is calculated from this model agrees, of course, with (10.13). Similarly, for the torque $\boldsymbol{T}$ on a dipole in an external field $\boldsymbol{E}_a$, one finds

$$\boldsymbol{T} = \boldsymbol{p} \times \boldsymbol{E}_a. \tag{10.14}$$

A corollary to the above results is a useful expression for the mutual potential energy of two (rigid) point dipoles. If we consider that dipole 1 produces the external field Φ_1 which acts on dipole 2, the mutual potential energy W_{12} of the dipoles is from (10.11),

$$W_{12} = \boldsymbol{p}_2 \cdot \nabla\Phi_1,$$

where Φ_1 is, of course, evaluated at the position of the second dipole. Without loss of generality, we can take the first dipole of moment $\boldsymbol{p}_1$ at the origin of a coordinate system and pointing in the z-direction, and the second dipole of arbitrary moment $\boldsymbol{p}_2$ a distance r away. Then

$$\Phi_1 = \frac{\boldsymbol{p}_1 \cdot \boldsymbol{r}}{r^3} = -\boldsymbol{p}_1 \cdot \nabla\left(\frac{1}{r}\right),$$

and we have

$$W_{12} = -\boldsymbol{p}_2 \cdot \nabla\left(\boldsymbol{p}_1 \cdot \nabla\left(\frac{1}{r}\right)\right) = \boldsymbol{p}_2 \cdot \left(\frac{1}{r^3}\nabla(\boldsymbol{p}_1 \cdot \boldsymbol{r}) + (\boldsymbol{p}_1 \cdot \boldsymbol{r})\nabla\left(\frac{1}{r^3}\right)\right).$$

With

$$\nabla(\boldsymbol{p}_1 \cdot \boldsymbol{r}) = \boldsymbol{p}_1 \qquad \text{and} \qquad \nabla\left(\frac{1}{r^3}\right) = -\frac{3\boldsymbol{r}}{r^5}$$

we have the useful form:

$$W_{12} = \frac{\boldsymbol{p}_1 \cdot \boldsymbol{p}_2}{r^3} - \frac{3(\boldsymbol{p}_1 \cdot \boldsymbol{r})(\boldsymbol{p}_2 \cdot \boldsymbol{r})}{r^5}. \tag{10.15}$$

10.4 RIGID CURRENT DISTRIBUTION IN APPLIED MAGNETIC FIELD

We turn now to a calculation of the forces on, and energy of, a rigid current distribution in an applied magnetic field. The concept of rigid distribution has both a mechanical *and* an electrical connotation: we first suppose that such a distribution is one that maintains its shape but we *also* suppose that the current it carries remains constant. For current in a wire, as we have remarked (and will discuss in more detail later), this means that work must be done by a battery or generator to keep the current constant; this work will *not* be considered here. We shall calculate only the work done by the *external* forces on the constant current loop.

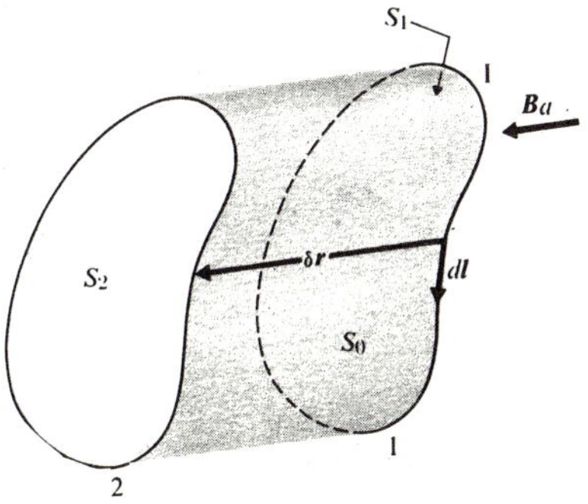

Fig. 10.1 Current loop in an applied field $\mathbf{B}_a$.

Instead of dealing with an arbitrary distribution $\boldsymbol{J}(\boldsymbol{r})$, it will be easier to begin with a filamentary current loop; the formulas for the general distribution can then be synthesized by superposition. Figure 10.1 shows such a loop carrying current I in an applied field $\boldsymbol{B}_a$. We suppose it is translated bodily from some initial position 1 to some final position 2 and that the translation is described by a vector $\delta\boldsymbol{r}$. The work δW_a done *on* the element of current $I\,d\boldsymbol{l}$ to translate it, i.e., the work done *against* the force $(I/c)d\boldsymbol{l} \times \boldsymbol{B}_a$ is

$$\delta W_a = -\frac{I}{c}(d\boldsymbol{l} \times \boldsymbol{B}_a)\cdot \delta\boldsymbol{r} = \frac{I}{c}(d\boldsymbol{l} \times \delta\boldsymbol{r}) \cdot \boldsymbol{B}_a.$$

From Fig. 10.1, $(\boldsymbol{dl} \times \delta \boldsymbol{r})$ is an inward-pointing element of area $\boldsymbol{ds}$ for the shaded ribbon-like surface S_0 shown, and

$$(\boldsymbol{dl} \times \delta \boldsymbol{r}) \cdot \boldsymbol{B}_a = \boldsymbol{B}_a \cdot \boldsymbol{ds}$$

is the *inward* flux of $\boldsymbol{B}_a$ through that surface. Then the work ΔW_a done on the loop as a whole is

$$\Delta W_a = \frac{I}{c} \int_{S_0} \boldsymbol{B}_a \cdot \boldsymbol{ds}.$$

Since $\nabla \cdot \boldsymbol{B}_a$ is zero, the integral in this expression is the same as the *outward** flux of $\boldsymbol{B}_a$ through the end surfaces S_1 and S_2 that span the loop; that is,

$$\Delta W_a = -\frac{I}{c} \int_{S_2} \boldsymbol{B}_a \cdot \boldsymbol{ds} + \int_{S_1} \boldsymbol{B}_a \cdot \boldsymbol{ds}.$$

With

$$\phi_{1,2} = \int_{S_{1,2}} \boldsymbol{B}_a \cdot \boldsymbol{ds}$$

and with $\Delta\phi = \phi_2 - \phi_1$ we have

$$\Delta W_a = -\frac{I}{c} \Delta\phi. \tag{10.16}$$

Since I is constant, Eq. (10.16) can be integrated to find the energy required to bring a rigid current distribution up from infinity. If we assume that $\boldsymbol{B}_a$ vanishes at infinity, and thus that the flux vanishes there also, we have for the energy W_a of the current loop in the applied field:

$$W_a = -\frac{I}{c} \int \boldsymbol{B}_a \cdot \boldsymbol{ds}. \tag{10.17}$$

The force $\boldsymbol{F}$ on the loop is

$$\boldsymbol{F} = \frac{I}{c} \oint \boldsymbol{dl} \times \boldsymbol{B}_a. \tag{10.18}$$

The force $\boldsymbol{F}$ on a rigid current distribution $\boldsymbol{J}(\boldsymbol{r})$ can therefore be found by considering such a distribution to be a sum of filamentary loops:

$$\boldsymbol{F} = \frac{1}{c} \int \boldsymbol{J}(\boldsymbol{r}) \times \boldsymbol{B}_a \, dv. \tag{10.19}$$

Not much more can be said in general about Eqs. (10.17) and (10.19); they are simply integrals to be evaluated. They can be expanded, however, for slowly varying applied fields; this is similar to the electrostatic case. We consider only the

* Recall the convention on page 122 for the *positive* normal.

lowest-order term of the expansion of Eq. (10.17). In terms of an origin centered in the distribution, we have $B_a(r) \approx B_a(0)$. Therefore,

$$W_a = -\frac{I}{c}\boldsymbol{B}_a(\mathbf{0}) \cdot \int d\boldsymbol{s}. \qquad (10.20)$$

With the definition (8.13) of magnetic moment $\boldsymbol{m}$, we see that, to lowest order, $W_a \approx W_{\text{dip}}$ where the *magnetic dipole energy* W_{dip} is

$$W_{\text{dip}} = -\boldsymbol{m} \cdot \boldsymbol{B}_a. \qquad (10.21)$$

This is analogous to the electrostatic result. Since the force on the dipole in an inhomogeneous field can be obtained by differentiating its potential energy, and since the potential energy (10.21) has the same form as for an electric dipole with $\boldsymbol{p} \to \boldsymbol{m}$ and $\boldsymbol{E}_a \to \boldsymbol{B}_a$, we conclude that the *force* $\boldsymbol{F}_{\text{dip}}$ on a *magnetic dipole* is

$$\boldsymbol{F}_{\text{dip}} = (\boldsymbol{m} \cdot \nabla)\boldsymbol{B}_a. \qquad (10.22)$$

10.5 FORCE ON CONDUCTORS

An important problem is that of the force on a charged conductor. This force is, of course, the same as that on the charges that reside on the surface since these charges are bound to the conductor. The surface charge distribution is not a rigid one, in general; it will change as the conductor is brought up from infinity. Nonetheless, Eq. (10.10) for the force could be used with the distribution $\rho(\boldsymbol{r})$ now being a surface one, since this equation simply says that the net force is the sum of the external forces acting on each charge element. To apply this, we must have an expression for the density distribution *and* the external force. But there is a surprisingly easier way. We derive a formula which allows the force to be calculated from the surface charge density *only*.

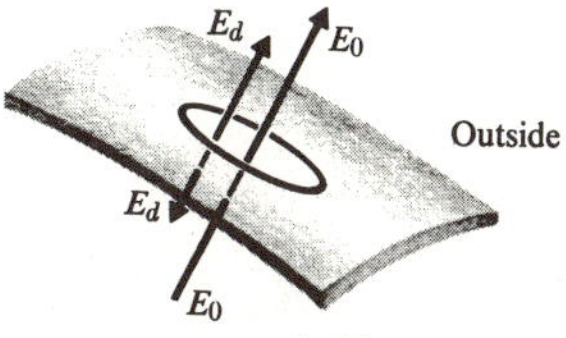

Fig. 10.2 Portion of conductor with small circle inscribed on surface.

Figure 10.2 shows a conductor with a surface charge σ, which may be a function of position. As we have seen, this surface charge distributes itself in such a way that the total electric field outside the surface is normal to it at every point and the total electric field vanishes in the interior. We focus attention on some point P which we assume is the center of a small circle inscribed on the surface. Whatever the curvature of the surface near P, we can always take the circle small enough so that the disk it defines is essentially plane. Consider now the total field $\boldsymbol{E}_t$ just outside P. This is made up of the field $\boldsymbol{E}_d$ due to the charge on the disk and the field $\boldsymbol{E}_0$ due to all other charges on or off the conductor. We have found, as shown in Eq. (3.53), that the field $\boldsymbol{E}_d$ has the same magnitude on each side of the disk, but reverses direction on crossing the disk. On the other hand, the field $\boldsymbol{E}_0$, due to the other charges, will be continuous through the disk. These fields are sketched in Fig. 10.2. Since the total field just *inside* the conductor is zero, it must be that the magnitudes of $\boldsymbol{E}_d$ and $\boldsymbol{E}_0$ are the same there, but that they point in opposite directions. Just *outside* the disk $\boldsymbol{E}_d$ and $\boldsymbol{E}_0$ are in the same direction, and since their magnitudes are the same, we have

$$\boldsymbol{E}_t = \boldsymbol{E}_d + \boldsymbol{E}_0 = 2\boldsymbol{E}_0.$$

The disk cannot act on itself; therefore, the force per unit area on the disk is the charge per unit area times the field due to all the other charges *except* those on the disk. Thus, we have the basic formula:

$$\text{Force/unit area} = \sigma \boldsymbol{E}_0 = \frac{\sigma \boldsymbol{E}_t}{2} = 2\pi\sigma^2, \tag{10.23}$$

with the understanding that this force density is normal to the conductor at the point where it is calculated.

10.6 DIELECTRIC MATTER IN APPLIED ELECTRIC FIELD

We now discuss the force on, and energy of, a dielectric body in an applied field $\boldsymbol{E}_a$, and we begin by considering force.

The net force on a sample of dipolar matter is the sum of the forces on the individual dipoles, and the force on one of these is given by Eq. (10.13). We can label the dipoles by an index i, and write the force $\boldsymbol{F}$ on an assemblage as:

$$\boldsymbol{F} = \sum_i (\boldsymbol{p}_i \cdot \nabla)\boldsymbol{E}_a(\boldsymbol{r}_i). \tag{10.24}$$

We shall show that this force can be calculated to a good approximation by assuming that the field $\boldsymbol{E}_a$ acts on the equivalent surface and volume polarization charges discussed in Chapter 6. The first step is to consider that the dipoles of the assemblage are smeared out to form a continuous distribution, as was done in Chapter 6 for calculating the exterior field, and to determine how the force on this model of continuous matter differs from the force on the actual assemblage. We shall see that it differs only negligibly; given this fact, the result quoted for the force in terms of polarization charges will follow directly.

Dipole matter aside for the moment, consider the force on a charge distribution in terms of its multipole moments. Equation (10.13) gives the force due to the *dipole moment*; it depends on the products of dipole moment components with the first derivatives of the field. We have not derived an expression for the force due to the quadrupole moment, but it is easy to see that it will depend on products of the quadrupole moments with the *second* derivatives of the field. For a distribution of typical linear dimension a containing total charge q, the dipole moment of the distribution will generally have components of order qa and the quadrupole moment tensor will have components of order qa^2. If the applied field is due to charge sources a mean distance R away, the field components will vary, apart from angular factors, as $1/R^2$ near the distribution. Successive derivatives of the field will introduce additional powers of $1/R$. Then the force due to the quadrupole moment will differ from the dipolar force by an extra power of a (from the definition of the moment) and an extra factor $1/R$ (from the gradient of the field). In short, the force will be smaller by a factor of order a/R, and if a is a microscopic length and R a macroscopic one, the quadrupolar force will be negligible.

Now we return to the problem of the force on dipolar matter. The continuous model with which we propose to replace dipolar matter will have the same *dipole* moment in each cell as does the actual matter, but the quadrupole moments of a cell of the model will differ from the actual quadrupole moment (if any). In an external field then, characterized as above by the scale R, the force on the *continuous* distribution $\boldsymbol{P}(\boldsymbol{r})$ will differ from the sum of forces on the original dipoles by terms of order a/R. For the ordinary external field, these will be negligible. For this case then, the summation in Eq. (10.24) is replaced by an integral, and the force on the body to order a/R is

$$\boldsymbol{F} = \int (\boldsymbol{P} \cdot \nabla)\, \boldsymbol{E}_a(\boldsymbol{r})\, dv.$$

Now we use the theorem from vector analysis that if $\boldsymbol{C}$ and $\boldsymbol{D}$ are two vector fields defined in a volume V that is bounded by a surface S and if $\boldsymbol{n}$ is the outward normal to S, then

$$\int_V (\boldsymbol{C} \cdot \nabla)\boldsymbol{D}\, dv = \int_S (\boldsymbol{n} \cdot \boldsymbol{C})\boldsymbol{D}\, ds - \int_V (\nabla \cdot \boldsymbol{C})\boldsymbol{D}\, dv.$$

We replace $\boldsymbol{C}$ by $\boldsymbol{P}$ and $\boldsymbol{D}$ by $\boldsymbol{E}_a$ and, with $\sigma_p = \boldsymbol{n} \cdot P$ as the density of surface polarization charge and $\rho_p = -\nabla \cdot \boldsymbol{P}$ as the density of volume polarization charge, we find

$$\boldsymbol{F} = \int_S \sigma_p \boldsymbol{E}_a\, ds + \int_V \rho_p \boldsymbol{E}_a\, dv. \tag{10.25}$$

This is the desired result: the force on the body is *as if* the external field acted on the polarization charges.

We turn now to the problem of the energy of a sample of dielectric matter in an applied field and begin by discussing the energy a *single* polarizable element (molecule) acquires when it is polarized by such a field. We shall assume the law of polarization of Eq. (6.6); the induced moment $\boldsymbol{p}$ is related to the applied field by

$$\boldsymbol{p} = \alpha \boldsymbol{E}_a. \tag{10.26}$$

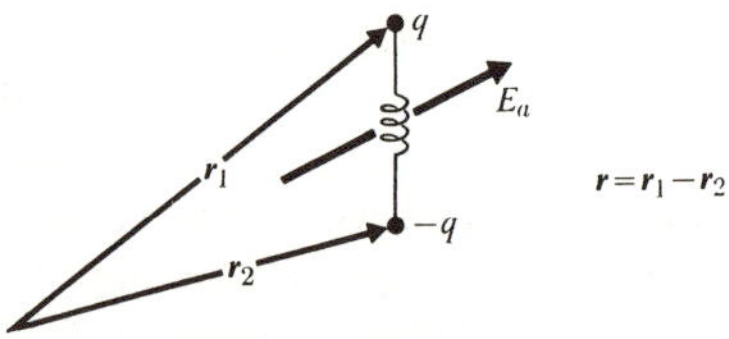

Fig. 10.3 Model of polarizable molecule in external field.

It will sometimes be convenient to think in terms of a quite concrete model of the dipolar molecule. As such a model, we can consider two charges, q and $-q$, that can be separated against spring-like forces as shown in Fig. 10.3. The results will be independent of the model, but the model will help clarify the discussion. The energy of the dipole in the field $\boldsymbol{E}_a$ is the work required to bring the molecule from its unpolarized state at infinity to some polarized and oriented state in the field. This work is done on the two charges by the forces that act on them: the applied field $\boldsymbol{E}_a$ which acts on each of the two charges; and the spring-like force which also acts on the charges, but only as a function of their *relative displacement*. Consider first the work done by the applied field. The work done on one of the charges is just the magnitude of the charge times the potential ϕ_a of the field $\boldsymbol{E}_a$ at that charge. The work W_a done on both charges is then $q\phi_a(\boldsymbol{r}_1) - q\phi_a(\boldsymbol{r}_2)$. If the relative displacement $\boldsymbol{r} = \boldsymbol{r}_1 - \boldsymbol{r}_2$ is small, this work is approximately $q\boldsymbol{r} \cdot \nabla\phi_a$. Now $q\boldsymbol{r}$ is the dipole moment $\boldsymbol{p}$ of the charge pair, and $\nabla\phi_a$ is the negative of the electric field, so that we have

$$W_a = -\boldsymbol{p} \cdot \boldsymbol{E}_a. \tag{10.27}$$

This result is, of course, identical with the general result in (10.11).

Now we calculate the work done by the spring-like force $\mathscr{F}$ that acts between the two charges. Consider the process whereby the molecule is polarized adiabatically (very slowly). Then $\mathscr{F}$ must be such that it just balances the external force that $\boldsymbol{E}_a$ exerts on each charge, i.e., $\mathscr{F} = q\boldsymbol{E}_a$. To calculate the work W_s done by $\mathscr{F}$, first consider a small displacement $d\boldsymbol{r}_1$ and $d\boldsymbol{r}_2$ of the charges. Since $\mathscr{F}$ acts equally and oppositely on the two charges, the work dW_s in this displacement is

$$dW_s = \mathscr{F} \cdot (d\boldsymbol{r}_1 - d\boldsymbol{r}_2) = \frac{\mathscr{F}}{q} \cdot (q\, d\boldsymbol{r}) = \frac{\mathscr{F}}{q} \cdot d\boldsymbol{p}.$$

With the assumption (10.26) and with $\mathscr{F}/q = \boldsymbol{E}_a$, we can write dW_s as

$$dW_s = \alpha \boldsymbol{E}_a \cdot d\boldsymbol{E}_a = \frac{\alpha}{2} d|\boldsymbol{E}_a|^2.$$

Integrating this equation, we find that the *internal*, or *springlike*, energy, W_s, is

$$W_s = \frac{\alpha}{2} \boldsymbol{E}_a^2 = \tfrac{1}{2}\boldsymbol{p} \cdot \boldsymbol{E}_a. \tag{10.28}$$

The total energy W of the dipole in the external field is therefore

$$W = W_a + W_s.$$

Consider now the energy of a *medium* of polarizable dipoles in an external field. This energy is the sum of the energies of the individual dipoles in that field, plus the sum of the mutual potential energies of the dipoles, that is, their energies in the fields of each other.

In detail then, the energy of formation of the dipoles (the sum of the spring-like energies) is obtained from Eq. (10.28) by summing over all dipoles. However, we must remember that the field on the ith dipole is $\boldsymbol{E}_a$, *plus* the field at i due to all the other dipoles. Call this latter field $\boldsymbol{E}_d(i)$. Then the energy W_S which represents the internal or spring-like energy of the medium is

$$W_S = \frac{1}{2}\sum_i \boldsymbol{p}_i \cdot (\boldsymbol{E}_a(i) + \boldsymbol{E}_d(i)). \tag{10.29}$$

To this we must add the electrostatic energy of the dipoles due to the applied fields and to their mutual fields. The energy W_A which is due to the applied field is a sum of terms like (10.27);

$$W_A = -\sum_i \boldsymbol{p}_i \cdot \boldsymbol{E}_a. \tag{10.30}$$

Next, the *mutual* energy of any *two* dipoles, say the ith and jth, can be thought of in one of three ways: as the energy of the ith dipole in the field $\boldsymbol{E}_d^j(i)$ of the jth, which we find from (10.27) to be $-\boldsymbol{p}_i \cdot \boldsymbol{E}_d^j(i)$; as the energy of the jth dipole in the field of the ith, which is $-\boldsymbol{p}_j \cdot \boldsymbol{E}_d^i(j)$; or as one-half the *sum* of these two energies. From this last point of view, the mutual *interaction* energy W_I of the assemblage is

$$W_I = -\frac{1}{2}\sum_i {\sum_j}' \boldsymbol{p}_i \cdot \boldsymbol{E}_d^j(i), \tag{10.31}$$

where the prime on the sum means that terms with $i = j$ are excluded. Now the previously defined field $\boldsymbol{E}_d(i)$, which is the field at the ith dipole due to all other dipoles, is represented by the sum over j in Eq. (10.31),

$$\boldsymbol{E}_d(i) = \sum_j \boldsymbol{E}_d^j(i).$$

Thus Eq. (10.31) can be written

$$W_I = -\frac{1}{2}\sum_i \boldsymbol{p}_i \cdot \boldsymbol{E}_d(i). \tag{10.32}$$

The total energy W_P of the polarized medium is the sum of the three terms W_S, W_A, and W_I. From Eqs. (10.29), (10.30), and (10.32), we find that

$$W_P = -\frac{1}{2}\sum_i \boldsymbol{p}_i \cdot \boldsymbol{E}_a(i). \tag{10.33}$$

The sum can be replaced by an integral under the same conditions we have discussed before and, hence, will not elaborate on now. To the approximation involved in this replacement, the energy of the polarized medium in the external field $\boldsymbol{E}_a$ is

$$W_P = -\frac{1}{2}\int \boldsymbol{P}\cdot\boldsymbol{E}_a\,dv. \tag{10.34}$$

This result can be interpreted as if there were an energy *density* w_P given by

$$w_P = -\tfrac{1}{2}\boldsymbol{P}\cdot\boldsymbol{E}_a.$$

Equation (10.34) is usually derived by beginning with the expression

$$W = \frac{1}{8\pi}\int \boldsymbol{E}\cdot\boldsymbol{D}\,dv$$

for the "energy" of a polarized medium. The microscopic derivation above bypasses this usual basis and shows once more the superfluity of the vector $\boldsymbol{D}$.

10.7 PERMEABLE MATTER IN APPLIED MAGNETIC FIELD

A next natural subject is the question of the force on, and energy of, a permeable body in an applied field. We shall, however, be able to discuss only the problem of force. The question of the energy of a permeable medium involves assembling an ensemble of currents, a special kind of dipolar currents to be sure, but currents nonetheless. As we remarked in Section 10.2, the process of assemblage introduces forces (Faraday's law of induction) that we have not discussed. We defer the question of energy then to Section 11.3. However, we can discuss the problem of force now.

We found in Section 8.5 that the external field of dipolar magnetic matter can be calculated either from a set of equivalent magnetic charges or a set of equivalent currents. From the above result for dielectrics, we might then expect that the force on such a body in an external field could be calculated *as if* the external field acted on the equivalent charges or *as if* it acted on the equivalent currents. We shall see that both of these expectations are correct.

Without any calculation, it is obvious that the force can be calculated from the equivalent charges, by comparison with the electrostatic case. This is because the force on a sample of permeable matter is the sum of the forces on the individual dipoles and the force on *one* of the dipoles has the same form as that for the electrostatic case, with $\boldsymbol{p}$ replaced by $\boldsymbol{m}$, and $\boldsymbol{E}_a$ replaced by $\boldsymbol{B}_a$. The arguments for replacing the discrete distribution by a continuous one can be taken over directly to the magnetic case. The force can then be calculated, to the approximation we hâve discussed, as that on the continuous distribution of magnetization $\boldsymbol{M}(\boldsymbol{r})$, by means of the formula

$$\boldsymbol{F} = \int (\boldsymbol{M} \cdot \nabla)\boldsymbol{B}_a(\boldsymbol{r})\, dv. \tag{10.35}$$

This integral expression can now be transformed much as was its electrostatic counterpart to find, analogously to Eq. (10.25),

$$\boldsymbol{F} = \int_S \sigma_m \boldsymbol{B}_a\, ds + \int_V \rho_m \boldsymbol{B}_a\, dv, \tag{10.36}$$

where $\sigma_m = \boldsymbol{M} \cdot \boldsymbol{n}$ and $\rho_m = -\nabla \cdot \boldsymbol{M}$ are the densities of equivalent surface and volume "magnetic charges."

The formula (10.35) can also be put into a form which can be interpreted as if the external field acts on the equivalent surface and volume currents discussed in Section 8.5. To show this, we quote a theorem from vector analysis. If $\boldsymbol{C}$ and $\boldsymbol{D}$ are two vector functions, and if V is a volume bounded by a closed surface S, with normal $\boldsymbol{n}$, then

$$-\int_S (\boldsymbol{n} \times \boldsymbol{C}) \times \boldsymbol{D}\, ds + \int_V (\nabla \times \boldsymbol{C}) \times \boldsymbol{D}\, dv$$

$$= \int_V \left[(\boldsymbol{C} \cdot \nabla)\boldsymbol{D} + \boldsymbol{C} \times (\nabla \times \boldsymbol{D}) - \boldsymbol{C}\nabla \cdot \boldsymbol{D} \right] dv.$$

We apply this to (10.35) with $\boldsymbol{C}$ replaced by $\boldsymbol{M}$ and $\boldsymbol{D}$ by $\boldsymbol{B}_a$. Now $\nabla \cdot \boldsymbol{B}_a = 0$ everywhere and $\nabla \times \boldsymbol{B}_a = 0$ inside the volume since there are no sources of $\boldsymbol{B}_a$ there. Then

$$\boldsymbol{F} \equiv \int_V (\boldsymbol{M} \cdot \nabla)\boldsymbol{B}_a\, dv = -\int_S (\boldsymbol{n} \times \boldsymbol{M}) \times \boldsymbol{B}_a\, ds + \int_V (\nabla \times \boldsymbol{M}) \times \boldsymbol{B}_a\, dv. \tag{10.37}$$

On comparing this with Eqs. (8.38) and (8.39) and with the force law (7.7), we see that it is as if the field $\boldsymbol{B}_a$ acted on the equivalent surface current $\boldsymbol{J}_s = c(\boldsymbol{M} \times \boldsymbol{n})$ and volume current $\boldsymbol{J}_v = c(\nabla \times \boldsymbol{M})$.

PROBLEMS

1. Prove Earnshaw's theorem: A charged particle in a static electric field cannot be in stable equilibrium under the action of only that field.
2. A charge q is a distance a from the center of a dielectric sphere of radius R, and dielectric constant ε. Show that the force F between the charge and the sphere is attractive and of magnitude

$$F = (\varepsilon - 1)q^2 \sum_{l=1}^{\infty} \frac{l(l+1)R^{2l+1}}{(l\varepsilon + l + 1)a^{2l+3}}.$$

3. Show that the dipole–dipole interaction energy (10.15) can be written as

$$W_{12} = -\frac{p_1 p_2}{r^3}(2\cos\theta_1\cos\theta_2 - \sin\theta_1\sin\theta_2\cos\varphi),$$

where θ_1 and θ_2 are the angles between the two dipoles and the vector $\boldsymbol{r}$ joining them, and φ is the angle between the planes formed by $\boldsymbol{p}_1$ and $\boldsymbol{r}$ and by $\boldsymbol{p}_2$ and $\boldsymbol{r}$.

4. Prove Thomson's theorem: Charge placed on a system of conductors will so distribute itself that the energy of the resultant electrostatic field is a minimum.

5. The half-space $z < 0$ is filled with matter of dielectric constant ε. An electric dipole of moment $\boldsymbol{p}$ is at $z = d$, and points toward the dielectric face. Find the force on the dipole.

6. A spherical volume of radius a contains a uniform distribution of rigidly fixed charge. The volume is cut into two hemispheres which are then separated along the diameter perpendicular to their plane faces, which remain parallel. Assuming that d is the distance between these faces, find the force between the hemispheres in an expansion in powers of a/d.

7. A point charge is a distance d from the plane face of an infinite perfect conductor. Use Eq. (10.23) to calculate the force on the conductor.

8. Two square loops of side a each carrying current I are placed with their edges parallel, and perpendicular to the line joining the center of the loops. The distance between centers along this line is d. Show that for $d \gg a$, and the currents in the same direction, the force between the loops is an attraction of magnitude $(6I^2/c^2)(a/d)^4$. This expression can be considered to be the first term in a power series in a/d. Find the next term.

9. A hemisphere of radius a is uniformly filled with charge of density ρ_0. Show that the energy of assemblage W_E is

$$W_E = \frac{4\pi^2\rho_0^2 a^5}{5}\left(\frac{1}{3} + \frac{1}{2}\sum_{n=1}^{\infty}\frac{(-\frac{1}{2})_n(-\frac{1}{2})_n}{2_n n!}\right),$$

where $b_n = b(b+1)\cdots(b+n-1)$. See B. C. Carlson and J. P. Morley, "Multipole Expansion of Coulomb Energy," *Am. J. Phys.* 30, 209 (1962).

10. A uniformly charged ring of radius a, total charge Q, is in the plane $z = 0$ with center at the origin. A dielectric sphere of radius b, dielectric constant ε, has its center at $z = c$ on the z-axis. Find the force on the sphere.

11. A bar magnet is external to a permeable sphere as in the problem of Section 8.5. Find the force on the sphere.

12. Show that the torque $\boldsymbol{T}$ on a dipole of moment $\boldsymbol{p}$ in an external field $\boldsymbol{E}_a$ is $\boldsymbol{T} = \boldsymbol{p} \times \boldsymbol{E}_a$.

13. Derive Eq. (10.22), $\boldsymbol{F} = (\boldsymbol{m}\cdot\nabla)\boldsymbol{B}_a$, for the force on a magnetic dipole in an external field $\boldsymbol{B}_a$ from the general formula

$$\boldsymbol{F} = \frac{1}{c}\int \boldsymbol{J}(\boldsymbol{r}') \times \boldsymbol{B}_a\, dv'.$$

REFERENCES

Böttcher, C. J. F., *Theory of Electric Polarization*, Elsevier, Amsterdam (1952).

This book has the standard derivations of the formulas for electric and magnetic energies in terms of the formalism involving ***D*** and ***H***.

Brown, W. F. Jr., *Micromagnetics,* New York, Wiley (Interscience) (1963).

Brown, W. F. Jr., *Magnetoelastic Interactions*, Springer, Berlin (1966).

The frequently confused problem of the energy of electric and magnetic dipolar matter has been greatly clarified by Professor W. F. Brown, Jr. in a considerable series of publications of which these two are examples. They contain sets of references to earlier works.

Landau, D. L., and Lifshitz, E. M., *Electrodynamics of Continuous Media*, Addison-Wesley, Reading, Mass. (1960).

Another standard work that discusses force and energy in electric and magnetic dipolar matter.

11
TIME-VARYING FIELDS

11.1 INTRODUCTION

Electrostatics deals with purely static charge distributions, that is, of charges that are (on the average) motionless. The electric fields they produce are, of course, independent of time. Magnetostatics deals with the fields of charges in motion —that is, of currents, but of *steady* or *stationary* currents whose magnitude at any point in space is *independent* of time. The magnetic fields of such currents are then also *independent* of time. Now however, we want to calculate the effect of charges that move in an arbitrary manner, and hence of fields which may change with time.

In free space—in the presence of only charge or current sources, but not of matter—there are four basic equations for the static fields: two equations define the divergence and curl of $\boldsymbol{E}$ and two others do similarly for $\boldsymbol{B}$. For the present, more general case, the two curl equations will be supplemented by time dependent terms, one of which was found experimentally by Michael Faraday, and the other of which was postulated on theoretical grounds by James C. Maxwell. The four (supplemented) equations are *Maxwell's equations* for free space.

One point is more delicate for the present case than for statics: the identification of the frames of reference in which $\boldsymbol{E}$ and $\boldsymbol{B}$ are defined. Although we have not stressed the fact, there is, of course, a frame of reference implicitly associated with the definition of any electric field. For the field is defined by the force on a test charge, the force is defined by an acceleration, and the acceleration is defined with respect to a specific frame. A similar remark applies to the definition of a magnetic field. In statics, however, the question of choosing a frame almost answers itself. Given a configuration of charges with fixed relative positions in electrostatics, one chooses a frame at rest with respect to the configuration, so that the field is indeed a static one. Or given a system of currents, one can choose a frame in magnetostatics in which they constitute a *stationary* system. In the present time-varying case, neither of these criteria for choice necessarily applies.

Although all the quantities in Maxwell's equations must refer to a well-defined frame of reference, this does not mean that there is only one such possible frame. There is an infinite set of alternate frames, the so-called *inertial frames*, in which the equations are valid. The relation between quantities in one such frame and those in another is a subject for the special theory of relativity, treated in Chapter 12, where the concept of inertial frames is discussed more fully. In the present chapter, we shall simply presuppose a single inertial frame with respect to which fields, charges, and currents are defined.

11.2 FARADAY'S LAW OF INDUCTION

The first generalization to time-varying fields is embodied in the *law of induction.* This celebrated law was deduced by Faraday as a consequence of a long series of experiments that we shall not describe here in detail. Qualitatively, the law relates the change with time of a magnetic field to the production of an electric field. More specifically, Faraday found that if the magnetic flux through a loop or coil of wire changed, either by moving a magnet in the vicinity or by changing the size of the loop in a fixed magnetic field, or both, an electric field was created in the loop; the field manifested itself by producing a current. Quantitatively, he found that the line integral of electric field around the loop was proportional to the time rate of change of the magnetic flux through the loop. If, for example, we consider a loop of fixed shape, his results are summarized by

$$\oint \boldsymbol{E} \cdot d\boldsymbol{l} \propto \int \frac{\partial \boldsymbol{B}}{\partial t} \cdot d\boldsymbol{s}. \tag{11.1}$$

All empirical evidence shows that this relation between a changing magnetic field and an electric field is a property of the fields in space: the wire, and the current flowing in it, simply serve as a practical way to make the relation (11.1) manifest.

If Eq. (11.1) is written with a constant of proportionality, there are two noteworthy facts about the constant, which is easily verified to have the dimension of reciprocal velocity in Gaussian units. First, it is found empirically that this reciprocal velocity is $1/c$, where c is the velocity of light. This result is perhaps not startling to the reader of this book, who undoubtedly knows that light and electromagnetism are connected. But when this fact was found empirically, in the middle of the nineteenth century, by the very precise experiments of Wilhelm Weber and Rudolf Kohlrausch, it evoked the utmost excitement. We must remember that at the beginning of Faraday's researches there was no suspicion that electric induction had any connection with a theory of light. In fact, it had only been a short time since the three previously separated subjects of electricity, magnetism, and optics had been partially united by the discoveries of Oersted that "electricity" in motion produces a magnetic field. But, given even this partial union, electricity and magnetism on the one hand and optical phenomena on the other were considered to be quite separate. The unexpected appearance of the velocity of light in the law of induction seemed clearly to be more than a remarkable numerical coincidence, but betokened something deeper. It was left to Maxwell to uncover this deeper significance in a way that will be discussed later.

The second point about the proportionality constant is that it contains a negative sign. This embodies *Lenz's law*: the change of flux through the loop induces an electromotive force* and hence, a current, which is such that the additional flux produced by the induced current opposes the original flux. Faraday's law in its integral form is then

$$\oint \boldsymbol{E} \cdot d\boldsymbol{l} = -\frac{1}{c}\int \frac{\partial \boldsymbol{B}}{\partial t} \cdot d\boldsymbol{s}. \tag{11.2}$$

The law can be expressed in differential form by using Stokes' theorem to rewrite the left-hand side. Then, either by equating the integrands or by imagining Eq. (11.2) applied to an infinitesimal loop, we find

$$\nabla \times \boldsymbol{E} = -\frac{1}{c}\frac{\partial \boldsymbol{B}}{\partial t} \tag{11.3}$$

which is the generalization of the static equation $\nabla \times E = 0$.

It is worth noting that Faraday's law is frequently taken to be a more general one that can also be applied in calculating the emf in loops of *changing* shape. The effects of changing shape are, it can be shown, included by taking a *total* derivative on the right-hand side of (11.2). This alternate version of Faraday's law, also known as the "flux rule," is then with ϕ the magnetic flux

$$\text{emf} = -\frac{1}{c}\frac{d}{dt}\int \boldsymbol{B} \cdot d\boldsymbol{s} = -\frac{1}{c}\frac{d\phi}{dt}. \tag{11.4}$$

A more detailed discussion of the flux rule is given in Section 11.5.

11.3 ENERGY OF STATIONARY CURRENTS AND OF PERMEABLE MATTER

We can now return to the two questions whose answers are dependent on knowledge of Faraday's law. These are the problems of calculating the energy required to assemble a stationary distribution of currents, and of calculating the energy of a sample of permeable matter in an external field.

First, given a stationary current distribution $\boldsymbol{J}(\boldsymbol{r})$, we seek the energy of assemblage, or, in other words, the amount of energy required to bring it into being. It is useful here to recall the calculation of the energy of assemblage of a charge distribution. This energy was found, in effect, by assuming that the distribution was broken up into a large number N of small charge elements and calculating the energy required to assemble these elements into the distribution. By taking N large enough, the energy of formation of the elements themselves could be neglected (as

* The integral on the left of (11.1), $\oint \boldsymbol{E} \cdot d\boldsymbol{l}$, is frequently called the *electromotive force* or *emf*. Since the dimensions of the integral are not those of force there are those who object to this terminology and prefer to call it the *electromotance* or, in the MKS system, the *voltage*. We shall, on the few occasions that we name it, use the conventional, if illogical, term *emf*.

proportional to N) by comparison with the energy comprised in the $N(N-1)/2$ mutual interactions. Similarly, given a stationary current distribution, we can imagine that the distribution is broken into N current filaments, and that these filaments are dispersed to infinity. The energy of assemblage is then the energy required to assemble the filaments subject to their $N(N-1)/2$ mutual interactions.

An obvious starting point is to calculate the energy required to bring a pair of filaments, or current loops, into proximity from initial, widely-separated positions. Consider such a pair with currents I_1 and I_2 in the two loops. We suppose that the loops are brought together in such a way that these currents remain *constant*. There are then three forces that do work to be considered in calculating the energy. First, there is what might be called the Biot–Savart force. The filaments are brought to a certain position relative to one another; we can then consider that one of the loops (either one) is fixed and that it generates a magnetic field that acts on, or is, so to speak, an applied field for, the other. Equation (10.16) for the energy of a rigid current distribution in an applied field, can then be used. For example, if we assume that loop 2 provides the applied field in which loop 1 moves, then we find that the work W_{BS} done against the Biot–Savart force is

$$W_{BS} = -\frac{I_1}{c}\phi_1, \tag{11.5}$$

where ϕ_1 is the flux through loop 1 in its final position due to the field of loop 2. We assume here that loop 2 produces no flux through loop 1 when they are widely separated. However, it can equally well be assumed that loop 1 is fixed and provides the applied field for loop 2 in which case the energy W_{BS} is

$$W_{BS} = -\frac{I_2}{c}\phi_2. \tag{11.6}$$

We can use either (11.5) or (11.6), but it is convenient to take W_{BS} as half the sum of these two

$$W_{BS} = -\frac{1}{2c}(I_1\phi_1 + I_2\phi_2). \tag{11.7}$$

Equation (11.7) is, however, not the whole story. In addition we must take into account the forces that are generated according to Faraday's law. For as loop 2 with its field moves closer to 1, the flux through 1 will change and by Faraday's law there will be an emf in loop 1 that will tend to change the current. To keep the current constant, we must then apply a *counter* emf of the same magnitude but opposite sign, say, by means of a battery. The energy supplied by this *counter* emf must be included as part of the energy required to bring the loops together. For either of the loops, let dW_F/dt be the rate at which work W_F (subscript F for Faraday) is done by this counter emf. Then, with $\boldsymbol{E}$ the field due to the changing $\boldsymbol{B}$,

$$\frac{dW_F}{dt} = I\int(-\boldsymbol{E}).\,d\boldsymbol{l} = \frac{I}{c}\frac{d\phi}{dt}.$$

We integrate this equation to find the general expression for the work ΔW_F done by the battery when the flux changes by $\Delta\phi$:

$$\Delta W_F = \frac{I}{c}\Delta\phi. \tag{11.8}$$

Then the amount of work done by the battery attached to 1 in keeping its current constant is

$$W_{F1} = \frac{I_1}{c}\phi_1$$

with a similar result for W_{F2}. Adding W_{BS}, W_{F1}, and W_{F2}, we conclude that the energy W_2 required to bring the pair of loops together is

$$W_2 = \frac{1}{2c}(I_1\phi_1 + I_2\phi_2). \tag{11.9}$$

We can now return to the original problem of finding the energy W_N required to assemble the N loops that are considered to make up a general current distribution. Generalizing (11.9), we have

$$W_N = \frac{1}{2c}\sum_{i=1}^{N} I_i\phi_i. \tag{11.10}$$

This equation can be rewritten in terms of the vector potential. If S_i is a surface spanning the ith filament and if $d\boldsymbol{l}_i$ is an element of length along that filament, then

$$\phi_i = \int_{S_i} \boldsymbol{B}\cdot d\boldsymbol{s}_i = \oint \boldsymbol{A}\cdot d\boldsymbol{l}_i. \tag{11.11}$$

Let $d\sigma_i$ be the cross-sectional area of the filament. Then

$$\boldsymbol{J}\,d\sigma_i\,dl_i = I_i\,d\boldsymbol{l}_i. \tag{11.12}$$

Putting (11.11) and (11.12) into (11.10) and changing the sum to an integral, we have for W_M, the energy of assemblage of a current distribution,

$$W_M = \frac{1}{2c}\int \boldsymbol{A}\cdot\boldsymbol{J}\,dv \tag{11.13}$$

which is the formula previously quoted as (10.7).

Next we consider the problem of the energy required to establish a sample of permeable matter in a fixed applied field. This problem is very similar to that of calculating the energy of a sample of dielectric matter in an applied electric field, but there are two differences. First, the concept of a fixed applied magnetic field is not so simple as that of a fixed applied electric field. Second, the magnetic dipoles of permeable matter are rather different from the electric dipoles we have discussed. We consider these differences now; once they are understood, it is relatively easy to modify the discussion for dielectric matter to apply to the present case.

The first difference is this: For dielectric matter a *fixed* applied field can be imagined to be one for which the source charges are held immobile by some kind of fanciful charge glue. There can then be no motion of these charges, and hence no work can be done on them by the electric fields of dipolar matter. The sources of magnetic field are, however, charges in motion, that is, currents. Elements of magnetic dipolar matter that are being brought near these currents create a changing magnetic field which induces an electric field that *necessarily* does work on them. To keep the currents constant, that is, to deal with a *fixed* applied field, a counter emf must be applied, and the work done by this emf must be included as part of the energy of assemblage of the permeable sample.

The difference between electric and magnetic dipoles manifests itself in several ways. For magnetic matter there are two kinds of microscopic dipoles, *diamagnetic* and *paramagnetic*, and the analogs of the spring-like internal restoring forces that act in electric dipoles are of a different nature for each of these two kinds. Also, the moment $\boldsymbol{m}_i$ that can be associated with the ith dipole is, for paramagnets, not an intrinsic one, but a mean or statistical one. It is necessary to understand the nature of the internal restoring forces, since part of the energy of dipolar matter is comprised in the energy of formation of the dipoles which represents the work done against these internal forces.

We now discuss the calculation of this work, or energy, for magnetic dipoles. As we have remarked, for a paramagnetic dipole, $\boldsymbol{m}_i$ is an average moment that is achieved only in opposition to the thermal motion, that is, to the randomizing effect of the interaction and collisions of the dipoles with one another. This thermal motion, (or more precisely, the momentum transferred per unit time due to it) acts like a restoring force, against which work must be done to establish the moment. For diamagnetic dipoles, the concept of internal forces is even more artificial since diamagnetism is essentially quantum mechanical. Nonetheless it is clear, if we refer to the calculation by means of Eq. (10.28) of the energy of formation of electric dipoles, that this energy is independent of the nature of the internal forces and is determined by the assumption that the induced dipole moment is proportional to the inducing field: $\boldsymbol{p} = \alpha\boldsymbol{E}$. Given this assumption, the energy of formation of a single dipole is $\frac{1}{2}\boldsymbol{p}\cdot\boldsymbol{E}$. We make the analogous assumption for magnetic dipoles, that the moment $\boldsymbol{m}$ is proportional to the inducing field $\boldsymbol{m} = \beta\boldsymbol{B}$, and assume the analogous result, that the energy of formation of the dipole is $\frac{1}{2}\boldsymbol{m}\cdot\boldsymbol{B}$.

We return now to the problem of the energy of a sample of permeable matter in an applied field. We can consider this energy as the sum of W_{FF}, the energy required to assemble the matter in an *assumed fixed field*, and W_S, the energy that must be supplied to the *sources* to keep the field fixed, in fact. The calculation of W_{FF} can be thrown back on the results for dielectric matter. The calculation of W_S requires new considerations.

Consider first the calculation of W_{FF}. An expression for this quantity can be written immediately by analogy with the dielectric case since there is the straightforward connection pointed out above between energies of formation (and also between energies of orientation) of electric and magnetic dipoles: Simply replace $\boldsymbol{p}$ by $\boldsymbol{m}$ and $\boldsymbol{E}$ by $\boldsymbol{B}$ in the various energy formulas. For permeable matter in an assumed fixed field $\boldsymbol{B}_a$ we have then, from Eq. (10.33) for dielectric matter,

$$W_{FF} = -\frac{1}{2}\sum_i \boldsymbol{m}_i \cdot \boldsymbol{B}_a. \tag{11.14}$$

Consider next the calculation of W_S. We can think of the source of the applied field $\boldsymbol{B}_a$ as a large current loop a that carries current I_a. As dipoles are brought into the proximity of the source loop a in the process of assembling a sample of permeable matter, the flux that these dipoles produce in a will change, generating an emf. We want to calculate the work that is done by the battery that produces the balancing, or counter, emf. We can begin by first considering that a *single* dipole is brought in the neighborhood of the loop. Since the work done on the source loop is obviously independent of the kind of dipole that produces the flux, it will be convenient to think of this dipole itself as a current loop d, carrying current I_d and producing a field $\boldsymbol{B}_d$. Now the general formula (11.8) gives the work W_c done in keeping the sources of a magnetic field constant. Applied to the present case, it is

$$W_c = \frac{I_a}{c}\int_a \boldsymbol{B}_d \cdot ds.$$

It follows from the equality of (11.5) and (11.6), that

$$I_a \int_a \boldsymbol{B}_d \cdot ds = I_d \int_d \boldsymbol{B}_a \cdot ds. \tag{11.15}$$

Given Eq. (11.15), W_c can be rewritten as

$$W_c = \frac{I_d}{c}\int_d \boldsymbol{B}_a \cdot ds.$$

The integral in this equation can be written as $(I_d \boldsymbol{B}_a/c) \cdot \int \boldsymbol{n}\, ds$ where $\boldsymbol{n}$ is the vector normal to the surface spanning the loop. From the definition (8.13) of the magnetic moment $\boldsymbol{m}$ of such a loop, we have

$$W_c = \boldsymbol{m} \cdot \boldsymbol{B}_a. \tag{11.16}$$

Equation (11.16) refers to a single dipole. For an assemblage constituting permeable matter, the work W_S done on the source is then

$$W_S = \sum_i \boldsymbol{m}_i \cdot \boldsymbol{B}_a.$$

The total energy required to assemble a sample of permeable matter is then $W = W_{FF} + W_S$ or

$$W = \tfrac{1}{2}\sum_i \boldsymbol{m}_i \cdot \boldsymbol{B}_a.$$

Replacing the sum by an integral with the usual justification, we conclude that the energy of a sample of permeable matter with magnetization density $\boldsymbol{M}(\boldsymbol{r})$ in an applied field $\boldsymbol{B}_a$ is

$$W = \tfrac{1}{2} \int \boldsymbol{M} \cdot \boldsymbol{B}_a \, dv \tag{11.17}$$

which is a standard result.

11.4 DISPLACEMENT CURRENT AND THE MAXWELL EQUATIONS

We discuss now a famous generalization due to Maxwell, of the equations we have considered. This generalization is essentially a postulate, but it is perhaps better understood if we look at it in the context of the physics of Maxwell's time. To this end, consider first the magnetostatic equation $\nabla \times \boldsymbol{B} = 4\pi \boldsymbol{J}/c$. Taking the divergence of this equation, we find, since the divergence of any curl is zero, that

$$\nabla \cdot \boldsymbol{J} = 0. \tag{11.18}$$

Now the fact that the divergence of a vector field is zero means that the field lines close on themselves. For example, the equation $\nabla \cdot \boldsymbol{B} = 0$ corresponds to the fact that the lines of B form closed loops. Equation (11.18) then implies that current flow must be in closed paths.

Current does flow in closed loops when static fields are being considered, but not when fields vary in time. If, for example, a condenser is being charged by connecting the two plates to a battery with a wire, we would say that there is a current in the wire, but no current in the space between the plates of the condenser. However, Maxwell's conception of this process was somewhat different. To understand his viewpoint, we recall that Faraday had discovered, and of course Maxwell knew about, the phenomenon of polarization, or the separation of charges, in a dielectric in an electric field. Charges must move to separate, and since charges in motion constitute a current, the process of polarization thus implies the existence of a current, which can be called a *polarization current.* We must also remember that it was considered that "electric actions" were transmitted not through empty space, but through an essentially material ether, which was not different except in the quantitative values of the parameters that described it, from, say, dielectric matter. Since polarization currents were known to exist in ordinary matter, it was not unnatural for Maxwell to assume that similar currents could exist in the ether. Maxwell considered that, in the charging of a condenser, the wires of the condenser and the ether between the plates formed a *continuous material circuit*, in all parts of which current could exist. The possibility then existed of salvaging, in the following way, what appears to have been considered the basic equation (11.18). From $\nabla \cdot \boldsymbol{E} = 4\pi\rho$ we have

$$\frac{\partial \rho}{\partial t} = \frac{1}{4\pi} \frac{\partial}{\partial t} \nabla \cdot \boldsymbol{E},$$

and on putting this into the continuity equation $\nabla \cdot \boldsymbol{J} + \partial\rho/\partial t = 0$, we get

$$\nabla \cdot \left(\boldsymbol{J} + \frac{1}{4\pi}\frac{\partial \boldsymbol{E}}{\partial t}\right) = 0. \tag{11.19}$$

The second term, $(1/4\pi)(\partial \boldsymbol{E}/\partial t)$, has the proper dimensions to be considered a current density; it was called by Maxwell the *displacement current.* Now Eq. (11.19) is of the form of (11.18). If then we consider that the total current consists of the true current *plus* the displacement current, we see that the total current always flows in closed loops. However, we should remember that the displacement current is not a current in the ordinary sense, and does not correspond to a flow of charge.

Whatever the cogency of the above discussion may appear to be today, Maxwell was sufficiently inspired by the equivalence of the displacement current and ordinary current in Eq. (11.19) to generalize the differential Ampere's law, $\nabla \times \boldsymbol{B} = 4\pi \boldsymbol{J}/c$, in a similar way. He assumed that for time-varying fields the current on the right-hand side of this equation for $\nabla \times \boldsymbol{B}$ should be the *total* current, including the displacement current. His generalization of this equation then reads

$$\nabla \times \boldsymbol{B} = \frac{4\pi \boldsymbol{J}}{c} + \frac{1}{c}\frac{\partial \boldsymbol{E}}{\partial t}. \tag{11.20}$$

In the last analysis, Eq. (11.20) is a new postulate, but it has been thoroughly confirmed, as we shall see. Among other things, the addition of the displacement current to $\boldsymbol{J}$ has the prediction of electromagnetic waves for a corollary.

Equation (11.20), plus the two divergence equations, plus the differential expression of Faraday's law, are frequently called *Maxwell's equations.* We assemble them here for reference.

$$\text{a) } \nabla \cdot \boldsymbol{E} = 4\pi\rho \qquad \text{b) } \nabla \cdot \boldsymbol{B} = 0$$

$$\text{c) } \nabla \times \boldsymbol{E} = -\frac{1}{c}\frac{\partial \boldsymbol{B}}{\partial t} \qquad \text{d) } \nabla \times \boldsymbol{B} = \frac{4\pi \boldsymbol{J}}{c} + \frac{1}{c}\frac{\partial \boldsymbol{E}}{\partial t} \tag{11.21}$$

Equations (11.21) might more properly be called the *microscopic* Maxwell equations, as opposed to the original Maxwell equations, which were written in terms of $\boldsymbol{D}$ and $\boldsymbol{H}$, as well as $\boldsymbol{E}$ and $\boldsymbol{B}$, and which purported to hold in the presence of matter.

11.5 THE LORENTZ FORCE DENSITY

Maxwell's equations permit, in principle at least, the calculation of the fields $\boldsymbol{E}$ and $\boldsymbol{B}$ from arbitrary sources. Since these fields are important mainly because of their action on charges, the foundations of electromagnetic theory are completed by a prescription for this action that is given by the *Lorentz force density* $\boldsymbol{f}$. Given a charge distribution of density $\rho(\boldsymbol{r})$, moving with velocity $\boldsymbol{v}(\boldsymbol{r})$ with respect to an inertial frame in which there are fields $\boldsymbol{E}$ and $\boldsymbol{B}$, the force density $\boldsymbol{f}$ (force/unit volume) acting on the distribution is given by the Lorentz formula

$$\boldsymbol{f} = \rho\left(\boldsymbol{E} + \frac{\boldsymbol{v} \times \boldsymbol{B}}{c}\right). \tag{11.22}$$

Equation (11.22) is a postulate, but it is illuminating to see its origin. The first term in it essentially extends the definition of $\boldsymbol{E}$, as the force exerted on a unit charge, to time-varying fields. The second term is the essence of the postulate; it generalizes the magnetostatic results on the force between two loops of stationary currents. As we have seen, this force can be calculated as if one filament produces a field $\boldsymbol{B}$ which exerts a force $d\boldsymbol{F}$ on a small current element $I\,d\boldsymbol{l}$ of the second by means of the Biot–Savart law $d\boldsymbol{F} = I\,d\boldsymbol{l} \times \boldsymbol{B}/c$. Lorentz assumed that the current in a wire was due to the motion of individual, microscopic, charged particles (and this was a bolder assumption in his time than in ours). Formally then, from $I = dq/dt$, we have $I\,d\boldsymbol{l} = dq\,d\boldsymbol{l}/dt$. Interpreting $d\boldsymbol{l}/dt$ as the velocity $\boldsymbol{v}$ of the charge dq, the force $d\boldsymbol{F}$ on this charge dq in motion is (from the Biot-Savart law)

$$d\boldsymbol{F} = dq\frac{\boldsymbol{v} \times \boldsymbol{B}}{c},$$

and we see here the origin of (11.22). Although the force-density formula was inspired by the results of experiments on ensembles of charges that constituted stationary currents, it can be considered as confirmed for general distributions of charges in arbitrary motion.

An important application of Eq. (11.22) is to a point charge. This is discussed in some detail in Section 14.9, but here we briefly anticipate some aspects of the results. Suppose that $\rho(\boldsymbol{r})$ in Eq. (11.22) describes a small distribution (point particle) that moves in the presence of an applied field $\boldsymbol{E}_a$ and $\boldsymbol{B}_a$. If the total charge in the distribution is q, one might think that the force on the electron is given by $q[\boldsymbol{E} + (\boldsymbol{v} \times \boldsymbol{B}/c)]$, but this is not quite right. The reason it is not right emerges when one looks more carefully at the concept of a point charge, which must be considered as a limiting case of a small, but finite, distribution. Now for any distribution (small, but finite, ones being no exception), the fields $\boldsymbol{E}$ and $\boldsymbol{B}$ that act on each infinitesimal element of the distribution will be the *total field*, that is, the sum of any applied field and the field (self-field) due to all the other charge elements of the distribution. It turns out that the net or resultant self-field acting on the distribution as a whole does *not* integrate to zero, even in the limit that the distribution shrinks down to a point charge. In short, it must be considered that even for a point charge there is a self-force $\boldsymbol{F}_s$ due to the mutual interaction of the various parts of the distribution, and thus that the total force $\boldsymbol{F}$ is $\boldsymbol{F}_s$ plus the force $\boldsymbol{F}_a$ due to the applied field:

$$\boldsymbol{F} = \boldsymbol{F}_a + \boldsymbol{F}_s.$$

However, for practical purposes, we can often consider the self-force to be small and neglect it, and include only the external forces. In this case, the force on a charged particle can be taken to be approximately:

$$F = q\left(E_a + \frac{v \times B_a}{c}\right), \tag{11.23}$$

a form in which it is often seen.

The generalized Faraday flux law mentioned in Section 11.2 is closely connected with Eq. (11.23). To see this, suppose first that a particle of *unit* charge moves in applied fields E_a and B_a in a closed orbit that thereby defines a curve C. We neglect the self-force on the particle so the *applied* fields are also the *total* fields E and B. Let S be a surface spanning the curve. Taking the curl of Eq. (11.23), forming $\int \nabla \times F \cdot ds$, and using Stokes' theorem, we have

$$\int_C F \cdot dl = -\frac{1}{c}\int_S \frac{\partial B}{\partial t} \cdot ds + \frac{1}{c}\int_C (v \times B) \cdot dl. \tag{11.24}$$

Consider next a purely mathematical theorem that involves a closed curve C_0 whose shape may be a function of time, $C_0 = C_0(t)$, in a time dependent magnetic field $B = B(t)$. The shape $S_0(t)$ of any surface spanning the curve is of course also a function of time. Now the flux $\phi = \int B \cdot ds$ of B through the surface will be time dependent both because B is a function of time, *and* because the shape is changing. If we form $d\phi/dt$, it can be shown that*

$$-\frac{1}{c}\frac{d\phi}{dt} = -\frac{1}{c}\int_{S_0} \frac{\partial B}{\partial t} \cdot ds + \frac{1}{c}\int_{C_0} (v \times B) \cdot dl. \tag{11.25}$$

We see that the right-hand side of Eq. (11.24) is formally identical to the right-hand side of Eq. (11.25).

Consider now a loop of wire containing electrons at rest with respect to it. Suppose that the loop is changing shape, and focus on a segment that moves with velocity v with respect to some inertial frame. If the loop is also in a changing magnetic field B, defined with respect to that *same* frame, there will be a force on the electrons due to the E field produced by the changing B, *and* a force due to the fact that the electrons in the segment have velocity v. The sum of these forces, *per unit charge on the electron* can be identified with F of Eq. (11.24) and the shape of the loop with C. The line integral of F around the loop is given by the right-hand side of Eq. (11.24), which is identical with the right-hand side of Eq. (11.25). Equating the left hand sides of these two equations and *defining* the emf in the wire as

$$\text{emf} = \int_C F \cdot dl,$$

* See Problem 11.8.

we have

$$\text{emf} = -\frac{1}{c}\frac{d\phi}{dt}. \tag{11.26}$$

Equation (11.26) is the *flux rule*, the generalized form of the Faraday law of induction that was mentioned in Section 11.2. It is unexceptionable if the emf is interpreted for what it is, namely, the line integral, at a given instant of time, around a loop having the shape of the wire at that instant, of a force (effectively an electric field) defined in the same inertial frame that defines $\boldsymbol{B}$ and $\boldsymbol{v}$. Equation (11.26) may not, however, be especially useful. The concept of emf in a wire is particularly convenient when it involves the line integral of an electric field defined in a frame *stationary* with respect to the wire. But if the wire is changing shape, there is no single stationary frame with respect to which all parts of the wire are at rest. Because of such reservations we have chosen not to take the flux law as the primary law of induction, but rather have chosen the Faraday law of Eq. (11.2), wherein all quantities refer to a single unambiguous frame. A discussion of various questions connected with the flux law, and particularly those connected with the difficulties associated with the definition of an emf in a moving frame, is given by Scanlon (R).

11.6 AUXILIARY POTENTIALS AND GAUGE TRANSFORMATIONS

As with static fields, it is frequently more convenient not to deal with $\boldsymbol{E}$ and $\boldsymbol{B}$ directly, but with potentials from which they can be derived. We discuss these potentials now.

Even for time-varying fields it is still true that $\nabla \cdot \boldsymbol{B} = 0$ so we can write

$$\boldsymbol{B} = \nabla \times \boldsymbol{A}, \tag{11.27}$$

where $\boldsymbol{A}$ is of course now a function of time. But since $\nabla \times \boldsymbol{E} \neq 0$, there is no simple scalar potential for $\boldsymbol{E}$. On putting (11.27) into (11.3) we see, however, that

$$\nabla \times \left(\boldsymbol{E} + \frac{1}{c}\frac{\partial \boldsymbol{A}}{\partial t}\right) = 0, \tag{11.28}$$

so we can try to write $\boldsymbol{E} + (1/c)(\partial \boldsymbol{A}/\partial t)$ as the gradient of some scalar function Φ,

$$\boldsymbol{E} + \frac{1}{c}\frac{\partial \boldsymbol{A}}{\partial t} = -\nabla\Phi. \tag{11.29}$$

The quantities $\boldsymbol{A}$ and Φ, both time dependent, are the generalized vector and scalar potentials. However, they are not uniquely determined by Eqs. (11.27) and (11.28). There are other potentials, which differ from $\boldsymbol{A}$ and Φ by a so-called *gauge transformation*, that give the same results for the fields. A gauge transformation is the operation of replacing $\boldsymbol{A}$ by $\boldsymbol{A} + \nabla\Gamma$, and Φ by $\Phi - (1/c)(\partial\Gamma/\partial t)$, where Γ is an arbitrary function of $\boldsymbol{r}$ and t. Consider then the fields calculable from the gauge-transformed potentials, $\boldsymbol{A}'$ and Φ', where

$$\boldsymbol{A}' = \boldsymbol{A} + \nabla\Gamma, \qquad \Phi' = \Phi - \frac{1}{c}\frac{\partial\Gamma}{\partial t}.$$

We have

$$\begin{aligned} \boldsymbol{E}' &= -\frac{1}{c}\frac{\partial \boldsymbol{A}'}{\partial t} - \nabla\Phi' = -\frac{1}{c}\frac{\partial \boldsymbol{A}}{\partial t} - \frac{1}{c}\frac{\partial}{\partial t}\nabla\Gamma - \nabla\Phi + \nabla\frac{1}{c}\frac{\partial\Gamma}{\partial t} \\ &= -\frac{1}{c}\frac{\partial \boldsymbol{A}}{\partial t} - \nabla\Phi = \boldsymbol{E} \end{aligned}$$

and

$$\boldsymbol{B}' = \nabla \times (\boldsymbol{A} + \nabla\Gamma) = \nabla \times \boldsymbol{A} = \boldsymbol{B}.$$

Thus the gauge-transformed potentials yield the same fields as the original potentials As we shall see, this arbitrariness is not necessarily a disadvantage, but it can be usefully exploited.

The potentials $\boldsymbol{A}$ and Φ are particularly useful in that equations can be derived which relate them to integrals over the sources $\boldsymbol{J}$ and ρ; then $\boldsymbol{E}$ and $\boldsymbol{B}$ can be found by differentiation. The first step to this end is to find the differential equations satisfied by $\boldsymbol{A}$ and Φ. We begin by substituting $\boldsymbol{B}$ and $\boldsymbol{E}$ from (11.27) and (11.29) into the Maxwell equation for $\nabla \times \boldsymbol{B}$. Using $\nabla \times (\nabla \times \boldsymbol{A}) = \nabla\nabla \cdot \boldsymbol{A} - \nabla^2 \boldsymbol{A}$, we find

$$-\nabla^2\boldsymbol{A} + \frac{1}{c^2}\frac{\partial^2\boldsymbol{A}}{\partial t^2} + \nabla\left(\nabla\cdot\boldsymbol{A} + \frac{1}{c}\frac{\partial\Phi}{\partial t}\right) = \frac{4\pi\boldsymbol{J}}{c}. \tag{11.30}$$

Similarly, from $\nabla \cdot \boldsymbol{E} = 4\pi\rho$, we find

$$\nabla^2\Phi + \frac{1}{c}\frac{\partial(\nabla\cdot\boldsymbol{A})}{\partial t} = -4\pi\rho. \tag{11.31}$$

Now $\nabla \times \boldsymbol{A}$ is specified by (11.27), but $\nabla \cdot \boldsymbol{A}$ is still at our disposal. Two choices are commonly made. First, we can require that

$$\nabla\cdot\boldsymbol{A} + \frac{1}{c}\frac{\partial\Phi}{\partial t} = 0, \tag{11.32}$$

whereupon Eqs. (11.30) and (11.31) become

$$\begin{aligned} \nabla^2\boldsymbol{A} - \frac{1}{c^2}\frac{\partial^2\boldsymbol{A}}{\partial t^2} &= -\frac{4\pi\boldsymbol{J}}{c}, \\ \nabla^2\Phi - \frac{1}{c^2}\frac{\partial^2\Phi}{\partial t^2} &= -4\pi\rho. \end{aligned} \tag{11.33}$$

Equation (11.32) is called the *Lorentz condition*, and it is this that we shall usually employ. If this condition is chosen, one is said to be working in the *Lorentz gauge.* The second choice, in which one simply sets

$$\nabla \cdot \boldsymbol{A} = 0, \tag{11.34}$$

defines the *Coulomb, transverse* or *radiation gauge.* Then Eqs. (11.30) and (11.31) become

$$\nabla^2 \boldsymbol{A} - \frac{1}{c^2}\frac{\partial^2 \boldsymbol{A}}{\partial t^2} - \frac{\nabla}{c}\frac{\partial \Phi}{\partial t} = -\frac{4\pi \boldsymbol{J}}{c}, \qquad \nabla^2 \Phi = -4\pi\rho. \tag{11.35}$$

From these equations, we note first that Φ is calculated from an integral formally identical to the Coulomb superposition integral of electrostatics,

$$\Phi(\boldsymbol{r}, t) = \int \frac{\rho(\boldsymbol{r}', t)\, dv'}{|\boldsymbol{r} - \boldsymbol{r}'|},$$

and this circumstance accounts for the origin of the name *Coulomb gauge.*

In discussing the Coulomb gauge, it is convenient to decompose $\boldsymbol{J}$ into $\boldsymbol{J}_t$, its transverse part ($\nabla \cdot \boldsymbol{J}_t = 0$), and $\boldsymbol{J}_l$, its longitudinal part ($\nabla \times \boldsymbol{J}_l = 0$), as prescribed by Helmholtz' theorem, discussed in Appendix C. With

$$\boldsymbol{J} = \boldsymbol{J}_l + \boldsymbol{J}_t,$$

we have from Appendix C:

$$\boldsymbol{J}_l = -\frac{\nabla}{4\pi}\int \frac{\nabla' \cdot \boldsymbol{J}}{|\boldsymbol{r} - \boldsymbol{r}'|}\, dv'. \tag{11.36}$$

Using the continuity equation in (11.36), we find

$$\frac{1}{c}\nabla \frac{\partial \Phi}{\partial t} = \frac{4\pi \boldsymbol{J}_l}{c}$$

and with this equation in (11.35) for $\boldsymbol{A}$, we have

$$\nabla^2 \boldsymbol{A} - \frac{1}{c^2}\frac{\partial^2 \boldsymbol{A}}{\partial t^2} = -\frac{4\pi \boldsymbol{J}_t}{c}. \tag{11.37}$$

In calculating $\boldsymbol{E}$ and $\boldsymbol{B}$ in this gauge, that part of $\boldsymbol{E}$ calculable from $\nabla\Phi$ will drop off at least as fast as $1/r^2$. As we shall see the fields of radiation drop off as $1/r$. In calculating radiation fields we can then in effect set $\Phi = 0$ and use

$$\boldsymbol{E} = -\frac{1}{c}\frac{\partial \boldsymbol{A}}{\partial t} \quad \text{and} \quad \boldsymbol{B} = \nabla \times \boldsymbol{A}.$$

This gauge thus conveniently separates out that part of the sources (the transverse currents) that are responsible for radiation, hence the names *radiation gauge* and *transverse gauge.*

11.7 INTEGRAL FORM OF THE POTENTIALS

We turn now to the problem of finding $\boldsymbol{A}$ and Φ in terms of $\boldsymbol{J}$ and ρ. For this purpose the Lorentz gauge is the more convenient one. In this gauge we have the four equations in (11.33). These are all similar, and for definiteness we focus on that for Φ, which is an obvious generalization of the Poisson equation. This latter equation was nicely solved in electrostatics by use of the Green function $1/|\boldsymbol{r} - \boldsymbol{r}'|$ which led to the superposition integral for Φ in terms of the charge distribution. We shall adopt a similar technique here, first finding the Green function for equations of the form of (11.33) and then using this to construct a general solution of such equations as an integral. We seek then a Green function $G(\boldsymbol{r}, t; \boldsymbol{r}', t')$, where $\boldsymbol{r}', t'$ are passive parameters, that satisfies

$$\nabla^2 G - \frac{1}{c^2}\frac{\partial^2 G}{\partial t^2} = -4\pi\delta(\boldsymbol{r} - \boldsymbol{r}')\delta(t - t'). \tag{11.38}$$

It will be convenient to begin with the point source at the origin, i.e., with $\boldsymbol{r}' = t' = 0$, and so we define $G_0(\boldsymbol{r}, t) = G(\boldsymbol{r}, t; \boldsymbol{0}, 0)$, such that G_0 satisfies

$$\nabla^2 G_0 - \frac{1}{c^2}\frac{\partial^2 G_0}{\partial t^2} = -4\pi\delta(\boldsymbol{r})\delta(t). \tag{11.39}$$

Consider the possible dependence of G_0 on $\boldsymbol{r}$. Since the source is a point, we do not anticipate any angular dependence, but expect the solution to be a function only of $r = |\boldsymbol{r}|$. The analogous solution, $1/r$, for the Poisson equation, is also of this form. Using the Laplacian in spherical coordinates, but with its angular part set equal to zero, Eq. (11.39) becomes

$$\frac{1}{r^2}\frac{\partial}{\partial r}\left(r^2\frac{\partial G_0}{\partial r}\right) - \frac{1}{c^2}\frac{\partial^2 G_0}{\partial t^2} = -4\pi\delta(\boldsymbol{r})\delta(t). \tag{11.40}$$

It is easy to verify that a solution of this equation away from the origin ($r \neq 0$ and $t \neq 0$) is

$$G_0(\boldsymbol{r}, t) = \frac{f(t \pm r/c)}{r}, \tag{11.41}$$

where f is an *arbitrary* function. We shall discuss later how to choose between the plus and minus signs. We must now see if the form of f can be chosen so that Eq. (11.40) is satisfied "at the origin" as well; in other words, so that G_0 has in fact the integral properties implied by the δ-functions on the right-hand side. Consider the property connected with $\delta(\boldsymbol{r})$. This requires that if (11.40) is integrated over a small volume ΔV containing the origin, that

$$\int_{\Delta V}\frac{1}{r^2}\frac{\partial}{\partial r}\left(r^2\frac{\partial G_0}{\partial r}\right)dv - \frac{1}{c^2}\frac{\partial^2}{\partial t^2}\int_{\Delta V} G_0\, dv = -4\pi\delta(t). \tag{11.42}$$

With the G_0 of Eq. (11.41) the integrand of the first integral in (11.42) is singular:

thus the integral is an improper one. It can be properly redefined on recognizing that the integrand is $\nabla^2 G_0 = \nabla \cdot \nabla G_0$. By the divergence theorem, the volume integral of the divergence can be converted to a surface integral over the small bounding surface ΔS (which is ultimately allowed to shrink to zero), and which we assume to be spherical. The details here are very similar to those discussed somewhat more fully in Appendix B, in connection with the meaning of $\nabla^2(1/r) = -4\pi\delta(\boldsymbol{r})$. With this conversion from volume to surface integral, the δ function condition in (11.42) becomes

$$\int_{\Delta S} (\nabla G_0) \cdot d\boldsymbol{s} - \frac{1}{c^2}\frac{\partial^2}{\partial t^2}\int_{\Delta V} G_0 \, dv = -4\pi\delta(t).$$

Now we substitute the explicit form (11.41) into this equation. With

$$\nabla G_0 = -\left(\frac{f}{r^2} \pm \frac{1}{c}\frac{f'}{r}\right)\hat{\boldsymbol{r}},$$

where prime denotes differentiation with respect to the argument, we have, with r_0 as the radius of the small sphere,

$$\left(-\frac{f[t \pm r_0/c]}{r_0^2} \mp \frac{f'[t \pm r_0/c]}{cr_0}\right)4\pi r_0^2 - \frac{1}{c^2}\int_{\Delta V}\frac{f'' \, dv}{r} = -4\pi\delta(t).$$

Letting $r_0 \to 0$, we note that only the term in f remains in the left-hand side of this equation and we conclude that

$$f(t) = \delta(t).$$

With this identification, we have

$$G_0(\boldsymbol{r}, t) = \frac{\delta(t \pm r/c)}{r}$$

and so we can construct $G(\boldsymbol{r}, t; \boldsymbol{r}', t')$ as

$$G(\boldsymbol{r}, t; \boldsymbol{r}', t') = \frac{\delta(t - t' \pm |\boldsymbol{r} - \boldsymbol{r}'|/c)}{|\boldsymbol{r} - \boldsymbol{r}'|}. \tag{11.43}$$

This Green function in turn enables us to perform the integration of Eqs. (11.33) so that the solution for Φ, for example, is

$$\Phi(\boldsymbol{r}, t) = \int \frac{\rho(\boldsymbol{r}', t \pm |\boldsymbol{r} - \boldsymbol{r}'|/c)}{|\boldsymbol{r} - \boldsymbol{r}'|} \, dv'. \tag{11.44}$$

The solution with the minus sign, involving $t - (|\boldsymbol{r} - \boldsymbol{r}'|/c)$, is called the *retarded solution* and that with the plus sign, the *advanced solution*. One usually chooses the retarded solution since, with it, the potential Φ at time t is determined by the behavior of the charges at times *previous* to t. The advanced solutions imply that the potential at time t is determined by what charges will be doing at later times, and this is unphysical. A solution like (11.44), and similar reasoning, applies to each of the rectangular components of $\boldsymbol{A}$. We have then the two expressions for the *retarded potentials* Φ and $\boldsymbol{A}$:

$$\Phi(\boldsymbol{r}, t) = \int \frac{\rho(\boldsymbol{r}', t - |\boldsymbol{r} - \boldsymbol{r}'|/c)}{|\boldsymbol{r} - \boldsymbol{r}'|} \, dv' \tag{11.45}$$

$$\boldsymbol{A}(\boldsymbol{r}, t) = \frac{1}{c} \int \frac{\boldsymbol{J}(\boldsymbol{r}', t - |\boldsymbol{r} - \boldsymbol{r}'|/c)}{|\boldsymbol{r} - \boldsymbol{r}'|} \, dv'. \tag{11.46}$$

These equations for $\boldsymbol{A}$ and Φ formally solve the problem of finding the fields produced by an arbitrary charge and current distribution.

11.8 ELECTROMAGNETIC WAVES

Consider next the potentials and fields, not as related to their sources, $\boldsymbol{J}$ and ρ, but as existing in free space, i.e., in regions where $\boldsymbol{J}$ and ρ are zero. From Eqs. (11.33) we see that Φ and the rectangular components of $\boldsymbol{A}$ satisfy the wave equation

$$\begin{aligned} \nabla^2\Phi - \frac{1}{c^2}\frac{\partial^2\Phi}{\partial t^2} &= 0 \\ \nabla^2\boldsymbol{A} - \frac{1}{c^2}\frac{\partial^2\boldsymbol{A}}{\partial t^2} &= 0. \end{aligned} \tag{11.47}$$

Moreover, since $\boldsymbol{E}$ and $\boldsymbol{B}$ are linearly related to $\boldsymbol{A}$ and Φ, each of the rectangular coordinates of $\boldsymbol{E}$ and $\boldsymbol{B}$ also satisfies the wave equation

$$\nabla^2\boldsymbol{E} - \frac{1}{c^2}\frac{\partial^2\boldsymbol{E}}{\partial t^2} = 0, \qquad \nabla^2\boldsymbol{B} - \frac{1}{c^2}\frac{\partial^2\boldsymbol{B}}{\partial t^2} = 0. \tag{11.48}$$

For future reference, we note here that if harmonic time dependence is assumed in any of these equations, say that for Φ, so that

$$\Phi(\boldsymbol{r}, t) = \chi(\boldsymbol{r}) \begin{cases} \sin \omega t \\ \cos \omega t, \end{cases}$$

then the amplitude function χ satisfies the equation

$$\left(\nabla^2 + \frac{\omega^2}{c^2}\right)\chi = 0 \tag{11.49}$$

which is called the *Helmholtz equation.*

The wave equation is ubiquitous; in this section we study some of the properties of its solutions. First, this equation can, of course, be written in various coordinate systems—Cartesian, spherical, cylindrical, etc.—and the corresponding solutions are *plane, spherical,* or *cylindrical waves.* The function $f(t - r/c)/r$ that was introduced in the last section is a simple example of an (outgoing) spherical wave, although there are other examples with more complicated angular dependence. Somewhat analogously, the simplest outgoing wave in cylindrical coordinates ρ, φ can be shown asymptotically (ρ large) to be of the form $g(t - \rho/c)/\sqrt{\rho}$, where g is an arbitrary function. But in this section we shall consider only plane waves in any detail. Plane waves are interesting in their own right, and also because they are approximations over a limited region of space to the more general spherical and cylindrical waves.

Any of the wave equations above describe, in general, waves that are functions of all four variables x, y, z, t. For simplicity, we shall first consider waves along the x-axis, for which there is dependence only on x and t. If then $u(x, t)$ is Φ or any of the rectangular components of $\boldsymbol{E}$, $\boldsymbol{B}$ or $\boldsymbol{A}$, we have

$$\frac{\partial^2 u}{\partial x^2} - \frac{1}{c^2}\frac{\partial^2 u}{\partial t^2} = 0. \tag{11.50}$$

It is easily verified that a general solution of this equation is of the form

$$u(x, t) = v(x - ct) + w(x + ct),$$

where v and w are arbitrary functions, *not* necessarily sinusoids. The function $v(x - ct)$ represents a wave propagating to the *right* with velocity c and $w(x + ct)$ represents one propagating to the left. An important property of *electromagnetic plane waves* is that the vectors $\boldsymbol{E}$ and $\boldsymbol{B}$ are perpendicular to each other *and* to the direction of propagation. This property is called the *transversality* of the wave. To prove it, we use the Coulomb gauge: $\Phi = 0$ and $\nabla \cdot \boldsymbol{A} = 0$. We take the direction of propagation to be the x-direction. Then $\nabla \cdot \boldsymbol{A} = 0$ yields $\partial A_x/\partial x = 0$, and from the wave equation $\partial^2 A_x/\partial t^2 = 0$, or $\partial A_x/\partial t =$ constant. But

$$E_x = -\frac{1}{c}\frac{\partial A_x}{\partial t}$$

shows that E_x is also constant so that, at most, there can be a static component of $\boldsymbol{E}$ along the propagation direction; since this does not refer to a wave motion, we shall neglect it. Thus $\boldsymbol{E}$ is perpendicular to the x-axis and we can take it to be along the z-direction. From $\boldsymbol{B} = \nabla \times \boldsymbol{A}$, we then find that $\boldsymbol{B}$ is perpendicular to both the x- and z-directions, and this completes the proof of the transversality of the wave.

An important special kind of wave is the plane wave of (angular) frequency ω. It is one for which the time dependence is harmonic, i.e., as $\sin \omega t$ or $\cos \omega t$. Consider first one dimension. If harmonic time dependence is assumed in (11.50), it is easy to verify that four solutions of it can be found:

$$\sin\frac{\omega x}{c}\sin \omega t, \qquad \sin\frac{\omega x}{c}\cos \omega t, \qquad \cos\frac{\omega x}{c}\sin \omega t, \qquad \cos\frac{\omega x}{c}\cos \omega t.$$

Although these solutions do not have the form of traveling waves, we can form four traveling wave solutions of (11.50) by taking linear combinations of them. With

$k = \omega/c$, these solutions are $\sin(kx \pm \omega t)$ and $\cos(kx \pm \omega t)$ and they represent sinusoidal disturbances traveling to the left or right (plus or minus sign, respectively). The dimensions of k are, of course, those of inverse length. Writing

$$k = 2\pi/\lambda, \tag{11.51}$$

we see that at any time t the disturbance is periodic over a length λ, which is defined to be the *wavelength*.

In dealing with harmonic waves, it is frequently convenient to use the *complex exponential notation*. Thus one-dimensional plane waves can be written in the form

$$u(x, t) = \text{Re}\colon Ae^{i(kx \pm \omega t)}$$

where the amplitude A can be complex and where Re: means "real part of." The wave $\cos(kx \pm \omega t)$ is given in this way of writing by setting $A = 1$ and the wave $\sin(kx \pm \omega t)$ by setting $A = -i$. Frequently it is expedient to do away with the necessity of continually writing "Re:." To this end, we simply drop it from the equations and write, for example,

$$u(x, t) = Ae^{i(kx \pm \omega t)}, \tag{11.52}$$

where, by convention, "real part of" is understood.

Consider next a harmonic plane wave amplitude $u(\boldsymbol{r}, t)$ in three dimensions, a solution of the wave equation

$$\nabla^2 u - \frac{1}{c^2}\frac{\partial^2 u}{\partial t^2} = 0.$$

Let $\boldsymbol{k}$ be a vector with components k_x, k_y, k_z and with magnitude k given by $k^2 = k_x^2 + k_y^2 + k_z^2 = \omega^2/c^2$. Then the functions $\sin(\boldsymbol{k} \cdot \boldsymbol{r} - \omega t)$ and $\cos(\boldsymbol{k} \cdot \boldsymbol{r} - \omega t)$ are solutions of this equation. The quantity $(\boldsymbol{k} \cdot \boldsymbol{r} - \omega t)$ is the *phase* of the wave and at a given time it is constant over a plane perpendicular to the vector $\boldsymbol{k}$. These two functions are said to describe *harmonic waves* propagating in the direction $\hat{\boldsymbol{k}}$. Using the convention above on taking real parts, we write such harmonic waves as

$$u(\boldsymbol{r}, t) = Ae^{i(\boldsymbol{k} \cdot \boldsymbol{r} - \omega t)}, \tag{11.53}$$

where A is a *complex* constant.

Electromagnetic plane waves are characterized by a field direction as well as a direction of propagation $\hat{\boldsymbol{k}}$. The electric and magnetic fields $\boldsymbol{E}$ and $\boldsymbol{B}$ associated with such a wave can, subject to certain relations, be taken as

$$\begin{aligned} \boldsymbol{E} &= \hat{\boldsymbol{e}} E_0 e^{i(\boldsymbol{k} \cdot \boldsymbol{r} - \omega t)} \\ \boldsymbol{B} &= \hat{\boldsymbol{b}} B_0 e^{i(\boldsymbol{k} \cdot \boldsymbol{r} - \omega t)} \end{aligned} \tag{11.54}$$

so that $\boldsymbol{E}$ and $\boldsymbol{B}$ are in the directions of the unit vectors $\hat{\boldsymbol{e}}$ and $\hat{\boldsymbol{b}}$. The relations mentioned arise of course from the fact that $\boldsymbol{E}$ and $\boldsymbol{B}$ are not independent, but must satisfy Maxwell's equations. For example, from the two divergence equations applied to (11.54) we find that

$$\hat{\boldsymbol{e}} \cdot \boldsymbol{k} = \hat{\boldsymbol{b}} \cdot \boldsymbol{k} = 0$$

which shows that both $\boldsymbol{E}$ and $\boldsymbol{B}$ are perpendicular to the direction of propagation. This comes as no surprise since we have previously proved the transversality of electromagnetic waves. A further relation is given by either of the two curl equations; for example, $\nabla \times \boldsymbol{E} + (1/c)(\partial \boldsymbol{B}/\partial t) = 0$ applied to (11.54) yields

$$(\hat{\boldsymbol{k}} \times \hat{\boldsymbol{e}})E_0 = \hat{\boldsymbol{b}}B_0$$

with $\hat{\boldsymbol{k}}$ the unit vector along $\boldsymbol{k}$. This equation is solved by first equating the magnitudes of the left- and right-hand sides, i.e., by setting

$$E_0 = B_0.$$

Given this equality, we obtain the vectorial condition:

$$\hat{\boldsymbol{k}} \times \hat{\boldsymbol{e}} = \hat{\boldsymbol{b}}.$$

This last equation shows that $\boldsymbol{E}$ and $\boldsymbol{B}$ are not only both perpendicular to $\boldsymbol{k}$ but to each other so that $\hat{\boldsymbol{k}}$, $\hat{\boldsymbol{e}}$, and $\hat{\boldsymbol{b}}$ form a mutually orthogonal right-handed triad of unit vectors.

Consider now the directional or *polarization* properties of $\boldsymbol{E}$. If a wave is propagating in direction $\hat{\boldsymbol{k}}$, we have seen that $\boldsymbol{E}$ must lie in a plane perpendicular to $\boldsymbol{k}$. In this plane, we set up a two-dimensional coordinate system x_1, x_2, with unit vectors $\hat{\boldsymbol{x}}_1$ and $\hat{\boldsymbol{x}}_2$ along the axes. Since $\boldsymbol{E}$ can have any direction in the plane, a general expression for it, using the convention for taking real parts, is

$$\boldsymbol{E} = (\hat{\boldsymbol{x}}_1 E_1 + \hat{\boldsymbol{x}}_2 E_2)e^{i(\boldsymbol{k}\cdot\boldsymbol{r} - \omega t)},$$

where E_1 and E_2 are arbitrary *complex* constants. There are then various polarization possibilities, depending on the nature of these constants. Both E_1 and E_2 are characterized by a magnitude and a phase, $E_1 = \mathscr{E}_1 e^{i\varphi_1}$, $E_2 = \mathscr{E}_2 e^{i\varphi_2}$. To take the simplest case, suppose that the phases are identical, $\varphi_1 = \varphi_2 = \varphi$, but the magnitudes are arbitrary. Then

$$\boldsymbol{E} = (\hat{\boldsymbol{x}}_1 \mathscr{E}_1 + \hat{\boldsymbol{x}}_2 \mathscr{E}_2)e^{i(\boldsymbol{k}\cdot\boldsymbol{r} - \omega t + \varphi)}$$

and on taking real parts, we see that the actual electric field is

$$\boldsymbol{E}_{\text{actual}} = (\hat{\boldsymbol{x}}_1 \mathscr{E}_1 + \hat{\boldsymbol{x}}_2 \mathscr{E}_2)\cos(\boldsymbol{k}\cdot\boldsymbol{r} - \omega t + \varphi).$$

Such a wave is said to be linearly polarized in the direction $\hat{\boldsymbol{x}}_1 \mathscr{E}_1 + \hat{\boldsymbol{x}}_2 \mathscr{E}_2$. By contrast, suppose that the magnitudes E_1 and E_2 are equal, but that the phases differ by $\pi/2$. Take for example $\mathscr{E}_1 = \mathscr{E}_2 = \mathscr{E}_0$, $\varphi_1 = 0$, $\varphi_2 = -\pi/2$. Then

$$\boldsymbol{E} = \mathscr{E}_0(\hat{\boldsymbol{x}}_1 e^{i(\boldsymbol{k}\cdot\boldsymbol{r} - \omega t)} + \hat{\boldsymbol{x}}_2 e^{i(\boldsymbol{k}\cdot\boldsymbol{r} - \omega t - \pi/2)}) \tag{11.55}$$

and the actual electric field, found by taking real parts, is

$$\boldsymbol{E}_{\text{actual}} = \mathscr{E}_0[\hat{\boldsymbol{x}}_1 \cos(\boldsymbol{k} \cdot \boldsymbol{r} - \omega t) + \hat{\boldsymbol{x}}_2 \sin(\boldsymbol{k} \cdot \boldsymbol{r} - \omega t)].$$

A wave with $\boldsymbol{E}$ of the form (11.55) is said to be *circularly polarized.* The electric field vector sweeps around like the hand of a clock, with frequency ω. If the rotation is clockwise to an observer facing the oncoming wave, the wave is said to be *right circularly polarized.* More generally, by taking both the magnitudes and phases of E_1 and E_2 to be different, one arrives at *elliptically polarized waves.*

11.9 CONSERVATION LAWS

We know from mechanics that a mutually interacting system of (neutral) particles which is not acted on by external forces moves in such a way that its energy and momenta (linear and angular) remain constant. If, however, the particles are charged so that electromagnetic fields are necessarily present, the thought experiment of Chapter 1 shows that the total mechanical energy and momentum of such a system is *not* necessarily constant. As we have indicated, the conservation laws for energy and momentum can be retained if it is assumed that these quantities can be associated with electromagnetic fields as well as with matter. In this section, these generalized conservation laws are derived on the assumption that Newton's laws govern the motion of the point charged particles acted on by the electromagnetic field. As the reader probably knows, Newton's laws are only approximations to what are now considered to be the correct laws of relativistic point mechanics. For small velocities, the relativistic laws become Newtonian, and we can neglect "relativistic effects." This neglect must not be too cavalier, however; we shall later see examples involving momentum conservation where relativistic effects enter somewhat unexpectedly.

We shall begin by recalling the derivation of the conservation laws in Newtonian mechanics; the comparison with the electromagnetic case is illuminating. These conservation laws, for energy, momentum, and angular momentum, are based on Newton's third law, on the equality of action and reaction forces. In other words, these laws can be derived using the assumption that the force that one particle exerts on a second is equal (in magnitude) and opposite (in direction) to the force the second exerts on the first. Thus, consider two mutually interacting particles that exert, say, gravitational forces on one another. Let them have masses m_1 and m_2 and move with velocities $\boldsymbol{v}_1$ and $\boldsymbol{v}_2$. The rate of change of momentum $\boldsymbol{P}$ of the *system* of two particles is

$$\frac{d\boldsymbol{P}}{dt} = m_1 \frac{d\boldsymbol{v}_1}{dt} + m_2 \frac{d\boldsymbol{v}_2}{dt}.$$

From Newton's second law, we have

$$\frac{d\boldsymbol{P}}{dt} = \boldsymbol{F}_1 + \boldsymbol{F}_2,$$

where $\boldsymbol{F}_1$ is the force on 1 due to 2, and $\boldsymbol{F}_2$ is the force on 2 due to 1. But if these forces are equal and opposite, $\boldsymbol{F}_1 = -\boldsymbol{F}_2$, then $d\boldsymbol{P}/dt = 0$, or

$$\boldsymbol{P} = \text{constant}.$$

The argument is easily generalized to a system of N interacting particles. Similar arguments, found in texts on classical mechanics, demonstrate the conservation of energy and angular momentum for such a system. All of these classical arguments depend on the fact of the equality of action and reaction forces. However, we have seen in the simple thought experiment of Chapter 1 that this equality breaks down when electromagnetic radiation is present. For it is implied in the above derivation that the equation $\boldsymbol{F}_1 + \boldsymbol{F}_2 = 0$ holds at every instant of time. This is certainly not the case with the interaction of that thought experiment, in which one particle exerts a force on the other *only* after an electromagnetic pulse has traveled from the first to the second. As another example in which action does not equal reaction, consider the magnetic forces exerted on each other by two moving charges,for example, the two charges 1 and 2 in the mutually perpendicular trajectories shown in Fig. 11.1. Now it is plausible that for the sake of calculating their magnetic fields we can think of these charges as current elements and use the Biot–Savart law (7.8), with $q\boldsymbol{v}$ replacing $I\,d\boldsymbol{l}$, for the field due to such elements. With this assumption,* we see immediately that q_2 exerts no force on q_1 since the field of q_2 vanishes at the place of q_1. On the other hand, the field of q_1 does not vanish at q_2 and it does exert a transverse force on that charge: The magnitudes of the action and reaction forces are not equal. Therefore, the total momentum of this system of two particles is not conserved even though no external forces act on the system. It is only by including the momentum of the electromagnetic field that the conservation law for momentum can be retained.

Fig. 11.1 Example showing inequality of action and reaction forces.

We turn then to the conservation laws that include the electromagnetic field. These laws have the form of mathematical theorems that are given a physical interpretation. The derivation of these theorems is unequivocal, but the physical interpretation is not quite so. We shall consider first the several mathematical theorems and then discuss their interpretation in more detail.

We start with the theorem that corresponds to the conservation of energy. Consider a system of charged particles moving inside a volume and constituting a current. These charges and the associated current produce electromagnetic fields $\boldsymbol{E}$ and $\boldsymbol{B}$; we shall assume that there are no other (external) sources of these fields.

* The discussion in Chapter 14 of the fields of a point particle will confirm this assumption. for slowly moving charges.

For simplicity, suppose that the distribution is dense enough to be effectively continuous. Let ρ be the charge density at a point and $\boldsymbol{u}$ the velocity there so the current density $\boldsymbol{J}$ is $\boldsymbol{J} = \rho\boldsymbol{u}$. The element of charge contained in dv will be acted on by a force $\boldsymbol{F} = \boldsymbol{f}\,dv$, where $\boldsymbol{f}$ is the Lorentz force density of Eq. (11.22). If $w_m\,dv$ is the mechanical energy of the element, so that w_m represents the mechanical energy *density*, we have

$$\frac{\partial(w_m\,dv)}{\partial t} = \boldsymbol{F}\cdot\boldsymbol{u} = \rho\left(\boldsymbol{E} + \frac{\boldsymbol{u}\times\boldsymbol{B}}{c}\right)\cdot\boldsymbol{u}\,dv = \boldsymbol{E}\cdot\boldsymbol{J}\,dv, \tag{11.56}$$

since $(\boldsymbol{u}\times\boldsymbol{B})\cdot\boldsymbol{u} = 0$; the magnetic field does no work. The relation

$$\nabla\cdot(\boldsymbol{E}\times\boldsymbol{B}) = \boldsymbol{B}\cdot\nabla\times\boldsymbol{E} - \boldsymbol{E}\cdot\nabla\times\boldsymbol{B}$$

becomes, on substituting for $\nabla\times\boldsymbol{E}$ and $\nabla\times\boldsymbol{B}$ from Maxwell's equations,

$$\nabla\cdot(\boldsymbol{E}\times\boldsymbol{B}) = \boldsymbol{B}\cdot\left(-\frac{1}{c}\frac{\partial\boldsymbol{B}}{\partial t}\right) - \boldsymbol{E}\cdot\left(\frac{4\pi\boldsymbol{J}}{c} + \frac{1}{c}\frac{\partial\boldsymbol{E}}{\partial t}\right).$$

On integrating this equation over a volume V_0, bounding surface S_0, it becomes

$$\int_{V_0}\nabla\cdot\left(\frac{c(\boldsymbol{E}\times\boldsymbol{B})}{4\pi}\right)dv + \frac{1}{8\pi}\frac{\partial}{\partial t}\int_{V_0}(|\boldsymbol{E}|^2 + |\boldsymbol{B}|^2)\,dv + \int_{V_0}\boldsymbol{E}\cdot\boldsymbol{J}\,dv = 0. \tag{11.57}$$

We recall that for static fields $1/8\pi\int|\boldsymbol{E}|^2\,dv$ was interpretable as the energy stored in the electric field, or as the energy of assemblage of a charge distribution, and the similar integral $1/8\pi\int|\boldsymbol{B}|^2\,dv$ as the energy required to assemble a stationary current distribution. It is as if $1/8\pi|\boldsymbol{E}|^2$ and $1/8\pi|\boldsymbol{B}|^2$ represented the *densities* of electric and magnetic energies. For the present case, the fields are time varying. But we are free, if only for dimensional reasons, to define the *field energy density* w_f, without any implications at the moment as to its meaning, by

$$w_f = \frac{1}{8\pi}(|\boldsymbol{E}|^2 + |\boldsymbol{B}|^2).$$

The physical meaning, or lack of physical meaning, of w_f will be discussed later. If we also define the *Poynting vector* $\boldsymbol{S}$ by

$$\boldsymbol{S} = \frac{c}{4\pi}(\boldsymbol{E}\times\boldsymbol{B}), \tag{11.58}$$

then by using the divergence theorem and Eq. (11.56), we can write Eq. (11.57) as

$$\frac{\partial}{\partial t}\int_{V_0}(w_f + w_m)\,dv + \int_{S_0}\boldsymbol{S}\cdot d\boldsymbol{s} = 0. \tag{11.59}$$

This is the conservation theorem we sought. The Poynting vector has the dimensions (energy/area $\times$ time), the dimensions of what one might call an *energy current density*. Then Eq. (11.59), resembles a continuity equation in integral form that relates the rate of change of energy in a volume to the energy flow through the bounding surface. Later we shall discuss the validity of this interpretation.

We now derive an analogous theorem for momentum. Let the mechanical momentum carried by the moving charge element $\rho\, dv$ be given by $\boldsymbol{p}_m\, dv$, so that $\boldsymbol{p}_m$ is the momentum *density*. Then, from Newton's second law, we have

$$\frac{\partial \boldsymbol{p}_m}{\partial t} = \left(\rho \boldsymbol{E} + \frac{1}{c}\boldsymbol{J} \times \boldsymbol{B}\right). \tag{11.60}$$

We transform this by substituting for ρ and $\boldsymbol{J}$ from Maxwell's equations, to find

$$\frac{\partial \boldsymbol{p}_m}{\partial t} = \frac{1}{4\pi}\left[\boldsymbol{E}(\nabla \cdot \boldsymbol{E}) + \frac{1}{c}\boldsymbol{B} \times \frac{\partial \boldsymbol{E}}{\partial t} - \boldsymbol{B} \times (\nabla \times \boldsymbol{B})\right]\cdot \tag{11.61}$$

Now we use

$$\boldsymbol{B} \times \frac{\partial \boldsymbol{E}}{\partial t} = -\frac{\partial}{\partial t}(\boldsymbol{E} \times \boldsymbol{B}) + \boldsymbol{E} \times \frac{\partial \boldsymbol{B}}{\partial t},$$

add $\boldsymbol{B}\nabla \cdot \boldsymbol{B} = 0$ to the right-hand side of (11.61), and integrate to get

$$\int_{V_0} \frac{\partial \boldsymbol{p}_m}{\partial t} + \frac{\partial}{\partial t}\int_{V_0} \frac{\boldsymbol{E} \times \boldsymbol{B}}{4\pi c}\, dv = \frac{1}{4\pi}\int_{V_0} [\boldsymbol{E} \cdot (\nabla \boldsymbol{E}) + \boldsymbol{B}(\nabla \cdot \boldsymbol{B})$$

$$-\boldsymbol{E} \times (\nabla \times \boldsymbol{E}) - \boldsymbol{B} \times (\nabla \times \boldsymbol{B})]\, dv. \tag{11.62}$$

The right-hand side can be transformed, using for example the vector identity

$$\tfrac{1}{2}\nabla(\boldsymbol{B} \cdot \boldsymbol{B}) = (\boldsymbol{B} \cdot \nabla)\boldsymbol{B} + \boldsymbol{B} \times (\nabla \times \boldsymbol{B}).$$

Then

$$\boldsymbol{B}(\nabla \cdot \boldsymbol{B}) - \boldsymbol{B} \times (\nabla \times \boldsymbol{B}) = \boldsymbol{B}(\nabla \cdot \boldsymbol{B}) + (\boldsymbol{B} \cdot \nabla)\boldsymbol{B} - \tfrac{1}{2}\nabla B^2.$$

The right-hand side of this last equation is interpretable as the divergence of a dyadic, with $\mathscr{I}$ the unit dyadic,

$$\boldsymbol{B}(\nabla \cdot \boldsymbol{B}) + (\boldsymbol{B} \cdot \nabla)\boldsymbol{B} - \tfrac{1}{2}\nabla B^2 = \nabla \cdot (\boldsymbol{BB} - \tfrac{1}{2}\mathscr{I} B^2).$$

We can treat the terms in $\boldsymbol{E}$ in (11.62) similarly and if we define the dyadic $\mathscr{T}$ by

$$\mathscr{T} = \frac{1}{4\pi}\left[\boldsymbol{EE} + \boldsymbol{BB} - \frac{\mathscr{I}}{2}(E^2 + B^2)\right],$$

then the right-hand side of (11.62) can be written as $\int_{V_0} \nabla \cdot \mathscr{T}\, dv$. We use the divergence theorem,

$$\int_{V_0} \nabla \cdot \mathscr{T}\, dv = \int_{S_0} \mathscr{T} \cdot d\boldsymbol{s},$$

and in the same spirit as in the definition of field energy density we define the *field momentum density* $\boldsymbol{p}_f$ by

$$\boldsymbol{p}_f = \frac{\boldsymbol{E} \times \boldsymbol{B}}{4\pi c}. \tag{11.63}$$

Then Eq. (11.62) becomes what we call the *momentum conservation theorem*:

$$\frac{\partial}{\partial t} \int_{V_0} (\boldsymbol{p}_m + \boldsymbol{p}_f)\, dv + \int_{S_0} \mathscr{T} \cdot d\boldsymbol{s} = 0. \tag{11.64}$$

Since $\mathscr{T}$ is a dyadic, Eq. (11.64) stands for three equations, each of which has the form of the integral continuity equation for a momentum component (mechanical plus electromagnetic) in which the rate of change of that component in a volume appears to be balanced by the momentum that "flows" through the enclosing surface.

A conservation theorem for angular momentum can be derived quite similarly. We define the mechanical angular momentum density $\boldsymbol{l}_m$ by $\boldsymbol{l}_m = \boldsymbol{r} \times \boldsymbol{p}_m$. The time rate of change of $\boldsymbol{l}_m$ is given by the torque density $\boldsymbol{r} \times \boldsymbol{f}$, where $\boldsymbol{f}$ is the Lorentz force density,

$$\frac{\partial \boldsymbol{l}_m}{\partial t} = \boldsymbol{r} \times \left(\rho \boldsymbol{E} + \frac{\boldsymbol{J} \times \boldsymbol{B}}{c} \right). \tag{11.65}$$

The right-hand side of (11.65) can be written in terms of the fields only, by substituting for ρ and $\boldsymbol{J}$ from Maxwell's equations. It can then be shown, analogously to the derivation of (11.64), that if the *field angular momentum density* $\boldsymbol{l}_f$ is defined by

$$\boldsymbol{l}_f = \frac{1}{4\pi c} \boldsymbol{r} \times (\boldsymbol{E} \times \boldsymbol{B}), \tag{11.66}$$

then

$$\frac{\partial}{\partial t} \int_{V_0} (\boldsymbol{l}_m + \boldsymbol{l}_f)\, dv + \int_{S_0} \mathscr{M} \cdot d\boldsymbol{s} = 0, \tag{11.67}$$

where the dyadic $\mathscr{M}$ is

$$\mathscr{M} = \mathscr{T} \times \boldsymbol{r}.$$

Tentatively, the interpretation of Eq. (11.67) is analogous to the interpretation of Eq. (11.64) for linear momentum.

We consider now the physical interpretation of the three conservation laws (11.59), (11.64), and (11.67). We shall discuss separately three different possibilities for the time variation of fields: stationary, quasi-stationary, and generally time-varying fields. Stationary fields are time independent, electrostatic, and magnetostatic ones. Quasi-stationary fields are discussed in some detail in Section 13.3, but for the present purpose we simply say that they are generated by charges and currents that vary very slowly with time.

To investigate the three conservation laws in the stationary case, we begin with Eq. (11.64) for momentum conservation. Since the fields are time independent, $\partial \boldsymbol{p}_f/\partial t = 0$, but the term $\int(\partial \boldsymbol{p}_m/\partial t)\,dv$ does not vanish, because it is simply, by Eq. (11.60), the volume integral of the Lorentz force. It is clearer to write this term as such and the momentum conservation law becomes

$$\int_{V_0} \rho\left(\boldsymbol{E} + \frac{\boldsymbol{u} \times \boldsymbol{B}}{c}\right) dv = -\int_{S_0} \mathscr{T} \cdot d\boldsymbol{s}. \tag{11.68}$$

The volume integral on the left-hand side of this equation represents the electromagnetic force on the charges in V_0 due to the fields produced by all the charges. The real interest in this equation lies in the right-hand side which gives an alternate prescription for this force in terms of a surface integral. Now it is just this kind of surface integral that is encountered in elasticity theory in calculating the force that one part of a stressed elastic medium exerts on another. The result (11.68) is important historically in that, in Maxwell's time, it seemed to support the view that electromagnetic actions were due to electrical "stresses" in a medium, the electromagnetic ether. Historical reasons aside, the result is of no great importance practically, since the volume integral is usually as easy to work with as the surface integral. It must also be recognized that Eq. (11.68) gives only part of the force acting on a distribution, that part having electric and magnetic origin. There may, and in fact must, be other forces. For example, we have found in electrostatics that a charge distribution cannot be in equilibrium under the influence of only electrical forces (Earnshaw's theorem); there must be other forces that hold the charges in place. In general, when there are stationary currents, those flowing in wires for example, there exist nonelectric forces due to batteries or generators that are not included in the above discussion. Quite similar results apply to the angular momentum conservation law (11.67) in the stationary case. Setting $(\partial/\partial t)\int \boldsymbol{l}_f\,dv$ to zero and recognizing that $\int(\partial \boldsymbol{l}_m/\partial t)\,dv$ is the electromagnetic torque on the charges in V_0 due to electric and magnetic forces, Eq. (11.67) shows how this torque may be calculated by a surface integral. Here again there must be other torques beside the electromagnetic ones if the distribution is stationary. Finally, the energy law is of the same nature as the momentum laws. It relates $\int \boldsymbol{E} \cdot \boldsymbol{J}\,dv = \int(\partial w_m/\partial t)\,dv$, the rate that the field does work on the charges in a volume, to a surface integral. However, since there are other forces that do work, this gives only a partial picture of energy conservation, so we shall not consider it further.

We now apply the three conservation theorems to quasi-stationary fields. These differ importantly from stationary fields, for which we had to imagine that forces other than electromagnetic ones acted. This need not be the case for time-varying fields in general, and for quasi-stationary ones in particular. We can then think of the system to which we apply the theorems as consisting of an ensemble of charged particles that interact by means of purely electromagnetic forces.

The properties of quasi-stationary fields are discussed in later chapters in some detail, but we shall briefly anticipate here some of the results. For *stationary* fields, the *two* time derivatives in Maxwell's equations, $\partial \boldsymbol{E}/\partial t$ and $\partial \boldsymbol{B}/\partial t$, are zero. *Quasi-stationary* fields are defined by the approximation of equating *one* time derivative, $\partial \boldsymbol{E}/\partial t$, but not the other, to zero. This asymmetry in the treatment of the two time derivatives is not inconsistent. Since the displacement current $(1/c)(\partial \boldsymbol{E}/\partial t)$ appears in the Maxwell equation

$$\nabla \times \boldsymbol{B} = \frac{4\pi \boldsymbol{J}}{c} + \frac{1}{c}\frac{\partial \boldsymbol{E}}{\partial t},$$

it can be neglected if it is small *compared* with $4\pi \boldsymbol{J}/c$. On the other hand, $\partial \boldsymbol{B}/\partial t$ appears by itself in

$$\nabla \times \boldsymbol{E} = -\frac{1}{c}\frac{\partial \boldsymbol{B}}{\partial t},$$

so a similar approximation cannot be made with it. Because the displacement current is responsible for the existence of electromagnetic waves, its neglect effectively means neglect of such radiation.

To characterize quasi-stationary fields more quantitatively, consider a charge–current distribution that is of typical dimension a and whose time dependence is either periodic with angular frequency ω_0, or can be Fourier analyzed into a frequency spectrum whose dominant frequency is ω_0. Then if $\omega_0 a \ll c$ (slowly varying fields), it is shown in Chapter 13 that for distances r for which $\omega_0 r \ll c$, the *spatial dependence* of the field is just the spatial dependence of static electric and magnetic fields. This implies that $\boldsymbol{E}$ drops off at least as fast as $1/r^2$, and $\boldsymbol{B}$ at least as fast as $1/r^3$. If, in addition, $r \gg a$, these static fields will be negligible, compared to their values in the neighborhood of the origin. In short, throughout a sphere of radius $r_0 \approx \varepsilon c/\omega_0$, where ε is small compared with unity, but $r_0 \gg a$, the fields behave like static fields, and are negligible at the surface of the sphere; the radius r_0 is effectively "infinite" for quasi-stationary fields. We shall apply the conservation theorems to a spherical volume V_0 whose radius is of the order of r_0.

We consider first the energy theorem (11.59). Over the "infinite" sphere of radius r_0, the Poynting vector $(c/4\pi)(\boldsymbol{E} \times \boldsymbol{B})$ will be proportional to $1/r_0^5$, whereas the surface area is proportional to r_0^2. Thus for sufficiently large r_0 (sufficiently small ω_0), the surface term involving $\boldsymbol{S}$ will effectively vanish. Then Eq. (11.59) becomes just

$$\int_{V_0} (w_m + w_f)\, dv = \text{constant}. \tag{11.69}$$

Now $\int w_m\, dv$ is the kinetic energy $\frac{1}{2}\Sigma_i m v_i^2$ of the particles, so we can write this equation as

$$\tfrac{1}{2}\sum_i m v_i^2 + \frac{1}{8\pi}\int (|\boldsymbol{E}|^2 + |\boldsymbol{B}|^2)\, dv = \text{constant}. \tag{11.70}$$

As we have seen in electrostatics, the integral $(1/8\pi)\int |\boldsymbol{E}|^2\,dv$ can also be written as $\Sigma_{j<i}\,\Sigma_i\,(q_iq_j/r_{ij})$, the sum of the mutual Coulomb potential energies of the particles. So we find that Eq. (11.70) is simply the ordinary energy conservation law stating that the sum of the kinetic and potential energies of the charges plus the magnetic field energy (the energy of assemblage of the currents) is constant.

In the case of the momentum conservation theorem, we take the volume V_0 to be the same as that for the energy theorem and again argue that the surface integral vanishes so that (11.64) becomes

$$\int_{V_0} (\boldsymbol{p}_m + \boldsymbol{p}_f)\,dv = \text{constant.} \tag{11.71}$$

Equation (11.71) has the same conceptual basis as (11.69) although it is perhaps less familiar to think of momentum stored in field than it is to think of energy in the field. Nonetheless, the analogy between energy and momentum is quite complete. We have found that electrostatic energy can be expressed in terms of the field as $(1/8\pi)\int |\boldsymbol{E}|^2\,dv$ *or* in terms of the charge as $\frac{1}{2}\int \rho\Phi\,dv$. There is a similar duality with momentum. For a natural definition of the field momentum $\boldsymbol{P}_f$ is the integral of the field momentum density,

$$\boldsymbol{P}_f = \int \boldsymbol{p}_f\,dv = \frac{1}{4\pi c}\int \boldsymbol{E}\times\boldsymbol{B}\,dv \tag{11.72}$$

in which case $\boldsymbol{P}_f$ is defined by, and associated with, the field. But the momentum can be equally well expressed in terms of the charge density and vector potential. To see this, we first consider a simple example in which the charge–current distribution consists of a single point charge in the presence of some arbitrary current. In general

$$\boldsymbol{E} = -\boldsymbol{\nabla}\Phi - \frac{1}{c}\frac{\partial \boldsymbol{A}}{\partial t},$$

but for the present quasi-stationary case we can neglect $(\partial\boldsymbol{A}/\partial t)$ with respect to Φ. Substituting $E = -\boldsymbol{\nabla}\Phi$ into (11.72), we have

$$\boldsymbol{P}_f = -\frac{1}{4\pi c}\int \boldsymbol{\nabla}\Phi\times\boldsymbol{B}\,dv = \frac{1}{4\pi c}\int \Phi\boldsymbol{\nabla}\times\boldsymbol{B}\,dv = \frac{1}{c^2}\int \Phi\boldsymbol{J}(\boldsymbol{r})\,dv.$$

If the point charge q is at $\boldsymbol{R}_0$, then $\Phi \approx q/|\boldsymbol{r}-\boldsymbol{R}_0|$ and

$$\boldsymbol{P}_f = \frac{q}{c^2}\int \frac{\boldsymbol{J}(\boldsymbol{r})}{|\boldsymbol{r}-\boldsymbol{R}_0|}\,dv.$$

The integral, in the present approximation, is just c times the vector potential at $\boldsymbol{R}_0$ so

$$\boldsymbol{P}_f = \frac{q}{c}\boldsymbol{A}(\boldsymbol{R}_0).$$

For a general charge distribution $\rho(\boldsymbol{r})$, this equation obviously generalizes to

$$\boldsymbol{P}_f = \frac{1}{c}\int \rho(\boldsymbol{r}')\boldsymbol{A}(\boldsymbol{r}')\,dv'. \tag{11.73}$$

This equation for the field momentum as an integral over the charge distribution is the obvious analog of $\frac{1}{2}\int \rho\Phi\,dv$ for the energy as an integral over the charge distribution.

One basic difference exists between the energy and the momentum theorems that makes the momentum theorem less relevant to the real world. To show this, we consider a simple thought experiment. Suppose a point charge is outside an infinitely long solenoid with constant current, at a distance d from the solenoid axis. There is, of course, no $\boldsymbol{B}$ field external to the solenoid. Now if the current changes with time so that the flux through the solenoid also changes, there will be an $\boldsymbol{E}$ field which is, in fact, in the φ direction and is given by $E_\varphi = -\dot{\Phi}/2\pi dc$. A force will act on the charge but there will be no corresponding force on the solenoid. If the *system* of solenoid plus charge were in a box in which there was some automatic timing device to start or change the current in the solenoid, the charge would be set in motion after the device went off and the center of mass of the system would begin moving, even though no external force had acted on it.

Now it can be shown,* as a consequence of quite general relativistic theorems, that such behavior is impossible. We have a paradox. The paradox is resolved by recognizing that in the discussion above we have tacitly neglected relativistic effects which, in fact, make for a force on the solenoid. Since we have not yet discussed the theory of relativity, we shall merely outline the resolution of the paradox, which can be effected in at least two different ways. First, in the argument above that the charge exerts no force on the solenoid we took for granted the assumption that any small element of the current-carrying wire is neutral. This neutrality would seem to be obvious for, say, a model of electrons of a given charge density flowing in a uniform background of positive charge of the same density. But relativity theory modifies this conclusion, and predicts that the charge density of the electrons in motion will be slightly different from that density when the electrons are at rest. This comes about because the charge density is defined as the quantity of charge (an invariant as discussed in Chapter 12) per unit *volume*, and, as the reader may know, lengths (and volumes) at rest differ from lengths (and volumes) in motion by the so-called Lorentz contraction. In short, according to relativity theory, the charge density of the flowing electrons will not exactly compensate for that of the

* See S. Coleman and J. H. Van Vleck, *Phys. Rev.*, **171**, 1370 (1968) for a discussion of this point as well as for references to previous discussions of this paradox. Independently of this paper, the reader may be familiar with a famous derivation by Einstein of the mass–energy equivalence $E = mc^2$, which derivation involves the reflection of photons traveling between two mirrors fixed in a box, and is predicated on the impossibility of moving the box as a whole by this process.

positive background charge. Locally the wire of the solenoid will appear charged, even though the solenoid is uncharged as a whole. This local unbalance of charge density turns out to be proportional to $(u/c)^2$, where u is the velocity of the electrons. Since the solenoid is charged locally, an external point charge will exert forces on the elements of the wire, and these forces do not integrate to zero, but rather yield a net force on the solenoid.

A second way of resolving the paradox is to remark that to be precise in calculating the force on a small volume element of the wire, we should first calculate the force on each of the electrons and *then* average this force over a small volume. Moreover, and this is the essential point, we should use the exact relativistic mechanics rather than the approximate Newtonian mechanics. Now in relativistic mechanics the equation of motion of an electron involves not $m_0(d\boldsymbol{u}/dt)$, where m_0 is the electron (rest) mass, but rather involves the time rate of change of the relativistic momentum

$$\frac{d}{dt}\left(\frac{m_0\boldsymbol{u}}{\sqrt{1-u^2/c^2}}\right).$$

The relativistic and nonrelativistic results differ by terms of order $(u/c)^2$. Taking these terms into account, it can be shown* that there is a force on the solenoid of just the correct magnitude to keep the center of mass of the system at rest. The paradox is thereby resolved.

We have seen above that one cannot consistently use the momentum conservation theorem in the quasi-stationary case if the motion of the particles is treated nonrelativistically. A somewhat different way of seeing this is to observe that the field momentum is proportional to $1/c^2$. In Eq. (11.73), for example, there is an external factor $1/c$, and $\boldsymbol{A}$ in the integrand is proportional to $1/c$. Then if we consider the balance of electromagnetic and mechanical momentum, we must have the mechanical momentum correct at least to terms of order $1/c^2$; this means we must use the relativistic expression for it.

The angular momentum conservation theorem in the quasi-static case is similar to that for linear momentum. It simply becomes, with $\boldsymbol{L}_f = \int \boldsymbol{l}_f\, dv$ and similarly for $\boldsymbol{L}_m$,

$$\boldsymbol{L}_m + \boldsymbol{L}_f = \text{constant}.$$

However, it, too, is really inconsistent if the mechanical momentum is defined in terms of Newtonian rather than relativistic mechanics so we shall not consider it further here.

We now consider the conservation theorems for general time-varying fields. We shall focus almost exclusively on the energy theorem (11.59), although similar remarks pertain to the two momentum theorems. We can apply the energy

---

* *Ibid.*

theorem to an arbitrarily small volume, which means, in effect, that we can equate the integrands. Rewriting the surface integral as a volume integral, we obtain

$$\nabla \cdot \boldsymbol{S} + \frac{\partial w}{\partial t} = 0. \tag{11.74}$$

Here $w = w_f + w_m$ is the combined electromagnetic and mechanical energy density. Aside from the approximation involved in using Newtonian mechanics in its derivation, there is no question about the validity of (11.74). As an equation relating mechanical energy to fields and their time derivatives, it is as sound as the Maxwell equations on which it is straightforwardly based. On the other hand, Eq. (11.74) has historically had certain difficulties and obscurities associated with it. These arise from attempting to *interpret* it in terms of energy and its flow. The equation does have the formal appearance of a continuity equation relating the rate of change of the energy in a small volume to the flow of energy, as given by the divergence of the purported energy current $\boldsymbol{S}$. This interpretation leads to certain traditional difficulties we now discuss.

The first dilemma arises on trying to infer, from Eq. (11.74) for the *divergence* of $\boldsymbol{S}$, the meaning of $\boldsymbol{S}$ itself. If it is held that the form of (11.74) entitles one to think of $\boldsymbol{S}$ as describing an energy current density, it must be recognized that if $\boldsymbol{S}$ is replaced by $\boldsymbol{S} + \boldsymbol{S}'$, where $\nabla \cdot \boldsymbol{S}' = 0$ but $\boldsymbol{S}'$ is otherwise an arbitrary vector field, that $\boldsymbol{S} + \boldsymbol{S}'$ also satisfies Eq. (11.74). On the other hand the flow pattern of the purported energy flow may be markedly different for $\boldsymbol{S} + \boldsymbol{S}'$ than for $\boldsymbol{S}$. Therefore, Eq. (11.74) does not predict a unique energy flow.

A second classic dilemma arises on assuming that energy flows wherever $\boldsymbol{S} = (c/4\pi)(\boldsymbol{E} \times \boldsymbol{B})$ is different from zero. Energy flow is thereby predicted in purely static electric and magnetic fields, for example in the field generated by a point charge in the presence of a bar magnet; a momentum density $\boldsymbol{p}_f = (1/4\pi c)(\boldsymbol{E} \times \boldsymbol{B})$ is similarly predicted. But the notion of flow of energy or of momentum in static fields is perplexingly unintuitive.

We now discuss these two dilemmas. Equation (11.74) is conventionally considered to relate the change of energy localized in a small volume to the flow of energy through the bounding surface. The analogous equation for momentum, linear or angular, is similarly supposed to relate the change of momentum localized in a volume to the flow of momentum through a bounding surface. The key concepts we want to focus on here are *localization* and *flow*, since they are at the root of the trouble. First, localization. To begin generally suppose some physical quantity Q is defined as a volume integral, say $Q = \int q(\boldsymbol{r})\, dv$. We are tempted to consider $q(\boldsymbol{r})$ as the "density of Q," but this is not always meaningful. The reason is that integrands are not unique: they can be added to, subtracted from, and, in general, transformed in an infinite number of ways that may leave the integral itself unchanged. Such transformed integrands may, and usually do, differ markedly in their spatial dependences. Clearly, they cannot all represent the density of Q, and we do not lack examples to illustrate this point. For example, think of the energy of assemblage W_E of a charge distribution, as given by Eqs. (10.4) *or* (10.5); namely, $\frac{1}{2}\int \rho\Phi\, dv$ *or* $(1/8\pi)\int |\boldsymbol{E}|^2\, dv$. If we consider a thin shell of charge, say, the "energy density" $\frac{1}{2}\rho\Phi$ describes energy which is in the shell and nowhere else and the "energy density" $(1/8\pi)|E|^2$ claims that the energy is everywhere but with only a small fraction in the shell. A similar situation obtains for the electromagnetic momentum, which when expressed in terms of the field by means of Eq. (11.72) is purportedly spread throughout space and when expressed in terms of the charge by means of Eq. (11.73) appears to be localized within the charge.

The question of the flow of electromagnetic energy is closely tied to that of its localization. For matter, or for charge, there is no ambiguity in the concept of flow. *Flow* implies that a quantity of stuff that is at one point at one time will be at another point later and at intermediate points for intermediate times. Such flow can be measured for say, a liquid or gas; localized parts of a liquid can be dyed, for example, or radioactive tracers can be used to follow the flow of a gas. But since one cannot identify the electromagnetic energy in a volume, one can scarcely verify that this same energy has moved someplace else. From this viewpoint, the concept of flow of electromagnetic energy is a tenuous one.

On the basis of these remarks, it is hard to see how the usual interpretation of Eq. (11.74), which relates the flow of electromagnetic energy as given by the Poynting vector to the change of localized energy, can be taken seriously. On the other hand the Poynting vector (usually) "works." One can not argue with its utility. It works in all kinds of applications, in physics and engineering, in the calculation of antenna patterns, of the radiation from point charges, etc., etc. It must be basically correct, in some sense. Our thesis will be that the concept of the flow of electromagnetic energy *is* correct, in a loose sense. Thus, when *energy flow* is used as a kind of vague shorthand description, it is satisfactory; but when one thinks of it in too close analogy to material flow, its looseness becomes apparent and leads to dilemmas of the kind we have described.

The discussion of this thesis for the electromagnetic field can be aided by analogies drawn with a simple mechanical system. Consider two pendula, 1 and 2, that are coupled by a spring that joins them. This system is a crude model for two charged particles that are coupled by the electromagnetic field. The system can be started by holding pendulum 2 fixed in its equilibrium position while displacing pendulum 1 and setting it in motion. To accomplish this displacement and beginning motion, we would have to apply a force to the pendulum 1, let it act over a certain distance, and then release the bob. The *work* done on the bob, and hence the *energy* given it in such a displacement, is of course calculable as the integral of the net force on the bob and the displacement. Given the initial conditions, the subsequent motion of this coupled system is well known: Pendulum 1 has a certain initial amplitude of motion. After a period of time, it will come to rest and pendulum 2 will be in motion with the full amplitude that pendulum 1 had. Still later pendulum 1 will regain this amplitude, and so it goes, periodically.

There is one crucial phrase in the above discussion. We have said that, in the initial displacement of pendulum 1, energy was given to its bob. This is not precise. In displacing the bob, we are not giving energy to *it*, but since the bob is coupled by a spring to the second bob, we are doing work on, and giving energy to, the *system as a whole*. Thus it is the energy of the whole system that is conserved, not the "energy" of the bob.

Although the language of "giving energy to the bob" is loose, it is illuminating to see how far we can carry it before getting into trouble. In this language, it is only natural, when pendulum 1 and 2 have changed roles (when 2 has attained the amplitude that 1 had and 1 is at rest), to say that energy has appeared in pendulum 2. And if we think of energy as the material substance it is *not*, the only way of describing how energy got from the first bob to the second is to say that it flowed from 1 to 2. In fact, this language is roughly descriptive of the facts. If we described to a person with some grasp of the concept of energy the behavior of the system in terms of a periodic *flow* of energy from one pendulum to the other, he would probably infer correctly the qualitative behavior we have described above. However, our layman would encounter difficulties when he began to examine the concept of flow more closely, by analogy with the flow of charge or matter. What is the path the energy takes? Does it wind round and round along the helical coils of the spring? If so, given a second spring of some other material but the same spring constant, so that the length along the helix is different, why does the longer path of energy flow not manifest itself in a different behavior of the system? On trying to answer questions like these, we see that the concept of energy flow does not replace the correct and detailed picture of the physical mechanism: the initial displacement of the one pendulum stretches the spring, causing a force to act on the second, as a result of which force the second pendulum is set into motion.

Now we carry this viewpoint over to the electromagnetic field. In a rough sense we may often say that electromagnetic energy flows. Take the typical example of a radio transmitter that emits energy which, we say, flows through space to a radio receiver where it is absorbed. This is a fine qualitative picture, good as far as it goes. On pushing it further, however, one arrives at the difficulties and dilemmas that introduced this subject. The reason for this is that the concept of energy flow is really a summary of a whole set of physical processes. Namely, a generator of some kind drives electrons against resistive forces in a transmitting antenna. As a result, the electrons accelerate and emit electromagnetic waves, which propagate and impinge on a receiving antenna. Work is done on the electrons in the receiving antenna and they are set in motion, that is, acquire energy. For *brevity* then, we may say that energy has flowed from the transmitter to receiver; but we must remember that this merely summarizes a complicated process in which it is the field that really flows or propagates. It is not an invitation to consider that the flow of energy resembles the flow of a fluid in any detailed way.

Since the concept of the flow of energy is a loose one and ultimately not even a consistent one, it is natural to look at this looseness as the source of the difficulties mentioned above in connection with the Poynting vector. The first such difficulty is the nonuniqueness of $\boldsymbol{S}$ because of the possibility of adding to it a divergenceless vector field $\boldsymbol{S}'$. Although this dilemma is not resolvable on the basis of the energy theorem, one knows the answer empirically: that $\boldsymbol{S}' = 0$, since in all practical work this assumption is made, and, as we have remarked, a considerable body of results that agree with experiment have been calculated using this assumption. But we can also conclude *a priori* that $\boldsymbol{S}' = 0$ by looking in more detail at what is meant operationally by the flow of energy in any particular case, i.e., by looking at what is measured. Consider for example the many correct calculations that have been made of the power radiated from radio transmitters, from so-called antenna arrays. The quantity we calculate that gives agreement with experiment is the Poynting vector in the radiation zone, i.e., the Poynting vector at "infinity," as a function of angles measured with respect to the transmitter. But if the Poynting theorem had never been discovered, the same angular distributions could be predicted by analyzing the mechanism of detection. To select one from the many possible cases, a typical detector might be a bolometer that measures the heating effect of the electric field incident on it, which effect turns out to be proportional to $|\boldsymbol{E}|^2$, the square of the electric field at the bolometer. The angular distribution of $|\boldsymbol{E}|^2$ in the radiation zone is the same as that for $\boldsymbol{S}$ itself, not $\boldsymbol{S}$ as supplemented by a divergence-free $\boldsymbol{S}'$. This modest inquiry into the mechanism of the purported energy flow is then enough to dispose of the ambiguity of the additional divergence in the Poynting vector, and indeed to dispose of the need for the Poynting vector itself, if one so wishes.

This operational point of view toward the flow of energy resolves the dilemma of energy (or other physical quantities) flowing in static fields. Our thesis is that flow of energy is a shorthand description of a process in which the motion of a charge generates a field, which propagates to another charge, sets it in motion, and thereby transfers "energy" to it. Then it is clear that there can be no flow of energy if fields are static and do not propagate. This conclusion is not meant to invalidate our previous conclusions that static and quasi-static fields can contain energy. If we integrate over all space, thereby foregoing the attempt to localize energy (or momentum), we may consider the field to contain these quantities, but we cannot think of them as localized, or as flowing, in any precise way.

PROBLEMS

1. For two-dimensional fields, all quantities are independent of one coordinate, say the z-coordinate. Show that Maxwell's equations uncouple for such fields and become two independent sets, the first relating E_z, B_x, and B_y to J_z, and the second relating B_z, E_x, and E_y to J_x and J_y.

2. Show that if F is an arbitrary function, then

   $F(x \pm ct)$ are solutions of the wave equation in one dimension;

   $\dfrac{F(t \pm r/c)}{r}$ are solutions in three dimensions;

   $F(t \pm \rho/c)/\sqrt{\rho}$ are asymptotic solutions (for the cylindrical coordinate $\rho \gg 1$) in two dimensions.

3. Show that the continuity equation follows from Maxwell's equations.

4. The only component of $\boldsymbol{E}$ of a certain electromagnetic field is E_y given by

$$E_y = \begin{cases} E_0 \cos \omega(z/c - t), & z > 0 \\ E_0 \cos \omega(z/c + t), & z < 0. \end{cases}$$

   Find the current source that produces this field.

5. Once in every lifetime one is tempted to interpret

$$\nabla \times \nabla \times \boldsymbol{E} = \nabla(\nabla \cdot \boldsymbol{E}) - \nabla^2 \boldsymbol{E}$$

   in other than Cartesian coordinates. Try it in spherical coordinates:

$$\nabla^2 E_r \stackrel{?}{=} -(\nabla \times \nabla \times \boldsymbol{E})_r + \nabla_r(\nabla \cdot \boldsymbol{E}).$$

6. Show by direct differentiation that the $\boldsymbol{A}$ and Φ given by (11.45) and (11.46) satisfy the Lorentz condition.

7. With the assumption of harmonic time dependence

$$\psi(\boldsymbol{r}, t) = \Phi(\boldsymbol{r}) \begin{cases} \sin \omega t \\ \cos \omega t \end{cases}$$

   the wave equation becomes the Helmholtz equation $(\nabla^2 + k^2)\Phi = 0$. Show that

$$\int_C f(\alpha) e^{ik\rho \cos(\varphi + \alpha)} \, d\alpha$$

   is a solution in cylindrical coordinates ρ, φ, where $f(\alpha)$ is arbitrary, as is the contour C in the complex plane.

8. Prove Eq. (11.25). One way to proceed is to use the definition

$$\frac{d\phi}{dt} = \lim_{\tau \to 0} \frac{\phi(t + \tau) - \phi(t)}{\tau}$$

   and the fact that $\nabla \cdot \boldsymbol{B} = 0$.

9. Derive the angular momentum conservation theorem (11.67).

10. Verify that Eq. (11.68) for the force on a charge distribution in terms of the stress tensor correctly gives the Coulomb law between two point charges. Use the infinite plane perpendicularly bisecting the line between the charges as the surface for integration.

11. In a neutral medium for which Ohm's law, $\boldsymbol{J} = \sigma \boldsymbol{E}$, holds, show that

$$\nabla^2 \begin{pmatrix} \boldsymbol{E} \\ \boldsymbol{B} \end{pmatrix} - \frac{4\pi\sigma}{c^2} \frac{\partial}{\partial t} \begin{pmatrix} \boldsymbol{E} \\ \boldsymbol{B} \end{pmatrix} = 0$$

if the time variation of the fields is such that $\partial \boldsymbol{E}/\partial t \ll \boldsymbol{J}$ (displacement current negligible with respect to conduction current).

12. Solve Eq. (11.39) for the Green function $G_0(\boldsymbol{r}, t)$ by Fourier transformation, that is, by writing

$$G_0(\boldsymbol{r}, t) = \frac{1}{(2\pi)^2} \int g(\boldsymbol{k}, \omega) e^{i(\boldsymbol{k} \cdot \boldsymbol{r} + \omega t)} \, d\boldsymbol{k} \, d\omega,$$

using (11.39) to find $g(\boldsymbol{k}, \omega)$ and then integrating. Note that this seemingly straightforward technique leads to an integral with a singular integrand and this integral must be carefully redefined to yield a Green function with the correct physical properties.

13. Discuss in as much detail as you can the motion of an uncharged sphere, made of a conductor obeying Ohm's law, and moving in a uniform magnetic field, subject only to the Lorentz force on the electrons in it. See H. B. Rosenstock and T. A. Chubb, *Am. J. Phys.* **24**, 413 (1956) and the references cited therein.

14. A stationary current distribution consists of N loops of wire of negligible cross section, with current I_i in the ith loop. Show that the magnetic energy of assemblage W_M, given in general by Eq. (11.13) can be written

$$W_M = \frac{1}{2} \sum_{i=1}^{N} I_i^2 L_i + \sum_{i=1}^{N} \sum_{j>i}^{N} I_i I_j M_{ij}$$

where L_i, a *coefficient of self induction*, is the magnetic flux through the ith loop due to unit current in it, and M_{ij}, a *coefficient of mutual induction*, is the magnetic flux through the ith loop due to unit current in the jth.

15. The fields of a "small" but arbitrary current distribution can be found from $\boldsymbol{A}$ and Φ by expanding $|\boldsymbol{r} - \boldsymbol{r}'|$ in their integrands in a Taylor series about the point $x' = y' = z' = 0$, as in the multipole expansion for static fields. Show that to lowest order (dipole approximation) in this expansion, the nonvanishing field components of the current distribution $\boldsymbol{J}(\boldsymbol{r}, t) = \boldsymbol{J}_0(\boldsymbol{r}) \cos \omega t$ are

$$E_r = \frac{2p \cos\theta}{r^2} \left[\frac{\cos(kr - \omega t)}{r} + k \sin(kr - \omega t) \right],$$

$$E_\theta = \frac{p \sin\theta}{r} \left[\left(\frac{1}{r^2} - k^2 \right) \cos(kr - \omega t) + \frac{k}{r} \sin(kr - \omega t) \right],$$

$$B_\varphi = -\frac{pk^2}{r} \left[\cos(kr - \omega t) - \frac{1}{kr} \sin(kr - \omega t) \right],$$

where $\boldsymbol{p} = \int \boldsymbol{J}_0 \, dv$, $k = \omega/c$, and the z-axis is along $\boldsymbol{p}$.

REFERENCES

Scanlon, P. J., R. N. Henriksen, and J. R. Allen, "Approaches to Electromagnetic Induction," *Am. J. Phys.* **37**, 698 (1969).

This article critically reviews the pedagogical literature on the subject of Faraday's law and the flux rule, and thereby contains many examples on the calculation of induced emf.

Whittaker, E. A., *History of the Theories of Aether and Electricity*, Thomas Nelson, London (1951).

The standard history, with a chapter on the work of Faraday, and one on the work of Maxwell.

12
ELECTRODYNAMICS AND RELATIVITY

This chapter discusses the special theory of relativity, especially as it relates to electromagnetism. To keep a proper balance, we have been brief in the development of relativity theory *per se* and have not, for example, dwelled on the "paradoxes" connected with it. However, we have tried to provide a complete, if condensed, treatment of the theory, with major emphasis on the electromagnetic applications.

At its inception, the special theory of relativity appeared to imply a revolutionary overthrow of established concepts of physics. Nonetheless, like most new physical theories, it shared much with the theory it replaced. The latter theory has been called *Galilean relativity*. For a better perspective on the special theory, we begin by discussing Galilean relativity and the associated concepts of *inertial frames* and *absolute time*.

12.1 NEWTON'S SECOND LAW AND INERTIAL FRAMES

Galilean relativity is closely connected with Newton's second law of motion

$$\boldsymbol{F} = m\boldsymbol{a}. \tag{12.1}$$

We begin then with a brief discussion of the conceptual basis of this law, which involves three quantities (or concepts). We assume that acceleration $\boldsymbol{a}$ needs no explanation except to remark that it is measured with respect to some well-defined reference frame. The other two quantities, *force* $\boldsymbol{F}$, and *mass* m (more properly, inertial rest mass) do require more discussion, but we need to discuss only one or the other, since if one of them is defined independently of Eq. (12.1), then that equation *defines* the other. We take the concept of mass as logically prior to that of force, and *define* the mass of a body, or rather the mass ratio of two bodies, by means of experiments involving slow collisions.* Briefly, if two bodies, 1 and 2, first approach each other along the line joining them with (vectorial) velocities v_1 and $-v_2$, respectively, then collide, and finally part along the same line with velocities $v_1 + \Delta v_1$, $-(v_2 + \Delta v_2)$, we *define* the relative masses m_1 and m_2 to be such that $m_1 v_1 - m_2 v_2 = m_1(v_1 + \Delta v_1) - m_2(v_2 + \Delta v_2)$. This equation implies that

$$\frac{m_1}{m_2} = \frac{\Delta v_2}{\Delta v_1}. \tag{12.2}$$

---

* For more details on this approach to defining mass, see Jammer, Max, *Concepts of Mass*, Cambridge University Press (1961).

Taking the mass m_1 as an arbitrary standard, we can determine the masses m_2, $m_3, \ldots, m_n$ of any of a set of bodies.

Next, given some entity (such as a compressed spring) that can "exert a force," we *define* the magnitude of that force quantitatively by means of Newton's second law. That is, this magnitude is taken to be the product of a *known* mass and the *measurable* acceleration it imparts to that mass. The importance of Newton's second law is that, given *any* mass, it enables one to characterize a *set* of forces and measure their magnitudes $F_1, F_2, \ldots, F_n$. Once this characterization is made, the law then enables us to calculate the acceleration of any *other* mass m_i of the set, when acted on by any force F_j. Thus, suppose that we had l springs of various strengths, and n different bodies. To find the acceleration produced by *any* of the springs acting on *any* of the bodies would require, without Newton's law, a number ln of experiments. With it, only $l + n - 1$ need be done: $n - 1$ collision experiments determine the masses, and l acceleration measurements with a *single* mass determine the forces.

We return now to the question of reference frames. In the discussion above, it has been tacitly assumed that the reference frame is one for which Newton's laws can be verified, i.e., for which in the example, $l + n - 1$ measurements do imply ln results. If now the frame is put on a large turntable and rotated, keeping everything else as before, then equation (12.1) will not be satisfied. The equation that *will* be satisfied is one in which there are certain supplementary terms due to so-called *centrifugal* and *Coriolis* accelerations. There are, then, reference frames in which Newton's law is valid and ones in which it is not. The frames in which it is valid are called *inertial frames* or *inertial coordinate systems*; they play a central role in the theory of special relativity.* Given one inertial frame, an infinite number of others can be derived from it by means of so-called *Galilean transformations*. We discuss these transformations now and examine more closely the concept of *absolute time* that is inherent in them and that was overthrown by the special theory.

12.2 GALILEAN RELATIVITY AND ABSOLUTE TIME

Galilean transformations are certain mathematical (as opposed to physically verifiable) relations between the coordinates x, y, z and the time t of a point or particle in one coordinate system and x', y', z', t' for the *same* point or particle in a system moving with uniform velocity with respect to it. Figure 12.1 illustrates an unprimed system and a primed system that move in the relative x-direction with velocity v. Suppose that the relation between the x-coordinate and the x'-coordinate is the "obvious" one, $x' = x - vt$, that the y and z coordinates are unchanged on

* Another common way of defining inertial frames is to observe that Newton's law predicts that a body subject to no net force will move with a constant velocity. If then, in a given coordinate frame, a body subject to no force does move uniformly, that frame is said to be an inertial one. The definition is somewhat circular, however, since the criterion that a body not be acted on by a force is that it move uniformly.

transformation, and that the times, t and t', are identical. These various relations constitute the *Galilean transformation*:

$$\begin{aligned} x' &= x - vt, \\ y' &= y, \\ z' &= z, \\ t' &= t. \end{aligned} \tag{12.3}$$

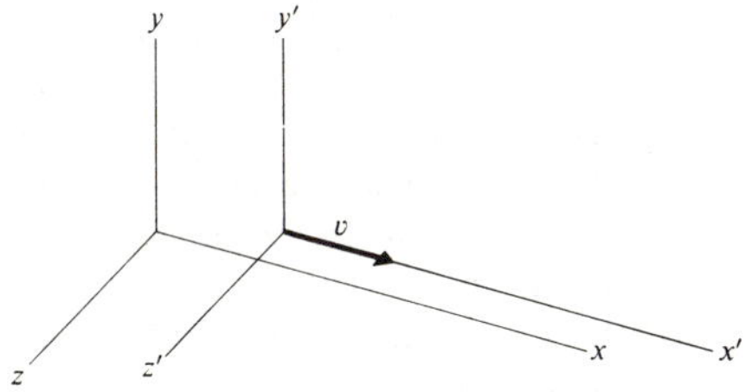

Fig. 12.1 Coordinate systems with relative velocity v.

The equation $t' = t$ is innocent in appearance, but let us consider it in more detail. It says that the measure of time is the same in the primed and the unprimed system. This time, common to the two systems and, of course, to any of the infinite number of others in relative uniform motion, is Newton's *absolute time*. This concept of absolute time is, roughly, that of some super master clock whose readings can be transmitted instantly to any or all experimenters in any or all uniformly moving frames of reference. However, no such clock exists, so we shall have to define absolute time from a more operational viewpoint. We proceed in three steps: First, we define time at *one* point of a given reference frame; second, we define a time common to *different points* of the *same* reference frame; and third, we define a time common to *uniformly moving* frames as well. This last is Newton's absolute time, operationally defined.

First, at a single point in space, time is measured by a *clock*, a device that returns periodically to some given condition. The device may be a pendulum or other mechanical or electrical or atomic oscillator; it may be any mechanism that departs from, and then returns to, some identifiable state. The quantitative measure of the *time* at that point is the number of repetitions of that state which have occurred since an arbitrary time origin.

To compare time measurements between *different* points of the same reference frame, we must somehow set up clocks at those points. There are various possibilities. For example, we might synchronize many identical clocks at the origin, i.e., set them to read identically, and then carry each of them to a point in space. In practice, one difficulty with this method is that there is no guarantee that the process of transporting the clocks does not affect their readings; equally, we cannot be sure that the physical conditions at the various points of space are precisely identical to those at the origin. Of course, we are not now really concerned with the truly practical experimental details and are permitted to idealize. But the slightest difference between the running of the clock at a point and that of the clock at the origin will eventually make an arbitrarily large difference between their readings. The high degree of idealization that is involved in this method is then rather uncomfortable.

We stay much closer to reality, although still an idealized one, by beginning with an ensemble of identical clocks which are *first* transported to various points of space and are then synchronized (and their calibration checked) by *signals* which are sent from one to another. This is the method we shall adopt. First, we must decide on a mode of signaling; we shall, of course, not choose light waves since the whole theory of relativity deals with the unexpected properties of the propagation of light. Think then in terms of some kind of acoustic signal, say, since the propagation of sound is presumably well understood. Moreover, since sound requires a *medium* in which to propagate, this choice will help to clarify an important contrast to the propagation of light which does *not* require such a medium. Given then a reference frame, fixed in a medium (assumed uniform and isotropic), we set up a clock at the origin O and carry another identical one to some other point P. We synchronize the two clocks by first arranging to have some kind of acoustic wave emitted from the origin and by then contriving to have this wave reflected from the clock at P. At the instant the wave is generated, we set the clock at the origin to zero. We record the time, call it $2t_0$, at which the reflected wave comes back from the point. Assuming that the wave takes the same time to go to P as it does to return to O, we conclude that the wave takes time t_0 to arrive at P. We then alert the observer at P to the fact that another signal will be sent, say at time T on the clock at O, and instruct him to set his clock at time $T + t_0$ when that signal is received. The clock at P is thereby *synchronized* with that at the origin. As many clocks at as many points as we like can be so synchronized, and a network is thereby constructed that can reasonably be said to define time throughout that reference frame. We shall call this time the *frame time*.

Now we shall show how *frame times* can be compared for different frames that are moving uniformly with respect to the stationary one discussed above, thus defining a universal or *absolute* time. This absolute time will be defined with the help of the Galilean transformations, so we shall show somewhat more precisely that the concepts of absolute time and of Galilean transformations are self-consistent.

Given some *moving* frame with its network of clocks fixed in it, suppose that we want to synchronize one of these clocks, say that at Q, with the clock at the origin of the stationary system. In general, this can be done by the same kind of acoustic signaling used in defining frame time, but the details are somewhat more complicated. The acoustic signal travels at a certain velocity in the medium, i.e., with respect to the reference frame fixed in that medium, but in the moving frame it will appear to travel at a different velocity. Assuming that the Galilean transformation is correct, we can correct for this. For concreteness, think of the two reference frames moving along their relative x-axis, as in Fig. 12.1, so that the Galilean transformation is that of Eq. (12.3). Then this transformation between positions of points can be made into one between velocities of such points. That is, if we differentiate (12.3) with respect to assumed absolute time (here is where the self-consistency enters) and write $u' = dx'/dt$, and $u = dx/dt$, we find that

$$u' = u - v. \tag{12.4}$$

This formula relates the velocities of any point, for example, the point that is the intersection of the acoustic wave front with the x-axis. If u is the sound velocity in the stationary system, then u' will be the velocity in the primed system. But with this velocity known, we can synchronize clocks in the moving system with those in the stationary system. For example, if an acoustic signal is sent out from the origin at $t = 0$ (at which time the origins of the two systems coincide), it will arrive at a point x'_0 after an interval $t'_0 = x'_0/u'$. If the clock at x'_0 is then set to t'_0 when the signal arrives at it, it will be synchronized with the clock at the origin of the rest frame and, hence, with all the other clocks in the rest frame. Proceeding in this way, we can compare any clock in any frame with any other clock and so arrive at a set of clocks that defines a *universal* time, which can be considered to exist at all points of all reference frames moving uniformly with respect to one another.

With these concepts, we return to Newton's equations and the *principle of Galilean relativity*. Consider the equations for the motion of an ensemble of particles, with vector positions $\boldsymbol{r}_1, \boldsymbol{r}_2, \ldots, \boldsymbol{r}_n$, which we suppose move subject to forces which are derivable from potentials $V(|\boldsymbol{r}_i - \boldsymbol{r}_j|)$ that act between the particles. The equation of motion of the ith particle is

$$m_i \frac{d^2\boldsymbol{r}_i}{dt^2} = -\sum_j \nabla_i V(|\boldsymbol{r}_i - \boldsymbol{r}_j|). \tag{12.5}$$

If these equations are transformed to any other Galilean system through use of the transformation (12.3), we have $|\boldsymbol{r}_i - \boldsymbol{r}_j| = |\boldsymbol{r}'_i - \boldsymbol{r}'_j|$ and $d^2\boldsymbol{r}_i/dt^2 = d^2\boldsymbol{r}'_i/dt'^2$. In the primed system, the equations will have the same *form* as in the unprimed:

$$m_i \frac{d^2\boldsymbol{r}'_i}{dt'^2} = -\sum_j \nabla_i V(|\boldsymbol{r}'_i - \boldsymbol{r}'_j|). \tag{12.6}$$

This *form-invariance* of Newton's equations under Galilean transformations is called the principle of *Galilean relativity*. It will be the analogous *form-invariance* of Maxwell's equations under *Lorentz transformations* that is central to the theory of *special relativity*.

12.3 THE LORENTZ TRANSFORMATION

Galilean transformations do not apply to the physical world except as an approximation. They are superseded by the Lorentz transformation. We begin with some history. There have been many theories of light through the ages, including a rather complete early corpuscular theory. Toward the middle of the nineteenth century, however, wave theories achieved ascendancy. At first these were patterned after the theories of elastic waves in continuous matter and they culminated in the theory by Maxwell which we have presented, and which was enormously successful. But it raised one important question which vexed physicists for decades. What were the properties of the medium that carried electromagnetic waves? It was difficult in the mechanistic context of the time to think of vibrations, or waves, without thinking of some medium which actually vibrated and in which the wave itself was but a special pattern of vibration. Thus, for acoustic waves in air, the medium is the ensemble of air molecules and the wave is a periodic longitudinal compression and rarefaction of that medium. Similarly, an elastic wave in an (assumed) continuous solid is a periodic displacement, both longitudinal and transverse, of the actual physical elements of the solid. It is the solid that vibrates; a wave is a special spatio-temporal kind of vibration of the solid. Moreover, for such waves there is a *preferred frame*, one which is unique and distinct. This frame is, of course, the one fixed in the medium.

For an electromagnetic wave, then, the question was: What is the nature of the vibration and of the medium that supports it? A semantic "answer" was given: the medium was called the electromagnetic ether. But the giving of a name did not answer physical questions. Where is the ether situated with respect to absolute space? What are its material properties? What is its effect on experiments done in frames moving with respect to it?

At the end of the last century many questions like these were asked, and partial answers given. There was hypothesis and counter hypothesis, experiment and counter experiment—a recounting of which would constitute an appreciable portion of the history of physics of that time. Finally, after a crucial experiment of Albert Michelson and Edward Morley, what is now considered to be the truth of the matter emerged. The mechanistic model is too narrow. *There is no ether; there is no single preferred frame for propagation of electromagnetic waves with velocity c; this velocity is the same for all inertial frames in uniform relative motion.* This universality of the velocity c contradicts the law of addition of velocities that the Galilean transformation implies. According to that transformation, if the velocity of light is c in one frame, it should be $c \pm v$ in a frame moving with uniform velocity v with respect to it. But, experimentally, this is not the way the world works. We can only conclude that some of the assumptions that underlie the Galilean transformation are not correct. One of these assumptions was the existence of an absolute time; it is this assumption that will have to be discarded.

Not all aspects of time measurement we have discussed need to be revised. As in the previous section, we can, for a given frame, still define a *frame time t* in terms of a network of clocks fixed in it as well as a *frame time t'* for any frame moving uniformly with respect to it. But we can no longer identify t with t'; *each frame has* *its private time*. We consider now the generalization of the Galilean transformation between x, y, z, t and x', y', z', t' subject to this condition.

In deducing this connection, it is natural to use the striking experimental result that the velocity of light must be the same in the two systems. In applying this, we begin with some simplifying assumptions.* We assume first that the transformation is a linear one. Second, we note, referring to Fig. 12.1, that the contradiction between the Galilean velocity addition formula and the experimental fact of the constancy of the light velocity has nothing to do with the transformation of the y and z coordinates. We assume then that these are the same as for the Galilean transformation: $y = y'$ and $z = z'$. The basic assumption is then that x and t are linear functions of x' and t':

$$x' = ax + bt, \tag{12.7}$$

$$t' = dx + ft, \tag{12.8}$$

It is also assumed here that at the instant when the two origins coincide ($x = x' = 0$), the clock at each origin is set to read zero ($t = t' = 0$). In other words, the time origins coincide when the space origins coincide. The transformation deduced under this condition is called the *homogeneous* one. We can eliminate one of the constants in (12.7) by noting that the origin of the primed system, $x' = 0$, is a point described in the unprimed frame by $x = vt$. This shows that $b = -va$, and we have

$$x' = a(x - vt). \tag{12.9}$$

We now use the fact that the velocity of light is the same in the two systems. This means that if a spherical light pulse is sent out from the origin at $t = t' = 0$, the equation of the wavefront as a function of time is given by

$$x^2 + y^2 + z^2 = c^2t^2 \tag{12.10}$$

or by

$$x'^2 + y'^2 + z'^2 = c^2t'^2. \tag{12.11}$$

Both these equations must hold simultaneously. Substituting t' from (12.8) and (x') from (12.9) into (12.11), we find that

$$a^2(x - vt)^2 + y^2 + z^2 = c^2(dx + ft)^2$$

or, on rearranging, that

$$(a^2 - c^2d^2)x^2 - 2(va^2 + c^2fd)xt + y^2 + z^2 = (c^2f^2 - a^2v^2)t^2.$$

If this is to be of the form (12.10), the coefficients of x^2, t^2, and xt in it can be equated to the same coefficients in (12.10) to obtain

$$c^2f^2 - a^2v^2 = c^2,$$

$$a^2 - c^2d^2 = 1,$$

$$va^2 + c^2fd = 0.$$

* For a detailed discussion of the rationale of these assumptions, see any standard text, e.g., Moller (R).

Eliminating first d from the second and third of these equations and then a^2, we find that $f^2 = 1/(1 - v^2/c^2)$, whence the first equation yields $a^2 = f^2$, and the third yields d^2. Taking the positive square root of f and a so the transformation reduces to the Galilean one in the limit $v \to 0$, we find

$$\begin{aligned} f &= 1/\sqrt{1 - v^2/c^2}, \qquad a = f; \\ d &= -\,v/(c^2\sqrt{1 - v^2/c^2}). \end{aligned} \tag{12.12}$$

Substitution of these results in (12.8) and (12.9) then yields the *Lorentz transformation equations*

$$\begin{aligned} x' &= \frac{x - vt}{\sqrt{1 - v^2/c^2}}, \\ y' &= y, \\ z' &= z, \\ t' &= \frac{t - \dfrac{v}{c^2}x}{\sqrt{1 - v^2/c^2}}. \end{aligned} \tag{12.13}$$

With the common abbreviations

$$\beta = v/c, \qquad \gamma = \frac{1}{\sqrt{1 - v^2/c^2}} \tag{12.14}$$

the transformation becomes

$$\begin{aligned} x' &= \gamma(x - vt), \\ y' &= y, \\ z' &= z, \\ t' &= \gamma(t - \beta x/c). \end{aligned} \tag{12.15}$$

The *inverse* transformation, that is from x to x', etc., is, of course, obtained from Eqs. (12.13) by changing the roles of primed and unprimed variables, and reversing the sign of v:

$$\begin{aligned} x &= \gamma(x' + vt'), \\ y &= y', \\ z &= z', \\ t &= \gamma(t' + \beta x'/c). \end{aligned} \tag{12.16}$$

12.4 POINT EVENTS

Although the Lorentz transformation was known prior to Einstein, it had been looked upon as a formal and mathematical one. Einstein elevated it to a more profound statement about the connection between space and time in the physical world. This was done, in part, by his *postulate of special relativity.* We have seen that the velocity of light is the same, and the equation of a wavefront is of the same form, in all uniformly moving inertial frames. Einstein generalized these facts to all physical phenomena by postulating, in short, that *all relatively moving inertial frames are equivalent: the laws of physics take the same form in all of them.* The full meaning of the postulate will have to be made more exact. The concepts of "equivalent" frames and the exact meaning of laws "of the same form" are not very precise as they stand. For the moment, we observe only that the postulate means that the equations of the Lorentz transformation have to be interpreted more broadly. As we have derived them, the variables x, y, z, t refer to points on the wavefront of a spherical pulse of light. But the postulate implies that the transformations have a more general interpretation: they relate coordinates and time of *any* phenomenon in one frame to coordinates and time in another.

To clarify the last sentence, we introduce the concept of a *point event.* A point event is a "happening" that takes place (to a sufficiently good approximation) at a single point x, y, z, at one time instant t. A point event is then an idealization, much as is a point charge. It is a happening which takes place over a duration of time which is very short, and throughout an extension of space which is very small, when compared to the other times and lengths that may be of interest. An explosion, the winking on or off of a flashlight, the radioactive decay of a nucleus, may, under suitable circumstances, be considered point events. If a rod is laid against a meter stick, the reading at the left end of the rod, at a given time, constitutes an event*; the reading at some other time constitutes another and *different* event, even if the rod is stationary and the x-coordinate remains unchanged. All physical laws and quantities can ultimately be analyzed as relations between events. A moving particle, for example, is characterized by position x_1 at t_1, x_2 at t_2, etc., or, in short, by a sequence of events (x_1, t_1), (x_2, t_2), etc. Of course, this sequence has an infinite number of members but this fact is irrelevant to the basic idea. The concept of point events applies also to Galilean relativity, but is less important there in the presence of an absolute time.

It is implicit in the assumptions above about point events that observers in different coordinate systems can identify the *same* event. Thus, if one observer sees an explosion at x, t, another observer can identify the *same* explosion, although he may assign different coordinates x', t', to it. The Lorentz transformation gives the coordinates (space and time) of any point event in one frame in terms of the coordinates of the *same* event in another.

---

* We shall frequently simply use the word *event* to signify a point event.

12.5 THE INVARIANT INTERVAL

The Lorentz transformation completely describes the relation between two events in different frames. It is, however, sometimes more convenient not to deal with the transformation itself, but with the closely related concept of the *invariant interval* between the two events. If the events have coordinates (x_1, y_1, z_1, t_1) and (x_2, y_2, z_2, t_2), the interval s_{12} is defined by

$$s_{12}^2 = (x_1 - x_2)^2 + (y_1 - y_2)^2 + (z_1 - z_2)^2 - c^2(t_1 - t_2)^2. \tag{12.17}$$

The interval is *form-invariant*; if it is Lorentz-transformed to a primed frame, it assumes the form (12.17), but with all variables primed. For the special transformation (12.13), for which y- and z-coordinates are unchanged, the form-invariance of the interval is expressed by the equation

$$(x_1 - x_2)^2 - c^2(t_1 - t_2)^2 = (x_1' - x_2')^2 - c^2(t_1' - t_2')^2. \tag{12.18}$$

The meaning of this equation between primed and unprimed variables is that it becomes an identity if the primed variables are introduced on the left-hand side by means of Eqs. (12.13). The fact of the invariance of the interval involves no physics that is not in the Lorentz transformation: that transformation was derived from Eqs. (12.10) and (12.11) which are simply statements of the invariance of the interval for the special case when the point events it refers to are successive positions of a point on a wavefront of light. However, sometimes it is easier to think in terms of the interval than in terms of the transformation.

Consider now some of the properties of the interval. First, s_{12}^2 can be either positive or negative. For example, Eq. (12.17) shows that if the two events occur at the same time, i.e., are *simultaneous* in the unprimed system, then

$$s_{12}^2 = (x_1 - x_2)^2 + (y_1 - y_2)^2 + (z_1 - z_2)^2 > 0. \tag{12.19}$$

If they occur at the same place, or are *colocal*, then

$$s_{12}^2 = -c^2(t_1 - t_2)^2 < 0. \tag{12.20}$$

In general, two events in a given frame will be neither simultaneous nor colocal, but the squared interval will be either positive or negative. Positive squared intervals are called *spacelike*. If an interval is spacelike, there is a frame in which the events it refers to are simultaneous. Similarly, if the squared interval for two events is negative, or *timelike*, there exists a frame in which they are colocal.

Consider, for simplicity, events whose y and z coordinates are zero. The *spatial separation* between two events with coordinates (x_1, t_1) and (x_2, t_2) in some unprimed frame is defined by $|x_1 - x_2| \equiv \sqrt{(x_1 - x_2)^2}$. The spatial separation in a primed frame is similarly $|x_1' - x_2'|$. Suppose now the interval is spacelike. We can then find an unprimed frame in which the events are simultaneous. For this case,

$$|x_1 - x_2| = ((x_1' - x_2')^2 - c^2(t_1' - t_2')^2)^{1/2} < |x_1' - x_2'|. \tag{12.21}$$

The spatial separation between two events is least in the frame in which they are simultaneous. Similarly, the temporal or *time separation* between two events, is defined by $|t_1 - t_2|$ in the unprimed frame. Suppose the interval is timelike and let the unprimed frame be the one in which the events are colocal; then

$$|t_1 - t_2| = \left((t_1' - t_2')^2 - \frac{(x_1'^2 - x_2'^2)}{c^2}\right)^{1/2} < |t_1' - t_2'|. \tag{12.22}$$

The time separation between two events is least in the frame in which they are colocal.

The only exception to the fact that an interval must be positive or negative occurs when the interval refers to the propagation of light. Let the event x, y, z, t be the position of a point on a spherical wavefront that was emitted from the origin at time $t = 0$ [point event (0, 0, 0, 0)]. The interval between the events is

$$s^2 = x^2 + y^2 + z^2 - c^2t^2 = 0.$$

This equation between x, y, z, t defines a "conical surface" in space time that is frequently called the *light cone.*

12.6 PROPER TIME

The invariant interval serves to define an important physical quantity: the *proper time* of a moving particle. Suppose a particle moves with uniform velocity $\boldsymbol{u}$ with respect to an unprimed frame, and that a primed frame is *attached* to the particle. The spatial coordinates x_1', y_1', z_1' of the particle at time t_1' define the event (x_1', y_1', z_1', t_1'), and the *same* spatial coordinates (since the particle is fixed in the primed frame) at time t_2' define the event (x_1', y_1', z_1', t_2'). These two events become the events (x_1, y_1, z_1, t_1) and (x_2, y_2, z_2, t_2) in the unprimed system. With $\Delta x = x_1 - x_2$, etc., we have, from the invariance of the interval,

$$(\Delta x)^2 + \Delta y^2 + (\Delta z)^2 - c^2(\Delta t)^2 = -c^2(\Delta t')^2$$

or

$$\Delta t' = \Delta t\left(1 - \frac{(\Delta x)^2 + (\Delta y)^2 + (\Delta z)^2}{c^2(\Delta t)^2}\right)^{1/2}.$$

The right-hand side of this equation is an invariant, since it is essentially the expression for the invariant interval. In the limit that Δx, Δy, Δz and Δt go to zero, $\Delta x/\Delta t$, $\Delta y/\Delta t$, $\Delta z/\Delta t$ become the velocity components u_x, u_y, u_z of the particle and we can write dt' and dt instead of $\Delta t'$ and Δt. The quantity dt' is then given a special symbol and is called the *element of proper time* $d\tau$. Since $u^2 = u_x^2 + u_y^2 + u_z^2$,

$$d\tau = \sqrt{1 - \frac{u^2}{c^2}}\, dt. \tag{12.23}$$

Equation (12.23) is for a particle moving with uniform velocity $\boldsymbol{u}$. If a particle moves with arbitrary and varying velocity, its trajectory can be divided into segments over each of which the velocity is approximately constant. The proper time for each segment can be defined by (12.23) and, by integration, we can define the proper time interval τ between two points

$$\tau = \int_{t_1}^{t_2} \sqrt{1 - \frac{u^2(t)}{c^2}}\, dt. \tag{12.24}$$

From its derivation, the proper time interval represents the time recorded by a clock moving with the particle, if one assumes that the acceleration of the clock attached to the particle has no effect on its rate. The proper time interval τ is obviously smaller than the time interval $t_2 - t_1$ in the frame with respect to which the particle is assumed to move.

This last observation is the basis of the famous *twin paradox*. We can think of the rate of biological processes, for example the rate of aging, as a kind of clock. Consider now two identical twins, A and B. Suppose that B leaves A to embark on some long round-trip space voyage at large velocity. From A's viewpoint, the time elapsed on the biological "aging clock" of B will be less than the time elapsed on his own: B will be younger than A. The paradox enters because of a spurious assumed symmetry between A and B. It is argued that since "all motion is relative," one might equally have considered that A had traveled, leaving B at rest, whereupon A should be younger than B. The resolution of the paradox lies in the fact that the assumed symmetry between A and B does not really exist. The times t_1 and t_2 in (12.24) apply to an observer in an inertial frame. But for A and B to mutually depart and then return, one or the other must accelerate and thereby move as if attached to a noninertial frame. If, for example, A's frame is inertial, B's is, for at least part of the time, necessarily noninertial. There is no contradiction, since the special theory claims symmetry only between two observers who are both moving in inertial frames.

12.7 KINEMATIC CONSEQUENCES OF THE LORENTZ TRANSFORMATION

The Lorentz transformation has implications for the relationships between all physical quantities in different frames: length, time, velocity, charge, field strengths, etc. In this section we lump together, as kinematical implications, those that can be discussed in terms of the transformation only. By contrast, the transformation of electric and magnetic fields is best discussed only after introducing the concepts of *charge invariance* and of *tensors*.

Physical quantities are defined in terms of point events. Before discussing the specific events that define specific physical quantities, we consider some consequences of the Lorentz transformation that apply to all point events, whatever their interpretation. Let (x_1, t_1) and (x_2, t_2) be the coordinates* of two events, 1

* For simplicity, we consider pairs of events for which the y- and z-coordinates remain unchanged, but similar results apply to more general event pairs.

and 2, in some unprimed frame. If the coordinates of these same events are (x_1', t_1') and (x_2', t_2') in a primed frame, we have from (12.15):

$$\begin{aligned} x_1' - x_2' &= \gamma(x_1 - x_2) - \gamma v(t_1 - t_2) \\ t_1' - t_2' &= \gamma(t_1 - t_2) - \frac{\gamma v}{c^2}(x_1 - x_2). \end{aligned} \tag{12.25}$$

In general, then, $x_1 - x_2 \neq x_1' - x_2'$ and $t_1 - t_2 \neq t_1' - t_2'$. Suppose now that the two events occur at the same time, i.e., are *simultaneous* in the unprimed frame:

$$t_1 - t_2 = 0.$$

We see from (12.25) that they will *not* be simultaneous in the primed system, but rather that

$$t_1' - t_2' = \gamma \frac{v}{c^2}(x_2 - x_1).$$

This effect is called the *relativity of simultaneity*. There is, of course, an analogous effect for spatial differences. Suppose that two events occur at the same place (are *colocal*) in the unprimed system:

$$x_1 - x_2 = 0.$$

The events will, in general, be spatially separated in the primed system:

$$x_1' - x_2' = -\gamma v(t_1 - t_2).$$

Many of the paradoxes that are perennially popular in relativity arise from forgetting that simultaneity and colocality are indeed different for different frames.

We now turn to the problem of using specific point events to define physical quantities. There are two aspects to consider: the definition of a quantity in one frame; and the problem of the transformation of that quantity from the one frame to another, i.e., the question of what the quantity "becomes" in a relatively moving primed frame. This last question can be the source of much confusion. The difficulty, as we shall see, is that there is more than one way of defining transformed quantities and it is essential that these ways not be confused with one another.

First, we consider the definition of physical quantities in a single frame. All such quantities are functions of the coordinates* of one or more point events. It will be general enough to consider only quantities that are functions $p(x, t)$ of the coordinates* x, t of a single point event, and quantities that are functions $P(x_1, t_1; x_2, t_2)$ of the coordinates x_1, t_1 and x_2, t_2 of two point events. An example of the first kind of quantity might be the *phase* φ of a wave

$$p(x, t) \equiv \varphi = 2\pi\left(\frac{x}{\lambda} - vt\right). \tag{12.26}$$

* For the sake of simplicity, we shall usually consider "one-dimensional" events for which the y- and z-coordinates play no role.

An example of the second kind might be, if x_1, t_1 and x_2, t_2 refer to two successive positions of a particle, the mean velocity $\bar{u}$ along the x-axis

$$P(x_1, t_1; x_2, t_2) \equiv \bar{u} = \frac{x_2 - x_1}{t_2 - t_1}, \tag{12.27}$$

or similarly the mean acceleration, or the instantaneous velocity in the limit that x_2 approaches x_1, etc.

Having defined such quantities we must consider the second problem of transforming them to the primed variables of a relatively moving frame. There are, as we have remarked, two methods of defining this transformation. We illustrate the first method with reference to the physical quantity $p(x, t)$ that is a function of one point event. We define the *same* quantity in the new frame by simply substituting the Lorentz transformation equations (12.15) into $p(x, t)$. That is, we take the *same* physical quantity in the new frame to be *defined by* the function $p'(x', t')$, where

$$p'(x', t') = p\left(\gamma(x' + vt'), \gamma\left(t' + \frac{v}{c^2}x'\right)\right). \tag{12.28}$$

Obviously, the functional form of p' will not, in general, be the same as the functional form of p, but the quantity will have a definite value at a given point of space-time, whether that point is called x, t or x', t'. An analogous definition can be used to transform quantities $P(x_1, t_1; x_2, t_2)$ that are functions of the coordinates of two point events. That is, the same quantity in the primed frame is defined by $P'(x'_1, t'_1; x'_2, t'_2)$, where

$$P'(x'_1, t'_1, x'_2, t'_2) = P\left(\gamma(x'_1 + vt'_1), \gamma\left(t'_1 + \frac{v}{c^2}x'_1\right); \gamma(x'_2 + vt'_2), \gamma\left(t'_2 + \frac{v}{c^2}x'_2\right)\right).$$

By contrast, consider now the second possibility for defining transformed quantities, in terms of the function $P(x_1, t_1; x_2, t_2)$ of two point events. Let x'_1, t'_1 be the coordinates in a primed frame of the event (x_1, t_1), and similarly for (x'_2, t'_2). In the primed frame, we define the "same" quantity as $P(x_1, t_1; x_2, t_2)$ by forming the *same function* $P(x'_1, t'_1; x'_2, t'_2)$ *of the new variables*. However, the *numerical value* of $P(x_1, t_1; x_2, t_2)$ will *not*, in general, be the same as the numerical value of $P(x'_1, t_1; x'_2, t_2)$. Only a few functions have this property of invariance; these include the squared interval s^2, the coordinate "scalar product"

$$x_1x_2 + y_1y_2 + z_1z_2 - c^2t_1t_2,$$

and, of course, any functions of these functions. But, usually, the numerical values of $P(x_1, t_1; x_2, t_2)$ and $P(x'_1, t'_1; x'_2, t'_2)$ will be different, so they cannot properly be called the same quantity. For the sake of terminology, we may say that if $P(x_1, t_1; x_2, t_2)$ defines the "quantity Q in the unprimed frame," then $P(x'_1, t'_1; x'_2, t'_2)$ defines the "quantity Q in the primed frame"; however, these are *not* the same quantities.

To summarize the two kinds of definitions, we may say that the first defines the *same* quantity in two frames but that the *functional* form of the definition differs between the frames. The second defines two quantities which have the same functional form in the two frames, but which cannot reasonably said to be the same quantity since they do not have the same value at a given space-time point. We shall call quantities defined according to the second definition *similar* quantities. Both definitions are useful and choosing between them is purely a question of convenience and suitability to the problem at hand. A more extended discussion of this point is given by Gamba (R).

Consider some examples. Suppose that there is a wave amplitude ψ of some kind, for instance, an acoustic pressure amplitude or one component of an electromagnetic field, given in some unprimed system by

$$\psi = Ae^{i(\boldsymbol{k}\cdot\boldsymbol{r} - \omega t)}$$

and characterized by the wave vector $\boldsymbol{k}$ and frequency ω. We usually want to know the *same* complex amplitude in another system, i.e., the first definition is the more relevant. If we transform ψ by the algorithm (12.28) of that definition, it becomes

$$\psi = Ae^{i(\boldsymbol{k}'\cdot\boldsymbol{r}' - \omega' t')},$$

where

$$\begin{aligned} k'_x &= \gamma\left(k_x - \frac{v\omega}{c^2}\right), \\ k'_y &= k_y, \\ k'_z &= k_z, \\ \omega' &= \gamma(\omega - vk_x). \end{aligned} \tag{12.29}$$

The wave moves in a different direction with respect to the primed system than it does in the unprimed; this is the phenomenon of *aberration*. The frequency ω' also differs from ω; this is called the *Doppler effect*.

We apply these formulas to the propagation of light* for which case

$$k \equiv |\boldsymbol{k}| = \omega/c \qquad \text{and} \qquad k' \equiv |\boldsymbol{k}'| = \omega'/c.$$

Let the wave vector $\boldsymbol{k}$ in the unprimed frame make an angle θ with the x-axis, $k_x = k\cos\theta$. Then the equation for ω' is

$$\omega' = \gamma\omega(1 - \beta\cos\theta), \tag{12.30}$$

a formula for the *relativistic Doppler shift*. For $v \ll c$, it becomes a standard non-

* The present treatment is oversimplified if A is an electromagnetic field amplitude, since one must also consider how A transforms. However, the more exact treatment of Section 12.10 gives the same results as far as the transformation of the exponent of the wave, which is all that comes into question here.

relativistic formula. In contrast to the nonrelativistic case, however, the factor γ in (12.30) shows that there is a frequency shift even for $\theta = \pi/2$. This relativistic shift has been confirmed experimentally. If $\boldsymbol{k}'$ makes an angle θ' with the x-axis, the first equation in (12.29) becomes $k' \cos\theta' = \gamma k(\cos\theta - \beta)$. This can be combined with Eq. (12.30) for ω' to yield

$$\tan\theta' = \frac{\sin\theta}{\gamma(\cos\theta - \beta)}, \tag{12.31}$$

a well-known equation describing the *aberration of light*.

Now we consider examples of the transformation of quantities according to the second definition above. Many important physical quantities are customarily so defined: length intervals, time intervals, velocities, etc. We consider velocities first since the results somehow encounter less psychological resistance than do the analogous formulas for time and length intervals, although all three formulas are on the same conceptual footing. The mean x-component of velocity of a particle $\bar{u}_x$ is defined by a special choice of $P(x_1, t_1; x_2, t_2)$; namely,

$$\bar{u}_x = \frac{x_2 - x_1}{t_2 - t_1}.$$

The *similar* mean velocity $\bar{u}'_x$ is defined by

$$\bar{u}'_x = \frac{x'_2 - x'_1}{t'_2 - t'_1}.$$

Instead of dealing with mean velocities, it is more useful to discuss *instantaneous* velocities, u_x, u_y, u_z. The point of even mentioning mean velocities is to show clearly that they are, in fact, examples of the second mode of defining transformed quantities. In that case, so are instantaneous velocities, since they are the limits of mean velocities as t_1 approaches t_2.

We define the velocity components as

$$u_x = \frac{dx}{dt}, \qquad u_y = \frac{dy}{dt}, \qquad u_z = \frac{dz}{dt}, \tag{12.32}$$

and the similar components in a primed frame, $u'_x = dx'/dt'$, etc., and we seek the relation between the primed and unprimed quantities. From the Lorentz transformation equations (12.13), we have

$$dx' = \gamma(dx - v\,dt), \qquad dy' = dy, \qquad dz' = dz, \qquad dt' = \gamma\left(dt - \frac{v\,dx}{c^2}\right).$$

On rewriting dt' as $\gamma\,dt(1 - vu_x/c^2)$ and forming $u'_x = dx'/dt'$, etc., we find

$$u'_x = \frac{u_x - v}{1 - vu_x/c^2}, \qquad u'_y = \frac{u_y}{\gamma(1 - vu_x/c^2)}, \qquad u'_z = \frac{u_z}{\gamma(1 - vu_x/c^2)}, \tag{12.33}$$

and these are the desired formulas. As a sidelight on these formulas, let the velocity u_x in the unprimed system be that of light, $u_x = c$. Then

$$u'_x = \frac{c - v}{1 - v/c} = c.$$

Thus this equation reflects the basic postulate of the theory: a velocity c in one system appears as a velocity c in another moving uniformly with respect to it.

Consider next the transformation of time intervals. As an important example, we take the time interval between two events that are colocal in an unprimed system. For the sake of vividness, we may suppose that there is a time bomb which, together with its fuse, is small enough to be located at some point x_1 of an unprimed frame. If the fuse of the bomb is lighted at t_1, this generates the event (x_1, t_1); let the event (x_1, t_2) refer to its explosion. The quantity T, "the time in the unprimed frame between the lighting of the bomb and its explosion" is defined by

$$T = t_2 - t_1. \tag{12.34}$$

The *similar* quantity T', "the time in the primed frame between the lighting of the bomb and its explosion," is then

$$T' = t'_2 - t'_1. \tag{12.35}$$

From Eq. (12.13), we obtain

$$T' = T/\sqrt{1 - \beta^2}, \tag{12.36}$$

i.e., T' is larger than T. This phenomenon is called *time dilation* and it has been verified using, in effect, subnuclear bombs. Certain elementary particles called *μ-mesons* disintegrate into two other particles with a probability law governed by a mean lifetime. It is found that when these mesons are in motion with respect to a frame, the decay rate differs from that when they are at rest in the frame. This difference is explained quantitatively by the phenomenon of time dilation.

As a final example, consider the transformation of lengths. Suppose a rod is fixed along the x-axis of the primed system of Fig. 12.1. In the *unprimed* system, we measure the coordinates x_1, x_2 of the two ends of the rod at some common time t_0, thereby generating the point events $(x_1, t_0), (x_2, t_0)$. We *define* the quantity, "length L of the moving rod," by

$$L = x_2 - x_1. \tag{12.37}$$

We then define the *similar* quantity L_0, "length of the rod at rest," by

$$L_0 = x'_2 - x'_1. \tag{12.38}$$

From (12.15), we have $L_0 = L\gamma$, or

$$L = \sqrt{1 - v^2/c^2}\, L_0. \tag{12.39}$$

The "length of the moving rod" is *less* than the "length of the rod at rest." This effect is called the *Lorentz–Fitzgerald contraction.*

We have remarked that the three transformations above, of velocities, time intervals, and length, are on the same conceptual footing but that, at first sight, the Lorentz contraction often seems most "paradoxical." The reason has nothing to do with physics, but with psychology. First, we are accustomed to the fact that velocities differ in relatively moving systems; velocities transform even in Galilean relativity, so it is perhaps no surprise that they transform in special relativity. The fact that time dilates is perhaps more surprising, although we tend not to make too much of the fact since our intuition about time is less assured than it is about length. Time has always been "mysterious." But length! Everyone *knows* that a body *has* a length and it should not then make any difference how it is measured. One should always get the same answer. This is incorrect, of course. A body has properties defined by functions of point events, and if we define quantities as noninvariant functions, as we do in Eqs. (12.34) and (12.37), it is not astonishing that they have different values in different systems.

One final remark. One might think that the Lorentz transformation contraction could be observed, either visually or by photographing it. Science fiction writers have assumed as much for years and supposed that if one looked at a fat man with relativistic velocities, he would appear to be skinny. This is not true. The reason is not that Eq. (12.39) is wrong, but that the two point events it refers to are not directly relevant to the picture-making process or to seeing. The point events involved in vision, say, are those corresponding to light quanta that strike different parts of the eye at the same time. These are different point events than those of Eqs. (12.37) and (12.38) that define the length of the rod. A discussion of this subject is given by Terrell (R).

12.8 INVARIANTS, FOUR-VECTORS, TENSORS

The concepts of *invariants* and *invariance* have emerged in various contexts, and with various meanings, in the previous discussion. They take on still new meanings in connection with the *four-vectors* and *tensors* that we are about to discuss. As an introduction to these new meanings, we shall, at the expense of some redundancy, review the previous somewhat scattered discussion.

A quantity is invariant when it "has the same value" in different coordinate systems. The number $\pi = 3.14159\ldots$, and the number $e = 2.718\ldots$ are, trivially, invariants by this definition. Physical quantities can be similarly invariant. For example, we can define a physical quantity by a measurement that yields a number in a certain frame and then, by convention, use that *same* number in other frames. Almost trivially, then, the quantity is invariant. Proper time and rest mass are examples of such invariants. It is important in discussing invariants to separate clearly the physical invariance from the mathematical description of it. Consider, for example, a temperature distribution $T(x, y)$ on the surface of a plane plate, where x and y are coordinates with respect to some system inscribed on the plate. Now the *physical* distribution is the same, i.e., the temperature at a given point is what it is, no matter how the distribution is described. But the *form* of the description will depend on the coordinate system chosen. If the distribution is described in terms of rotated coordinates $x' = x \cos\theta + y \sin\theta$, $y' = -x \sin\theta + y \cos\theta$, or more briefly $x = x(x', y')$, $y = y(x', y')$, then the function

$$T'(x', y') = T(x(x', y'), y(x', y'))$$

which describes the temperature in terms of the new variables will *not*, in general, be the same function of its variables as T is of its variables. However, there are functions for which the functional form of T *is* the same as that of T'. For example, if $T(x, y) = T_0 e^{-(x^2+y^2)}$, then $T'(x', y') = T_0 e^{-(x'^2+y'^2)}$, and T is the same function of its variables as T' is of its variables. Such functions are said to be *form invariant*; we have already encountered a form invariant function, the invariant interval of Eq. (12.18). Form invariance comprises two separate concepts: the invariance of a physical quantity *and* the invariance of the form of the mathematical description of it. Differential expressions can be similarly form invariant. In the Galilean transformation of Eq. (12.3), we have $dx/dt = (dx'/dt') + v$ so that dx/dt is not form invariant. On the other hand, d^2x/dt^2 *is* form invariant:

$$d^2x/dt^2 = d^2x'/dt'^2.$$

In general, an expression is form invariant under transformation from unprimed to primed variables if the labor of the transformation is redundant, i.e., if the result could be obtained simply by putting primes on the variables of the original expression.

The concepts above of *scalar* invariants are not general enough to describe the kind of invariance involved in the *vectorial* Maxwell equations. The generalizations that are needed are called *four-vectors* and *tensors*. Four-vectors are, in turn, generalizations of ordinary two- and three-dimensional vectors. For introducing this generalization, however, the elementary notion of a vector as a "directed line segment," or some equivalent rough definition, is not precise enough. Therefore we begin by discussing the familiar vectors in two and three dimensions from a viewpoint in which the generalization to four-vectors and to tensors will be more transparent.

Two dimensions will be as good as three for our exposition. Consider, as in Fig. 12.2, a point P with coordinates x_1, x_2. We draw the line from the origin to P and call this line the *displacement OP*. The coordinates x_1 and x_2, which we write as the ordered pair (x_1, x_2), are called the *components* of this displacement. Now consider how the *same* displacement would be characterized in the primed coordinate system of Fig. 12.2 which is rotated with respect to the unprimed system by an angle θ'. The components of the displacement OP will be (x'_1, x'_2) where

$$\begin{aligned} x'_1 &= x_1 \cos\theta' + x_2 \sin\theta', \\ x'_2 &= -x_1 \sin\theta' + x_2 \cos\theta'. \end{aligned} \tag{12.40}$$

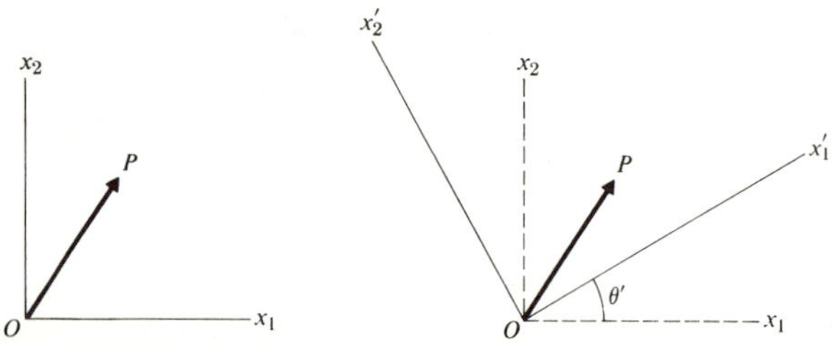

Fig. 12.2 The displacement OP in the unprimed system and the same displacement represented in the primed system.

In still a third system, characterized by a rotation angle θ'', the *same* displacement will be defined by the ordered pair (x_1'', x_2''). There are, of course, an infinite number of different coordinate systems of this kind and, hence, an infinite number of sets of the form (x_1, x_2), (x_1', x_2'), (x_1'', x_2''), (x_1''', x_2'''), ... The mathematical concept of a vector is this: *a vector is the abstract entity to which all of these coordinate sets refer.* Each coordinate system describes the *displacement* by a different pair of numbers or in a different language, so to speak. The common entity described by all the languages is the displacement *vector*; it is almost the Platonic idea common to all the representations.

This idea is so basic that we rephrase it in terms of an analogy. When the American says "house," the Frenchman says "maison," and the German says "Haus," etc., each is referring to the same idea in his own language. It is the *concept* house (as opposed, say, to the word "house" in the English language) that represents the common denominator of all the words for it in all possible languages. In the same way, a vector is the common denominator of all of its representations in all different coordinate systems.

To generalize this example, we first rewrite Eq. (12.40) as the linear transformation

$$x'_\mu = \sum_{\nu=1}^{2} a_{\mu\nu} x_\nu, \qquad \mu = 1, 2, \tag{12.41}$$

where the coefficients $a_{\mu\nu}$ are elements of the matrix $\{a_{\mu\nu}\}$

$$\{a_{\mu\nu}\} = \begin{pmatrix} \cos\theta' & \sin\theta' \\ -\sin\theta' & \cos\theta' \end{pmatrix}. \tag{12.42}$$

The transformation (12.41) is an *orthogonal* one, i.e., the coefficients $a_{\mu\nu}$ satisfy the *orthogonality relations*

$$\sum_{\mu=1}^{2} a_{\mu\nu} a_{\mu\lambda} = \delta_{\nu\lambda}. \tag{12.43}$$

With this displacement vector as a model, we now define a more general vector. Suppose two numbers (A_1, A_2) are associated with a certain coordinate system x_1, x_2. In another coordinate system, x'_1, x'_2, related to the first by Eqs. (12.41) we define a pair (A'_1, A'_2) by

$$A'_\mu = \sum_{\nu=1}^{2} a_{\mu\nu} A_\nu, \qquad \mu = 1, 2. \tag{12.44}$$

We define a similar set (A''_1, A''_2) in the coordinate system x''_1, x''_2, etc., and so end up with the infinite number of sets $(A_1, A_2), (A'_1, A'_2), (A''_1, A''_2), \ldots$. The vector $\boldsymbol{A}$ is now defined as the entity which *exists* independently of a given system, but is *described* in any system by the ordered pair appropriate to it. The components of a vector then transform like the coordinates of a point. A vector is a construct which is independent of any particular coordinate system; for this reason, vectors are ideally suited for representing physical laws that are independent of the coordinate system.

The vector concept can obviously be generalized to three dimensions, but we shall bypass this to focus on vectors in four-dimensional space-time, or *four-vectors.* We begin by rewriting the Lorentz transformation: time is put on a more equal footing with other coordinates by introducing the variable $x_4 = ict$, and by relabeling variables according to

$$x_1 = x, \qquad x_2 = y, \qquad x_3 = z, \qquad x_4 = ict. \tag{12.45}$$

Then the Lorentz transformation (12.15) can be written

$$x'_\mu = \sum_{\nu=1}^{4} a_{\mu\nu} x_\nu, \qquad \mu = 1, 2, 3, 4, \tag{12.46}$$

where again the coefficients $a_{\mu\nu}$ can be represented by a matrix $\{a_{\mu\nu}\}$

$$\{a_{\mu\nu}\} = \begin{pmatrix} \gamma & 0 & 0 & iv\gamma/c \\ 0 & 1 & 0 & 0 \\ 0 & 0 & 1 & 0 \\ -iv\gamma/c & 0 & 0 & \gamma \end{pmatrix}. \tag{12.47}$$

It is easy to verify that the following *orthogonality relation* holds:

$$\sum_{\mu=1}^{4} a_{\mu\nu} a_{\mu\lambda} = \delta_{\nu\lambda}. \tag{12.48}$$

An equation like (12.48) is also satisfied by coefficients of a general Lorentz transformation, i.e., one for which two frames move in an *arbitrary* relative direction. In the form (12.46), the Lorentz transformation is seen to resemble closely the simple transformation (12.41) in two dimensions. The analog of the rotation angle θ' is the parameter β which can take any value from 0 to 1.

Now consider four numbers (A_1, A_2, A_3, A_4) that are defined in the unprimed system, four others (A'_1, A'_2, A'_3, A'_4) that are defined in the primed system corresponding to a given β, four others $(A''_1, A''_2, A''_3, A''_4)$ in still another system, etc. If the numbers in the primed and unprimed systems are related by equations like

$$A'_\mu = \sum_{\nu=1}^{4} a_{\mu\nu} A_\nu, \qquad \mu = 1, 2, 3, 4, \tag{12.49}$$

they define, in the sense discussed above, a *four-vector*.

There are important form invariant scalars associated with four-vectors. A useful one is the dot product,* $\boldsymbol{A} \cdot \boldsymbol{B}$, of two four-vectors

$$\boldsymbol{A} \cdot \boldsymbol{B} = \sum_\mu A_\mu B_\mu. \tag{12.50}$$

From (12.49) and the orthogonality relation (12.48), the invariance follows directly:

$$\boldsymbol{A}' \cdot \boldsymbol{B}' = \sum_\mu A'_\mu B'_\mu = \sum_\mu \sum_\nu \sum_\lambda a_{\mu\nu} a_{\mu\lambda} A_\nu A_\lambda = \sum_\lambda A_\lambda B_\lambda = \boldsymbol{A} \cdot \boldsymbol{B}. \tag{12.51}$$

The special case $\boldsymbol{A} = \boldsymbol{B}$ shows that the *squared magnitude* $\Sigma_\mu A_\mu^2$ of a four-vector is invariant. If the four-vector $\boldsymbol{A}$ is the coordinate four-vector, $(\boldsymbol{r}, ict)$, the invariance of the magnitude is just the invariance of the interval:

$$x^2 + y^2 + z^2 - c^2t^2 = x'^2 + y'^2 + z'^2 - c^2t'^2.$$

In discussing the differential invariants associated with four-vectors, it is sometimes convenient, by analogy with the three-dimensional ∇ operator, to define the four-dimensional operator $\Box$ as a *symbolic vector* with components

$$\left(\frac{\partial}{\partial x_1}, \frac{\partial}{\partial x_2}, \frac{\partial}{\partial x_3}, \frac{\partial}{\partial x_4}\right).$$

The *divergence*, div $\boldsymbol{A}$, of a four-vector $\boldsymbol{A}$ is then defined by

$$\operatorname{div} A \equiv \Box \cdot \boldsymbol{A} = \sum_\mu \frac{\partial A_\mu}{\partial x_\mu}. \tag{12.52}$$

Much in the same way that we showed the dot product to be invariant, we can show this divergence to be a differential invariant

$$\sum_\mu \frac{\partial A_\mu}{\partial x_\mu} = \sum_\mu \frac{\partial A'_\mu}{\partial x'_\mu}. \tag{12.53}$$

If in (12.53) we let $A_\mu = \partial\phi/\partial x_\mu$, we get the four-dimensional Laplacian, or the *Dalembertian*, $\Box^2\phi$, defined by

$$\Box^2\phi \equiv \sum_\mu \frac{\partial^2\phi}{\partial x_\mu^2} = \sum_\mu \frac{\partial^2\phi}{\partial x_\mu'^2}. \tag{12.54}$$

* In the rest of this section, it will be understood that summations over the indices of four-vectors and tensors run from 1 through 4.

We now generalize these ideas to define *tensor*. Tensors can be considered in spaces of any number of dimensions but we shall discuss only those in the four-dimensional x_1, x_2, x_3, x_4 space. As we define a four-vector in one frame by a set of four quantities, written as a symbol with one index T_μ, so we define a tensor in one frame by a set of quantities that are written as symbols with multiple indices; for example, $T_{\mu\nu}$, $T_{\mu\nu\lambda}$, etc. Each of the indices takes on the values 1, 2, 3, 4; the number of indices defines the *rank* of the tensor. Thus $T_{\mu\nu}$ defines a tensor of the second rank and $T_{\mu\nu\lambda}$, a tensor of the third rank. An essential part of the tensor concept is that there are similar component sets $T'_{\mu\nu}$ and $T'_{\mu\nu\lambda}$ in arbitrary primed frames, and that the connection between the unprimed and primed sets is given by

$$T'_{\alpha\beta} = \sum_\mu \sum_\nu a_{\alpha\mu} a_{\beta\nu} T_{\mu\nu} \tag{12.55}$$

for *second rank* tensors, by

$$T'_{\alpha\beta\gamma} = \sum_\mu \sum_\nu \sum_\lambda a_{\alpha\mu} a_{\beta\nu} a_{\gamma\lambda} T_{\mu\nu\lambda} \tag{12.56}$$

for *third rank* tensors, etc. As with four-vectors, we consider that a tensor, for example of the second rank, is *described* by the components $T_{\mu\nu}$ in one frame, $T'_{\mu\nu}$ in a second, $T''_{\mu\nu}$ in a third, etc., but that the *tensor* $\mathbf{T}$ *itself** is the *entity underlying the component descriptions, and common to them all.* A four-vector is a tensor of the first rank, and a scalar function of x_1, x_2, x_3, x_4 can be considered as a tensor of rank zero. Tensors of any rank can be defined by obvious generalizations of the above definitions.

Consider now some general properties of tensors. Tensors can be *symmetric* or *antisymmetric* in pairs of indices according to whether their components change sign or not on interchange of the two indices of the pair. These symmetry properties can be quite complicated for high rank tensors, which may be symmetric in some pairs, antisymmetric in others, and without special properties in still others. We shall consider only tensors of the second rank. For the symmetric one,

$$T_{\mu\nu} = T_{\nu\mu}$$

for all μ and ν. Similarly, for the antisymmetric one,

$$T_{\mu\nu} = -T_{\nu\mu}.$$

An antisymmetric tensor in four dimensions has only six independent components.† Of the sixteen possible components, the four diagonal ones must vanish since $T_{\mu\mu} = -T_{\mu\mu}$ by definition and obviously only six of the remaining twelve are independent.

---

* We use **boldface** to indicate the tensor itself, as for vectors.

† Analogously, the antisymmetric tensor in three dimensions has only *three* independent components and is sometimes called an axial vector. Its three components transform as do the components of an ordinary (or polar) vector under rotations, but not under reflections.

There are two important processes for generating tensors from other tensors: the first is the formation of *outer products*, the second is *contraction*. The outer product of two tensors, $\boldsymbol{P}$ and $\boldsymbol{Q}$, whose components are $P_{\alpha\beta}$ and $Q_{\gamma\delta\varepsilon}$, is a tensor $\boldsymbol{R}$ with components $R_{\alpha\beta\gamma\delta\varepsilon}$ defined by

$$R_{\alpha\beta\gamma\delta\varepsilon} = P_{\alpha\beta}Q_{\gamma\delta\varepsilon}. \tag{12.57}$$

It is straightforward to show that $\boldsymbol{R}$ indeed has the correct transformation properties of a tensor. The definition (12.57) is extended in an obvious way to the outer product of two tensors of arbitrary rank. A tensor is *contracted* by setting two of its indices equal and then summing over the repeated index. For example, a contraction of the tensor $T_{\alpha\beta\gamma}$ is the quantity $\Sigma_\beta \Gamma_{\alpha\beta\beta}$. It is easy to show that the contraction of a tensor of rank r is a tensor of rank $r - 2$. As a case to exemplify these processes: if $\boldsymbol{A}$ and $\boldsymbol{B}$ are four-vectors, their outer product components, $A_\mu B_\lambda$, define a tensor of rank two. The contraction of this outer product is just the ordinary dot product $\Sigma_\mu A_\mu B_\mu$, a tensor of rank zero, or a scalar. A second example is the formation of a four-vector from the contracted outer product of a second rank tensor $F_{\alpha\beta}$ and a first rank tensor (four-vector) J_γ. We form the outer product $F_{\alpha\beta}J_\gamma$ and contract over the second and third indices to define $\boldsymbol{M}$, where

$$M_\alpha = \sum_\beta F_{\alpha\beta}J_\beta. \tag{12.58}$$

Then the quantities M_α define a four-vector; the relation of M_α and $M'_\mu = \Sigma_\nu F'_{\mu\nu}J'_\nu$ is readily shown to be

$$M'_\mu = \sum_\alpha a_{\mu\alpha}M_\alpha.$$

In a similar manner, we can define the four-divergence $\text{div}_\mu \boldsymbol{T}$ of a tensor $\boldsymbol{T}$ with components $T_{\mu\nu}$ by

$$\text{div}_\mu \boldsymbol{T} = \sum_\nu \frac{\partial T_{\mu\nu}}{\partial x_\nu}. \tag{12.59}$$

In the same way that the divergence of a four-vector is a scalar (a tensor of one rank less), so the divergence of a second rank tensor can be shown to be a four-vector (a tensor of one rank less).

12.9 COVARIANCE OF THE MAXWELL EQUATIONS

Einstein's postulate of the equivalence of inertial systems implies that all the equations of physics must be form invariant under Lorentz transformations. We now investigate this condition of form invariance as it applies to Maxwell's equations. There are at least two complementary ways to proceed. Most directly, we have Maxwell's equations that define $\boldsymbol{E}$ and $\boldsymbol{B}$ as a function of the coordinates x, y, z, t in some frame and the Lorentz transformation that relates these to the coordinates x', y', z', t' of another frame. It is then a simple procedure to transform the Maxwell equations to the new coordinates. Doing this, we find that we can satisfy the requirement that the equations retain the same form if we assume that the fields in the one frame are related to the fields in the other in a certain special way. For example, a purely electric field in one frame must be considered as an electric *and* magnetic field in a relatively moving frame.

A complementary approach is to show that Maxwell's equations can be written as relations between four-vectors and tensors and the differential invariants that can be formed from them. The equations are then obviously invariant in form, or *covariant*,* and the transformation laws of the tensors become the laws of transformations of the fields.

We undertake the first approach now. Consider, for example, the y-component of the Maxwell equation for $\nabla \times \boldsymbol{E}$,

$$(\nabla \times \boldsymbol{E})_y = \frac{\partial E_x}{\partial z} - \frac{\partial E_z}{\partial x} = -\frac{1}{c}\frac{\partial B_y}{\partial t}. \tag{12.60}$$

Performing the Lorentz transformation to primed variables x', y', z', t', we obtain, with Eqs. (12.13)

$$\frac{\partial E_z}{\partial x} = \frac{\partial E_z}{\partial x'}\frac{\partial x'}{\partial x} + \frac{\partial E_z}{\partial y'}\frac{\partial y'}{\partial x} + \frac{\partial E_z}{\partial z'}\frac{\partial z'}{\partial x} + \frac{\partial E_z}{\partial t'}\frac{\partial t'}{\partial x} = \gamma\left(\frac{\partial E_z}{\partial x'} - \frac{v}{c^2}\frac{\partial E_z}{\partial t'}\right),$$

and

$$\frac{\partial E_x}{\partial z} = \frac{\partial E_x}{\partial z'}, \qquad \frac{\partial B_y}{\partial t} = \gamma\left(\frac{\partial B_y}{\partial t'} - v\frac{\partial B_y}{\partial x'}\right).$$

With these results, Eq. (12.60) becomes

$$\frac{\partial E_x}{\partial z'} - \frac{\partial}{\partial x'}\left[\gamma\left(E_z + \frac{v}{c}B_y\right)\right] = -\frac{1}{c}\frac{\partial}{\partial t'}\left[\gamma\left(B_y + \frac{v}{c}E_z\right)\right]. \tag{12.61}$$

Form invariance requires that the transformed Eq. (12.61) should be of the form (12.60), but with primed† variables,

$$\frac{\partial E'_x}{\partial z'} - \frac{\partial E'_z}{\partial x'} = -\frac{1}{c}\frac{\partial B'_y}{\partial t'}; \tag{12.62}$$

but it is not immediately and obviously of this form. For example, $\partial/\partial x'$ operates on a linear combination of two terms in Eq. (12.61) whereas $\partial/\partial x$ in Eq. (12.60)

---

* In the literature, the term *covariant* is used with a wide variety of meanings, not all of them well defined. The word loosely means "unchanged in form" and it is, hence, sometimes taken to be a synonym for invariant. This is unfortunate since, as we have seen, the word invariant has a variety of shadings. In connection with general tensor analysis, covariant has a well-defined meaning, as in the phrase "covariant differentiation," but that does not concern us here. In general, we shall say an equation is *covariant* if it is one between four-vectors and tensors.

† Since the x, y, z-axes are respectively parallel to the x', y', z'-axes, we can write E'_x, etc., instead of $E'_{x'}$.

operates on only a single term. We can, however, effect the form invariance of (12.61) by identifying the term in it that $\partial/\partial x'$ operates on as E_z', the z-component of field as measured in the primed frame:

$$E_z' = \gamma\left(E_z + \frac{v}{c}B_y\right).$$

Similarly, if we write

$$B_y' = \gamma\left(B_y + \frac{v}{c}E_z\right)$$

we are, in fact, led to Eq. (12.62). We see that the requirement of form invariance is far from an empty one; it requires that fields lose their purely electric or magnetic identities on transformation. Similar treatment of the other Maxwell equations yields the formulas for the transformation of all the six components of $\boldsymbol{B}$ and $\boldsymbol{E}$:

$$\begin{aligned} E_x' &= E_x, & B_x' &= B_x, \\ E_y' &= \gamma\left(E_y - \frac{v}{c}B_z\right), & B_y' &= \gamma\left(B_y + \frac{v}{c}E_z\right), \\ E_z' &= \gamma\left(E_z + \frac{v}{c}B_y\right), & B_z' &= \gamma\left(B_z - \frac{v}{c}E_y\right). \end{aligned} \tag{12.63}$$

In these equations, the x-components of fields are those *parallel* to the x-axes in relative motion, and the y- and z-components are perpendicular thereto. The equations can be written more generally and more succinctly in terms of $\boldsymbol{E}_{||}$, $\boldsymbol{B}_{||}$, and $\boldsymbol{E}_\perp$, $\boldsymbol{B}_\perp$, the parallel and perpendicular components of the field

$$\begin{aligned} \boldsymbol{E}_{||}' &= \boldsymbol{E}_{||}, & \boldsymbol{B}_{||}' &= \boldsymbol{B}_{||}, \\ \boldsymbol{E}_\perp' &= \gamma\left(\boldsymbol{E}_\perp + \frac{\boldsymbol{v}\times\boldsymbol{B}}{c}\right), & \boldsymbol{B}_\perp' &= \gamma\left(\boldsymbol{B}_\perp - \frac{\boldsymbol{v}\times\boldsymbol{E}}{c}\right). \end{aligned} \tag{12.64}$$

We now turn to the problem of transformation of Maxwell's equations from the second point of view, which shows that they can be written as tensor equations defined in four-dimensional space-time. The equations then transform from frame to frame according to the laws of transformation of tensor components. These laws, of course, yield results identical to those above. We begin with the continuity equation, which is not one of Maxwell's equations *per se*, but is a consequence of them. Using notation parallel to the $x_1, x_2, x_3, x_4 = ict$ notation of Eq. (12.45), namely,

$$J_1 = J_x, \qquad J_2 = J_y, \qquad J_3 = J_z, \qquad J_4 = ic\rho, \tag{12.65}$$

the continuity equation takes the form

$$\sum_\mu \frac{\partial J_\mu}{\partial x_\mu} = 0. \tag{12.66}$$

In this form, the left-hand side resembles the divergence of a four-vector whose first three components are $\boldsymbol{J}$ and whose fourth component is $ic\rho$. We write this presumed four-vector as $(\boldsymbol{J}, ic\rho)$. We cannot, however, rely on the fact that a change of variables has induced this (possibly superficial) resemblance. A four-vector is defined by its transformation or invariance properties, and these must be investigated. Since $\boldsymbol{J} = \rho\boldsymbol{u}$, the presumed four-vector is $(\rho\boldsymbol{u}, ic\rho)$. To find its transformation properties, we must, since we have already defined how $\boldsymbol{u}$ transforms, consider how the charge density ρ transforms. Preliminary to that, we must treat the transformation of the *total charge* on a body. How is a measurement of the total charge on a body related to a similar measurement of the charge from a frame past which the body is moving with velocity v? The question is answered by the *principle of charge invariance*: the charge measured in the two cases will be the same. The principle can be taken to be either an emperical fact or a matter of definition. If we take it as an empirical fact, we note that experimental evidence for it comes from (among other places) the neutrality of atoms. It is known that the electron and proton, when each at rest, have the same magnitude (although opposite sign) of charge. It is also found that a hydrogen atom is electrically neutral. Since, in this atom, the electron can be considered to move with respect to the proton, and since its charge exactly balances the proton charge, the motion of the electron is shown to have no effect on the quantity of charge it carries.

Although we have characterized the principle of charge invariance as an empirical one, the characterization is incomplete in that we have begged the question of how charge is measured. The original definition of charge in electrostatics was made in terms of Coulomb's law and the forces between two charged particles that are *at rest* in some frame. It is then not clear how to apply this definition to charge that is in motion with respect to the frame. Charge invariance as an empirical principle implies, in fact, an alternate definition of the measure of the quantity of charge. Namely, we can reshuffle the logic of the development of electrostatics in Chapter 2 and take $\boldsymbol{E}$ as a primary quantity. The quantity of charge in a volume can then be *defined* via Gauss' law and the flux through a closed surface of the field $\boldsymbol{E}$ due to the charge. This definition can be extended to charge that is moving with respect to a given frame. A closed surface that encloses the charge and is fixed in the frame can be drawn, and the flux of $\boldsymbol{E}$ at a *given instant* through that surface can be taken as a measure of the charge contained within. It is on the basis of this tacit definition of quantity of charge in a moving frame that the principle of charge invariance holds.

We can adopt a second viewpoint: Given a charged body, we can measure its charge in a frame in which it is at rest by using the original definition that involves Coulomb's law. We can then, by decree, use this same number in moving systems so that charge is *defined* to be an invariant. In this viewpoint, charge is analogous to inertial rest mass, for which a number is defined by measurements in one system, and this same number is by convention used in other systems. Whichever point of view is adopted toward the principle of charge invariance (it might even simply be labeled a postulate), it will follow from the covariance of Maxwell's equations that the flux of $\boldsymbol{E}$ due to a charged body in a volume enclosed by a surface is the same whether the body is at rest or in motion with respect to the surface.

The question of how charge *density* (charge/volume) transforms is obviously closely connected to the question of the transformation of volumes, and hence lengths. Suppose a charge cloud is at rest in the primed system of Fig. 12.1. Enclose in it a small volume, a parallelepiped, which as measured from the *unprimed* system, has sides Δx, Δy, Δz. The lengths Δx, Δy, Δz are defined, as in the previous discussion of length measurements, by *simultaneous* measurements in the *unprimed* frame of the end points of Δx, etc. The charge density ρ in the unprimed frame is then the total charge Q (an invariant) in the volume divided by $\Delta x\Delta y\Delta z$, or $\rho = Q/\Delta x\Delta y\Delta z$. The similar quantity, the density ρ_0 in the primed (rest) frame, is then naturally taken to be Q divided by the volume $\Delta x_0\Delta y\Delta z$, where Δx_0 is the length Δx transformed to the primed system as discussed in Section 12.7. Since

$$\Delta x = \Delta x_0\sqrt{1 - v^2/c^2},$$

we see that ρ is related to ρ_0 by

$$\rho = \frac{\rho_0}{\sqrt{1 - v^2/c^2}}. \tag{12.67}$$

In the unprimed frame, the moving cloud will constitute a current in the x-direction, with x-component J_x given by

$$J_x = \rho v = \frac{\rho_0 v}{\sqrt{1 - \beta^2}}. \tag{12.68}$$

In the rest system, the current J_{0x} is zero and the density is just ρ_0. Using the expressions above for J_x and ρ, we then have

$$J_{0x}^2 - c^2\rho_0^2 = J_x^2 - c^2\rho^2.$$

But this invariance property is just the one that characterizes a four-vector, so without further ado, we assume that $(\boldsymbol{J}, ic\rho)$ constitutes a four-vector.

Consider now the inhomogeneous pair of Maxwell equations that involves $\boldsymbol{J}$ and ρ,

$$\nabla \times \boldsymbol{B} = \frac{1}{c}\frac{\partial \boldsymbol{E}}{\partial t} + \frac{4\pi\boldsymbol{J}}{c}, \qquad \nabla \cdot \boldsymbol{E} = 4\pi\rho, \tag{12.69}$$

and the possible tensorial nature of the fields. We have seen that, under Lorentz transformation, the fields $\boldsymbol{E}$ and $\boldsymbol{B}$ can transform into one another. Since there are six field-strength components, they cannot define a four-vector. The next possibility is that they define a four-tensor of second rank. Such a tensor has sixteen components in general, but we recall that the antisymmetric tensor of this kind has only six independent components. This fact suggests that the tensor we have to deal with is antisymmetric. The plausibility of this suggestion is strengthened if we write out explicitly the two equations (12.69) in four-dimensional notation in which B_x, B_y, B_z become B_1, B_2, B_3, and similarly for $\boldsymbol{E}$. They become

$$\begin{array}{lllll}
0 + & \dfrac{\partial B_3}{\partial x_2} - & \dfrac{\partial B_2}{\partial x_3} - & \dfrac{\partial(iE_1)}{\partial x_4} & = \dfrac{4\pi}{c} J_1, \\[2ex]
-\dfrac{\partial B_3}{\partial x_1} + & 0 + & \dfrac{\partial B_1}{\partial x_3} - & \dfrac{\partial(iE_2)}{\partial x_4} & = \dfrac{4\pi}{c} J_2, \\[2ex]
\dfrac{\partial B_2}{\partial x_1} - & \dfrac{\partial B_1}{\partial x_2} + & 0 - & \dfrac{\partial(iE_3)}{\partial x_4} & = \dfrac{4\pi}{c} J_3, \\[2ex]
\dfrac{\partial(iE_1)}{\partial x_1} + & \dfrac{\partial(iE_2)}{\partial x_2} + & \dfrac{\partial(iE_3)}{\partial x_3} + & 0 & = \dfrac{4\pi}{c} J_4.
\end{array}$$

The left-hand side of this equation is a kind of 4×4-matrix whose terms along the diagonal are zero (the diagonal terms of an antisymmetric tensor are zero). There is also a suggestive *antisymmetry* across the diagonal. If a given field component appears in row i, column j, then the *negative* of that component appears in row j, column i. With these hints, we are led to tentatively define the *field-strength tensor* $F_{\mu\nu}$ as one with components

$$\{F_{\mu\nu}\} = \begin{pmatrix} 0 & B_3 & -B_2 & -iE_1 \\ -B_3 & 0 & B_1 & -iE_2 \\ B_2 & -B_1 & 0 & -iE_3 \\ iE_1 & iE_2 & iE_3 & 0 \end{pmatrix} \tag{12.70}$$

We can then write the set of Maxwell equations (12.69) as

$$\sum_\nu \frac{\partial F_{\mu\nu}}{\partial x_\nu} = \frac{4\pi}{c} J_\mu. \tag{12.71}$$

The fact that we have been able to array the field-strength components in a suggestive pattern does not prove they form a tensor, since tensors are defined by their transformation properties. But the form of Eq. (12.71) supports the assumption that $\{F_{\mu\nu}\}$ represents a tensor. For if it is such, the left-hand side of Eq. (12.71) is the four-divergence of a tensor of rank two; in other words, it is a four-vector, consistent with the fact that the right-hand side is *known* to be a four-vector. We shall take this self-consistency as evidence that $\{F_{\mu\nu}\}$ is indeed a tensor and that Eq. (12.71) is a covariant expression of the two Maxwell equations. The ultimate test of our assumptions (and they pass this test) is, of course, the appeal to experiment and, in particular, to the transformation laws of the field that are predicted by the assumptions.

With the identification in (12.70), the other two Maxwell equations

$$\nabla \cdot \boldsymbol{B} = 0$$
$$\nabla \times \boldsymbol{E} = -\frac{1}{c}\frac{\partial \boldsymbol{B}}{\partial t} \tag{12.72}$$

can, it is easy to verify, be written

$$\frac{\partial F_{\mu\nu}}{\partial x_\lambda} + \frac{\partial F_{\lambda\mu}}{\partial x_\nu} + \frac{\partial F_{\nu\lambda}}{\partial x_\mu} = 0. \tag{12.73}$$

Here each of the indices, λ, μ, ν takes on the values 1, 2, 3, 4. Because of the antisymmetry of the $F_{\mu\nu}$, and the invariance properties under index permutation of Eq. (12.73), only a few of the possible choices for the set λ, μ, ν will be independent. Not surprisingly, there are only *four* such choices corresponding to the *four* equations comprised in (12.72). These can be taken to be the sets (1, 2, 3), (4, 2, 3), (4, 3, 1), (4, 1, 2).

The invariance properties of Eq. (12.73) are not particularly clear as it is written. It is not obvious that, in a primed system, it becomes an equation of the same form but with all variables primed. This form invariance is made more evident by writing the equation as the four-divergence of a certain tensor-like entity called a *pseudo tensor* (or tensor density) which we discuss now.

Formally, a tensor in one reference frame is merely a set of components: a symbol with multiple indices. We can think of *any* such symbol as representing the components of a tensor; the components in another frame are then *defined* by the tensor transformation laws. An example might be the Kronecker symbol, $\delta_{\mu\nu}$. It is defined in one system by $\delta_{\mu\nu} = 1$, $\mu = \nu$; $\delta_{\mu\nu} = 0$, $\mu \neq \nu$. These components then define, through use of the transformation law (12.55), the components in another system. The tensor so defined is symmetric. A generalized analog of this Kronecker-δ is the so-called *Levi–Civita symbol*, which has multiple indices and certain antisymmetry properties. We are presently concerned with the symbol having four indices, which is written $\varepsilon_{\lambda\mu\nu\xi}$. It is defined by

$$\varepsilon_{\lambda\mu\nu\xi} = \begin{cases} 0, \text{ if any two indices are equal} \\ 1, \text{ if } \lambda\mu\nu\xi \text{ is an even permutation of 1234} \\ -1, \text{ if } \lambda\mu\nu\xi \text{ is an odd permutation of 1234.} \end{cases} \tag{12.74}$$

If we think of $\varepsilon_{\lambda\mu\nu\xi}$ as the components in one system of a tensor, the components of another system are defined by the tensor transformation laws, such as (12.56). In the context of a more general discussion of vector analysis, this tensor is not quite of the same character as the ones we have treated. Such a discussion comprises cases where the coefficients $a_{\mu\nu}$ refer not only to the coordinate transformations (12.46) which are "rotations" in the x_1, x_2, x_3, x_4 hyperspace, but to *reflections* in which $x_1 \to -x_1$, etc. The Levi–Civita symbol is a tensor with respect to "rota-

tions," but not reflections. For this reason, it is sometime
For the present purposes, i.e., neglecting reflections, it ca
and we can use the various tensorial rules, for example, tha.
form then the following contracted outer product of the Levi–Civita
and the field-strength tensor:

$$F^*_{\lambda\mu} = \sum_\nu \sum_\xi \varepsilon_{\lambda\mu\nu\xi} F_{\nu\xi}. \tag{12.75}$$

It can be shown† that Eq. (12.73) can be written as the four-divergence of this *dual tensor* $F^*_{\mu\nu}$,

$$\sum_\nu \frac{\partial F^*_{\mu\nu}}{\partial x_\nu} = 0, \tag{12.76}$$

and consequently that it is form invariant under the transformations (12.46).

12.10 TRANSFORMATION OF ELECTROMAGNETIC QUANTITIES

Having established the tensorial properties of the fields, charge, and current, we can discuss some examples of how these quantities transform from one frame to another. Consider first the charge-current four-vector. From Eq. (12.49), we find that

$$\begin{aligned} J'_x &= \gamma\left(J_x - \frac{v}{c}\rho\right), \\ J'_y &= J_y, \\ J'_z &= J_z, \\ \rho' &= \gamma\left(\rho - \frac{vJ_x}{c^2}\right). \end{aligned} \tag{12.77}$$

These equations have interesting consequences. The last of them shows, for example, that a system of charges that is neutral ($\rho = 0$), but that carries a current, will appear from a moving system to have a local charge density ρ' that is different from zero. Such a system might consist of electrons of a certain average density moving in a wire which contains an equal density of fixed positive background charges. Also from the last equation, it is obvious that the charge density ρ' will have a sign which depends on the sign of J_x. Thus in any current flowing in a closed loop J_x, and hence ρ', will be sometimes positive and sometimes negative; the net charge on the loop will always turn out to be zero, in accordance with the principle of charge invariance.

As a second example, we calculate the field of a point charge in uniform motion by transforming the field of a charge at rest. Let a charge q be at the origin of the primed system shown in Fig. 12.3. We want to calculate the field at the point P which is at $x = 0$, $y = b$, $z = 0$ in the unprimed system. The transformation of the

† Moller (R) has a fuller discussion of dual tensors.

eld components is given by Eqs. (12.63) with the primed and unprimed components interchanged and with v replaced by $-v$. Since $B'_x = B'_y = B'_z = 0$, we have

$$E_x = E'_x, \qquad B_x = 0,$$
$$E_y = \gamma E'_y, \qquad B_y = -\gamma \frac{v}{c} E'_z, \tag{12.78}$$
$$E_z = \gamma E'_z, \qquad B_z = \gamma \frac{v}{c} E'_y.$$

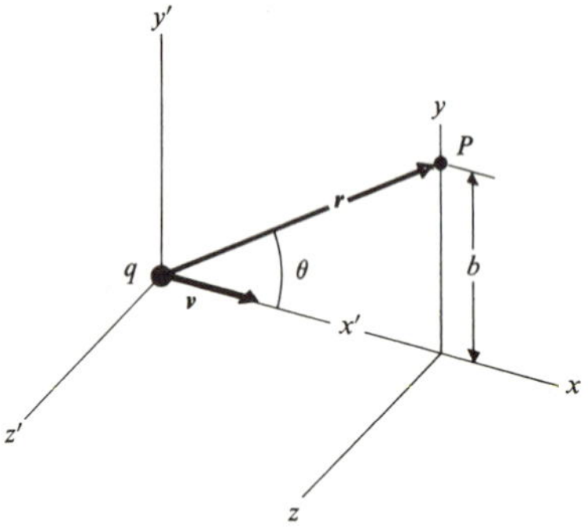

Fig. 12.3 Field of moving point charge.

To use these equations, we first find E'_x, E'_y, E'_z as functions of the primed variables x', y', z', t', and then express these variables in terms of the unprimed ones. The components of $\boldsymbol{E}'$ are, in general, with $r'^2 = (x'^2 + y'^2 + z'^2)^{1/2}$,

$$E'_x = \frac{qx'}{r'^3}, \qquad E'_y = \frac{qy'}{r'^3}, \qquad E'_z = \frac{qz'}{r'^3}. \tag{12.79}$$

From the Lorentz transformation, the space-time coordinates $(0, b, 0, t)$ of the field point P in the unprimed system correspond to the following special values of the primed variables,

$$x' = -\gamma v t, \qquad y' = b, \qquad z' = 0, \qquad t' = \gamma t.$$

Thus we have, from (12.78) and (12.79)

$$E_x = -\frac{q\gamma v t}{(b^2 + \gamma^2 v^2 t^2)^{3/2}},$$

$$E_y = \frac{q\gamma b}{(b^2 + \gamma^2 v^2 t^2)^{3/2}}, \tag{12.80}$$

$$B_z = \frac{q\gamma v b}{c(b^2 + \gamma^2 v^2 t^2)^{3/2}},$$

$$B_x = B_y = E_z = 0.$$

Let $\boldsymbol{r}$ be the vector from the charge to P in the unprimed system, so $\boldsymbol{r} = -vt\hat{\boldsymbol{x}} + b\hat{\boldsymbol{y}}$ and $r = (b^2 + v^2t^2)^{1/2}$. We note first that there is a magnetic field at P that for $\beta \ll 1$ is given by the Biot–Savart law $dB \approx (q\boldsymbol{v} \times \boldsymbol{r}/cr^3)$ since $\boldsymbol{v} \times \boldsymbol{r}$ is in the z-direction with magnitude vb in the present case. The two components E_x and E_y are quite different as functions of t. The longitudinal component E_x vanishes at the origin ($t = 0$) and is of opposite sign on either side of it. On the other hand, E_y is of one sign and has its maximum value at the origin. These functions are plotted in Fig. 12.4. An alternate way to describe the field is in terms of the angle θ, as measured in the unprimed system, between $\boldsymbol{r}$ and the x-axis. With $\cos\theta = -vt/r$ and $\sin\theta = b/r$ it is easy to verify that the field can be written as

$$\boldsymbol{E} = \frac{q(1-\beta^2)}{r^2(1-\beta^2\sin^2\theta)^{3/2}}\,\hat{\boldsymbol{r}}. \tag{12.81}$$

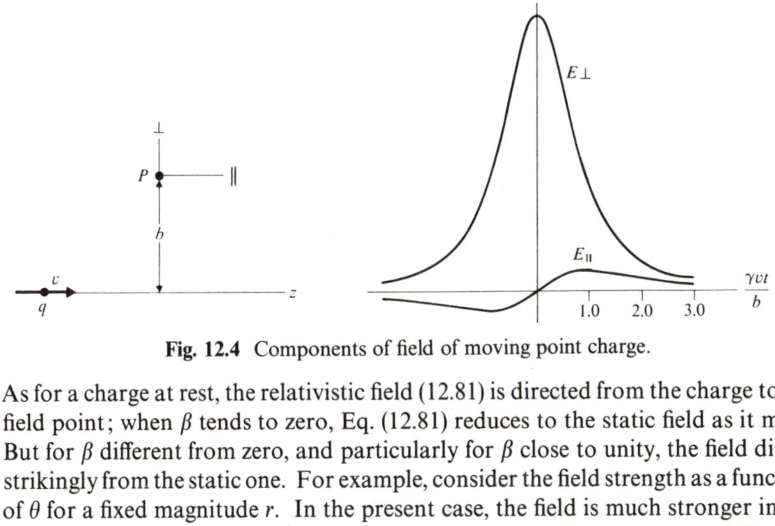

Fig. 12.4 Components of field of moving point charge.

As for a charge at rest, the relativistic field (12.81) is directed from the charge to the field point; when β tends to zero, Eq. (12.81) reduces to the static field as it must. But for β different from zero, and particularly for β close to unity, the field differs strikingly from the static one. For example, consider the field strength as a function of θ for a fixed magnitude r. In the present case, the field is much stronger in the direction perpendicular to the motion ($\theta \approx \pi/2$, $\sin\theta \approx 1$) than it is parallel to the motion ($\theta \approx 0$). In fact, at a given r, the ratio of these two strengths is

$$E(\pi/2)/E(0) = 1/(1-\beta^2)^{3/2};$$

for β close to unity, this ratio can be very large. This effect of the Lorentz transformation, of weakening the field in the direction of motion and enhancing it in the perpendicular direction, is sometimes called the *contraction of the field.*

12.11 RELATIVISTIC MECHANICS

Electrodynamics and mechanics cannot be divorced completely since the central concept of field is defined in terms of the mechanical force exerted on a charged particle. For completeness, we shall discuss briefly some of the modifications that relativity theory introduces into mechanics, but shall cover only a small part of this subject, which includes rigid body motion, elasticity theory, etc. We shall concentrate on the modifications of Newton's second law for a point particle, and on the revision of the momentum conservation law for colliding particles.

Relativity modifies mechanics more profoundly than it does electrodynamics. The reason is clear: Maxwell's equations are already relativistic, i.e., consistent with the Lorentz transformation. The task of the previous sections has been the comparatively trivial one not of generalizing these equations, but of rewriting them in a more convenient four-dimensional language. Newton's second law and the law of conservation of momentum are, by contrast, form-invariant under Galilean transformations and, hence, are *not* form-invariant under Lorentz transformations. They must be generalized to be so invariant.

The relativistic mechanical equations must be relations among invariants, four-vectors, and tensors. The only restriction we have in deducing them is that they must reduce to the Newtonian equations in the limit $v \ll c$. The technique is a kind of guesswork. We find reasonable invariant or four-vectorial analogs of the quantities that occur in the Newtonian equations and replace the quantities by their relativistic analogs. The process is somewhat arbitrary, but perhaps less so than might appear at first sight.

Newton's second law involves the acceleration, the time derivative of the velocity vector. We ask first then for a reasonable four-dimensional analog of the velocity vector. We have found that the position $\boldsymbol{r}$ and absolute time t of Galilean relativity are replaced by a single entity, the four-vector $(\boldsymbol{r}, ict)$, and $d\boldsymbol{r}$ by $(d\boldsymbol{r}, ic\,dt)$. The three-dimensional velocity vector $\boldsymbol{v}$ is obtained by dividing $d\boldsymbol{r}$ by dt, but if we divide the four-vector $(d\boldsymbol{r}, ic\,dt)$ by dt, the resultant entity $(d\boldsymbol{r}/dt, ic)$ is *not* a four-vector because dt is not an invariant. If, however, we divide $(d\boldsymbol{r}, ic\,dt)$ by an invariant, the result is a four-vector and the natural invariant choice to replace dt is, of course, $d\tau$, the element of proper time. As the analog of the velocity, we choose then the *velocity four-vector* $\boldsymbol{V}$ defined by

$$\boldsymbol{V} = \left(\frac{d\boldsymbol{r}}{d\tau}, ic\frac{dt}{d\tau}\right), \tag{12.82}$$

In ordinary mechanics, the momentum vector obtained by multiplying the three-velocity by the rest mass m_0 plays an important role. There is an obvious four-dimensional analog. We multiply the four-velocity by the rest-mass m_0 which is, of course, an invariant, to form the *four-momentum* $\boldsymbol{P}$,

$$\boldsymbol{P} = m_0\left(\frac{d\boldsymbol{r}}{d\tau}, ic\frac{dt}{d\tau}\right). \tag{12.83}$$

The four components P_μ of $\boldsymbol{P}$ are, with $d\tau = dt\sqrt{1 - \beta^2}$

$$P_i = m_0\frac{dx_i}{d\tau}, \qquad i = 1, 2, 3$$

$$P_4 = \frac{icm_0}{\sqrt{1 - \beta^2}}$$

Consider this important four-vector in more detail. The first three components have dimensions of momentum and when $\beta \to 0$ they become the three components of the ordinary momentum. These first three components of the four-momentum are then sometimes called the components of the *relativistic momentum*. We relabel them as $\mathfrak{p}_i$ and we have

$$\mathfrak{p}_i \equiv P_i = \frac{m_0}{\sqrt{1-\beta^2}}\frac{dx_i}{dt}, \qquad i = 1, 2, 3. \tag{12.84}$$

Equation (12.84) now resembles the nonrelativistic formula for momentum except that the rest-mass m_0 is replaced by the so-called *relativistic mass* m,

$$m = \frac{m_0}{\sqrt{1-\beta^2}}. \tag{12.85}$$

Thus the relativistic mass is *velocity dependent*. The relativistic mass is a much-used concept, but not a very fundamental one. As we see above, it merely introduces new terminology into the definition for four-momentum and associates $\sqrt{1-\beta^2}$ with m_0, rather than with dt.

Consider now the limit for small velocities of the fourth component in the four-momentum of (12.83). For $\beta \ll 1$, we have

$$P_4 = \frac{i}{c}\left(\frac{m_0c^2}{\sqrt{1-\beta^2}}\right) = \frac{i}{c}\left(m_0c^2 + \tfrac{1}{2}\, m_0v^2 + \cdots\right).$$

The terms in parentheses have the dimensions of energy and, partly for this reason, one traditionally *defines* the *relativistic energy* $\mathscr{E}$ by

$$\mathscr{E} = \frac{m_0c^2}{\sqrt{1-\beta^2}} = mc^2. \tag{12.86}$$

In the limit of small velocities,

$$\mathscr{E} \approx m_0c^2 + \tfrac{1}{2}mv^2.$$

The term m_0c^2 is called the *rest energy*. The term $\frac{1}{2}m_0v^2$ is, of course, just the ordinary kinetic energy. With the notations above, the four-momentum can be written

$$\boldsymbol{P} \equiv (\mathfrak{p}, i\mathscr{E}/c). \tag{12.87}$$

It should be well noted that *there is no physics* yet in any of the equations that follow (12.83). They involve merely a relabeling of terms in the original *definition* (12.83) of the momentum four-vector. But since we have not yet discussed any laws or equations that this four-vector obeys, we have not yet said anything about the physical world. We get new physical results by using the momentum four-vector to generalize a Newtonian law, that of the *conservation of momentum* of an isolated system. The nonrelativistic law follows from Newton's second law, combined with the equality of action and reaction prescribed by Newton's third law. It holds, say, for a system of particles for which the only forces that act are internal ones, directed along the lines between the particles; in particular, it is importantly applied to collisions between particles. If particles with (ordinary) momentum $\boldsymbol{p}^{(1)}\boldsymbol{p}^{(2)}\ldots\boldsymbol{p}^{(n)}$ collide among themselves and after the collision have momenta $\boldsymbol{p}'^{(1)}\boldsymbol{p}'^{(2)}\ldots\boldsymbol{p}'^{(m)}$, then*

$$\sum_i \boldsymbol{p}^{(i)} = \sum_j \boldsymbol{p}'^{(j)}. \tag{12.88}$$

The natural generalization of (12.88), and our hypothesis, is that *in the collision of particles, the total four-momentum of the particles is conserved*, i.e., the sum of the individual four-momenta before the collision equals that sum after the collision. If $\boldsymbol{P}^{(i)}$ and $\boldsymbol{P}'^{(i)}$ are the four-momenta of the ith particle before and after collision, the hypothesis is

$$\sum_i \boldsymbol{P}^{(i)} = \sum_j \boldsymbol{P}'^{(j)}. \tag{12.89}$$

Since this equation holds for each of the components of the four-momentum, an alternative, but not really novel, way of restating it is to say that *relativistic momentum* and *relativistic energy* are conserved in collisions. Equation (12.89) has famous and striking implications concerning the equivalence of inertial rest mass and energy. To take perhaps the simplest example, consider an "inverse" collision: the splitting of a single particle into two. Such a process is well known in subatomic physics; an example is the so-called decay of one meson into two others. Let the particle that decays have rest-mass $m_0^{(1)}$, and let the two decay products have masses $m_0^{(2)}$ and $m_0^{(3)}$. For simplicity, suppose also that the initial particle decays from rest, i.e., that its velocity $\boldsymbol{v}_1$ is zero, and let the velocities of the other two particles be $\boldsymbol{v}_2$ and $\boldsymbol{v}_3$. The equation for the fourth component of the four-momentum, the relativistic energy, is then

$$m_0^{(1)} = \frac{m_0^{(2)}}{\sqrt{1-\beta_2^2}} + \frac{m_0^{(3)}}{\sqrt{1-\beta_3^2}}. \tag{12.90}$$

We note a striking difference from Newtonian mechanics, according to which rest-mass would be conserved, i.e., which would imply that $m_0^{(1)} = m_0^{(2)} + m_0^{(3)}$. Here we see that, if v_2 and v_3 are nonzero, *rest-mass cannot be conserved*. In a sense, to compensate for the velocities of the final particles, their rest-masses must add up to less than the original rest-mass. In this process, rest-mass has been converted to velocity or energy, and this is the physical implication of the famous Eq. (12.86).

We consider now the problem of the relativistic generalization of Newton's second law for a particle acted on by a force $\boldsymbol{F}$. In doing this, we shall neglect the self-force that the particle experiences due to the mutual interaction of its own charge

---

* Note that the number of particles need not be the same before and after collisions since particles may stick together or a single particle may split into several.

elements. This is discussed more fully in Chapter 14. The nonrelativistic form of Newton's law relates force $\boldsymbol{F}$ and momentum $\boldsymbol{p}$ by

$$\boldsymbol{F} = \frac{d\boldsymbol{p}}{dt}.$$

To make the relativistic form resemble this as closely as possible and reduce to it in the limit of small velocities, it is natural to replace the right-hand side by the four-vector $d\boldsymbol{P}/d\tau$, the derivative with respect to proper time of the four-momentum, and to replace the left-hand side by some *four-vector force* $\mathfrak{F}$. The equation then becomes

$$\mathfrak{F} = \frac{d\boldsymbol{P}}{d\tau}.$$

Consider the possibilities for $\mathfrak{F}$ in terms of the kind of force which interests us, namely, that due to the electromagnetic field. Nonrelativistically, this force is characterized for a particle with charge q by the Lorentz force $q(\boldsymbol{E} + (\boldsymbol{v} \times \boldsymbol{B}/c))$. We ask first if there is some natural relativistic generalization, some four-vector that is simply related to the Lorentz force. Since this latter involves the field strengths and velocity components, we look for four-vectors involving the field-strength tensor $F_{\mu\nu}$ and the velocity four-vector whose components are V_μ. A little reflection shows that a natural candidate is the four-vector $\boldsymbol{f}$ with components f_μ given by

$$f_\mu = \frac{1}{c}\sum_\nu F_{\mu\nu}V_\nu. \tag{12.91}$$

It is easy to see that the first three components of $\boldsymbol{f}$ in the limit $v \ll c$ are those of the three-vector $\boldsymbol{E} + (\boldsymbol{v} \times \boldsymbol{B}/c)$, the Lorentz-force per unit charge. Since charge is an invariant the components qf_μ are also those of a four-vector. A natural hypothesis for the relativistic equation of motion is then

$$qf_\mu = \frac{dP_\mu}{d\tau}. \tag{12.92}$$

The first three components of Eq. (12.92) are

$$\frac{q(\boldsymbol{E} + (\boldsymbol{v} \times \boldsymbol{B}/c))}{\sqrt{1-\beta^2}} = \frac{d}{d\tau}\frac{m_0\boldsymbol{v}}{\sqrt{1-\beta^2}}, \tag{12.93}$$

or with $d\tau = \sqrt{1-\beta^2}\,dt$,

$$q\left(\boldsymbol{E} + \frac{\boldsymbol{v} \times \boldsymbol{B}}{c}\right) = \frac{d}{dt}\frac{m_0\boldsymbol{v}}{\sqrt{1-\beta^2}}. \tag{12.94}$$

The fourth component of (12.92) yields the equation

$$\frac{d\mathscr{E}}{dt} = q\boldsymbol{v}\cdot\boldsymbol{E}. \tag{12.95}$$

The correctness of Eqs. (12.94) and (12.95) has been verified in that they have been successfully used to calculate particle trajectories in many contexts; in particle accelerators, electrostatic and magnetostatic lens, etc., etc.

PROBLEMS

1. Show that the four-divergence of a second rank tensor, as defined by Eq. (12.59), is a four-vector.
2. Show that the sum of the diagonal components of a second rank tensor is an invariant.
3. Show that the outer product of two tensors, defined according to the rule of Eq. (12.57), is itself a tensor.
4. Prove that a contraction of a tensor of rank r yields a tensor of rank $r - 2$.
5. If $\boldsymbol{E}, \boldsymbol{B}$ and $\boldsymbol{E}', \boldsymbol{B}'$ constitute an electromagnetic field as measured in two frames in uniform relative motion, show that $E^2 - B^2$ and $\boldsymbol{E} \cdot \boldsymbol{B}$ are invariants. That is, show that
$$E^2 - B^2 = E'^2 - B'^2 \qquad \text{and} \qquad \boldsymbol{E} \cdot \boldsymbol{B} = \boldsymbol{E}' \cdot \boldsymbol{B}'.$$
6. A long thin wire with no net charge carries current I. Find the fields and current in a reference frame that moves with velocity v parallel to the wire.
7. Use the field of Eq. (12.81) for a moving charge q to show explicitly that Gauss' law is still satisfied for a charge in uniform motion. Thus, show that the flux of the field in the unprimed (rest) system through some convenient surface, such as a sphere, which is stationary in the rest system, yields the value $4\pi q$.
8. Verify that Eq. (12.73), the covariant form of the second set of Maxwell's equations, is equivalent to Eq. (12.76) which expresses the vanishing of the four-divergence of the dual tensor.
9. Show, or at least make plausible, the fact that $(\boldsymbol{A}, i\Phi)$ constitutes a four-vector.
10. Given that $(\boldsymbol{A}, i\Phi)$ is a four-vector, use this fact and the static Coulomb potential q/r to show that the potentials $\boldsymbol{A}, \Phi$ for a particle that moves with constant velocity v along the x-axis, passing through the origin at $t = 0$, are
$$A_x(\boldsymbol{r}, t) = \frac{qv}{c[(x - vt)^2 + (1 - \beta^2)(y^2 + z^2)]^{1/2}},$$
$$A_y = A_z = 0,$$
$$\phi(\boldsymbol{r}, t) = \frac{q}{[(x - vt)^2 + (1 - \beta^2)(y^2 + z^2)]^{1/2}}.$$
These are a special case (for constant velocity) of the general Lienard–Wiechert potentials that are discussed in Chapter 14.
11. An electric dipole of moment p points along the x-axis and moves with uniform velocity, passing through the origin at time $t = 0$. Find $\boldsymbol{E}$ and $\boldsymbol{B}$ at $x = 0$, $y = b$ as a function of time.
12. In a certain frame there are static uniform fields $\boldsymbol{E}$ and $\boldsymbol{B}$ orthogonal to one another. Show that there exists a primed frame in uniform relative motion for which the fields $\boldsymbol{E}'$ and $\boldsymbol{B}'$ are such that either $\boldsymbol{E}' = 0$ or $\boldsymbol{B}' = 0$.

13. In a certain frame $\boldsymbol{E} = \hat{\boldsymbol{x}}E_0$ and $\boldsymbol{B} = (E_0/2)\,(\hat{\boldsymbol{x}}\cos\theta + \boldsymbol{y}\sin\theta)$. Find a primed frame in uniform relative motion in which $\boldsymbol{E}'$ and $\boldsymbol{B}'$ are parallel.

14. A particle with charge q and rest mass m_0 moves along the x-axis and is acted on by a constant field of magnitude E_0 in the x-direction. If $x = \dot{x} = 0$ when $t = 0$, use the relativistic equations of motion, (12.94) and (12.95), to show that

$$x = \frac{m_0c^2}{qE_0}\left\{\left(1 + \left(\frac{qE_0t}{m_0c}\right)^2\right)^{1/2} - 1\right\}.$$

REFERENCES

Arzelies, H., *Relativistic Kinematics*, Pergamon, New York (1966).

A very complete and detailed study of relativistic kinematics with, as well, witty little discursions on life and literature.

Gamba, A., "Physical Quantities in Different Reference Systems According to Relativity," *Am. J. Phys.*, **35**, 83 (1967).

The discussion in our Section 12.7 of the relativistic definitions of physical quantities, and of the relations between quantities defined in different Lorentz frames, borrows from Gamba's article.

Landau, L. D., and E. M. Lifshitz, *The Classical Theory of Fields*, Addison-Wesley, Reading, Mass. (1961).

A treatment of the electromagnetic field which emphasizes its relativistic aspects from the beginning. The discussion is very physical and there is no mathematical formalism for its own sake.

Moller, C., *Theory of Relativity*, Oxford University Press (1952).

A standard text on the special and general theory.

Resnick, R., *Introduction to Special Relativity*, Wiley, New York (1968).

One of the clearest of the many introductions to special relativity.

Rosser, W. G. V., *Classical Electromagnetism via Relativity*, Plenum Press, New York (1968).

This book approaches Maxwell's equations from the Coulomb law and the transformations of special relativity as postulates. Many standard topics are treated from a fresh viewpoint.

Special Relativity Theory, Selected Reprints, American Institute of Physics, New York.

Sixteen reprints of pedagogical articles from various sources, chosen for their clarity or special interest. Ten of them refer to the twin paradox.

Terrell, J., *Phys. Rev.* **116**, 1041 (1959).

13
THE SUMMATION PROBLEM FOR TIME-HARMONIC CURRENTS

We turn once more to a summation problem, which includes those for both electrostatics and magnetostatics as special cases. Equations (11.45) and (11.46) give the potentials Φ and $\boldsymbol{A}$ in terms of integrals over an assumed known current and charge distribution $\boldsymbol{J}$ and ρ; Eqs. (11.9) and (11.10) then give the fields. The summation problem is that of evaluating the integrals and the corresponding fields in a convenient and practical form. There is little that can be said in detail for currents $\boldsymbol{J}(\boldsymbol{r}, t)$ which are arbitrary functions of space and time. Such perfectly arbitrary distributions are simply too general to be tractable. Of the two special cases we shall consider, the first is that of *time-periodic* or *harmonic* currents in which the *spatial dependence* of the currents is arbitrary but the *time dependence* is *simple harmonic* with angular frequency ω; that is, as $\sin \omega t$ or $\cos \omega t$. Harmonic time variation is less restrictive than it might appear since an arbitrary time dependence of current can be synthesized by a Fourier series or integral; we can think of the study of harmonic currents as the study of one such Fourier component. Harmonic currents give rise to, among other kinds of waves, the harmonic plane waves discussed in Section 11.8. The second case of the summation problem, to be discussed in Chapter 14, is that of the *point current*, which is generated by the motion of a *point charge*, e.g., a proton in a synchrotron, or a charged particle in cosmic rays.

13.1 THE EQUATIONS OF HARMONIC VARIATION

For a harmonic current, any component of $\boldsymbol{J}$, say J_x, can be written in the form

$$J_x(\boldsymbol{r}, t) = (\text{Function of } \boldsymbol{r})\begin{Bmatrix}\sin \omega t \\ \cos \omega t\end{Bmatrix}. \tag{13.1}$$

From the linearity of Maxwell's equations, it follows that all the quantities that enter them that depend on $\boldsymbol{r}$ and t, such as $\rho(\boldsymbol{r}, t)$, $B_z(\boldsymbol{r}, t)$ etc., also vary harmonically with time in the same way as (13.1). We can express this time dependence in complex exponential notation by writing for any such function $\boldsymbol{\Gamma}(\boldsymbol{r}, t)$,

$$\boldsymbol{\Gamma}(\boldsymbol{r}, t) = \text{Re}: \boldsymbol{\Gamma}_\omega(\boldsymbol{r})e^{-i\omega t}, \tag{13.2}$$

where, as in Section 11.8, Re: means "real part of." It is unhandy to carry around continually the symbol Re:, so we shall use Eq. (13.2) in the somewhat imprecise but more convenient form

$$\boldsymbol{\Gamma}(\boldsymbol{r}, t) = \boldsymbol{\Gamma}_\omega(\boldsymbol{r})e^{-i\omega t}, \tag{13.3}$$

where "real part of" is understood, and real parts are taken at the end of a calculation. No error results providing that operations are linear. $\boldsymbol{\Gamma}_\omega$ may be complex, $\boldsymbol{\Gamma}_\omega = \boldsymbol{\Gamma}_\omega^r + i\boldsymbol{\Gamma}_\omega^i$, thereby permitting a time dependence that is a general linear combination of sin ωt and cos ωt.

With the assumption (13.3), the various time dependent equations take on a special form. Maxwell's equations themselves become, with $k = \omega/c$:

$$\text{a) } \nabla \times \boldsymbol{E}_\omega = ik\boldsymbol{B}_\omega, \qquad \text{b) } \nabla \times \boldsymbol{B}_\omega = \frac{4\pi \boldsymbol{J}_\omega}{c} - ik\boldsymbol{E}_\omega,$$
$$\text{c) } \nabla \cdot \boldsymbol{E}_\omega = 4\pi\rho_\omega, \qquad \text{d) } \nabla \cdot \boldsymbol{B}_\omega = 0. \tag{13.4}$$

Of the four Maxwell equations in (13.4), only the two curl equations are independent for the present time harmonic case. To see this, we first observe that $\nabla \cdot (\nabla \times \boldsymbol{F}) = 0$, for an arbitrary vector $\boldsymbol{F}$. Thus Eq. (d) in Eqs. (13.4) follows from (a). Similarly, taking the divergence of (b), we have $0 = 4\pi\nabla \cdot \boldsymbol{J}_\omega - i\omega\nabla \cdot \boldsymbol{E}_\omega$. However, from the continuity equation, $\nabla \cdot \boldsymbol{J}_\omega = i\omega\, \rho_\omega$, so that (c) follows from (b) if the continuity equation is satisfied, and we shall assume this to be the case. Thus the two curl equations are the only independent ones. The important Eq. (11.46) for $\boldsymbol{A}$ and the Lorentz condition,

$$\nabla \cdot \boldsymbol{A} + \frac{1}{c}\frac{\partial \Phi}{\partial t} = 0,$$

become

$$\boldsymbol{A}_\omega = \frac{1}{c}\int \boldsymbol{J}_\omega(\boldsymbol{r}')\frac{e^{ik|\boldsymbol{r}-\boldsymbol{r}'|}}{|\boldsymbol{r}-\boldsymbol{r}'|}\,dv' \tag{13.5}$$

and

$$\nabla \cdot \boldsymbol{A}_\omega - ik\Phi_\omega = 0. \tag{13.6}$$

Using (13.6), we find that Eqs. (11.9) and (11.10) for $\boldsymbol{E}$ and $\boldsymbol{B}$ in terms of the potentials become equations for the fields in terms of $\boldsymbol{A}_\omega$ only:

$$\boldsymbol{E}_\omega = \frac{i}{k}\nabla(\nabla \cdot \boldsymbol{A}_\omega) + ik\,\boldsymbol{A}_\omega \tag{13.7}$$

$$\boldsymbol{B}_\omega = \nabla \times \boldsymbol{A}_\omega.$$

Equations (13.5) and (13.7) are, then, the formal solution to the summation problem. Given $\boldsymbol{J}_\omega$, we calculate $\boldsymbol{A}_\omega$ from (13.5) and the fields $\boldsymbol{E}_\omega$ and $\boldsymbol{B}_\omega$ from (13.7).

Before discussing the approximations that must be made to calculate (13.5) in practice, we discuss the Poynting vector for harmonic fields. For $\boldsymbol{E}$ and $\boldsymbol{B}$ represented in the form (13.2), we have

$$\boldsymbol{S} = \frac{c}{4\pi}(\mathrm{Re}: \boldsymbol{E}_\omega e^{-i\omega t}) \times (\mathrm{Re}: \boldsymbol{B}_\omega e^{-i\omega t})$$

$$= \frac{c}{4\pi}(\boldsymbol{E}_\omega^r \cos \omega t + \boldsymbol{E}_\omega^i \sin \omega t) \times (\boldsymbol{B}_\omega^r \cos \omega t + \boldsymbol{B}_\omega^i \sin \omega t).$$

Multiplying out gives four terms: one each involving $\sin^2\omega t$ and $\cos^2\omega t$ and two in $\sin \omega t \cos \omega t$. These last terms oscillate in time and their average value over a period is zero, whereas the first two terms average to $\frac{1}{2}$. If we define the *time-average Poynting vector* $\bar{\boldsymbol{S}}$ to be the Poynting vector $\boldsymbol{S}$ averaged over a period, we have

$$\bar{\boldsymbol{S}} = \frac{c}{8\pi}(\boldsymbol{E}_\omega^r \times \boldsymbol{B}_\omega^r + \boldsymbol{E}_\omega^i \times \boldsymbol{B}_\omega^i).$$

This equation is equivalent to

$$\bar{\boldsymbol{S}} = \frac{c}{8\pi} \mathrm{Re}: (\boldsymbol{E}_\omega \times \boldsymbol{B}_\omega^*), \tag{13.8}$$

where * means complex conjugate; we shall ordinarily use it in this last form.

13.2 OUTLINE OF APPROXIMATIONS

In this section we sketch various approximate methods for evaluating the integral (13.5). The different methods arise from the fact that there are essentially three lengths in the integral and different expansions can be used depending on the relative values of these lengths. These lengths are: the wavelength† $\lambda = 2\pi c/\omega$; the distance r to the field point; and the "size" a of the source. We mean by the latter a "typical" dimension of the radiating object; for example, the diameter if it is a sphere, the length if it is a linear antenna, etc. The approximation to the integral (13.5) will obviously depend on approximations in the expansion of the factor

$$\frac{e^{ik|\boldsymbol{r}-\boldsymbol{r}'|}}{|\boldsymbol{r} - \boldsymbol{r}'|}. \tag{13.9}$$

We shall always assume $r > r'$ and expand in powers of r'/r, i.e., we shall be interested in exterior fields. However, we could equally well make interior expansions in powers of r/r', as in electrostatics and magnetostatics.

There are two ways of undertaking the expansion of $e^{ik|\boldsymbol{r}-\boldsymbol{r}'|}/|\boldsymbol{r} - \boldsymbol{r}'|$. We could, for example write $|\boldsymbol{r} - \boldsymbol{r}'|$ as

$$r\sqrt{1 + \frac{r'^2}{r^2} - 2\frac{r'}{r}\cos \gamma}$$

† Strictly speaking we should call λ the wavelength only when the field it refers to has the mathematical form of a wave, but somewhat loosely we shall call it the wavelength in general.

and then expand this in powers of r'/r in the denominator and in the exponential in a more or less straightforward way. This is the simpler, if less systematic, approach; we shall adopt it first. It is satisfactory enough for the first few terms, but soon becomes tedious. The second, more systematic, method is discussed in Section 13.7. It involves an expansion in Bessel functions and Legendre polynomials that generalizes the familiar expansion of $1/|\boldsymbol{r} - \boldsymbol{r}'|$.

In performing the unsystematic expansion, it is important to remember that it is not the three lengths themselves, but their ratios, that are involved. For example, we can write (13.9) as

$$\frac{e^{ik\,r\sqrt{1+r'^2/r^2-2(r'/r)\cos\gamma}}}{r\sqrt{1+r'^2/r^2-2(r'/r)\cos\gamma}}$$

and the possibilities of an expansion of this function will depend on kr (the ratio $2\pi r/\lambda$) and the ratio r'/r. Thus, in addition to the quantity r'/r important in electrostatics and magnetostatics, there is the new parameter kr. The dependence of approximations on kr is sometimes more subtle than might appear at first sight. This is because kr appears in the exponential, and approximations in exponentials are particularly delicate. If the quantity $a + b$, where $a \ll b$, appears in some algebraic expression, it is usually permissible to write $a + b \approx b$. But in an exponent it is not necessarily true that $e^{i(a+b)} \approx e^{ib}$ for $a \ll b$. (Try $a = 10$ and $b = 10^6$.)

With this observation, we consider the expansion of (13.9) and begin with that for $e^{ik|\boldsymbol{r}-\boldsymbol{r}'|}$. We have for $r > r'$, with γ the angle between $\boldsymbol{r}$ and $\boldsymbol{r}'$:

$$k|\boldsymbol{r} - \boldsymbol{r}'| = kr\sqrt{1 + \frac{r'^2}{r^2} - 2\frac{r'}{r}\cos\gamma}$$

$$\approx kr\left[1 + \frac{1}{2}\left(\frac{r'^2}{r^2} - \frac{2r'}{r}\cos\gamma\right) - \frac{1}{8}\left(\frac{r'^2}{r^2} - \frac{2r'}{r}\cos\gamma\right)^2 + \cdots\right].$$

Keeping only the first term, we have

$$e^{ik|\boldsymbol{r}-\boldsymbol{r}'|} \approx e^{ikr + (ikr'/2)(r'/r) - ikr'\cos\gamma}$$

Suppose that $kr \ll 1$; since r' is always less than r, it follows that $kr' \ll 1$, and the exponential can be set equal to unity. This approximation is called the *quasi-stationary* one. For it, we have

$$\boldsymbol{A}_\omega = \frac{1}{c}\int\frac{\boldsymbol{J}_\omega(\boldsymbol{r}')\,dv'}{|\boldsymbol{r} - \boldsymbol{r}'|}, \qquad \begin{array}{c}\text{Quasi-stationary}\\ kr \ll 1,\\ ka \ll 1.\end{array} \tag{13.10}$$

The reason for the name is that (13.10) is formally like the expression (7.27) for the vector potential of stationary currents. The approximation is discussed in more detail in Section 13.3.

Next we consider the *multipole* or *long wavelength* expansion. Here we assume that $kr' \ll 1$, and also that r'/r is small enough so that $e^{ikr'^2/2r} \approx 1$ and $e^{-ikr'\cos\gamma} \approx 1 - ikr'\cos\gamma$. Then, in (13.9),

$$e^{ik|\boldsymbol{r}-\boldsymbol{r}'|} \approx e^{ikr}(1 - ikr'\cos\gamma).$$

Further expanding $1/|\boldsymbol{r} - \boldsymbol{r}'|$ in Legendre polynomials, we have, to first order in $P_l(\cos\gamma)$,

$$\frac{e^{ik|\boldsymbol{r}-\boldsymbol{r}'|}}{|\boldsymbol{r}-\boldsymbol{r}'|} \approx \frac{e^{ikr}}{r}\left[1 - \left(i - \frac{1}{kr}\right)kr'\cos\gamma + \cdots\right].$$

Then, to these approximations, we have

$$\boldsymbol{A}_\omega = \frac{e^{ikr}}{cr}\left[\int \boldsymbol{J}_\omega\, dv' - \left(ik - \frac{1}{r}\right)\int r'\boldsymbol{J}_\omega \cos\gamma\, dv'\right]. \qquad \begin{matrix}\text{Multipole}\\ ka \ll 1\\ r > a\end{matrix} \qquad (13.11)$$

The terms we have included comprise the *electric dipole*, *magnetic dipole*, and *electric quadrupole* terms. The reason for these names will become clear in Section 13.4, where the approximation is discussed in detail.

Finally, we consider the *wave-zone* or *radiation-zone* approximation, for which $r \gg a$. For this case,

$$\frac{e^{ik|\boldsymbol{r}-\boldsymbol{r}'|}}{|\boldsymbol{r}-\boldsymbol{r}'|} \approx \frac{e^{ikr}}{r}e^{-ikr'\cos\gamma}$$

so that

$$\boldsymbol{A}_\omega = \frac{e^{ikr}}{cr}\int \boldsymbol{J}_\omega e^{-ikr'\cos\gamma}\, dv', \qquad \begin{matrix}\text{Radiation zone}\\ r \gg a\end{matrix} \qquad (13.12)$$

In differentiating $\boldsymbol{A}_\omega$ to find the fields, we shall *also* have to use the condition $r \gg \lambda$, so that the radiation zone approximation really entails two conditions on r.

The approximations discussed above are sketched schematically in Fig. 13.1.

13.3 QUASI-STATIONARY APPROXIMATION

The expression for $\boldsymbol{A}_\omega$ in the quasi-stationary approximation is Eq. (13.10),

$$\boldsymbol{A}_\omega = \frac{1}{c}\int \frac{\boldsymbol{J}_\omega(\boldsymbol{r}')\, dv'}{|\boldsymbol{r}-\boldsymbol{r}'|}.$$

But this is formally like the expression for $\boldsymbol{A}$ in magnetostatics, where the current at a point is *stationary*, i.e., independent of time. This formal correspondence is the basis of the designation *quasi-stationary*. The essential difference between (13.10) and the magnetostatic $\boldsymbol{A}$ is that the present case is time dependent. We may say that the basic approximation in (13.10) is the complete neglect of retardation.

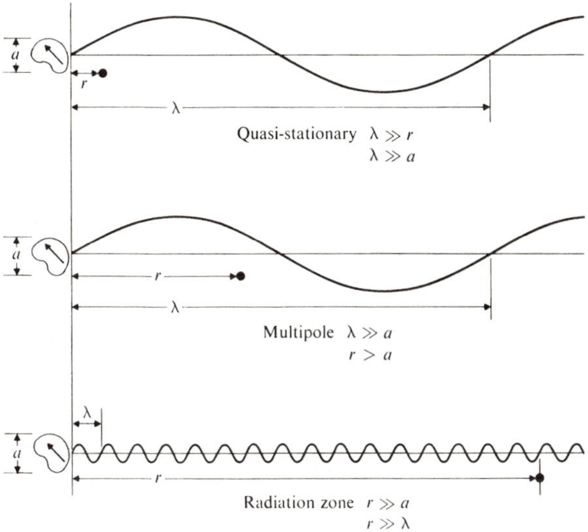

Fig. 13.1 Pictorialization of the approximations for evaluating the time harmonic $\boldsymbol{A}_\omega$. The condition $r \gg \lambda$ for the radiation zone has not yet been introduced into the text, but it will be needed later in calculating the fields.

That is, it corresponds to setting $e^{ik|\boldsymbol{r}-\boldsymbol{r}'|}$ in the integrand of (13.5) equal to unity. This factor, it will be recalled, comes from the fact that the retarded potentials are evaluated not at time t, but at the retarded time $(t - (|\boldsymbol{r} - \boldsymbol{r}'|/c))$. Neglecting the difference between t and the retarded time is equivalent to the condition $e^{ik|\boldsymbol{r}-\boldsymbol{r}'|} = 1$. From the differential point of view of Maxwell's equations, the approximation derives from neglecting $ik\boldsymbol{E}_\omega$ with respect to $4\pi\boldsymbol{J}_\omega/c$ in the equation

$$\nabla \times \boldsymbol{B}_\omega = 4\pi\boldsymbol{J}_\omega/c - ik\boldsymbol{E}_\omega.$$

Given this neglect,* the equations relating $\boldsymbol{J}_\omega$ to $\boldsymbol{A}_\omega$ and, hence, $\boldsymbol{B}_\omega$ to $\boldsymbol{A}_\omega$ are like those for the stationary case. Moreover, the electric field is also calculated as it was for the stationary case. Namely, the retarded potential Φ_ω becomes, on neglecting the retardation factor $e^{ik|\boldsymbol{r}-\boldsymbol{r}'|}$,

$$\Phi_\omega = \int \frac{\rho_\omega(\boldsymbol{r}')\,dv'}{|\boldsymbol{r} - \boldsymbol{r}'|}. \tag{13.13}$$

The charge density that enters (13.13) is calculated from the continuity equation by

* Although the quasi-stationary approximation derives from assuming that $ik\boldsymbol{E}_\omega \ll 4\pi\boldsymbol{J}_\omega/c$ and $ik\boldsymbol{A}_\omega \ll \nabla\Phi_\omega$, i.e., by assuming that k is small, we cannot completely characterize the approximation by saying that for it $k \approx 0$. In the equation $\nabla \times \boldsymbol{E}_\omega = ik\boldsymbol{B}_\omega$, one cannot set the right-hand side to zero without reverting to the completely stationary case.

means of $\rho_\omega = \nabla \cdot \boldsymbol{J}_\omega / i\omega$. Calculating $\boldsymbol{E}_\omega$ from the potentials by means of

$$\boldsymbol{E}_\omega = -\nabla \Phi_\omega + ik\boldsymbol{A}_\omega,$$

we can argue much as above that the term in $\boldsymbol{A}_\omega$ will be smaller than that in Φ_ω by the factor ka. Then $\boldsymbol{E}_\omega$ is calculated formally as it was in electrostatics.

We summarize this discussion: Given a harmonic current distribution of size a, there will be a charge distribution, calculable from the continuity equation for every instantaneous current configuration. If the wavelength $\lambda = 2\pi c/\omega$ is sufficiently large ($\lambda \gg a$) and if the field point is close enough to the distribution, $\lambda \gg r$, then the instantaneous $\boldsymbol{E}$ and $\boldsymbol{B}$ are calculated formally as they would be for the static case. Of course $\boldsymbol{E}$ and $\boldsymbol{B}$ vary with time, but $\boldsymbol{B}$ varies with the same phase as the currents and $\boldsymbol{E}$ with the same phase as the charge; all typical radiation effects are absent. The quasi-stationary approximation underlies circuit theory. For a frequency $f = 2\pi\omega$ of 60 cycles/sec, for example, the wavelength λ is about 8×10^5 m. One sees that the approximation is indeed excellent for most practical applications.

13.4 MULTIPOLE APPROXIMATION

The multipole or long wavelength approximation involves, as we have seen, an expansion in powers of $ka = 2\pi a/\lambda$. This approximation (or its quantum-mechanical analog) is much used in atomic and nuclear physics, where the conditions for its validity are often very well satisfied. For example, an atom with dimensions of the order of 10^{-8} cm may emit light with a wavelength of a few thousand Angstroms, for which case $ka \approx 10^{-3}$. It is important to be clear from the beginning that the present multipole approximation differs from the electrostatic and magnetostatic ones in a major way. In the two former expansions, the important parameter was a/r, the ratio of source size to distance to field point, and the fields of successively higher multipoles fell off with successively higher powers of a/r. In the present expansion, we shall find that *all* multipole fields drop off as $1/r$ in the radiation zone and the parameter that defines the relative importance of successive multipoles is not a/r, but ka.

We start with Eq. (13.11) for $\boldsymbol{A}_\omega$ and rewrite the integrand of the second term. Since

$$\boldsymbol{J}_\omega r' \cos\gamma = \boldsymbol{J}_\omega \frac{\boldsymbol{r}' \cdot \boldsymbol{r}}{r},$$

we can use the theorem $(\boldsymbol{A} \times \boldsymbol{B}) \times \boldsymbol{C} = (\boldsymbol{A} \cdot \boldsymbol{C})\boldsymbol{B} - \boldsymbol{A}(\boldsymbol{B} \cdot \boldsymbol{C})$ with $\boldsymbol{A} = \boldsymbol{r}'$, $\boldsymbol{B} = \boldsymbol{J}_\omega$, $\boldsymbol{C} = \boldsymbol{r}$ to find

$$\boldsymbol{J}_\omega(\boldsymbol{r}' \cdot \boldsymbol{r}) = \tfrac{1}{2}(\boldsymbol{r}' \times \boldsymbol{J}_\omega) \times \boldsymbol{r} + \tfrac{1}{2}((\boldsymbol{r}' \cdot \boldsymbol{r})\boldsymbol{J}_\omega + (\boldsymbol{r} \cdot \boldsymbol{J}_\omega)\boldsymbol{r}').$$

Substituting this into (13.11) we obtain the sum of three terms

$$\boldsymbol{A}_\omega = \boldsymbol{A}_\omega^{\text{ed}} + \boldsymbol{A}_\omega^{\text{md}} + \boldsymbol{A}_\omega^{\text{eq}},$$

where

$$\boldsymbol{A}_\omega^{\text{ed}} = \frac{e^{ikr}}{cr}\int \boldsymbol{J}_\omega\, dv', \tag{13.14}$$

$$\boldsymbol{A}_\omega^{\text{md}} = \frac{e^{ikr}}{r}\left(\frac{1}{r} - ik\right)\left(\frac{1}{2c}\int \boldsymbol{r}' \times \boldsymbol{J}_\omega\, dv'\right) \times \hat{\boldsymbol{r}}, \tag{13.15}$$

$$\boldsymbol{A}_\omega^{\text{eq}} = \frac{e^{ikr}}{2cr}\left(\frac{1}{r} - ik\right)\int\left[\frac{(\boldsymbol{r}'\cdot\boldsymbol{r})}{r}\boldsymbol{J}_\omega + \left(\frac{\boldsymbol{r}\cdot\boldsymbol{J}_\omega}{r}\right)\boldsymbol{r}'\right] dv'. \tag{13.16}$$

The superscripts on the left-hand sides of these equations stand for *electric dipole*, *magnetic dipole*, and *electric quadrupole*, respectively. We shall see shortly the origin of the names.

In the case of the term (13.14) for $\boldsymbol{A}_\omega^{\text{ed}}$, the name derives from the fact that the integral is related to the electric dipole moment. By integration by parts, and using the continuity equation, we find

$$\int \boldsymbol{J}_\omega(\boldsymbol{r}')\, dv' = -\int \boldsymbol{r}'(\nabla\cdot\boldsymbol{J}_\omega)\, dv' = -i\omega\int \boldsymbol{r}'\rho_\omega\, dv'. \tag{13.17}$$

But $\int \boldsymbol{r}'\rho_\omega\, dv'$ is the analog of the electrostatic dipole moment $\boldsymbol{p}$, and we denote it similarly,

$$\boldsymbol{p}_\omega = \int \boldsymbol{r}'\rho_\omega\, dv', \tag{13.18}$$

so Eq. (13.14) can be written as

$$\boldsymbol{A}_\omega^{\text{ed}} = -ik\boldsymbol{p}_\omega\frac{e^{ikr}}{r}. \tag{13.19}$$

The dipole approximation does *not* imply that the current is constant as a function of $\boldsymbol{r}'$. The current may have any space dependence, so long as the conditions in (13.11) are satisfied.

From (13.19), the fields can be calculated. Taking $\boldsymbol{p}_\omega$ along the z-axis, we have

$$\begin{aligned} A_{\omega r} &= -ikp_\omega\frac{e^{ikr}}{r}\cos\theta \\ A_{\omega\theta} &= ikp_\omega\frac{e^{ikr}}{r}\sin\theta. \end{aligned} \tag{13.20}$$

Then Eqs. (13.7) yield

$$\begin{aligned} E_{\omega r} &= 2p_\omega k^3 e^{ikr}\left(\frac{1}{(kr)^3} - \frac{i}{(kr)^2}\right)\cos\theta, \\ E_{\omega\theta} &= p_\omega k^3 e^{ikr}\left(\frac{1}{(kr)^3} - \frac{i}{(kr)^2} - \frac{1}{kr}\right)\sin\theta, \\ B_{\omega\varphi} &= -ip_\omega k^3 e^{ikr}\left(\frac{1}{(kr)^2} - \frac{i}{kr}\right)\sin\theta, \\ E_{\omega\varphi} &= B_{\omega r} = B_{\omega\theta} = 0. \end{aligned} \tag{13.21}$$

These expressions can be interpreted as special cases of the more general vector form

$$\boldsymbol{E}_\omega = k^2\left((\hat{\boldsymbol{r}} \times \boldsymbol{p}_\omega) \times \hat{\boldsymbol{r}} + [3\hat{\boldsymbol{r}}(\hat{\boldsymbol{r}}\cdot\hat{\boldsymbol{p}}_\omega) - \boldsymbol{p}_\omega]\left(\frac{1}{(kr)^2} - \frac{i}{kr}\right)\right)\frac{e^{ikr}}{r}$$

$$\boldsymbol{B}_\omega = k^2(\hat{\boldsymbol{r}} \times \boldsymbol{p}_\omega)\left(1 + \frac{i}{kr}\right)\frac{e^{ikr}}{r}.$$

Consider Eq. (13.21) in more detail. For small distances ($kr \ll 1$) the term $1/(kr)^3$ dominates $1/(kr)^2$ and $(1/kr)$; thus we have approximately

$$E_{\omega r} = \frac{2p_\omega \cos\theta}{r^3}, \qquad E_{\omega\theta} = \frac{p_\omega \sin\theta}{r^3},$$

$$B_{\omega\varphi} = -\frac{i\omega p_\omega \sin\theta}{cr^2}.$$

These electric field components are like those of a static dipole of moment p_ω. The magnetic field is that which would be calculated from the Biot-Savart law for a current element remembering that $-i\omega p_\omega = \int I_\omega\, dz' \approx I_\omega \Delta z$. These results are not surprising, for the condition $kr \ll 1$, in addition to $ka \ll 1$, is just the condition for the quasi-stationary approximation. At the other extreme, $kr \gg 1$, the observation point many wavelengths distant, we have approximately,

$$\begin{aligned} E_{\omega\theta} &= -k^2 p_\omega \sin\theta\,\frac{e^{ikr}}{r} \\ B_{\omega\varphi} &= -k^2 p_\omega \sin\theta\,\frac{e^{ikr}}{r} \end{aligned} \tag{13.22}$$

Moreover, the radial component $E_{\omega r}$ is smaller than these two components by the factor $1/kr$. We see then that $\boldsymbol{E}_\omega$ and $\boldsymbol{B}_\omega$ are (approximately) perpendicular to each other and to the radius vector to the point of observation; in other words, they have essentially the relations that they have in a plane wave. Moreover, their magnitudes are equal, as is the case for a plane wave.

The time average Poynting vector $\bar{S}$ is, from (13.8),

$$\bar{\boldsymbol{S}} = \frac{ck^4 p_\omega^2 \sin^2\theta}{8\pi r^2}\,\hat{\boldsymbol{r}}. \tag{13.23}$$

We see that there is maximum radiation at right angles to the dipole ($\theta = \pi/2$) and no radiation along its axis. Integrating (13.23) over a sphere of radius r yields dW/dt, the rate of radiation of energy:

$$dW/dt = 2\pi r^2 \int \bar{S}\sin\theta\, d\theta = \frac{ck^4 p_\omega^2}{3}. \tag{13.24}$$

This result is important in applications. In particular, the fact that the rate of energy loss is proportional to ω^4 and to p_ω^2 plays a role in many and diverse subjects.

Consider the magnetic dipole term (13.15). The origin of the name is clear if we remember that, within a factor, the integral in (13.15) is the time-dependent analog of the magnetic dipole moment $\boldsymbol{m}$ defined by Eq. (8.32). With a similar notation,

$$\boldsymbol{m}_\omega = \frac{1}{2c}\int \boldsymbol{r}' \times \boldsymbol{J}_\omega\, dv', \qquad 13.25)$$

Eq. (13.15) becomes

$$\boldsymbol{A}_\omega^{md} = \frac{e^{ikr}}{r}\left(\frac{1 - ikr}{r}\right)\boldsymbol{m}_\omega \times \hat{\boldsymbol{r}}. \tag{13.26}$$

Taking $\boldsymbol{m}_\omega$ in the z-direction we find for the fields

$$\begin{aligned}
E_{\omega\varphi} &= m_\omega k^3 e^{ikr}\left(\frac{i}{(kr)^2} + \frac{1}{kr}\right)\sin\theta,\\
B_{\omega r} &= 2m_\omega k^3 e^{ikr}\left(\frac{1}{(kr)^3} - \frac{i}{(kr)^2}\right)\\
B_{\omega\theta} &= m_\omega k^3 e^{ikr}\left(\frac{1}{(kr)^3} - \frac{i}{(kr)^2} - \frac{1}{kr}\right)\sin\theta,\\
E_{\omega r} &= E_{\omega\theta} = B_{\omega\varphi} = 0.
\end{aligned} \tag{13.27}$$

These fields can also be written in the vectorial form

$$\boldsymbol{E}_\omega = -k^2(\hat{\boldsymbol{r}} \times \boldsymbol{m}_\omega)\left(1 + \frac{i}{kr}\right)\frac{e^{ikr}}{r}$$

$$\boldsymbol{B}_\omega = k^2\left((\hat{\boldsymbol{r}} \times \boldsymbol{m}_\omega) \times \hat{\boldsymbol{r}} + [3\hat{\boldsymbol{r}}(\hat{\boldsymbol{r}}\cdot\hat{\boldsymbol{m}}_\omega) - \boldsymbol{m}_\omega]\left(\frac{1}{(kr)^2} - \frac{i}{kr}\right)\right)\frac{e^{ikr}}{r}.$$

The magnetic dipole field resembles that for an electrical dipole, if, in the latter, $\boldsymbol{E}_\omega$ is replaced by $\boldsymbol{B}_\omega$ and $\boldsymbol{B}_\omega$ by $-\boldsymbol{E}_\omega$. The only component of $\boldsymbol{E}_\omega$ is $E_{\omega\varphi}$ which is transverse to the radius vector from the origin, at all distances. For this reason a field configuration such as this is often called a *transverse electric*, abbreviated TE, wave (or mode). For a similar reason, the field of an electric dipole is often termed a *transverse magnetic*, abbreviated TM, mode. The rate of radiation dW/dt calculated as for the electric dipole, is

$$\frac{dW}{dt} = \frac{k^4 m_\omega^2}{3}. \tag{13.28}$$

We consider now the electric quadrupole term (13.16). It derives its name from the fact that the integral in (13.16), which involves $\boldsymbol{J}_\omega$ times x', y', z', can be transformed by means of the continuity equation to ones involving ρ_ω times x'^2, $x'y'$, etc; these latter integrals resemble those used in defining the static quadrupole moments. To see this, consider component by component the integral in (13.16):

$$\int [\boldsymbol{J}_\omega(\boldsymbol{r}'\cdot\boldsymbol{r}) + \boldsymbol{r}'(\boldsymbol{r}\cdot\boldsymbol{J}_\omega)]\,dv'. \tag{13.29}$$

The x-component is

$$2x\int J_{\omega x}x'\,dv' + y\int (y'J_{\omega x} + x'J_{\omega y})\,dv' + z\int (z'J_{\omega x} + x'J_{\omega z})\,dv'. \tag{13.30}$$

Now, if ψ is any scalar,

$$\nabla\cdot(\psi\boldsymbol{J}_\omega) = \boldsymbol{J}_\omega\cdot\nabla\psi + \psi\nabla\cdot\boldsymbol{J}_\omega.$$

Integrating this relation over a volume containing all the current and converting the left-hand side to a surface integral which vanishes because there is no current through it, we conclude that

$$-\int \boldsymbol{J}_\omega\cdot\nabla'\psi\,dv' = \int \psi\nabla'\cdot\boldsymbol{J}_\omega\,dv'.$$

By applying this theorem, where we choose ψ appropriately, to each of the three terms in (13.30), we achieve our result. For example, for the second term in (13.30), take ψ to be $x'y'$ whence $\nabla'\psi = \hat{\boldsymbol{x}}y' + \hat{\boldsymbol{y}}x'$ and the integral in this term is

$$\int \boldsymbol{J}_\omega\cdot(\hat{\boldsymbol{x}}y' + \hat{\boldsymbol{y}}x')\,dv' = -\int x'y'\nabla'\cdot\boldsymbol{J}_\omega\,dv' = -i\omega\int x'y'\rho_\omega\,dv'.$$

with transformations like these, the integral (13.29) can be put into the form

$$\int [\boldsymbol{J}_\omega(\boldsymbol{r}'\cdot\boldsymbol{r}) + \boldsymbol{r}'(\boldsymbol{r}\cdot\boldsymbol{J}_\omega)]\,dv' = -i\omega\int \rho_\omega\boldsymbol{r}'(\boldsymbol{r}'\cdot\boldsymbol{r})\,dv'. \tag{13.31}$$

Now we define the *quadrupole dyadic* $\mathfrak{Q}$ by

$$\mathfrak{Q} = \int \rho_\omega \mathbf{r}'\mathbf{r}'\, dv'$$

Written out in more detail $\mathfrak{Q}$ is of the form, $\mathfrak{Q} = Q_{xx}\hat{\mathbf{x}}\hat{\mathbf{x}} + Q_{xy}\hat{\mathbf{x}}\hat{\mathbf{y}} + \cdots Q_{zz}\hat{\mathbf{z}}\hat{\mathbf{z}}$ where the elements Q_{xx}, Q_{xy} etc., of the dyadic are,* for example,

$$Q_{xx} = \int \rho_\omega x'^2\, dv', \qquad Q_{xy} = \int \rho_\omega x'y'\, dv'.$$

With this definition of $\mathfrak{Q}$ the integral in (13.31) is

$$\int \rho_\omega \mathbf{r}'(\mathbf{r}' \cdot \mathbf{r})\, dv' = \mathfrak{Q} \cdot \mathbf{r},$$

and the quadrupolar potential $\mathbf{A}_\omega^{\text{eq}}$ can be written as

$$\mathbf{A}_\omega^{\text{eq}} = -\frac{ik}{2}\frac{e^{ikr}}{r}\left(\frac{1 - ikr}{r}\right)\mathfrak{Q} \cdot \hat{\mathbf{r}}. \tag{13.32}$$

The fields that are calculated from this can be quite complicated. They become relatively simple at large distances, i.e., in the radiation zone. In such cases, it is usually simplest to treat them with the radiation zone approximation of the next section, so we leave quadrupole fields with this.

13.5 RADIATION ZONE APPROXIMATION

The radiation zone approximation is directed toward calculating the fields of an arbitrary current distribution at large distances from it. We have seen already for the two special cases of the electric and magnetic dipoles that the distant fields are especially simple, in that they fall off as $1/r$. Moreover, only these far fields are needed for calculating the rate of radiated energy by means of the Poynting vector.

To begin, for r much greater than the source size a, $\mathbf{A}_\omega$ is, from (13.12):

$$\mathbf{A}_\omega = \frac{e^{ikr}}{cr}\int \mathbf{J}_\omega(\mathbf{r}')e^{-ikr'\cos\gamma}\, dv'. \tag{13.33}$$

The radiation zone approximation entails another condition in addition to the condition $r \gg a$: the assumption that $kr \gg 1$. In words, the field point is many source sizes *and* many wavelengths away from the current source. To see the consequences of the condition $kr \gg 1$, consider the factor e^{ikr}/r in Eq. (13.33). For k and r large, e^{ikr} varies rapidly with r, whereas $1/r$ varies slowly. In differentiating

* Note that the present convenient definitions of the "quadrupole moments" Q_{xx}, Q_{xy} etc. differ in minor ways from the definitions of the analogous moments in electrostatics.

$\boldsymbol{A}_\omega$ to find the fields, we can then consider $1/r$ as constant and simply differentiate e^{ikr}. For example, consider $\partial/\partial x\,(e^{ikr}/r)$. With $\partial r/\partial x = x/r$ it is

$$\frac{\partial}{\partial x}\frac{e^{ikr}}{r} = \left(\frac{e^{ikr}}{r^2} + \frac{ike^{ikr}}{r}\right)\frac{x}{r} \approx \frac{ikx}{r}\frac{e^{ikr}}{r}, \qquad kr \gg 1.$$

But this result is just what would have been obtained by assuming that $1/r$ was constant in the differentiation. It is also easy to see that in the radiation zone we can neglect the $\boldsymbol{r}$-dependence in the integrand of (13.33).

To use Eq. (13.7) for $\boldsymbol{E}_\omega$, we need to calculate first $\nabla \cdot \boldsymbol{A}_\omega$. In rectangular coordinates, when we apply the prescription just mentioned for differentiation, it is obvious that

$$\nabla \cdot \boldsymbol{A}_\omega = ik(\hat{\boldsymbol{r}} \cdot \boldsymbol{A}_\omega) \tag{13.34}$$

in terms of the unit vector $\hat{\boldsymbol{r}}$. Calculating $\nabla(\nabla \cdot \boldsymbol{A}_\omega)$ with the same prescription, we find that

$$\nabla(\nabla \cdot \boldsymbol{A}_\omega) = ik\hat{\boldsymbol{r}}(\nabla \cdot \boldsymbol{A}_\omega),$$

which, on using (13.34), becomes

$$\nabla(\nabla \cdot \boldsymbol{A}_\omega) = (ik)^2\hat{\boldsymbol{r}}(\hat{\boldsymbol{r}} \cdot \boldsymbol{A}_\omega).$$

From Eq. (13.7), we have then

$$\boldsymbol{E}_\omega = -ik[\hat{\boldsymbol{r}}(\hat{\boldsymbol{r}} \cdot \boldsymbol{A}_\omega) - \boldsymbol{A}_\omega]. \tag{13.35}$$

The expression in brackets has an interesting interpretation. Imagine that $\boldsymbol{A}_\omega$ is decomposed into two components, a component $\boldsymbol{A}_\omega^\perp$ perpendicular to direction of observation $\hat{\boldsymbol{r}}$, and a component $\boldsymbol{A}_\omega^{\parallel}$ parallel to it:

$$\boldsymbol{A}_\omega = \boldsymbol{A}_\omega^\perp + \boldsymbol{A}_\omega^{\parallel}.$$

The magnitude of $\boldsymbol{A}_\omega^{\parallel}$ is just $\hat{\boldsymbol{r}} \cdot \boldsymbol{A}_\omega$ so

$$\boldsymbol{A}_\omega^{\parallel} = \hat{\boldsymbol{r}}(\hat{\boldsymbol{r}} \cdot \boldsymbol{A}_\omega).$$

With this result, we see that the quantity in brackets in (13.35) is $\boldsymbol{A}_\omega^\perp$

$$\boldsymbol{A}_\omega^\perp = \hat{\boldsymbol{r}}(\hat{\boldsymbol{r}} \cdot \boldsymbol{A}_\omega) - \boldsymbol{A}_\omega$$

so that

$$\boldsymbol{E}_\omega = -ik\boldsymbol{A}_\omega^\perp. \tag{13.36}$$

The vector $\boldsymbol{A}_\omega$ is generated in Eq. (13.5) from the vector $\boldsymbol{J}_\omega$; in particular, $\boldsymbol{A}_\omega^\perp$ is generated from $\boldsymbol{J}_\omega^\perp$. Equation (13.36) is then

$$\boldsymbol{E}_\omega = -ik\frac{e^{ikr}}{cr}\int \boldsymbol{J}_\omega'^\perp e^{-ikr'\cos\gamma}\,dv'.$$

From $\boldsymbol{B}_\omega = \nabla \times \boldsymbol{A}_\omega$, we obtain easily:

$$\boldsymbol{B}_\omega = (\hat{\boldsymbol{r}} \times \boldsymbol{E}_\omega). \tag{13.37}$$

Here $\boldsymbol{J}_\omega^\perp$ is defined like $\boldsymbol{A}_\omega^\perp$,

$$\boldsymbol{J}_\omega^\perp = \hat{\boldsymbol{r}}(\hat{\boldsymbol{r}} \cdot \boldsymbol{J}_\omega) - \boldsymbol{J}_\omega.$$

Note that part of the angular dependence of the field is contained in $\boldsymbol{J}_\omega^\perp$, since the perpendicular component of current will differ for different directions in space.

With these expressions for $\boldsymbol{E}_\omega$ and $\boldsymbol{B}_\omega$, the time average Poynting vector $\bar{\boldsymbol{S}}$ is

$$\bar{\boldsymbol{S}} = \frac{k^2}{8\pi c r^2}\left|\int \boldsymbol{J}_\omega^\perp e^{-ikr'\cos\gamma}\, dv'\right|^2 \hat{\boldsymbol{r}}. \tag{13.38}$$

There is a useful alternative form of this equation. With $\boldsymbol{k} = k\hat{\boldsymbol{r}}$ it is readily shown that

$$\bar{\boldsymbol{S}} = \frac{1}{8\pi c r^2}\left|\int (\boldsymbol{k} \times \boldsymbol{J}_\omega) e^{-i\boldsymbol{k}\cdot\boldsymbol{r}'} dv'\right|^2 \hat{r} \tag{13.38a}$$

Equation (13.38) is a basic formula that enables us to calculate the "radiation pattern" i.e., the angular dependence of the Poynting vector in the radiation zone of any current distribution. It is very useful in physics and engineering.

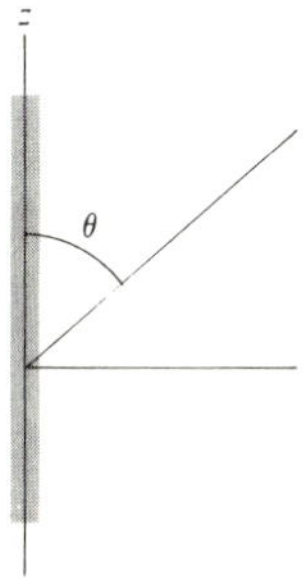

Fig. 13.2 Linear antenna.

We shall describe two applications of (13.38). The first is to linear antennas, i.e., antennas which consist of a thin conductor in which current flows, say, along the z-direction, as in Fig. 13.2. For this case, $\boldsymbol{r}'$ ranges only along the z'-axis and $\boldsymbol{J}_\omega\, dv'$ in (13.38) is replaced by $I_\omega\, dz'$. Moreover, $\boldsymbol{J}_\omega^\perp$ becomes $\boldsymbol{I}_\omega^\perp = -\hat{\boldsymbol{\theta}} I_\omega \sin\theta$, and $\cos\gamma$ is, to a good approximation, replaced by $\cos\theta$. We have then, for the only component of $\boldsymbol{E}_\omega$,

$$E_{\omega\theta} = \frac{ike^{ikr}}{cr} \sin\theta \int I_\omega(z') e^{-ikz'\cos\theta}\, dz'. \tag{13.39}$$

The only component of $\boldsymbol{B}_\omega$ is, of course, $B_{\omega\varphi}$. Note that one can look at (13.39) as giving the field due to an ensemble of dipoles of strength $I_\omega\, dz'$, each of which contributes to the field with a phase factor $e^{-ikz'\cos\theta}$. To apply this formula, suppose, as an example, that the current $I(z', t)$ is of the form

$$I(z', t) = I_0(z') \cos \omega t,$$

where $I_0(z')$ is some real amplitude function. Then

$$I_\omega(z') = I_0(z').$$

For the sake of being able to do the integrations in (13.39) explicitly, suppose further that the antenna extends from $-L/2$ to $L/2$ on the z'-axis and that

$$I_0(z') = \mathscr{I} \sin m\pi(z'/L + 1/2),$$

where m is an integer. Then the case $m = 1$ corresponds to a current distribution $\cos(\pi z'/L)$, the case $m = 2$, to the distribution $\sin(2\pi z'/L)$ etc. Note that the length L of the antenna has no necessary relation to the wavelength λ. The wavelength λ is determined only by the frequency of the generator that feeds the antenna. The length L is an independent parameter. However, for practical reasons, L is frequently chosen to be related to the wavelength. If L is taken to be an integral number of half-wavelengths

$$L = m\lambda/2,$$

then the case $m = 1$ is called a *half-wave* linear antenna, $m = 2$ is a *full-wave* one, and so on. For this case then, the integral we perform for evaluating $E_{\omega\theta}$ and/or the Poynting vector is

$$\int_{-L/2}^{L/2} I_\omega(z') e^{-ikz'\cos\theta}\, d\theta = \mathscr{I} \int_{-L/2}^{L/2} \sin\left(\frac{m\pi z'}{L} + \frac{m\pi}{2}\right) e^{-i(m\pi z'/L)\cos\theta}\, dz'.$$

The integral is elementary and leads to the result

$$\bar{S} = \frac{\mathscr{I}^2}{2\pi c r^2} \frac{\cos^2(m\pi \cos\theta/2)}{\sin^2\theta}, \qquad m = 1, 3, 5 \ldots$$

$$\bar{S} = \frac{\mathscr{I}^2}{2\pi c r^2} \frac{\sin 2(m\pi \cos\theta/2)}{\sin^2\theta}, \qquad m = 2, 4, 6 \ldots.$$

The radiation patterns are sketched in Fig. 13.3.

A second important application of (13.38) is to *arrays* of radiating elements. These elements might be the linear radiators we have just considered, constituting a linear antenna array. Or the elements might be atoms or molecules so that the array constituted a crystal lattice. For example, think of a linear antenna array consisting of a series of N identical radiators aligned along the z-axis, with the first centered at the origin, as in Fig. 13.4, and with distance a between them. As for a single linear antenna, the integral over $J_\omega^\perp$ in Eq. (13.38) now becomes an integral

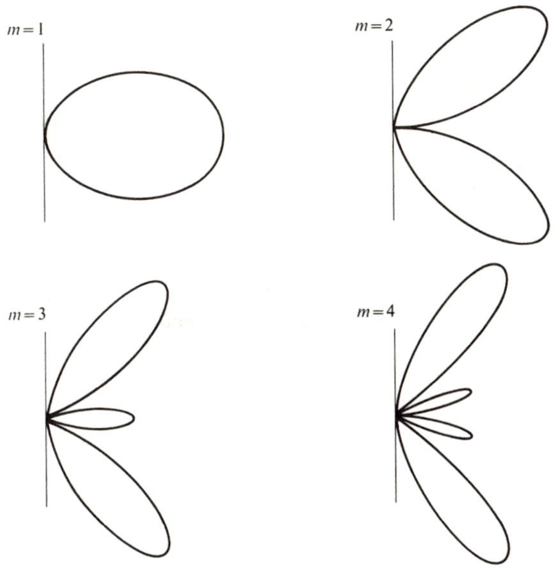

Fig. 13.3 Radiation patterns from linear antennas of length $m\lambda/2$ with current distributions varying as $\sin(2\pi z'/\lambda + m\pi/2)$.

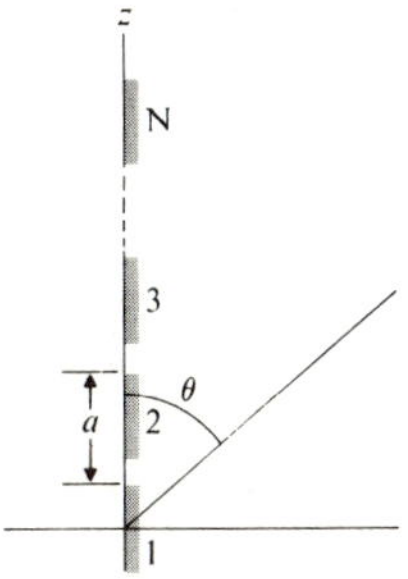

Fig. 13.4 Linear array of radiators.

over $I_\omega(z')$, so Eq. (13.39) is still valid, where the integral goes over all radiators. We have

$$E_{\omega\theta} = \frac{ike^{ikr}}{cr} F(\theta),$$

where

$$F(\theta) = \sin\theta \int_{\text{All radiators}} I_\omega(z')e^{-ikz'\cos\theta}\,dz' \tag{13.40}$$

The integral over all radiators in (13.40) is just the sum of integrals over individual radiators. Let $F_n(\theta)$ be the function obtained by integrating the right-hand side of (13.40) over the n-th radiator. Then

$$F_1(\theta) = \sin\theta \int_{\text{Radiator 1}} I_\omega(z')e^{-ikz'\cos\theta}\,dz'. \tag{13.41}$$

In the second radiator, displaced a distance a, the current is the same function of $z' - a$ as the current in the first is of z', so the function $F_2(\theta)$ is

$$F_2(\theta) = \sin\theta \int_{\text{Radiator 2}} I_\omega(z' - a)e^{-ikz'\cos\theta}\,dz' = e^{ika\cos\theta}\,F_1(\theta).$$

Similarly,

$$F_N(\theta) = e^{ik(N-1)a\cos\theta}\,F_1(\theta).$$

We have then

$$F(\theta) = F_1(\theta)\sum_{n=0}^{N-1} e^{ikna}.$$

The sum is a geometric progression, and is therefore readily evaluated:

$$\sum_{n=0}^{N-1} e^{ikna\cos\theta} = \frac{1 - e^{ikNa\cos\theta}}{1 - e^{ika\cos\theta}}.$$

Then from $\bar{S} = \dfrac{c}{8\pi}|E_{\omega\theta}|^2$ we have

$$\bar{S} = \frac{k^2}{8\pi c r^2}|F_1(\theta)|^2 \frac{\sin^2\left(\dfrac{Nka}{2}\cos\theta\right)}{\sin^2\left(\dfrac{ka}{2}\cos\theta\right)}.$$

Thus we have the important result that the radiation pattern of the array is given by a product of two factors: a function $|F_1(\theta)|^2$ characteristic of the individual element and a function $\sin^2((Nka/2)\cos\theta)/\sin^2((ka/2)\cos\theta)$ characteristic of the array. As N gets large, this last function becomes more sharply peaked as a function of θ. A similar result holds for two- and three-dimensional spatial arrays of identical elements. That is, the radiation pattern is the product of a function characterizing the element and a function characterizing the lattice structure of the array.*

* See Problem 13 at the end of this chapter.

13.6 TWO-DIMENSIONAL DISTRIBUTIONS

A special case of the time harmonic summation problem is that for *two-dimensional distributions.* As in electrostatics, these depend on only two coordinates, and are independent, say, of the coordinate z. They are then cylindrical* distributions in which the axis of the cylinder is parallel to the z-axis. Consider the vector potential $\boldsymbol{A}_\omega$ for this case. Since $\boldsymbol{J}_\omega$ is independent of z', the obvious first step in evaluating $\boldsymbol{A}_\omega$, as given by the integral of Eq. (13.5) is to integrate out the dependence on z' in the factor $e^{ik|\boldsymbol{r}-\boldsymbol{r}'|}/|\boldsymbol{r}-\boldsymbol{r}'|$ in the integrand. Since all field quantities are independent of z anyway, it is convenient to take the field point at $z = 0$. This factor becomes $e^{ik\sqrt{\sigma^2+z'^2}}/\sqrt{\sigma^2 + z'^2}$, where

$$\sigma = |\boldsymbol{\rho} - \boldsymbol{\rho}'|$$

and $\boldsymbol{\rho}$ and $\boldsymbol{\rho}'$ are vectors in a plane perpendicular to the cylinder axis. Then $\boldsymbol{A}_\omega$ is, with $dv' = ds'\,dz'$,

$$\boldsymbol{A}_\omega(\boldsymbol{\rho}) = \frac{1}{c}\int \boldsymbol{J}_\omega(\boldsymbol{\rho}')\,ds' \int_{-\infty}^{\infty} \frac{e^{ik\sqrt{\sigma^2+z'^2}}}{\sqrt{\sigma^2 + z'^2}}\,dz'.$$

The integral over z' can be done, but it is not elementary. It does arise in other contexts, however, and it defines† a *Hankel function*, designated by $H_0(k\sigma)$,

$$H_0(k\sigma) = -\frac{i}{\pi}\int_{-\infty}^{\infty} \frac{e^{ik\sqrt{\sigma^2+z^2}}}{\sqrt{\sigma^2 + z^2}}\,dz. \tag{13.43}$$

Using this, $\boldsymbol{A}_\omega(\boldsymbol{\rho})$ becomes

$$\boldsymbol{A}_\omega(\boldsymbol{\rho}) = \frac{i\pi}{c}\int \boldsymbol{J}_\omega(\boldsymbol{\rho}')H_0(k|\boldsymbol{\rho} - \boldsymbol{\rho}'|)\,ds'. \tag{13.44}$$

We now discuss some of the properties of H_0. Apart from (13.43), it is defined in terms of the Bessel and Neumann functions, J_0 and N_0, by

$$H_0(x) = J_0(x) + iN_0(x).$$

To see the physical meaning of H_0 in the present context, consider Eq. (13.44) for the case in which the current $\boldsymbol{J}_\omega$ is a *line source* of *unit strength* along the z-axis:

$$\boldsymbol{J}_\omega(\boldsymbol{\rho}') = \delta(\boldsymbol{\rho}')\hat{\boldsymbol{z}}. \tag{13.45}$$

Then $\boldsymbol{A}_\omega$ has only a z-component, which is, from (13.44),

$$A_{\omega z}{}^{\text{unit source}} = \frac{i\pi}{c}H_0(k\rho). \tag{13.46}$$

---

* Cylindrical is meant in the general sense, i.e., the cylinders are not necessarily circular in cross section.

† See for example, Morse and Feshbach (B), p. 1323. In fact, there are two Hankel functions, $H_0{}^{(1)}$ and $H_0{}^{(2)}$. That of Eq. (13.43) is really $H_0{}^{(1)}$, the Hankel function of the first kind, but we have dropped the superscript since we never deal with $H_0{}^{(2)}$.

Thus, $H_0(k\rho)$ is proportional to $A_{\omega_z}{}^{\text{unit source}}$. Now each component of $\boldsymbol{A}_\omega$ satisfies

$$(\nabla^2 + k^2)\boldsymbol{A}_\omega = -\frac{4\pi \boldsymbol{J}_\omega}{c}. \tag{13.47}$$

We see that $H_0(k\rho)$ is the basic cylindrical wave solution due to a line source along the z-axis, much as e^{ikr}/kr is the basic wave solution in three dimensions due to a point source at the origin. As a summation of *outgoing* spherical waves, H_0 must then represent an *outgoing cylindrical wave.* This is confirmed by looking at its asymptotic form. For large z, $J_0(z)$ approaches the value $\sqrt{2/\pi z}\cos(z - (\pi/4))$ and $N_0(z)$ approaches $\sqrt{2/\pi z}\sin(z - (\pi/4))$, so that

$$H_0(k\rho) \to \sqrt{\frac{2}{\pi k\rho}}\, e^{i(k\rho - \pi/4)}, \qquad k\rho \to \infty. \tag{13.48}$$

The factor $e^{ik\rho}$ shows that H_0 does indeed describe an outgoing wave.* Equation (13.44) is then interpretable as a superposition integral: the wave at $\boldsymbol{\rho}$ is the sum of all the outgoing cylindrical waves produced at the source points $\boldsymbol{\rho}'$. A further important property of H_0 is found by operating with $(\nabla^2 + k^2)$ on $A_{\omega_z}{}^{\text{unit source}}$ of (13.46), where ∇^2 is the two-dimensional Laplacian. Using (13.45) and (13.47), we conclude that

$$(\nabla^2 + k^2)H_0(k\rho) = 4i\delta(\boldsymbol{\rho}). \tag{13.49}$$

We can consider that in the last equation we have found the free-space Green function for the Helmholtz equation. That is, on defining

$$G_k(\boldsymbol{\rho}, \boldsymbol{\rho}') = -\frac{i}{4}H_0(k|\boldsymbol{\rho} - \boldsymbol{\rho}'|), \tag{13.50}$$

we see that

$$(\nabla^2 + k^2)G_k(\boldsymbol{\rho}, \boldsymbol{\rho}') = \delta(\boldsymbol{\rho} - \boldsymbol{\rho}'). \tag{13.51}$$

With the properties of H_0 now known to some degree, the two-dimensional integral (13.44) can be expanded and approximated much as was done in the previous sections for the three-dimensional integral. We shall not go into much detail, but simply mention two cases: a *multipole expansion*, and a *radiation zone* approximation. The multipole expansion might be generated by a straightforward Taylor series, but since there is neither much novelty nor much system in this, we shall bypass it. Rather we recall that in electrostatics a systematic multipole expansion was generated for two-dimensional charge distributions by using the expansion (3.48) for $\ln|\boldsymbol{\rho} - \boldsymbol{\rho}'|$. For the present case, a similar expansion, an

---

* If the split-off time dependence is taken to be $e^{+i\omega t}$, as it sometimes is, then the solution $H_0(k\rho) \equiv H_0^{(1)}(k\rho)$ must be replaced by $H_0^{(2)}(k\rho)$ to represent an outgoing wave, since $H_0^{(2)}$ contains the factor $e^{-ik\rho}$ in its asymptotic form.

addition theorem for $H_0(k|\boldsymbol{\rho} - \boldsymbol{\rho}'|)$, can be used. We discuss this now. As we have seen, $H_0(k|\boldsymbol{\rho} - \boldsymbol{\rho}'|)$ satisfies the Helmholtz equation (as a function of $\boldsymbol{\rho}$ say, for fixed $\boldsymbol{\rho}'$) and represents an outgoing cylindrical wave from a source at $\boldsymbol{\rho}'$. In terms of the cylindrical coordinates ρ, φ of $\boldsymbol{\rho}$ and coordinates ρ', φ' of $\boldsymbol{\rho}'$, we have

$$|\boldsymbol{\rho} - \boldsymbol{\rho}'| = (\rho^2 + \rho'^2 - 2\rho\rho' \cos(\varphi - \varphi'))^{1/2}.$$

$H_0(k|\boldsymbol{\rho} - \boldsymbol{\rho}'|)$ is then a function of ρ, ρ' and of $\varphi - \varphi'$, rather than φ and φ' separately. Assume some fixed ρ', and examine the expansion of H_0 for $\rho < \rho'$. Since H_0 satisfies the Helmholtz equation in cylindrical coordinates, since the range of H_0 includes the origin, and since H_0 is finite at the origin for fixed ρ', the cylindrical solutions of the Helmholtz equation that must be used are those that are finite at the origin. From Appendix D, these are

$$J_l(k\rho)e^{il\varphi}.$$

We conclude that for $\rho < \rho'$, H_0 can be expanded in the form

$$H_0(k|\boldsymbol{\rho} - \boldsymbol{\rho}'|) = \sum_{l=-\infty}^{\infty} A_l f_l(\rho') J_l(k\rho) e^{il(\varphi-\varphi')}, \qquad \rho < \rho', \tag{13.52}$$

where A_l and $f_l(\rho')$ are arbitrary so far. Consider now the opposite case, $\rho > \rho'$. Here similar arguments apply except that the expansion must behave asymptotically as an outgoing cylindrical wave; i.e., must contain the factor $e^{ik\rho}/\sqrt{\rho}$. The asymptotic form of the Hankel function of order l is

$$H_l(k\rho) \rightarrow \sqrt{\frac{2}{\pi k\rho}}\, e^{i(k\rho - \pi/2(l+1/2))}, \qquad \rho \rightarrow \infty \tag{13.53}$$

so that a Hankel function of any order satisfies this outgoing wave condition. The functions

$$H_l(k\rho)e^{il\varphi}$$

can therefore be taken as a basis set for the expansion for $\rho > \rho'$, which must then be of the form

$$H_0(k|\boldsymbol{\rho} - \boldsymbol{\rho}'|) = \sum_{l=-\infty}^{\infty} B_l g_l(\rho') H_l(k\rho) e^{il(\varphi-\varphi')}, \qquad \rho > \rho', \tag{13.54}$$

where B_l and $g_l(\rho')$ are arbitrary. But the original function is symmetric in ρ and ρ'. The expansion (13.52) for $\rho < \rho'$ must then become the expansion (13.54) if ρ and ρ' are interchanged. We conclude that $f_l = H_l(k\rho)$, $g_l = J_l(k\rho)$, and $A_l = B_l$. We can then write the expansion as

$$H_0(k|\boldsymbol{\rho} - \boldsymbol{\rho}'|) = \sum_{l=-\infty}^{\infty} B_l J_l(k\rho_<) H_l(k\rho_>) e^{il(\varphi-\varphi')},$$

where $\rho_<$ is the lesser of ρ and ρ', and $\rho_>$ is the greater. We determine the coefficients B_l by noting that this expansion becomes the known expansion (3.48) for $\ln|\boldsymbol{\rho} - \boldsymbol{\rho}'|$ when $k \to 0$. We find that $B_l = 1$, and so we can write

$$H_0(k|\boldsymbol{\rho} - \boldsymbol{\rho}'|) = \sum_{l=-\infty}^{\infty} J_l(k\rho_<)H_l(k\rho_>)e^{il(\varphi-\varphi')}. \tag{13.55}$$

This is the desired addition theorem and substituting it in (13.44) we have, for $\rho > a$, where a is the maximum value that ρ' assumes,

$$\boldsymbol{A}_\omega(\rho, \varphi) = \sum_{l=-\infty}^{\infty} \boldsymbol{\mu}_l H_l(k\rho)e^{il\varphi}.$$

The *moments* $\boldsymbol{\mu}_l$ are defined by

$$\boldsymbol{\mu}_l = \frac{i\pi}{c}\int \boldsymbol{J}_\omega(\boldsymbol{\rho}')J_l(k\rho')e^{-il\varphi'}\,ds'. \tag{13.56}$$

We leave examples of this expansion to the problem section.

For the radiation zone approximation we shall need to assume, as in the three-dimensional case, the conditions $k\rho \gg 1$ and $\rho \gg a$. We use the first of these to find, from Eq. (13.53),

$$H_0(k|\boldsymbol{\rho} - \boldsymbol{\rho}'|) \approx \sqrt{\frac{2}{\pi k|\boldsymbol{\rho} - \boldsymbol{\rho}'|}}\, e^{-\pi i/4}e^{ik|\boldsymbol{\rho} - \boldsymbol{\rho}'|}. \tag{13.57}$$

With the condition $\rho \gg \rho'$, this result can be further expanded much as for the three-dimensional case. That is, we set $1/|\boldsymbol{\rho} - \boldsymbol{\rho}'| \approx 1/\rho$ in the denominator of (13.57), and use in the exponent

$$e^{ik|\boldsymbol{\rho} - \boldsymbol{\rho}'|} \approx e^{ik(\rho - \rho'\cos(\varphi - \varphi'))}.$$

Then we obtain from (13.44):

$$\boldsymbol{A}_\omega \approx \frac{i\pi e^{-i\pi/4}}{c}\sqrt{\frac{2}{\pi k\rho}}\, e^{ik\rho}\int \boldsymbol{J}_\omega(\boldsymbol{\rho}')e^{-ik\rho'\cos(\varphi-\varphi')}\,ds'. \tag{13.58}$$

We can calculate relatively easily the radiation pattern of any two-dimensional distribution with this equation. Suppose, for example, that the current is in the z-direction only. It is easy to see that

$$E_{\omega z} = ikA_{\omega z} \approx -\sqrt{\frac{2\pi k}{\rho c^2}}\, e^{i(k\rho - \pi/4)}\int J_{\omega z}(\boldsymbol{\rho}')e^{-ik\rho'\cos(\varphi-\varphi')}\,ds'$$

and that, in the limit $k\rho \gg 1$, the only component of $\boldsymbol{B}_\omega$ is

$$B_{\omega\varphi} = -E_{\omega z}.$$

The time average Poynting vector $\bar{\boldsymbol{S}}$ is then given by

$$\bar{\boldsymbol{S}} = \frac{k}{4c\rho}\left|\int J_{\omega z}(\boldsymbol{\rho}')e^{-ik\rho'\cos(\varphi-\varphi')}\,ds'\right|^2\hat{\boldsymbol{\rho}} = \frac{k}{4c\rho}\left|\int J_{\omega z}(\boldsymbol{\rho}')e^{-i\boldsymbol{k}\cdot\boldsymbol{\rho}'}\,ds'\right|^2\hat{\boldsymbol{\rho}} \tag{13.59}$$

where $\boldsymbol{k} = k(\hat{\boldsymbol{x}}\cos\varphi + \hat{\boldsymbol{y}}\sin\varphi)$.

13.7 SYSTEMATIC MULTIPOLE EXPANSION

There are alternative approaches to the Taylor series expansion of Section 13.2 for making a multipole or long wavelength approximation. We shall sketch two of these, and then discuss in some detail a third, that combines the advantages of the two, although at some cost in algebraic complexity.

First, we recall that it was found in electrostatics that higher multipole field configurations could be generated by differentiation of lower ones; the field of a point charge yields a dipole field, while the differentiated dipole field yields a quadrupole, etc. There is a similar possibility for time harmonic fields. Equations (13.21) and (13.27) are the fields of electric and magnetic time harmonic dipoles; successive differentiations will generate other solutions that can be classified as electric and magnetic quadrupoles, octupoles, etc. This method soon gets unwieldy but it has one feature worth mentioning. We recall that the electric dipole field is a transverse magnetic (TM) one; the radial component $B_{\omega r}$ is zero. Similarly the magnetic dipole field is a transverse electric (or TE) mode. If we differentiate these dipole fields to form fields of higher multipoles, the TE or TM character is obviously maintained. This observation suggests that it should be possible to find a sequence of multipoles all of which can be classified as either TE or TM. The systematic expansion we later discuss will do just this.

The second possibility for a multipole expansion is the analog of the two-dimensional expansion that uses the addition theorem (13.55) for $H_0(k|\boldsymbol{\rho} - \boldsymbol{\rho}'|)$. Just as $H_0(k|\boldsymbol{\rho} - \boldsymbol{\rho}'|)$ and the function $1/|\boldsymbol{r} - \boldsymbol{r}'|$ of electrostatics have their addition theorems, so is there one for the function $e^{ik|\boldsymbol{r}-\boldsymbol{r}'|}/|\boldsymbol{r} - \boldsymbol{r}'|$ that appears in the integral for $\boldsymbol{A}_\omega$. In terms of the spherical Bessel and Hankel functions j_l and h_l, it is, with $r_>(r_<)$ as the greater (lesser) of r and r',

$$e^{ik|\boldsymbol{r}-\boldsymbol{r}'|}/|\boldsymbol{r} - \boldsymbol{r}'| = ik\sum_{l=0}^{\infty}(2l+1)j_l(kr_<)h_l(kr_>)P_l(\cos\gamma), \tag{13.60}$$

where γ is the angle between $\boldsymbol{r}$ and $\boldsymbol{r}'$. The proof of Eq. (13.60) is essentially identical to the proof of the addition theorem for $H_0(k|\boldsymbol{\rho} - \boldsymbol{\rho}'|)$ outlined in the last section and so we omit it.

We might now simply substitute this expansion in (13.5) for $\boldsymbol{A}_\omega$, and in fact, if we do, the first two terms for $l = 0$ and $l = 1$ simply reproduces the Taylor series result of Section 13.2. However, this straightforward technique has the disadvantage of not naturally yielding up the useful division into TE and TM modes that we have mentioned above. We turn then to the systematic multipole development which exploits the expansion (13.60), but *also* provides the division into modes. The angular parts of this multipole solution to the summation problem are called *vector spherical harmonics.* They are important in their own right, beyond the problem of relating fields to known sources, since they constitute an expansion basis for the angular dependence of fields of unknown sources, and are therefore important in discussing boundary value problems. For this and other reasons, it will be convenient to first classify and systematize the multipole modes in *free space*, without reference to their source, and only then find formulas which relate these modes to the currents that produce them.

Consider the time harmonic Maxwell equations (13.4) in free space. Eliminating $\boldsymbol{E}_\omega$ between the two curl equations, we find

$$\begin{aligned} \nabla \times (\nabla \times \boldsymbol{B}_\omega) - k^2\boldsymbol{B}_\omega &= 0, \\ \nabla \cdot \boldsymbol{B}_\omega &= 0, \\ \boldsymbol{E}_\omega &= i/k\nabla \times \boldsymbol{B}_\omega. \end{aligned} \tag{13.61}$$

Similarly, by eliminating $\boldsymbol{B}_\omega$, we find

$$\begin{aligned} \nabla \times (\nabla \times \boldsymbol{E}_\omega) - k^2\boldsymbol{E}_\omega &= 0, \\ \nabla \cdot \boldsymbol{E}_\omega &= 0, \\ \boldsymbol{B}_\omega &= -i/k\nabla \times \boldsymbol{E}_\omega. \end{aligned} \tag{13.62}$$

Consider the possible solutions of the first two equations in (13.61). Expanding $\nabla \times (\nabla \times \boldsymbol{B}_\omega)$ and using $\nabla \cdot \boldsymbol{B}_\omega = 0$, we obtain

$$(\nabla^2 + k^2)\boldsymbol{B}_\omega = 0, \qquad \nabla \cdot \boldsymbol{B}_\omega = 0. \tag{13.63}$$

Each *rectangular* component of $\boldsymbol{B}_\omega$ thus satisfies the scalar Helmholtz equation. Now the scalar Helmholtz equation for an amplitude ψ,

$$(\nabla^2 + k^2)\psi = 0, \tag{13.64}$$

has from Appendix D a set of solutions labeled by indices l and m. We denote a member of this set by ϕ_{lm}, so that $(\nabla^2 + k^2)\phi_{lm} = 0$. Explicitly,

$$\phi_{lm} = \gamma_l(kr)Y_{lm}(\theta, \varphi), \tag{13.65}$$

where γ_l is a spherical Bessel, Neumann, or Hankel function. We *cannot* however simply use these functions to attempt to write a solution of $(\nabla^2 + k^2)\boldsymbol{B}_\omega = 0$ in the form

$$\boldsymbol{B}_\omega = \sum_{l,m} A_{lm}\gamma_l(kr)Y_{lm}(\theta, \varphi),$$

where $\boldsymbol{A}_{lm}$ is an *arbitrary* vector coefficient, for there is no guarantee that the second equation, $\nabla \cdot \boldsymbol{B}_\omega = 0$, will be satisfied. However, we can satisfy this second equation and still use the solutions (13.65) of the scalar equation in the following way. Consider the vector function

$$(\boldsymbol{r} \times \nabla)\psi,$$

where ψ satisfies Eq. (13.64). Since $(\boldsymbol{r} \times \nabla)\psi = \nabla \times (\boldsymbol{r}\psi)$ and since the divergence of any curl is zero, we have $\nabla \cdot ((\boldsymbol{r} \times \nabla)\psi) = 0$. More importantly, $(\boldsymbol{r} \times \nabla)\psi$ satisfies the *vector* Helmholtz equation.

$$(\nabla^2 + k^2)(\boldsymbol{r} \times \nabla)\psi = 0, \tag{13.66}$$

i.e., each component satisfies (13.64). We can show this in a straightforward manner by taking (13.66) one component at a time and working out the differentiations. The function $(\boldsymbol{r} \times \nabla)\psi$ can then be taken as a solution of Eqs. (13.63) for $\boldsymbol{B}_\omega$. Moreover it follows from the cross product in $\boldsymbol{B}_\omega = (\boldsymbol{r} \times \nabla)\psi$ that, for this solution, $\boldsymbol{B}_\omega$ is perpendicular to $\boldsymbol{r}$, or is a *transverse magnetic* or TM field mode. This means that only $\boldsymbol{E}_\omega$ has a radial component. For this reason, a TM field mode is also called an *electric mode*, and is identified by a superscript e on $\boldsymbol{B}_\omega$ and $\boldsymbol{E}_\omega$, so that the fields of an electric mode are $\boldsymbol{B}^e_\omega$ and $\boldsymbol{E}^e_\omega$.

For compactness of notation at this point, it is convenient to introduce a *vector operator** $\boldsymbol{L}$, which is, in fact, except for a factor $\hbar$, the angular momentum operator* of quantum mechanics. We define $\boldsymbol{L}$ by

$$\boldsymbol{L} = -i(\boldsymbol{r} \times \nabla). \tag{13.67}$$

The operator $\boldsymbol{L}$ is useful not only for its compactness, but also because the angular momentum properties of fields are important in atomic and nuclear physics. Remembering (13.61), we can write the solution of (13.63) we have just discussed in terms of $\boldsymbol{L}$:

$$\boldsymbol{B}^e_\omega = \boldsymbol{L}\psi, \qquad \boldsymbol{E}^e_\omega = (i/k)\nabla \times (\boldsymbol{L}\psi). \tag{13.68}$$

Equations (13.62) for $\boldsymbol{E}_\omega$ can now be treated similarly. If a function χ satisfies $(\nabla^2 + k^2)\chi = 0$, then the function $\boldsymbol{L}\chi$ satisfies $(\nabla^2 + k^2)(\boldsymbol{L}\chi) = 0$, has zero divergence, and corresponds to a transverse electric field (TE) or a *magnetic* mode, in which only the magnetic field is radial. A solution of (13.62) can then be taken to be

$$\boldsymbol{E}^m_\omega = \boldsymbol{L}\chi, \qquad \boldsymbol{B}^e_\omega = -(i/k)\nabla \times (\boldsymbol{L}\chi). \tag{13.69}$$

The two potentials ψ and χ are called *Debye potentials.* Together they provide a

---

* An *operator* essentially embodies a rule for changing one function into another function. For example, the operator d/dx operating on any function $\psi(x)$ produces $d\psi/dx$; the operator $\sqrt{}$ produces $\psi^{1/2}$. A vector operator such as $(\boldsymbol{r} \times \nabla)$ is one which is equivalent to three scalar operators.

solution of the time harmonic Maxwell equations of the form

$$\begin{aligned} \boldsymbol{E}_\omega &= \boldsymbol{E}_\omega^e + \boldsymbol{E}_\omega^m \\ \boldsymbol{B}_\omega &= \boldsymbol{B}_\omega^e + \boldsymbol{B}_\omega^m \end{aligned} \tag{13.70}$$

or,

$$\begin{aligned} \boldsymbol{E}_\omega &= \boldsymbol{L}\chi + (i/k)\boldsymbol{\nabla} \times (\boldsymbol{L}\psi) \\ \boldsymbol{B}_\omega &= \boldsymbol{L}\psi - (i/k)\boldsymbol{\nabla} \times (\boldsymbol{L}\chi). \end{aligned} \tag{13.71}$$

Moreover, it can be shown* that Eqs. (13.71) constitute a *general* decomposition: any time harmonic electromagnetic field can be represented in this form. Using the fact that ψ and χ can be written as a linear combination of the elementary solutions (13.65), we have, with $\chi = \Sigma_{l,m} C_{lm}\phi_{lm}$ and $\psi = \Sigma_{l,m} D_{lm}\phi_{lm}$,

$$\begin{aligned} \boldsymbol{E}_\omega &= \sum_{l,m} C_{lm}\boldsymbol{L}\phi_{lm} + \frac{i}{k}\sum_{l,m} D_{lm}\boldsymbol{\nabla} \times (\boldsymbol{L}\phi_{lm}) \\ \boldsymbol{B}_\omega &= \sum_{l,m} D_{lm}\boldsymbol{L}\phi_{lm} - \frac{i}{k}\sum_{l,m} C_{lm}\boldsymbol{\nabla} \times (\boldsymbol{L}\phi_{lm}). \end{aligned} \tag{13.72}$$

As we have remarked, the choice of the radial function $\gamma_l(kr)$ in ϕ_{lm} as Bessel, Neumann, or Hankel will depend on the boundary conditions to be satisfied. If (13.72) represents the field external to sources that are bounded in space, then γ_l must be chosen as a Hankel function to correspond to outgoing waves at infinity, so that

$$\phi_{lm} = h_l(kr) Y_{lm}(\theta, \varphi). \tag{13.73}$$

Since the functions $\boldsymbol{L}\phi_{lm}$ are central to the solution of the vector wave equation, their angular parts are given the special name of *vector spherical harmonics*, designated $\boldsymbol{Y}_{lm}(\theta, \varphi)$ and defined by

$$\boldsymbol{Y}_{lm}(\theta, \varphi) = \frac{1}{\sqrt{l(l+1)}}\boldsymbol{L}\, Y_{lm}(\theta, \varphi). \tag{13.74}$$

They can be shown to satisfy the orthogonality relation

$$\int \boldsymbol{Y}^*_{l'm'} \cdot \boldsymbol{Y}_{lm}\, d\Omega = \delta_{ll'}\delta_{mm'}.$$

Equations (13.72) give the general form of the field of an arbitrary current distribution; the coefficients C_{lm} and D_{lm} will, of course, depend on the distribution. Now we shall relate these coefficients to the current $\boldsymbol{J}_\omega$ for the field exterior to bounded sources. The general technique is this. From (13.72) we can find expressions for $\boldsymbol{r} \cdot \boldsymbol{E}_\omega$ and $\boldsymbol{r} \cdot \boldsymbol{B}_\omega$, the radial components of the fields, in terms of the

* See the book by Papas (R) and references therein.

coefficients C_{lm} and D_{lm}. Independently, we can find $\boldsymbol{r}\cdot\boldsymbol{E}_\omega$ and $\boldsymbol{r}\cdot\boldsymbol{B}_\omega$ in terms of integrals over the current $\boldsymbol{J}_\omega$ from the field equations. By comparing these two expressions, C_{lm} and D_{lm} can be found in terms of the integrals. Consider first the calculation of $\boldsymbol{r}\cdot\boldsymbol{E}_\omega$. From (13.72) we have

$$\boldsymbol{r}\cdot\boldsymbol{E}_\omega = \sum_{l,m} C_{lm}\boldsymbol{r}\cdot(\boldsymbol{L}\phi_{lm}) + \frac{i}{k}\sum_{l,m} D_{lm}\boldsymbol{r}\cdot\nabla\times(\boldsymbol{L}\phi_{lm}).$$

For the factor multiplying C_{lm} in this expression, we have

$$\boldsymbol{r}\cdot(\boldsymbol{L}\phi_{lm}) = -i\boldsymbol{r}\cdot(\boldsymbol{r}\times\nabla)\phi_{lm} = 0$$

since $\boldsymbol{r}$ is perpendicular to $\boldsymbol{r}\times\nabla$. For the factor multiplying D_{lm}, we transform $\boldsymbol{r}\cdot\nabla\times(\boldsymbol{L}\phi_{lm}) = -i\boldsymbol{r}\cdot\nabla\times((\boldsymbol{r}\times\nabla)\phi_{lm})$ by means of the vector formula

$$\boldsymbol{A}\cdot\boldsymbol{B}\times(\boldsymbol{C}\times\boldsymbol{D}) = (\boldsymbol{A}\times\boldsymbol{B})\cdot(\boldsymbol{C}\times\boldsymbol{D}) \qquad \text{with } \boldsymbol{A}=\boldsymbol{C}=\boldsymbol{r} \text{ and } \boldsymbol{B}=\boldsymbol{D}=\nabla$$

to find $\boldsymbol{r}\cdot\nabla\times((\boldsymbol{r}\times\nabla)\phi_{lm}) = (\boldsymbol{r}\times\nabla)\cdot(\boldsymbol{r}\times\nabla)\phi_{lm}$. But since $(\boldsymbol{r}\times\nabla)\cdot(\boldsymbol{r}\times\nabla) = -L^2$ and since* $L^2\phi_{lm} = l(l+1)\phi_{lm}$, we have, with (13.73),

$$\boldsymbol{r}\cdot\boldsymbol{E}_\omega = \frac{1}{k}\sum_{l,m} l(l+1)D_{lm}h_l(kr)Y_{lm}(\Omega). \tag{13.75}$$

Similarly

$$\boldsymbol{r}\cdot\boldsymbol{B}_\omega = -\frac{1}{k}\sum_{l,m} l(l+1)C_{lm}h_l(kr)Y_{lm}(\Omega). \tag{13.76}$$

Now we calculate $\boldsymbol{r}\cdot\boldsymbol{E}_\omega$ and $\boldsymbol{r}\cdot\boldsymbol{B}_\omega$ in terms of the currents from the two Maxwell curl equations, (13.4a) and (13.4b). On taking the curl of either of these equations and substituting in the other, we find that

$$\nabla\times(\nabla\times\boldsymbol{E}_\omega) - k^2\boldsymbol{E}_\omega = \frac{4\pi ik\boldsymbol{J}_\omega}{c} \tag{13.77}$$

$$\nabla\times(\nabla\times\boldsymbol{B}_\omega) - k^2\boldsymbol{B}_\omega = \frac{4\pi\nabla\times\boldsymbol{J}_\omega}{c}. \tag{13.78}$$

From these equations it can be shown that

$$(\nabla^2 + k^2)(\boldsymbol{r}\cdot\boldsymbol{B}_\omega) = -\frac{4\pi\boldsymbol{r}\cdot(\nabla\times\boldsymbol{J}_\omega)}{c}. \tag{13.79}$$

To prove this, we use the vector identity

$$\boldsymbol{r}\cdot\nabla\times(\nabla\times\boldsymbol{F}) = 2\nabla\cdot\boldsymbol{F} + \boldsymbol{r}\cdot\nabla(\nabla\cdot\boldsymbol{F}) - \nabla^2(\boldsymbol{r}\cdot\boldsymbol{F})$$

---

* We quote the result without proof. It is proven in almost any text on quantum mechanics.

Applying this with $\boldsymbol{F} = \boldsymbol{B}_\omega$ to (13.78) and using the fact that $\nabla \cdot \boldsymbol{B}_\omega = 0$, we reproduce (13.79). Similarly, if we apply the identity with $\boldsymbol{F} = \boldsymbol{E}_\omega$ to (13.71), and then apply it with $\boldsymbol{F} = \boldsymbol{J}_\omega$, we find readily

$$(\nabla^2 + k^2)\left(\boldsymbol{r}\cdot\boldsymbol{E}_\omega + \frac{4\pi i}{\omega}\boldsymbol{r}\cdot\boldsymbol{J}_\omega\right) = \frac{4\pi}{i\omega}\boldsymbol{r}\cdot\nabla\times(\nabla\times\boldsymbol{J}_\omega). \tag{13.80}$$

Now Eq. (13.79) is of the form of the scalar Helmholtz equation

$$(\nabla^2 + k^2)f(\boldsymbol{r}) = -4\pi s(\boldsymbol{r})$$

whose solution that satisfies the radiation condition (outgoing waves at infinity) is

$$f(\boldsymbol{r}) = \int s(\boldsymbol{r}')\frac{e^{ik|\boldsymbol{r}-\boldsymbol{r}'|}}{|\boldsymbol{r}-\boldsymbol{r}'|}\,dv'.$$

Applying this result to the solution of Eq. (13.79) we have

$$\boldsymbol{r}\cdot\boldsymbol{B}_\omega = \frac{1}{c}\int \boldsymbol{r}'\cdot(\nabla'\times\boldsymbol{J}_\omega)\frac{e^{ik|\boldsymbol{r}-\boldsymbol{r}'|}}{|\boldsymbol{r}-\boldsymbol{r}'|}\,dv'.$$

Now we use the expansion (13.60) and the addition theorem for $P_l(\cos\gamma)$ to obtain

$$\boldsymbol{r}\cdot\boldsymbol{B}_\omega = \frac{4\pi ik}{c}\int \boldsymbol{r}'\cdot(\nabla'\times\boldsymbol{J}_\omega)\left(\sum_{l,m} j_l(kr')h_l(kr)Y_{lm}(\Omega)Y_{lm}^*(\Omega')\right)dv'.$$

Comparing this with (13.76), we find, for example, that

$$C_{lm} = -\frac{4\pi ik^2}{cl(l+1)}\int \boldsymbol{r}'\cdot(\nabla'\times\boldsymbol{J}_\omega)j_l(kr')Y_{lm}^*(\Omega')\,dv'.$$

For $\boldsymbol{r}\cdot\boldsymbol{E}_\omega$, using (13.80) outside the current distribution, so $\boldsymbol{r}\cdot\boldsymbol{J}_\omega$ is identically zero, we have similarly

$$\boldsymbol{r}\cdot\boldsymbol{E}_\omega = \frac{i}{\omega}\int \boldsymbol{r}'\cdot\nabla'\times(\nabla'\times\boldsymbol{J}_\omega)\frac{e^{ik|\boldsymbol{r}-\boldsymbol{r}'|}}{|\boldsymbol{r}-\boldsymbol{r}'|}\,dv',$$

whence

$$D_{lm} = -\frac{4\pi k}{cl(l+1)}\int \boldsymbol{r}'\cdot\nabla'\times(\nabla'\times\boldsymbol{J}_\omega)j_l(kr')Y_{lm}^*(\Omega')\,dv'.$$

PROBLEMS

1. The plane $x = 0$ constitutes an infinite current sheet of uniform harmonic current of surface density J_s flowing in the z-direction. Find $\boldsymbol{E}_\omega$ and $\boldsymbol{B}_\omega$ everywhere.

2. If, in Problem 1, the current was confined only to a strip of width $d(-d/2 < y < d/2)$, show that the magnitude of the Poynting vector at large distances is

$$\bar{S} = \frac{k\pi J_s^2 d^2}{4c\rho}\left(\frac{\sin\left(\frac{kd}{2}\sin\varphi\right)}{\frac{kd}{2}\sin\varphi}\right)^2$$

where φ is in the $x \cdot y$-plane, measured from the x-axis.

3. A uniformly distributed harmonic surface current of surface density J_s flows in the z-direction on a cylinder of radius a whose axis is the z-axis. Show that the only component of $\boldsymbol{E}_\omega$ is $E_{\omega z}$, given for $\rho > a$ by

$$E_{\omega z} = CH_0(k\rho).$$

Find the constant C, the form of $E_{\omega z}$ for $\rho < a$, and $\boldsymbol{B}_\omega$.

4. If ψ satisfies the Helmholtz equation, $(\nabla^2 + k^2)\psi = 0$, show that each of the rectangular components of $(\boldsymbol{r} \times \nabla)\psi$ also satisfies the Helmholtz equation.

5. Six time harmonic electric dipoles are spaced uniformly around the circumference of a circle of radius a, centered at the origin in the plane $z = 0$. One of the dipoles is at $x = 0$, $y = a$. Find the radiation pattern (Poynting vector at large distances) if: (a) the dipoles are oriented in the z-direction, (b) if they are oriented in the x-direction.

6. Two-dimensional time harmonic fields are those for which there is no z-dependence (although the fields may have *z-components*). Show that such fields are determined if $\boldsymbol{E}_{\omega z}$ and $\boldsymbol{B}_{\omega z}$ are known, by showing that outside the source ($\boldsymbol{J}_\omega = 0$)

$$-ik\boldsymbol{B}_\omega = \hat{z} \times \nabla E_{\omega z}$$

$$ik\boldsymbol{E}_\omega = \hat{z} \times \nabla B_{\omega z}.$$

7. A time-harmonic source a produces fields $\boldsymbol{E}_\omega^a$ and $\boldsymbol{B}_\omega^a$ and an independent source b produces fields $\boldsymbol{E}_\omega^b$ and $\boldsymbol{B}_\omega^b$. Show that at any point outside the sources

$$\nabla \cdot (\boldsymbol{E}_\omega^a \times \boldsymbol{B}_\omega^b) = \nabla \cdot (\boldsymbol{E}_\omega^b \times \boldsymbol{B}_\omega^a)$$

This relation is sometimes known as *Lorentz's lemma.*

8. *Isotropic* radiators exist for scalar waves. An example is an (idealized) *point source* of acoustic waves, or a pulsating sphere. Prove, or make plausible, that an isotropic antenna, i.e., a system of currents that radiates energy *uniformly* in all directions is impossible. One such proof is due to H. F. Mathis, *Proc. IRE* **39**, 970 (1951).

9. A spherical balloon carries a charge Q uniformly distributed on its surface. The balloon pulsates with frequency v and amplitude a, so that its radius r is given by $r = r_0 + a\sin(2\pi v t)$. Calculate the rate of radiation of electromagnetic energy.

10. A current $I = I_0 \sin(2\pi z/L) \sin\omega t$ flows on the z-axis between $z = 0$ and $z = L$. Find a good approximation for $\boldsymbol{E}$ and $\boldsymbol{B}$ for the following three cases, in which r is the distance to the field point, and $\lambda = 2\pi c/\omega$:

(a) $\lambda > 10^4 L$, $r > 10^4\lambda$

(b) $\lambda = L$, $r > 10^4\lambda$

(c) $L < r < 2L$, $\lambda > 10^4 L$.

11. Four infinite wires each carrying current $I = I_0 \sin \omega t$ in the same direction are parallel to the z-axis, and intersect the plane $z = 0$ at points with coordinates (x, y) given by (a, a), $(a, -a)$, $(-a, a)$, $(-a, -a)$. Find the Poynting vector at large distances.

12. A circular current loop of radius a carrying uniform time-harmonic current of amplitude I_0 is in the plane $z = 0$, centered at the origin. Show that the fields in the radiation zone are

$$E_{\omega\varphi} = -2\pi I_0 ka \frac{e^{ikr}}{r} \sum_{l=0}^{\infty} \frac{(4l+3)(2l)!j_{2l+1}(ka)P_{2l+1}^1(\cos\theta)}{2^{2l+1}l!(l+1)!}$$

$$B_{\omega\theta} = -E_{\omega\varphi}.$$

Hint: Use (13.36), the relation

$$e^{ikr\cos\gamma} = \sum_{l=0}^{\infty} i^l (2l+1) j_l(kr) P_l(\cos\gamma),$$

which can be derived from Eq. (13.60) and $P_s^1(0) = 0$, s even.

13. A linear radiating element, when at the origin and directed along the z-axis, has a radiation pattern $r^2\bar{S}$ given by $r^2\bar{S} = F(\theta)$. A three-dimensional antenna array is made by placing such identical elements at the points with coordinates $\boldsymbol{r}_n$ given by

$$\boldsymbol{r}_n = n_x \boldsymbol{a} + n_y \boldsymbol{b} + n_z \boldsymbol{c},$$

where $n_x = 0, 1, 2, \ldots, N_x$ and $n_y = 0, 1, 2, \ldots, N_y$ and $n_z = 0, 1, 2, \ldots, N_z$. Show that the radiation pattern for the array is proportional to $|FF_xF_yF_z|^2$, where

$$F_x(\theta, \varphi) = \frac{\sin(N_x ka \sin\theta\cos\varphi/2)}{\sin(ka\sin\theta\cos\varphi/2)}$$

$$F_y(\theta, \varphi) = \frac{\sin(N_y kb \sin\theta\sin\varphi/2)}{\sin(kb\sin\theta\sin\varphi/2)}$$

$$F_z(\theta, \varphi) = \frac{\sin(N_z kc\cos\theta/2)}{\sin(kc\cos\theta/2)}.$$

14. Show that the leading terms of the systematic multipole expansion of Section 13.7 yield the same results as the unsystematic Taylor series expansion of Section 13.2.

REFERENCES

Harrington, R. F., *Time Harmonic Electromagnetic Fields*, McGraw-Hill, New York (1961).

There is an abundance of detail in this thorough work on time harmonic fields and it contains a large variety of problems.

Jones, D. S., *The Theory of Electromagnetism*, Macmillan, New York (1964).

This generally useful and well-referenced book includes topics in radiation theory that we have not touched on, including helical antennas and the synthesis of prescribed antenna radiation patterns.

Papas, C., *Theory of Electromagnetic Wave Propagation*, McGraw-Hill, New York (1965).

Chapter 4 is devoted to the systematic multipole expansion discussed in our Section 13.7 and provides more detail than we have given.

Silver, S., Ed., *Microwave Antenna Theory and Design*, Vol. 12, Rad. Lab. Series, McGraw-Hill New York (1949).

This book is not recent but it contains a clear discussion of timeless subjects, such as reciprocity theorems, and the general theory of antenna array patterns.

14
POINT CURRENTS

14.1 INTRODUCTION

In this chapter we consider the field produced by *point currents*, i.e., by point charges which, under the influence of an external force, move in some orbit or trajectory. The external force may be of various kinds. For example, it might be a uniform magnetic field which bends a moving particle into a circular orbit (as in a particle accelerator); it might be the electric field of a passing electromagnetic wave which forces an atomic electron into oscillation (*Thomson scattering*); it might be the Coulomb field of a nucleus which deflects a passing electron or proton (*bremsstrahlung* or *X-radiation*). As we have remarked in Section 11.5 it is not strictly correct in any of these examples to suppose that the force on the particle is due to the external field only. In addition a small accelerated charge entity experiences a *self-force*. In the early sections of this chapter we shall neglect this self-force, and in fact generally neglect the details of how the applied forces produce a given motion. We shall simply suppose that the trajectory of the particle is given in advance. We leave the discussion of the question of the self-force, or *radiation reaction* as it is called, to the last section.

We assume that the trajectory of the particle of charge q, as sketched in Fig. 14.1, is some vector function of time, $\boldsymbol{r}_0(t')$. Just as the point charge at rest is described formally by a density $\rho(\boldsymbol{r}') = q\delta(\boldsymbol{r}')$ so the density of the moving point charge is

$$\rho(\boldsymbol{r}', t') = q\delta(\boldsymbol{r}' - \boldsymbol{r}_0(t')). \tag{14.1}$$

Since, in general, current density $\boldsymbol{J}$ is $\rho\boldsymbol{v}$, we have for such a charge

$$\boldsymbol{J}(\boldsymbol{r}', t') = q\boldsymbol{v}(t')\,\delta(\boldsymbol{r}' - \boldsymbol{r}_0(t')), \tag{14.2}$$

where $\boldsymbol{v}(t') = d\boldsymbol{r}_0/dt'$. When these expressions for ρ and $\boldsymbol{J}$ are put into Eqs. (11.45) and (11.46) for $\boldsymbol{A}$ and Φ the integrals can be done. The resultant point-charge potentials are called the *Lienard-Wiechert* potentials and from them explicit formulas for the fields can be found. These formulas are deceptively simple, however, since, to find the field at a space-time point $\boldsymbol{r},t$, they must be evaluated at an earlier time, the so-called *retarded time*, when the particle is at a certain *retarded point* on its trajectory. It is generally difficult to find useful exact expressions for the retarded time and approximations must usually be made to evaluate point-charge fields.

14.2 LIENARD-WIECHERT POTENTIALS

The retarded potentials Φ and $\boldsymbol{A}$ of Eqs. (11.45) and (11.46) become, on using Eqs. (14.1) and (14.2) for ρ and $\boldsymbol{J}$, and $\boldsymbol{\beta} = \boldsymbol{v}/c$,

$$\begin{aligned}\Phi(\boldsymbol{r}, t) &= q\int\frac{\delta\left(\boldsymbol{r}' - \boldsymbol{r}_0\left(t - \dfrac{|\boldsymbol{r} - \boldsymbol{r}'|}{c}\right)\right)dv'}{|\boldsymbol{r} - \boldsymbol{r}'|},\\ \boldsymbol{A}(\boldsymbol{r}, t) &= q\int\frac{\boldsymbol{\beta}\left(t - \dfrac{|\boldsymbol{r} - \boldsymbol{r}'|}{c}\right)\delta\left(\boldsymbol{r}' - \boldsymbol{r}_0\left(t - \dfrac{|\boldsymbol{r} - \boldsymbol{r}'|}{c}\right)\right)dv'}{|\boldsymbol{r} - \boldsymbol{r}'|}.\end{aligned} \tag{14.3}$$

In these equations the integration over dv' goes only over the region of space where the current is (or has been) nonvanishing so that any contribution to the integrals can come only from *somewhere* along the previous trajectory of the particle. But as we have remarked, there is in fact only *one point* on the whole trajectory, the retarded point, which, for a given space time field point $\boldsymbol{r}, t$ yields a contribution. We denote the *retarded point* by $\tilde{\boldsymbol{r}}$.* Since there is a contribution to the integrals in (14.3) only when the argument of the δ-function vanishes $\tilde{\boldsymbol{r}}$ is defined by

$$\tilde{\boldsymbol{r}} = \boldsymbol{r}_0\left(t - \frac{|\boldsymbol{r} - \tilde{\boldsymbol{r}}|}{c}\right). \tag{14.4}$$

For $\boldsymbol{r}$ and t fixed this intrinsic equation determines $\tilde{\boldsymbol{r}}$ in principle, although in practice it is usually intractable. Given $\tilde{\boldsymbol{r}}$, Eq. (14.4) shows that the *retarded time* $\tilde{t}$ associated with it, i.e. the time at which the particle was at the retarded point, is

$$\tilde{t} = t - \frac{|\boldsymbol{r} - \tilde{\boldsymbol{r}}|}{c} \tag{14.5}$$

so that $\tilde{\boldsymbol{r}} = \boldsymbol{r}_0(\tilde{t})$. Now $t - \tilde{t}$ is the time for the particle to get from the retarded point $\tilde{Q}$ shown in Fig. 14.1 to its position Q at t, which point we call the *present point*. And $|\boldsymbol{r} - \tilde{\boldsymbol{r}}|/c$ is the time for light to travel from the retarded point to the field point. The retarded point $\tilde{Q}$ is then *that point on the previous trajectory for which the time for the particle to get from $\tilde{Q}$ to Q is the same as for light to get from $\tilde{Q}$ to P.* It is easy to see qualitatively why there is only a single retarded point.†

---

* We shall generally use the tilde ($\tilde{\ }$) above a quantity to show that it refers to the retarded position or the retarded time. Similarly, when all the quantities in brackets or parentheses are to be evaluated at the retarded time we add the tilde as a subscript, thus: $(\ \ )_{\sim}$. For a quantity α, we also sometimes write $\alpha|_{\sim}$ as a convenient variant of $\tilde{\alpha}$.

† If the particle velocity v could be greater than the velocity of light c, the following argument breaks down. In effect, this can happen in a medium where the velocity of light c' can be less than c, in which case so-called *Cerenkov radiation* may be emitted. It is then shown, for example, in Clemmow and Dougherty (R), that there are *two* retarded points for a given field point.

Equation (14.5) is $t - \tilde{t} = |\boldsymbol{r} - \tilde{\boldsymbol{r}}|/c$. Considering $\tilde{Q}$ and $\tilde{t}$ as variables for the moment, we see in this equation that if $\tilde{Q}$ approaches Q, the left-hand side becomes very small and must eventually be *less than the right-hand side* which has a finite value in this limit. If, on the other hand, $\tilde{Q}$ is far back on the trajectory and very distant from Q, the distances $\tilde{Q}Q$ and $\tilde{Q}P$ will become approximately equal. Since the particle necessarily moves with a velocity $v < c$, the time $t - \tilde{t}$ it takes to move from $\tilde{Q}$ to Q will be *greater* than the time $|\boldsymbol{r} - \tilde{\boldsymbol{r}}|/c$ which light takes to traverse $\tilde{Q}P$. In short, the left-hand side of the equation will be *larger* than the right-hand side for this case. Thus, for some intermediate point, the left- and right-hand sides of (14.5) will be equal and this defines the retarded point.

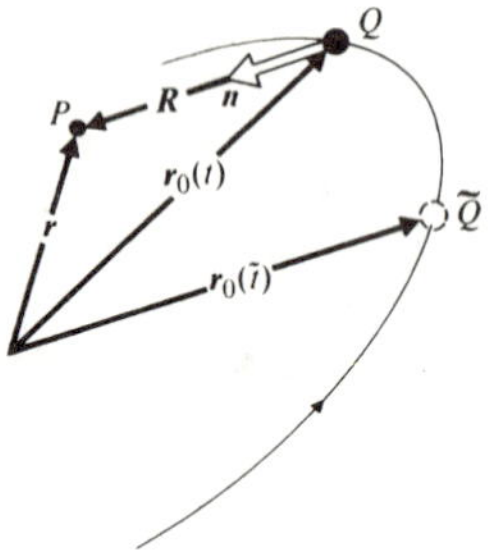

Fig. 14.1 Particle trajectory, with present point Q, retarded point $\tilde{Q}$ and the field point P.

Since the whole contribution to the integrals in (14.3) comes from the single retarded point, one might be tempted to write intuitively $\Phi = q/|\boldsymbol{r} - \tilde{\boldsymbol{r}}|$, and similarly for $\boldsymbol{A}$. However, this is incorrect. The mathematical reason is that the δ-functions in these equations are not simply of the form $\delta(\boldsymbol{r}' - \boldsymbol{r}_0)$ where $\boldsymbol{r}_0$ is a *constant*, but in fact $\boldsymbol{r}_0$ is a function of the integration variable: The simple integration formulas for δ-functions do not apply. The physical reason that intuition is incorrect, as we shall shortly see, is that different parts of a moving charge distribution contribute at different retarded times to the field at a given $\boldsymbol{r}, t$ and (perhaps surprisingly) this produces an effect that persists even in the limit when the distribution dimensions shrink to zero.

Consider first the mathematical problem of integrating the δ-functions in Eq. (14.3). To do this, we introduce a new variable $\boldsymbol{r}^*$ so that this equation involves a $\delta(\boldsymbol{r}^*)$ which can be integrated straightforwardly:

$$\boldsymbol{r}^* = \boldsymbol{r}' - \boldsymbol{r}_0(t - |\boldsymbol{r} - \boldsymbol{r}'|/c).$$

Let $\boldsymbol{R}$ be the vector from the particle to the field point $\boldsymbol{R}(t') = \boldsymbol{r} - \boldsymbol{r}_0(t')$ and let

$$\boldsymbol{n} = \boldsymbol{R}/R.$$

The volume elements dv^* and dv' are related by $dv^* = J\,dv'$ where J is the Jacobian of the transformation. On calculating it (see Problem 2 at the end of this chapter), we find

$$dv^* = s(t')\,dv',$$

where

$$s(t') = 1 - \boldsymbol{n}(t') \cdot \boldsymbol{\beta}(t') \tag{14.6}$$

and $t' = t - (|\boldsymbol{r} - \boldsymbol{r}'|/c)$. With this result, Eq. (14.3) can be readily integrated to yield the *Lienard-Wiechert potentials*:

$$\Phi = \frac{q}{(sR)_{\sim}}, \qquad \boldsymbol{A} = \frac{q\tilde{\boldsymbol{\beta}}}{(sR)_{\sim}} \tag{14.7}$$

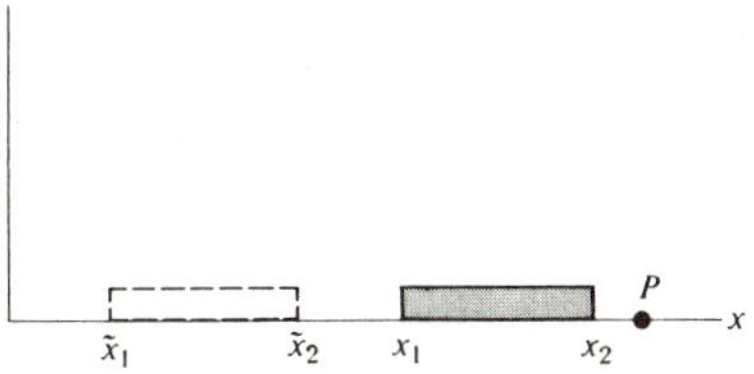

Fig. 14.2 One-dimensional illustration of origin of shrinkage factor $1/(1 - \boldsymbol{n} \cdot \boldsymbol{\beta})_{\sim}$.

Using a simple model, consider now the physical origin of the "shrinkage factor" $1/\tilde{s} \equiv 1/(1 - \boldsymbol{n} \cdot \boldsymbol{\beta})_{\sim}$ in (14.7). Suppose, as in Fig. 14.2, that a thin cylinder of charge is moving along the x-axis with velocity v. We want to calculate the field at x when the right end of the cylinder is at x_2, the left is at x_1, and $x_2 - x_1 = L$ is the length of the cylinder. The retarded point $\tilde{x}_2$ associated with a charge element at the right end of the cylinder is of course different from the retarded point $\tilde{x}_1$ associated with the left end. These points are defined by

$$\frac{x_1 - \tilde{x}_1}{v} = \frac{x - \tilde{x}_1}{c}, \qquad \frac{x_2 - \tilde{x}_2}{v} = \frac{x - \tilde{x}_2}{c}.$$

Subtracting the second equation from the first, we find, with $\tilde{L} = \tilde{x}_2 - \tilde{x}_1$, that

$$\tilde{L} = \frac{L}{1 - v/c}.$$

The *effective length* $\tilde{L}$ then differs by the factor $1/(1 - v/c)$ from the length L, and this effect persists even when $L \to 0$. It is easy to fill in the details that show that what we may call the *effective charge* of a moving cylinder differs by a similar factor; this illustrates the physical origin of the factor $1/\tilde{s}$ in (14.7).

14.3 FIELDS OF A MOVING CHARGE

We now use the Lienard-Wiechert potentials to calculate the fields of a point charge. We shall exhibit these fields in two forms: a more common and perhaps more convenient form due to A. Lienard and E. Wiechert, and a less familiar but nicely simple form due to O. Heaviside and R. Feynman. In using the Lienard-Wiechert potentials (14.7) to evaluate the fields, one must remember that although they may depend explicitly on $\boldsymbol{r}$ and t, they depend implicitly on these variables as well, since they are evaluated at the retarded time, a function of $\boldsymbol{r}, t$. A useful computational technique is to make the implicit dependence explicit by writing the potentials as integrals over a δ-function of the retarded time:

$$\Phi(\boldsymbol{r}, t) = q \int \frac{\delta\left(t' + \dfrac{R(t')}{c} - t\right)}{R(t')}\, dt',$$

$$\boldsymbol{A}(\boldsymbol{r}, t) = q \int \frac{\boldsymbol{\beta}(t')\delta\left(t' + \dfrac{R(t')}{c} - t\right)}{R(t')}\, dt', \tag{14.8}$$

where we recall that $R(t') = |\boldsymbol{r} - \boldsymbol{r}_0(t')|$. These integral representations can easily be shown (Problem 2) to be equivalent to (14.7), if we remember, on integrating, that the arguments of the δ-functions in them are nonlinear functions of the integration variable. One must then introduce a new variable t'',

$$t'' = t' + \frac{R(t')}{c} - t,$$

or equivalently use the formula from Appendix B

$$\int g(t')\,\delta\,(f(t'))\,dt' = \left.\frac{g(t')}{|df/dt'|}\right|_{f(t')=0} \tag{14.9}$$

To use (14.8) to calculate the fields, we must form

$$\boldsymbol{E} = -\nabla\Phi - \frac{1}{c}\frac{\partial \boldsymbol{A}}{\partial t} \qquad \text{and} \qquad \boldsymbol{B} = \nabla \times \boldsymbol{A}.$$

Consider first the evaluation of $\boldsymbol{E}$. Note that in operating with ∇ on Φ or on $\boldsymbol{A}$ we have, since they depend on x, y, z only through $R = ((x - x_0(t'))^2 + (y - y_0(t'))^2 + (z - z_0(t'))^2)^{1/2}$,

$$\nabla = \boldsymbol{n}\frac{\partial}{\partial R}\cdot \tag{14.10}$$

Then $\boldsymbol{E}$ is

$$\boldsymbol{E}(\boldsymbol{r}, t) = q\int\left(\frac{\boldsymbol{n}}{R^2}\delta\left(t' + \frac{R(t')}{c} - t\right) - \frac{\boldsymbol{n}}{cR}\delta'\left(t' + \frac{R(t')}{c} - t\right)\right) dt'$$
$$- \frac{q}{c}\frac{d}{dt}\int \frac{\boldsymbol{\beta}(t')\delta\left(t' + \dfrac{R(t')}{c} - t\right)}{R}\, dt',$$

where a prime on a δ-function means differentiation with respect to its *argument*. Since

$$\delta'\left(t' + \frac{R(t')}{c} - t\right) = -\frac{d}{dt}\delta\left(t' + \frac{R(t')}{c} - t\right),$$

we have

$$\boldsymbol{E}(\boldsymbol{r}, t) = q\int \frac{\boldsymbol{n}}{R^2}\delta\left(t' + \frac{R(t')}{c} - t\right)dt' + q\frac{d}{dt}\int \frac{(\boldsymbol{n} - \boldsymbol{\beta})}{cR}\delta\left(t' + \frac{R(t')}{c} - t\right)dt',$$

which becomes, on using (14.9) and $d/dt'(t' + R(t')/c - t) = s(t')$,

$$\boldsymbol{E} = q\left(\frac{\boldsymbol{n}}{sR^2}\right)_{\sim} + \frac{q}{c}\frac{d}{dt}\left(\frac{\boldsymbol{n} - \boldsymbol{\beta}}{sR}\right)_{\sim}. \tag{14.11}$$

It follows from (14.5) and the definition of s in (14.6) that

$$\tilde{s} = \frac{dt}{d\tilde{t}},$$

whence

$$\frac{1}{\tilde{s}} = \frac{d\tilde{t}}{dt} = 1 - \frac{1}{c}\frac{dR(\tilde{t})}{dt}.$$

Since

$$\boldsymbol{\beta}(t') = \frac{1}{c}\frac{d\boldsymbol{r}_0(t')}{dt'},$$

the quantity $\tilde{\boldsymbol{\beta}}$ that occurs in (14.11) is

$$\tilde{\boldsymbol{\beta}} = \frac{1}{c}\frac{d\boldsymbol{r}_0(\tilde{t})}{d\tilde{t}} = -\frac{1}{c}\frac{d}{d\tilde{t}}(R\boldsymbol{n})_{\sim} = -\frac{\tilde{s}}{c}\frac{d}{dt}(R\boldsymbol{n})_{\sim}.$$

Putting these last two expressions for $1/\tilde{s}$ and $\tilde{\boldsymbol{\beta}}$ into (14.11), we have*

$$\boldsymbol{E} = q\left\{\frac{\boldsymbol{n}}{R^2}\left(1 - \frac{1}{c}\frac{dR}{dt}\right) + \frac{1}{c}\frac{d}{dt}\left[\frac{\boldsymbol{n}}{R}\left(1 - \frac{1}{c}\frac{dR}{dt}\right) + \frac{1}{Rc}\frac{d}{dt}(R\boldsymbol{n})\right]\right\}_{\sim}$$

or

$$\boldsymbol{E} = q\left\{\frac{\boldsymbol{n}}{R^2} + \frac{R}{c}\frac{d}{dt}\left(\frac{\boldsymbol{n}}{R^2}\right) + \frac{1}{c^2}\frac{d^2\boldsymbol{n}}{dt^2}\right\}_{\sim}. \tag{14.12}$$

---

* Note that in (14.12) and the preceding equation the notation $(dF/dt)\sim$ means $dF(\tilde{t})/dt$ and *not* $dF(\tilde{t})/d\tilde{t}$. That is, the tilde ($\tilde{}$) outside the braces applies to the *arguments* of the functions inside and not to the variable of differentiation.

Equation (14.12) is the *Heaviside-Feynman* form for $\boldsymbol{E}$. To calculate $\boldsymbol{B} = \nabla \times A$, we use (14.10) and, much as in the calculation of $\boldsymbol{E}$, we find

$$\boldsymbol{B} = q\left(\frac{\boldsymbol{\beta} \times \boldsymbol{n}}{sR^2}\right)_{\sim} + \frac{q}{c}\frac{d}{dt}\left(\frac{\boldsymbol{\beta} \times \boldsymbol{n}}{sR}\right)_{\sim}$$

which on using

$$\left(\frac{\boldsymbol{\beta} \times \boldsymbol{n}}{s}\right)_{\sim} = -\left(\frac{R}{c}\frac{d\boldsymbol{n}}{dt} \times \boldsymbol{n}\right)_{\sim}$$

becomes

$$\boldsymbol{B} = q\left(\frac{\boldsymbol{n}}{Rc} \times \frac{d\boldsymbol{n}}{dt} + \frac{1}{c^2}\boldsymbol{n} \times \frac{d\boldsymbol{n}^2}{dt^2}\right)_{\sim}$$

or

$$\boldsymbol{B} = \tilde{\boldsymbol{n}} \times \boldsymbol{E}. \tag{14.13}$$

The *Lienard-Wiechert* form of the fields can be found by using $d/dt = (1/\tilde{s})/d/d\tilde{t}$ in (14.11) and

$$\frac{d}{d\tilde{t}}\left(\frac{\boldsymbol{n} - \boldsymbol{\beta}}{sR}\right)_{\sim} = \left(\frac{\dot{\boldsymbol{n}} - \dot{\boldsymbol{\beta}}}{sR} - \frac{\boldsymbol{n} - \boldsymbol{\beta}}{(sR)^2}(s\dot{R} + R\dot{s})\right)_{\sim},$$

where the dot means differentiation with respect to $\tilde{t}$. With this result, along with

$$\dot{R}|_{\sim} = -c(\boldsymbol{n} \cdot \boldsymbol{\beta})_{\sim},$$

$$\dot{\boldsymbol{n}}|_{\sim} = \frac{c}{R}(\boldsymbol{n}(\boldsymbol{n} \cdot \boldsymbol{\beta}) - \boldsymbol{\beta})_{\sim},$$

$$\dot{s}|_{\sim} = -(\boldsymbol{n} \cdot \dot{\boldsymbol{\beta}} + \boldsymbol{\beta} \cdot \dot{\boldsymbol{n}})_{\sim},$$

Eq. (14.11) can be rearranged into the Lienard-Wiechert form,

$$\boldsymbol{E} = q\left(\frac{(\boldsymbol{n} - \boldsymbol{\beta})(1 - \beta^2)}{s^3R^2} + \frac{\boldsymbol{n} \times ((\boldsymbol{n} - \boldsymbol{\beta}) \times \dot{\boldsymbol{\beta}})}{cs^3R}\right)_{\sim} \tag{14.14}$$

with $\boldsymbol{B}$ still given by (14.13).

The fields in both the forms (14.12) and (14.14) involve the velocity and acceleration of the particle. Both forms reduce, of course, to the Coulomb law $q\boldsymbol{n}/R^2$ for a particle at rest. Although the Heaviside-Feynman form is beautifully simple, the Lienard-Wiechert form exhibits explicitly the useful separation of the field into the *near* or *velocity field*, falling off as $1/R^2$, and the *far*, *radiation* or *acceleration* field, falling off as $1/R$. The radiation field then dominates at large distances and it is this that must be used to calculate energy loss by radiation.

From (14.14) the ratio of the magnitudes of radiation field and velocity field is of the order $(R\dot{v}/c^2)_\sim$ so the condition that the radiation field dominate is

$$\tilde{R}\tilde{\dot{v}} \gg c^2.$$

If $\dot{v}$ is strictly zero, there is, of course, no acceleration field at all, but if $\dot{v}$ is finite, however small, this condition can be satisfied. So it was with the analogous time harmonic condition $kr \gg 1$ for the radiation zone, which for arbitrarily small k could always be satisfied for large enough r.

14.4 UNIFORM MOTION

There are few nontrivial cases for which the Lienard-Wiechert fields can be calculated exactly. One such is for a charge in uniform motion. We discuss this now, mainly for pedagogical reasons, since the result will be the same as that derived in Chapter 12 by Lorentz transformation of the field of a point charge.

For uniform motion ($\dot{\boldsymbol{\beta}} = 0$), only the velocity term in (14.14) remains and

$$\boldsymbol{E} = q\left(\frac{(\boldsymbol{n} - \boldsymbol{\beta})(1 - \beta^2)}{s^3R^2}\right)_\sim. \tag{14.15}$$

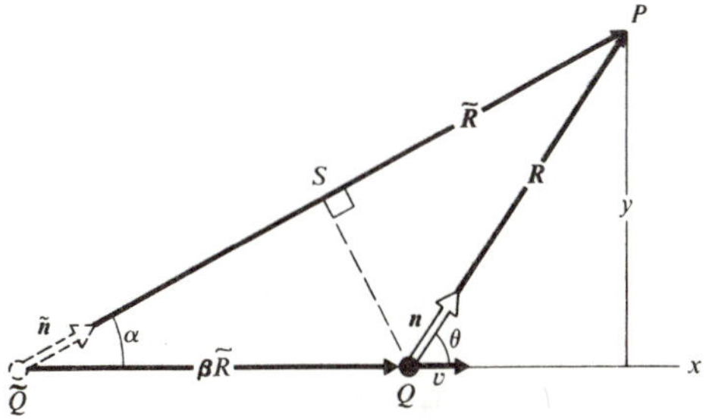

Fig. 14.3 Charge in uniform motion.

We suppose that the motion is along the x-axis, as in Fig. 14.3, and that we want the field at P when the particle is at Q. We must then express the retarded quantities in (14.15) in terms of present ones. We have obviously, since velocity is constant, that $\tilde{\boldsymbol{\beta}} = \boldsymbol{\beta}$. Moreover, when light travels the distance $\tilde{\boldsymbol{R}}$ (from $\tilde{Q}$ to P) the particle travels only $\boldsymbol{\beta}\tilde{R}$, the distance from $\tilde{Q}$ to Q. From the diagram, $\boldsymbol{\beta}\tilde{R} + \boldsymbol{R} = \tilde{\boldsymbol{R}}$ or $(\tilde{\boldsymbol{n}} - \boldsymbol{\beta}) = \boldsymbol{R}/\tilde{R}$. With this result in (14.15), the final quantity needed is $\tilde{s}\tilde{R}$. Now $\tilde{Q}S$ is of length $\tilde{\boldsymbol{n}}\cdot\boldsymbol{\beta}\tilde{R}$ whence SP is of length $(1 - \tilde{\boldsymbol{n}}\cdot\boldsymbol{\beta})\tilde{R} \equiv \tilde{s}\tilde{R}$. From the right triangle formed by SPQ,

$$(\tilde{s}\tilde{R})^2 + (\beta\tilde{R}\sin\alpha)^2 = R^2.$$

But $\tilde{R}\sin\alpha = y = R\sin\theta$, whence

$$\tilde{s}\tilde{R} = R(1 - \beta^2 \sin^2 \theta)^{1/2}.$$

Putting all these results together, we have

$$\boldsymbol{E} = q\frac{\boldsymbol{n}(1 - \beta^2)}{R^2(1 - \beta^2 \sin^2 \theta)^{3/2}}. \tag{14.16}$$

This agrees with the Lorentz transformation result of Eq. (12.81).

From the viewpoint of an observer at P who sees the charge whizzing by, the field (14.16) is time dependent. The time dependence, as seen by an observer at distance y from the line of motion, is found by setting $x = vt$ in (14.16) and using $\sin^2\theta = y^2/(x^2 + y^2)$. We find, with $\gamma = 1/\sqrt{1 - \beta^2}$,

$$E_x = -\frac{q\gamma vt}{(y^2 + \gamma^2 v^2 t^2)^{3/2}}, \qquad E_y = \frac{q\gamma y}{(y^2 + \gamma^2 v^2 t^2)^{3/2}}. \tag{14.17}$$

The time Fourier transforms of these fields are of considerable physical interest. We define the Fourier transform $\boldsymbol{E}_\omega$ by

$$\boldsymbol{E}_\omega(\boldsymbol{r}) = \frac{1}{2\pi}\int_{-\infty}^{\infty} \boldsymbol{E}(\boldsymbol{r}, t)e^{i\omega t}\,dt \tag{14.18}$$

with the inversion

$$\boldsymbol{E}(\boldsymbol{r}, t) = \int_{-\infty}^{\infty} \boldsymbol{E}_\omega(r)e^{-i\omega t}\,d\omega. \tag{14.19}$$

The Fourier transforms of Eq. (14.17) are then

$$\begin{aligned} E_{\omega x} &= -\frac{q}{2\pi\gamma v y}\int_{-\infty}^{\infty} \frac{e^{i\omega y\xi/\gamma v}\xi\,d\xi}{(1 + \xi^2)^{3/2}}, \\ E_{\omega y} &= \frac{q}{2\pi v y}\int_{-\infty}^{\infty} \frac{e^{i\omega y\xi/\gamma v}\,d\xi}{(1 + \xi^2)^{3/2}}. \end{aligned} \tag{14.20}$$

Even without evaluating the integrals in (14.20) explicitly, we can see their qualitative behavior. For small ω, $E_{\omega x}$ tends to zero since the integrand becomes an odd function of ξ. On the other hand $E_{\omega y}$ tends in this limit to the constant value $q/\pi v y$. For large ω $(\omega y/\gamma v \gg 1)$ the integrands of both integrals become rapidly varying functions and the values of both $E_{\omega x}$ and $E_{\omega y}$ tend rapidly to zero.

These qualitative features are confirmed by the express evaluation of the integrals in terms of Hankel functions of the first kind. One finds from Fourier transform tables

$$E_{\omega x} = \frac{q\omega}{2\gamma^2 v^2} H_0\left(\frac{i\omega y}{\gamma v}\right), \qquad E_{\omega y} = -\frac{q\omega}{2\gamma^2 v^2} H_1\left(\frac{i\omega y}{\gamma v}\right). \tag{14.21}$$

These functions have the limiting forms

$$H_n(x) \simeq \sqrt{\frac{2}{\pi x}}\, e^{i(x-(2n+1)\pi/4)}, \qquad x \to \infty;$$

$$\left.\begin{aligned} H_0 &\simeq -\frac{2i}{\pi}\left(\ln\frac{2}{x} - \ln 1.781\right) \\ H_1 &\simeq -\frac{2i}{\pi x} \end{aligned}\right\}, \quad x \to 0.$$

These asymptotic forms confirm the above qualitative results for $E_{\omega x}$ and $E_{\omega y}$: For small ω, $E_{\omega y}$ is a constant, $E_{\omega y} \simeq q/\pi v y$, whereas $E_{\omega x}$ goes as $\omega \ln \omega$ and so vanishes. For larger values of ω, the *integrals* in (14.20) tend to be comparable in value. However, there is a factor of $\sqrt{1-\beta^2}$ in front of the integral defining $E_{\omega x}$ that tends to make $E_{\omega x}$ *smaller* than $E_{\omega y}$ by at least this factor. For relativistic velocities, $E_{\omega x}$ is negligible and there is effectively only the transverse component $E_{\omega y}$. Since $\boldsymbol{B}_\omega$ is in the z-direction and has the same magnitude as $E_{\omega y}$, we see that the structure of each harmonic field component resembles that of a plane wave propagating in the x-direction, with $\boldsymbol{E}$ and $\boldsymbol{B}$ perpendicular to each other and to the direction of motion and with (for $\beta \approx 1$) the relative magnitudes of $\boldsymbol{E}$ and $\boldsymbol{B}$ having the same relation as for a plane wave. Therefore, if a charged particle passes by some system (for example an atom or nucleus), we can calculate the interaction between the particle and system approximately by first calculating the (usually easier) problem of interaction with a plane wave of arbitrary frequency, and then summing over all the frequencies of the equivalent plane waves as given above. The method is called the *method of virtual quanta* or *method of virtual photons* since in quantum theory each plane wave of frequency ω is associated with a *quantum* of light energy, or *photon*, of energy $h\omega/2\pi$.

14.5 APPROXIMATE EVALUATION OF THE FIELDS

Even given the Lienard-Wiechert or Heaviside-Feynman formulas, the problem of actually calculating the field of an arbitrarily moving point charge is not trivial, since these formulas are evaluated at the retarded time, and it is difficult to find useful explicit expressions for the retarded time in terms of the present time. Hence the formulas have to be evaluated approximately, and in fact the kinds of approximations that can be made are closely related to those for time-harmonic fields. Two important approximations in that case were the multipole expansion and the radiation zone approximation. The former was essentially a long wavelength (or low velocity) expansion, and the latter was an approximation based on the ratio of the size of the current distribution to the distance to the field point. We shall derive similar approximations for the present case.

Harmonic approximations involved the size a of the distribution and the present case will involve a similar length. Sometimes it will be immediately clear what this length is. For, say, a particle bound to an atom, a will obviously be the atomic radius. For problems in which the trajectory goes to infinity however, it is not so immediately obvious what a is. For such problems, we are usually not interested in the fields associated with the entire motion, but only those associated with the deflection or acceleration of the particle. This deflection (at least for forces of finite range) will take place over a region of space which can then be characterized by a length a. We shall assume in what follows that a region of interest of size a can be defined, and that the origin of the coordinate system is somewhere in it, as shown in Fig. 14.4.

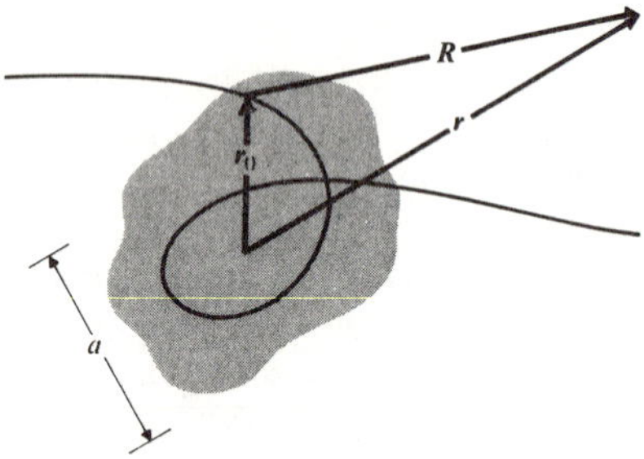

Fig. 14.4 Effective region of a trajectory (shaded), and associated coordinate system.

With this introduction consider the low velocity expansion of the Lienard-Wiechert fields. There are two places in which the approximation $\beta \ll 1$ enters. First, it can be made explicitly in Eq. (14.14) by setting $s \approx 1$ and $\boldsymbol{n} - \boldsymbol{\beta} \approx \boldsymbol{n}$. Second, and more importantly, approximations can be made for functions of the retarded time $\tilde{t} = t - (|\boldsymbol{r} - \tilde{\boldsymbol{r}}|/c)$, since $\tilde{\boldsymbol{r}} \equiv \boldsymbol{r}_0(\tilde{t})$ is small of order a. Consider any function f of $\tilde{t}$ (e.g. a component of the velocity or acceleration):

$$f\left(t - \frac{|\boldsymbol{r} - \tilde{\boldsymbol{r}}|}{c}\right).$$

It can be expanded in a Taylor series in $\tilde{\boldsymbol{r}}$. To first order, we obtain

$$f\left(t - \frac{|\boldsymbol{r} - \tilde{\boldsymbol{r}}|}{c}\right) \approx f\left(t - \frac{r}{c}\right) + \tilde{\boldsymbol{r}} \cdot \nabla f\left(t - \frac{r}{c}\right). \tag{14.22}$$

The ratio of the second term to the first in this expression is of the order $f'\tilde{r}/fc$, where prime means differentiation with respect to $t - r/c$. If f'/f is small, as it will be for $\beta \ll 1$, then

$$f(\tilde{t}) \approx f\left(t - \frac{r}{c}\right) \tag{14.23}$$

and the retardation is the *same* for all the points of present interest on the orbit.

The effect of retardation is then essentially trivial, although in a sense it may be very large.* The approximation (14.23) is called the *dipole approximation*, from its similarity to the first term of the multipole expansion for harmonic fields.

We now examine the approximation in more detail. The local or near field in (14.14) which goes as $1/R^2$ is of little practical interest, and we shall concentrate on the second, or radiation term, which goes as $1/R$. We shall consider it understood in discussing the *dipole approximation* that the *fields are evaluated at the time* $t - r/c$ and shall drop the explicit retardation signs. We also replace $\boldsymbol{R}$ by $\boldsymbol{r}$. Thus, the radiation field is

$$\boldsymbol{E} = \frac{q\boldsymbol{n} \times (\boldsymbol{n} \times \dot{\boldsymbol{\beta}})}{cr}.$$

If θ is the angle between $\boldsymbol{n}$ and $\dot{\boldsymbol{\beta}}$ and $\hat{\boldsymbol{\theta}}$ is the associated unit vector, then

$$\boldsymbol{E} = -\frac{q\dot{v}\sin\theta}{c^2 r}\hat{\boldsymbol{\theta}},$$

and the Poynting vector is

$$S = \frac{q^2}{4\pi}\frac{\dot{v}^2\sin^2\theta}{c^3 r^2}\hat{\boldsymbol{r}}. \tag{14.24}$$

Multiplying (14.24) by $r^2\, d\Omega = 2\pi r^2 \sin\theta\, d\theta$ and integrating over θ yields the well-known *Larmor formula* for the rate of energy loss of an accelerated particle with nonrelativistic velocity:

$$\frac{dW}{dt} = \frac{2}{3}\frac{q^2\dot{v}^2}{c^3}. \tag{14.25}$$

Another approximation of great practical importance is the *radiation zone* one, of which the dipole approximation above is a special case. In this approximation, no assumption is made about the magnitude of β, but it is assumed that R is large, $R \gg a$, so that $\boldsymbol{R} \approx \boldsymbol{r}$. Then only the second term in (14.14) need be kept. Moreover, in it we can approximate the retarded time using

$$\tilde{t} \equiv t - \frac{|\boldsymbol{r} - \tilde{\boldsymbol{r}}|}{c} \approx t - \frac{r}{c} + \frac{\boldsymbol{n}\cdot\tilde{\boldsymbol{r}}}{c}, \tag{14.26}$$

where $\hat{\boldsymbol{r}}$, the unit vector along $\boldsymbol{r}$, has been replaced by $\boldsymbol{n}$. The electric field $\boldsymbol{E}$ in the radiation zone is then, with (14.26) assumed,

$$\boldsymbol{E} = \frac{q}{cr}\left(\frac{\boldsymbol{n} \times [(\boldsymbol{n} - \boldsymbol{\beta}) \times \dot{\boldsymbol{\beta}}]}{s^3}\right)_{\sim}. \tag{14.27}$$

* For example, if a particle starts from rest at $t = 0$, there will be no signal felt at a distance r for what may be the very long time $T = r/c$. But, for almost all practical purposes, this time lag is of no concern. If, say, an atomic system emits radiation it is usually inconsequential that there is a time delay in recording this radiation on a photographic plate.

From (14.27) the magnitude S of the Poynting vector can be shown to be, after some vector algebra based mainly on $\boldsymbol{A} \times (\boldsymbol{B} \times \boldsymbol{C}) = \boldsymbol{B}(\boldsymbol{A} \cdot \boldsymbol{C}) - \boldsymbol{C}(\boldsymbol{A} \cdot \boldsymbol{B})$,

$$S = \frac{q^2}{4\pi c r^2}\left(\frac{\dot{\boldsymbol{\beta}}^2}{(1 - \boldsymbol{n}\cdot\boldsymbol{\beta})^4} + \frac{2(\boldsymbol{n}\cdot\dot{\boldsymbol{\beta}})(\boldsymbol{\beta}\cdot\dot{\boldsymbol{\beta}})}{(1 - \boldsymbol{n}\cdot\boldsymbol{\beta})^5} - \frac{(1 - \beta^2)(\boldsymbol{n}\cdot\dot{\boldsymbol{\beta}})^2}{(1 - \boldsymbol{n}\cdot\boldsymbol{\beta})^6}\right)_{\sim}. \tag{14.28}$$

This expression for the rate at which energy is radiated is ultimately connected with the rate at which the particle loses energy, or has work done on it. The connection is not trivial however, since the radiated intensity at time t is associated with the behavior of the particle at time $\tilde{t}$, and the differentials dt and $d\tilde{t}$ of field time and retarded time are not equal: $dt = d\tilde{t}(1 - \boldsymbol{n}\cdot\tilde{\boldsymbol{\beta}})$. If, for example, a particle of velocity $\boldsymbol{\beta}$ is impulsively accelerated over a very short time interval τ and is then impulsively brought to rest, a pulse of radiation will appear at the field point at a time $t = r/c$ and it will be of duration $\tau(1 - \boldsymbol{n}\cdot\tilde{\boldsymbol{\beta}})$. The total energy lost by the particle will be the same as the total energy ultimately radiated, but the energy lost *per unit time* by the charge will differ from the energy radiated per unit time by the factor $(1 - \boldsymbol{n}\cdot\tilde{\boldsymbol{\beta}})$. In general then, what one may call the *rate of energy loss of the charge, due to radiation through solid angle* $d\Omega$, is obtained by multiplying (14.28) by $r^2(1 - \boldsymbol{n}\cdot\tilde{\boldsymbol{\beta}})$.

Frequently, one is not interested in the *rate* at which energy is radiated into the solid angle $d\Omega$, but in the total amount of energy, $dW/d\Omega$, so radiated. This quantity is found by multiplying (14.28) by r^2 and integrating with respect to time:

$$\frac{dW}{d\Omega} = r^2 \int S\, dt. \tag{14.29}$$

Since S in (14.28) is expressed in terms of the retarded time, it is useful to introduce the retarded time as an integration variable by means of $dt = \tilde{s}\, d\tilde{t}$ to find

$$\frac{dW}{d\Omega} = \frac{q^2}{4\pi c}\int \left(\frac{\dot{\boldsymbol{\beta}}^2}{s^3} + \frac{2(\boldsymbol{n}\cdot\dot{\boldsymbol{\beta}})(\boldsymbol{\beta}\cdot\dot{\boldsymbol{\beta}})}{s^4} - \frac{(1 - \beta^2)(\boldsymbol{n}\cdot\dot{\boldsymbol{\beta}})^2}{s^5}\right)_{\sim} d\tilde{t}. \tag{14.30}$$

In this expression, the tilde on $\tilde{t}$ is not essential any longer since, in it, $\tilde{t}$ is just an integration variable. In using (14.30), the integration goes over all time if the motion is nonperiodic and the energy radiated is finite. For periodic motion we integrate over a period T and divide by T to find $d\overline{W}/d\Omega$, the *time average rate of energy loss per steradian*:

$$\frac{d\overline{W}}{d\Omega} = \frac{q^2}{4\pi c T}\int_0^T \left(\frac{\dot{\boldsymbol{\beta}}^2}{s^3} + \frac{2(\boldsymbol{n}\cdot\dot{\boldsymbol{\beta}})(\boldsymbol{\beta}\cdot\dot{\boldsymbol{\beta}})}{s^4} - \frac{(1 - \dot{\beta}^2)(\boldsymbol{n}\cdot\dot{\boldsymbol{\beta}})^2}{s^5}\right) dt. \tag{14.31}$$

14.6 FREQUENCY-ANGLE DISTRIBUTION OF THE RADIATED ENERGY

An important characteristic of the angular distributions of the radiated energy, $dW/d\Omega$ or $d\overline{W}/d\Omega$, is the spectrum of frequencies inherent in them. We consider first nonperiodic motion. In the radiation zone, $\boldsymbol{E}$ and $\boldsymbol{B}$ are perpendicular and have the same magnitude. Thus, on reverting to (14.29) for $dW/d\Omega$, we have

$$\frac{dW}{d\Omega} = \frac{cr^2}{4\pi}\int_{-\infty}^{\infty} |\boldsymbol{E}_r|^2\, dt,$$

where $\boldsymbol{E}_r$ is the radiation zone field. Using the Fourier transform $\boldsymbol{E}_\omega$ of $\boldsymbol{E}_r$ as defined in (14.18), we obtain

$$\frac{dW}{d\Omega} = \frac{cr^2}{4\pi}\int_{-\infty}^{\infty} dt \left(\int_{-\infty}^{\infty} \boldsymbol{E}_\omega^* e^{i\omega t}\, d\omega\right) \cdot \left(\int_{-\infty}^{\infty} \boldsymbol{E}_{\omega'} e^{-i\omega' t}\, d\omega'\right).$$

Once the order of integration has been changed, the integral over dt can be done by using $\int_{-\infty}^{\infty} e^{i(\omega-\omega')t}\, dt = 2\pi\delta(\omega - \omega')$ and we find that

$$\frac{dW}{d\Omega} = \frac{cr^2}{2}\int_{-\infty}^{\infty} |\boldsymbol{E}_\omega|^2\, d\omega. \tag{14.32}$$

The integral here goes over all frequencies, positive as well as negative. But since $\boldsymbol{E}_r$ is a real quantity, it follows from (14.18) that $\boldsymbol{E}_\omega = \boldsymbol{E}^*_{-\omega}$, and hence that $|\boldsymbol{E}_\omega|^2 = |\boldsymbol{E}_{-\omega}|^2$. In (14.32) we can then integrate over positive frequencies and multiply by two. We now define $I(\omega)$, the *intensity distribution in angle and frequency of the radiated energy*, by

$$\frac{dI(\omega)}{d\Omega} = cr^2|\boldsymbol{E}_\omega|^2 \tag{14.33}$$

so that Eq. (14.32) can be written as

$$\frac{dW}{d\Omega} = \int_0^{\infty} \frac{dI(\omega)}{d\Omega}\, d\omega.$$

The quantity $I(\omega)$ is important in that it is often a convenient one to measure.

To find $I(\omega)$, we must calculate $\boldsymbol{E}_\omega$ defined by Eq. (14.18). With $\boldsymbol{E}$ the radiation zone field of Eq. (14.27), we have

$$\boldsymbol{E}_\omega = \frac{q}{2\pi cr}\int_{-\infty}^{\infty} \left(\frac{\boldsymbol{n} \times [(\boldsymbol{n} - \boldsymbol{\beta}) \times \dot{\boldsymbol{\beta}}]}{(1 - \boldsymbol{n}\cdot\boldsymbol{\beta})^3}\right)_{\sim} e^{i\omega t}\, dt.$$

It is obviously easier to introduce $\tilde{t}$ as an integration variable by means of $\tilde{t} = t - r/c + \boldsymbol{n}\cdot\boldsymbol{r}_0(\tilde{t})/c$. With $dt = (1 - \boldsymbol{n}\cdot\tilde{\boldsymbol{\beta}})\, d\tilde{t}$, we have

$$\boldsymbol{E}_\omega = \frac{qe^{i\omega r/c}}{2\pi cr}\int_{-\infty}^{\infty} \left(\frac{\boldsymbol{n} \times [(\boldsymbol{n} - \boldsymbol{\beta}) \times \dot{\boldsymbol{\beta}}]}{(1 - \boldsymbol{n}\cdot\boldsymbol{\beta})^2}\right)_{\sim} e^{i\omega(\tilde{t} - \boldsymbol{n}\cdot\boldsymbol{r}_0(\tilde{t})/c)}\, d\tilde{t}. \tag{14.34}$$

Since $\tilde{t}$ is now simply an integration variable in this equation, it is permissible to write t rather than $\tilde{t}$. In other words, the retarded time does not really figure in (14.34). We simply put into it the known trajectory $\boldsymbol{r}_0(t)$ and $\boldsymbol{\beta}(t)$ etc. and do the integral. Given (14.34), we find the intensity distribution in the radiation zone to be, from (14.33)

$$\frac{dI(\omega)}{d\Omega} = \frac{q^2}{4\pi^2 c}\left|\int_{-\infty}^{\infty} \frac{\boldsymbol{n} \times [(\boldsymbol{n} - \boldsymbol{\beta}) \times \dot{\boldsymbol{\beta}}]}{(1 - \boldsymbol{n}\cdot\boldsymbol{\beta})^2} e^{i\omega(t - \boldsymbol{n}\cdot\boldsymbol{r}_0(t)/c)}\, dt\right|^2. \tag{14.35}$$

As an important special case, Eq. (14.35) can be calculated in the *dipole approximation.* If the velocity of the particle is small, $\beta \ll 1$, then $\boldsymbol{n}\cdot\boldsymbol{r}_0(t)$ is weakly time dependent, so that $e^{-i\omega\boldsymbol{n}\cdot\boldsymbol{r}_0(t)/c}$ simply becomes an essentially constant phase factor which can be taken outside the integrand. The frequency angle spectrum is then,

$$\frac{dI(\omega)}{d\Omega} = \frac{q^2}{4\pi^2 c}\left|\int_{-\infty}^{\infty} \boldsymbol{n} \times (\boldsymbol{n} \times \dot{\boldsymbol{\beta}}) e^{i\omega t}\, dt\right|^2. \tag{14.36}$$

Equation (14.35) can be put into an alternate form that is often more convenient. It is easy to show that the integrand of (14.35), exclusive of the exponential, is a perfect differential

$$\frac{\boldsymbol{n} \times [(\boldsymbol{n} - \boldsymbol{\beta}) \times \dot{\boldsymbol{\beta}}]}{(1 - \boldsymbol{n}\cdot\boldsymbol{\beta})^2} = \frac{d}{dt}\left[\frac{\boldsymbol{n} \times (\boldsymbol{n} \times \boldsymbol{\beta})}{(1 - \boldsymbol{n}\cdot\boldsymbol{\beta})}\right].$$

Then an integration by parts converts (14.35) to

$$\frac{dI(\omega)}{d\Omega} = \frac{q^2\omega^2}{4\pi^2 c}\left|\int_{-\infty}^{\infty} \boldsymbol{n} \times (\boldsymbol{n} \times \boldsymbol{\beta}) e^{i\omega(t - \boldsymbol{n}\cdot\boldsymbol{r}_0(t)/c)}\, dt\right|^2. \tag{14.37}$$

We have assumed here that $\boldsymbol{\beta}$ vanishes at the limits of integration, but this assumption is not true for some common cases, and (14.37) is, of course, then not valid. For example, if a particle moves uniformly from $t = -\infty$ to $t = 0$, is accelerated briefly near $t = 0$, and then continues to move uniformly, $\boldsymbol{\beta}$ does not vanish at the endpoints and the original form of (14.37) must be used.

For a periodic trajectory with period $T = 2\pi/\omega_0$, the time average of the energy, or the *power*, radiated per steradian, $d\overline{W}/d\Omega$, is defined by

$$\frac{d\overline{W}}{d\Omega} = \frac{cr^2}{4\pi T}\int_0^T |\boldsymbol{E}_r|^2\, dt. \tag{14.38}$$

Now $\boldsymbol{E}_r$ is expanded in Fourier *series*

$$\boldsymbol{E}_r(r, t) = \frac{1}{2\pi}\sum_{l=-\infty}^{\infty} \boldsymbol{E}_l(r) e^{il\omega_0 t}. \tag{14.39}$$

This representation of the field, as a series of *harmonic components* of frequency $l\omega_0$ will imply that energy is radiated only in these frequencies. To see this, put (14.39) into (14.38) and transform the integral in much the same way as led to (14.37). Changing the resulting sum over positive and negative frequencies to twice the sum over positive ones, we find that $d\overline{W}/d\Omega$ reduces to

$$\frac{d\overline{W}}{d\Omega} = \sum_{l=0}^{\infty} \frac{dP_l}{d\Omega}, \tag{14.40}$$

where $dP_l/d\Omega$, the *power per steradian radiated in the lth harmonic*, is

$$\frac{dP_l}{d\Omega} = \frac{q^2 l^2 \omega_0^4}{(2\pi)^3 c} \left| \int_0^{2\pi/\omega_0} \boldsymbol{n} \times (\boldsymbol{n} \times \boldsymbol{\beta}) e^{il\omega_0(t - \boldsymbol{n}\cdot\boldsymbol{r}_0(t)/c)}\, dt \right|^2. \tag{14.41}$$

This formula can be put into a simpler form for periodic and bounded motion. In this case, the vector $\boldsymbol{n}$ from the charge in its orbit to a very distant field point is essentially constant in time, although $\boldsymbol{\beta}$ is a function of time. Think of the squared integral in (14.41) as the dot product of two vector integrals, one over dt and one over dt'. In that dot product, we can use

$$[\boldsymbol{n} \times (\boldsymbol{n} \times \boldsymbol{\beta}(t))] \cdot [\boldsymbol{n} \times (\boldsymbol{n} \times \boldsymbol{\beta}(t'))] = (\boldsymbol{\beta}(t) \times \boldsymbol{n}) \cdot (\boldsymbol{\beta}(t') \times \boldsymbol{n}).$$

Then (14.41) can be put in the useful form:

$$\frac{dP_l}{d\Omega} = \frac{q^2 l^2 \omega_0^4}{(2\pi)^3 c} \left| \int_0^{2\pi/\omega_0} \boldsymbol{\beta} \times \boldsymbol{n} e^{il\omega_0(t - \boldsymbol{n}\cdot\boldsymbol{r}_0(t)/c)}\, dt \right|^2. \tag{14.42}$$

14.7 LINEAR ACCELERATION

In this and the next section, we give examples to illustrate some of the results of the previous sections. First, if a particle moves along a line with variable velocity, the acceleration is also directed along the line. For such linear acceleration, $\boldsymbol{\beta} \times \dot{\boldsymbol{\beta}} = 0$ in the Lienard-Wiechert formula (14.14), and the radiation field becomes

$$\boldsymbol{E} = \frac{q}{cr} \left(\frac{\boldsymbol{n} \times (\boldsymbol{n} \times \dot{\boldsymbol{\beta}})}{s^3} \right)_{\sim}. \tag{14.43}$$

Consider this first in the low-velocity approximation for which $s \approx 1$. To find the fields, we must first specify the variation of $\boldsymbol{\beta}$ with time. For the sake of a simple example, suppose that a particle moves uniformly, except for a small segment of its track in which the acceleration $\dot{v}$ is constant, $\dot{v} = w$. A pulse of radiation will then be emitted and a field will be felt at some distant point over some limited time interval. To apply (14.43) to find $\boldsymbol{E}$ *during* that interval, we first observe that the vector $\boldsymbol{n} \times (\boldsymbol{n} \times \dot{\boldsymbol{\beta}})$ lies in the plane containing $\boldsymbol{n}$ and $\boldsymbol{\beta}$, and is perpendicular to $\boldsymbol{n}$. Let θ be the angle between $\boldsymbol{n}$ and the track segment, and let $\hat{\boldsymbol{\theta}}$ be the unit vector in the direction of increasing θ. Then the radiation zone field is, since the difference between retarded and nonretarded quantities in (14.43) is now inconsequential,

$$\boldsymbol{E} = \frac{qw \sin\theta}{c^2 r} \hat{\boldsymbol{\theta}},$$

and the magnitude of the Poynting vector is

$$S = \frac{q^2 w^2 \sin^2\theta}{4\pi c^3 r^2}. \tag{14.44}$$

The radiation pattern is the $\sin^2\theta$ one characteristic of the dipole approximation.

Consider, for contrast, the analogous relativistic case. Here the factor $s = (1 - \boldsymbol{n}\cdot\boldsymbol{\beta})$ plays a crucial role. The electric field is

$$\boldsymbol{E} = \frac{q}{c^2 r}\left(\frac{\dot{v}\sin\theta}{(1-\beta\cos\theta)^3}\right)_{\sim} \hat{\boldsymbol{\theta}}. \tag{14.45}$$

We take the same example as above, of constant acceleration $\dot{v} = w$ over a short distance. The time variation of the field at the field point will depend on the time dependence of β. If, however, for the sake of an example we assume that the velocity change over the short interval of acceleration is small, we can approximate $\tilde{\beta}$ in (14.45) by some average value β_a. We have then for the Poynting vector, during the time the pulse is at the field point,

$$S = \frac{q^2 w^2 \sin^2\theta}{4\pi c^3 r^2 (1-\beta_a\cos\theta)^6}. \tag{14.46}$$

Compared with (14.44), there is an extra factor here of $1/(1-\beta_a\cos\theta)^6$. For values of β_a close to unity, this factor pushes the radiation pattern toward the forward direction, as sketched in Fig. 14.5, drastically modifying it. In other examples of radiation from highly relativistic particles, a similar factor usually plays a central role in determining the angular dependence of the emitted radiation.

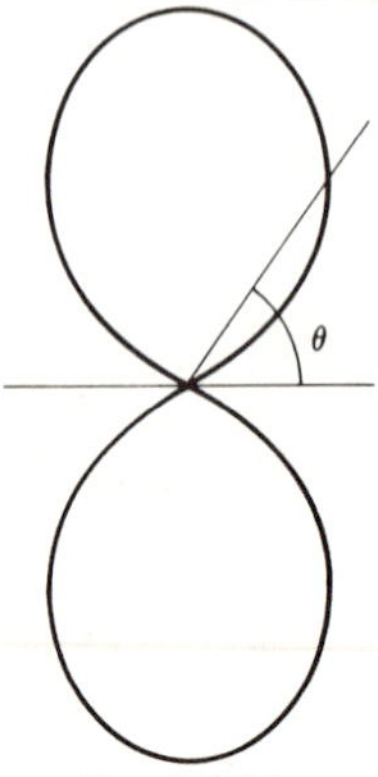

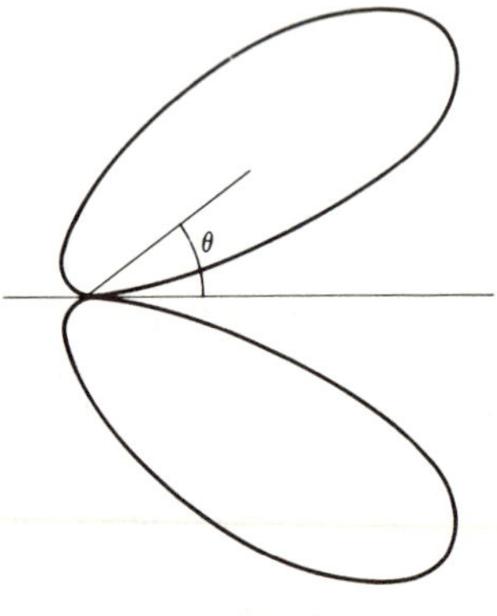

Fig. 14.5 Schematic nonrelativistic and relativistic radiation patterns for brief linear acceleration.

Consider next an example of the frequency-angle spectrum of radiated energy. For illustration, we take this spectrum in the dipole approximation of Eq. (14.36).

Suppose that the acceleration is a slowing down in which, after uniform motion up to $t = 0$, β varies subsequently as $\beta_0 e^{-t/t_0}$. Then

$$|\boldsymbol{n} \times (\boldsymbol{n} \times \dot{\boldsymbol{\beta}})| = \frac{\beta_0}{t_0} e^{-t/t_0} \sin\theta, \qquad t > 0$$

and Eq. (14.36) gives

$$\frac{dI(\omega)}{d\Omega} = \frac{q^2\beta_0^2 \sin^2\theta}{4\pi^2 c(1 + \omega^2 t_0^2)}.$$

This frequency spectrum is approximately constant for small ω and drops off for ω appreciably larger than $1/t_0$. A spectrum that is qualitatively of this kind will be produced by any nonrelativistic deceleration (or acceleration) characterized by a time t_0.

An important example of linear acceleration is that induced in an electron by the electric field of an incident light wave. Consider such an electron which we suppose to be weakly bound; i.e., an electron for which the atomic binding force is small compared to that due to the field of the incident wave. Moreover, we suppose that the velocity of the forced vibration of the electron is small compared to c, so that only the electric field of the incident wave needs to be considered and not the $\boldsymbol{v} \times \boldsymbol{B}/c$ force associated with its magnetic field. Let the incident wave travel in the z-direction and be polarized in the x-direction so that

$$\boldsymbol{E}_\omega = \hat{\boldsymbol{x}}\, E_0 e^{ikz}.$$

Assuming E_0 to be real, we find that the equation of motion of the electron situated at the origin is

$$m\dot{v} = eE_0 \cos\omega t.$$

We find simply, for $d\overline{W}/d\Omega$, the time average energy radiated per steradian,

$$\frac{d\overline{W}}{d\Omega} = \frac{cE_0^2}{8\pi}\left(\frac{e}{mc^2}\right)^2 \sin^2\Theta, \tag{14.47}$$

where Θ is the angle between the x-axis and the field point. This result is frequently expressed in terms of a *differential scattering cross section*, $d\sigma/d\Omega$ which essentially normalizes (14.47) to unit incident energy flux. It is defined by

$$\frac{d\sigma}{d\Omega} = \frac{\text{Radiated energy/(sec} \times \text{steradian)}}{\text{Incident energy/(sec} \times \text{cm}^2)}.$$

The incident energy/(sec × cm²) is $cE_0^2/8\pi$, so that

$$\frac{d\sigma}{d\Omega} = \left(\frac{e^2}{mc^2}\right)^2 \sin^2\Theta. \tag{14.48}$$

This process of induced radiation is called *Thomson scattering*.

14.8 UNIFORM CIRCULAR MOTION

As a second practical example, we consider a charged particle that moves with uniform angular velocity in a circular orbit. This motion gives rise to so-called *synchrotron radiation*, since it is important in the design of synchrotrons and other particle accelerators. This kind of radiation has also, however, been found to be associated with other physical processes, such as astrophysical ones. The two important properties of the radiation we discuss are the angular distribution of the average energy, as given by Eq. (14.31), and of the power in the lth harmonic, $dP_l/d\Omega$ defined by (14.42).

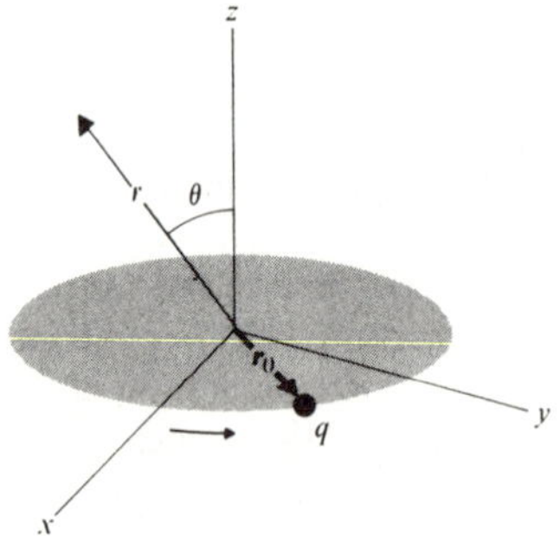

Fig. 14.6 Coordinates associated with uniform circular motion.

The coordinates associated with the charge q moving with speed $v = \omega_0 a$ in a circle of radius a are shown in Fig. 14.6. The position $\boldsymbol{r}_0$ of the charge is

$$\boldsymbol{r}_0(t) = a(\hat{\boldsymbol{x}} \cos \omega_0 t + \hat{\boldsymbol{y}} \sin \omega_0 t) \tag{14.49}$$

so that

$$\begin{aligned} \boldsymbol{\beta}(t) &= \frac{\omega_0 a}{c}(-\hat{\boldsymbol{x}} \sin \omega_0 t + \hat{\boldsymbol{y}} \cos \omega_0 t) \\ \dot{\boldsymbol{\beta}}(t) &= -\frac{\omega_0^2}{c}\, \boldsymbol{r}_0(t). \end{aligned} \tag{14.50}$$

The field point is located by the vector $\boldsymbol{r}$ in the x, z-plane, with spherical coordinates r, θ. The normal $\boldsymbol{n}$ from the charge to the field point is, in the radiation zone ($r \gg a$), identical with $\hat{\boldsymbol{r}}$ and is

$$\boldsymbol{n} = \hat{\boldsymbol{x}} \sin \theta + \hat{\boldsymbol{z}} \cos \theta.$$

Then, for use in (14.31) we have $s = 1 + \beta \sin \theta \sin \omega_0 t$, $\boldsymbol{\beta} \cdot \dot{\boldsymbol{\beta}} = 0$, $\dot{\beta}^2 = \omega_0^2 \beta^2$ and $(\boldsymbol{n} \cdot \dot{\boldsymbol{\beta}})^2 = \omega_0^2 \beta^2 \sin^2 \theta \cos^2 \omega_0 t$. With these, and with the integration variable $\varphi = \omega_0 t$, we can put Eq. (14.31) in the following form:

$$\frac{d\overline{W}}{d\Omega} = \frac{cq^2\beta^4}{8\pi^2 a^2}\int_0^{2\pi} \frac{(1-\beta^2)\cos^2\theta + (\beta + \sin\theta\sin\varphi)^2}{(1+\beta\sin\theta\sin\varphi)^5}\,d\varphi.$$

The integral is elementary but tedious, and we find

$$\frac{d\overline{W}}{d\Omega} = \frac{cq^2\beta^4}{8\pi a^2(1-\beta^2\sin^2\theta)^{7/2}}\left[1 + \cos^2\theta - \frac{\beta^2}{4}(1+3\beta^2)\sin^4\theta\right]. \quad (14.51)$$

There is a striking difference here between the nonrelativistic and the relativistic motion. For the former case, $\beta \ll 1$, Eq. (14.51), becomes

$$\frac{d\overline{W}}{d\Omega} = \frac{q^2 a^2 \omega_0^4}{8\pi c^3}(1 + \cos^2\theta). \quad (14.52)$$

This can be interpreted as the radiated power due to the two nonrelativistic harmonic oscillators to which Eq. (14.49) shows the motion is equivalent. The full amplitude of the y-oscillation is seen at the field point, whereas that in x is seen to be shortened by the factor $\cos\theta$, whence the factor $(1 + \cos^2\theta)$ in (14.52). The angular distribution (14.52) is then a slowly varying one. For the relativistic case, $\beta \approx 1$, on the other hand, the factor $(1 - \beta^2\sin^2\theta)^{-7/2}$ in (14.51) peaks the radiation sharply around $\theta = \pi/2$. This effect is reminiscent of the difference between the nonrelativistic and relativistic cases for linear acceleration.

Consider next the angular distribution of the power in the lth harmonic, $dP_l/d\Omega$, as given by (14.42). It requires the integral $\boldsymbol{\Gamma}$ defined by

$$\boldsymbol{\Gamma} = \int_0^{2\pi/\omega_0} (\boldsymbol{\beta} \times \boldsymbol{n}) e^{il\omega_0(t - \boldsymbol{n}\cdot\boldsymbol{r}_0(t)/c)}\,dt.$$

Since $\boldsymbol{\beta} \times \boldsymbol{n}$ is normal to $\boldsymbol{n}$ we can consider that it has a θ-component and a φ-component (the latter identical with the y-component), and these are easily verified to be

$$(\boldsymbol{\beta} \times \boldsymbol{n})_\theta = \frac{\omega_0 a}{c}\cos\omega_0 t,$$
$$(\boldsymbol{\beta} \times \boldsymbol{n})_\varphi = \frac{\omega_0 a}{c}\cos\theta\sin\omega_0 t. \quad (14.53)$$

Then $\boldsymbol{\Gamma}$ also has components Γ_θ and Γ_φ in terms of which

$$\frac{dP_l}{d\Omega} = \frac{q^2\omega_0^4 l^2}{(2\pi)^3 c}(|\Gamma_\theta|^2 + |\Gamma_\varphi|^2). \quad (14.54)$$

Using (14.53) and the integration variable $\alpha = \omega_0 t$, we have

$$\Gamma_\theta = \frac{a}{c}\int_0^{2\pi} \cos\alpha\, e^{il(\alpha - \beta\sin\theta\cos\alpha)}\,d\alpha,$$

$$\Gamma_\varphi = \frac{a\cos\theta}{c}\int_0^{2\pi} \sin\alpha\, e^{il(\alpha - \beta\sin\theta\cos\alpha)}\,d\alpha.$$

The integrals here can be expressed in terms of Bessel functions. From Appendix D, we have

$$J_l(x) = \frac{i^{-l}}{2\pi}\int_0^{2\pi} e^{i(l\alpha - x\cos\alpha)}\,d\alpha,$$

$$J_{l+1} + J_{l-1} = \frac{2l}{x}J_l(x),$$

whence

$$\Gamma_\theta = \frac{2\pi a}{c}\,i^{l+1}J_l'(l\beta\sin\theta),$$

$$\Gamma_\varphi = \frac{2\pi a}{c}\,i^l\,\frac{J_l(l\beta\sin\theta)}{\beta\sin\theta}\cos\theta.$$

These results in (14.54) finally yield

$$\frac{dP_l}{d\Omega} = \frac{q^2\omega_0^4a^2l^2}{2\pi c^3}\left[(J_l'(l\beta\sin\theta))^2 + \frac{\cot^2\theta}{\beta^2}(J_l(l\beta\sin\theta))^2\right]. \tag{14.55}$$

We briefly discuss the evaluation of (14.55). The most interesting practical case is the relativistic one $\beta \sim 1$, since Eq. (14.55) has shown in this case that the radiated intensity peaks sharply near $\theta = \pi/2$. At $\theta = \pi/2$, (14.55) becomes

$$\left.\frac{dP_l}{d\Omega}\right|_{\theta=\pi/2} = \frac{q^2\omega_0^4a^2l^2}{2\pi c^3}(J_l'(l\beta))^2. \tag{14.56}$$

The effective values of l can be large and what is needed to evaluate (14.56) are asymptotic expressions for the Bessel functions that are valid when the order and argument are large and equal. With such expressions, it can be shown* that the frequency spectrum peaks at about $\omega_c = l_c\omega_0$ where

$$l_c \approx \frac{3}{(1-\beta^2)^{3/2}}.$$

14.9 POINT PARTICLES AND RADIATION REACTION

We have discussed point currents that are due to the motion of point particles, but have left rather cloudy both the idealization, and the difficulties with that idealization, that are implicit in the concept of a point particle. A first and obvious difficulty is that of stability. A static charge distribution containing total charge Q and confined to some radius a has, whatever the detailed nature of the distribution, a Coulomb self-energy, or energy of assemblage, of order Q^2/a. In the limit of a

* See, for example, Clemmow and Dougherty (R).

point charge, $a \to 0$, this energy becomes infinite. If we avoid this difficulty by taking a to be small but finite, we must recognize that a charge distribution cannot be in equilibrium under purely electric forces (Earnshaw's theorem). We shall bypass these difficulties, without really resolving them, by simply assuming that there *are* other forces which can produce a small rigid charge distribution. We shall first assume it is of uniform density and is spherical of radius a. Later we consider the limit $a \to 0$.

Suppose that such a charge distribution moves in an external field. Each volume element of the distribution constitutes a little charge dq in motion, which produces a field that acts on every other element. As we have emphasized, and as the Lienard-Wiechert formula reconfirms, the law of equality of action and reaction forces does not hold for charges in motion. Thus, the force an element dq_1 exerts on another element dq_2 is not equal and opposite to the force that dq_2 exerts on dq_1. Consequently, it is possible that the summation of the forces on the individual charge elements yields a *net* force $\boldsymbol{F}_s$ on the distribution, the *self-force*. Such a summation is not easy to do for an arbitrary distribution. If, however, the distribution is small, we can make a Taylor-series expansion in functions of the retarded time, analogous to that in the dipole approximation of Eq. (14.22). Such a calculation was first done by Abraham and Lorentz. The calculation is somewhat lengthy, however, and we shall simply quote the result here; a heuristic derivation of it is given below.

The Abraham-Lorentz formula involves powers of the radius a coupled with successive time derivatives of the velocity $\dot{\boldsymbol{v}}$, $\ddot{\boldsymbol{v}}$ etc. According to it, $\boldsymbol{F}_s$ is given for a particle of charge e that we shall henceforth call an *electron*, by

$$\boldsymbol{F}_s = -\frac{4U}{3c^2}\dot{\boldsymbol{v}} + \frac{2e^2}{3c^3}\ddot{\boldsymbol{v}} + O(a),$$

where

$$U = \frac{1}{2}\int \frac{\rho(\boldsymbol{r})\rho(\boldsymbol{r}')}{|\boldsymbol{r} - \boldsymbol{r}'|}\, dv' \tag{14.57}$$

is the electrostatic self-energy mentioned above in connection with the stability problem, and $O(a)$ stands for terms that are proportional to positive powers of a. Note that there is no term proportional to $\boldsymbol{v}$. A particle in uniform motion, like a particle at rest, suffers no self-force. This result is, of course, a direct consequence of the Lorentz transformation law for forces. If the electron is in the presence of an *applied field* which exerts a force $\boldsymbol{F}_a$ (in addition to the force $\boldsymbol{F}_s$), the equation of motion is

$$m_0\dot{\boldsymbol{v}} = \boldsymbol{F}_a - \frac{4U}{3c^2}\dot{\boldsymbol{v}} + \frac{2e^2}{3c^3}\ddot{\boldsymbol{v}} + O(a),$$

where m_0 is the "mass," or more properly the *inertial mass* of the electron. Now the coefficient $4U/3c^2$ obviously has the dimensions of mass; it is called the *electromagnetic mass*, m_e, of the electron,

$$m_e = \frac{4U}{3c^2}.$$

The equation of motion is then

$$(m_0 + m_e)\dot{\boldsymbol{v}} = \boldsymbol{F}_a + \frac{2e^2}{3c^3}\ddot{\boldsymbol{v}} + O(a). \tag{14.58}$$

If this equation is assumed to be correct and is applied to an electron of *measured mass m*, we must assume that the coefficient of $\dot{\boldsymbol{v}}$ in it should be identified with m, that is, that neither m_0 nor m_e are measurable separately, but rather that only their sum

$$m = m_0 + m_e \tag{14.59}$$

has meaning.

We shall assume (14.59), but it makes for still other difficulties. Namely, Eq. (14.58) is structure dependent: The term in it symbolized by $O(a)$ in fact depends on the detailed nature of the charge distribution that is taken to represent the electron. We could bypass this ambiguity by letting $a \to 0$, but then we encounter the other difficulty that U and hence m_e become infinite. We have a dilemma. We shall not go into the details of its resolution except to say that the infinity incurred by making $a \to 0$ can be made less troublesome than might appear at first sight. It can be dealt with* by "renormalization," in such a way that one can let $a \to 0$ and still retain (14.59). We shall assume that this renormalization has been effected, and that $a \to 0$, so that our final equation of motion is

$$m\dot{\boldsymbol{v}} = \boldsymbol{F}_a + \frac{2e^2}{3c^3}\ddot{\boldsymbol{v}}. \tag{14.60}$$

Consider now the heuristic derivation, mentioned above, of (14.60). If an accelerated particle emits radiation, its energy necessarily changes, much as its energy would change if a force acted on it. The effect of radiation is then roughly equivalent to a force we call $\boldsymbol{F}_r$, the force of *radiation reaction*. To fix $\boldsymbol{F}_r$, we require that the work done by this force on the particle is the negative of the energy that is radiated according to the Larmor law (14.25). Applying this criterion between two time intervals t_1 and t_2, we have

$$\int_{t_1}^{t_2} \boldsymbol{F}_r \cdot \boldsymbol{v}\, dt = -\frac{2e^2}{3c^3}\int_{t_1}^{t_2} \dot{\boldsymbol{v}} \cdot \dot{\boldsymbol{v}}\, dt.$$

---

* See the book by Rohrlich (R).

Integration by parts on the right-hand side gives

$$\int_{t_1}^{t_2} \boldsymbol{F}_r \cdot \boldsymbol{v}\, dt = \frac{2e^2}{3c^3}\int_{t_1}^{t_2} \ddot{\boldsymbol{v}} \cdot \boldsymbol{v}\, dt - \frac{2e^2}{3c^3}(\dot{\boldsymbol{v}} \cdot \boldsymbol{v})\Big|_{t_1}^{t_2}.$$

If the motion is periodic so that $\dot{\boldsymbol{v}} \cdot \boldsymbol{v}$ is the same at t_1 and t_2, or if it is such that $\dot{\boldsymbol{v}} \cdot \boldsymbol{v} = 0$ at t_1 and t_2, we argue that we can equate the integrands of the integrals in this equation to get

$$\boldsymbol{F}_r = \frac{2e^2}{3c^3}\ddot{\boldsymbol{v}} \tag{14.61}$$

which is the result contained in Eq. (14.60). The weakness of this argument is clear. The times t_1 and t_2 must be such that $\dot{\boldsymbol{v}} \cdot \boldsymbol{v}$ has the special properties mentioned; the interval from t_1 to t_2 is not arbitrary. On the other hand, the *integrands* of two equal *integrals* can be equated only if there is equality over arbitrary intervals.

Whatever the rigor in the derivation of $\boldsymbol{F}_r$, we return to the equation of motion (14.60) in which it is involved. There is, in addition to the difficulties discussed above, still another—that of the so-called *runaway solution.* If the applied force $\boldsymbol{F}_a$ in Eq. (14.60) is zero, then

$$m\dot{\boldsymbol{v}} = \frac{2e^2}{3c^3}\ddot{\boldsymbol{v}}.$$

A solution of this equation is given by

$$\boldsymbol{v} = \boldsymbol{v}_0 + \boldsymbol{v}_1 e^{t/\tau}, \tag{14.62}$$

where

$$\tau = \frac{2e^2}{3mc^3} \tag{14.63}$$

and $\boldsymbol{v}_0$ and $\boldsymbol{v}_1$ are vectorial constants. Equation (14.62) predicts that a particle will self-accelerate and, starting with a finite velocity, may end up with an infinite one, even though no external force acts. This difficulty can be countered by assuming that the force of radiation reaction applies only when there is an applied field that acts as well, and only when the loss of energy due to the radiation reaction is small compared to the total energy of the particle. For example, if a particle in periodic motion loses only a small fraction of its energy per period by radiation, we assume that the radiation reaction force gives a good qualitative picture of the physics involved. This qualitative picture is in fact confirmed and made quantitative by *quantum electrodynamics*, the extension of the present classical theory to the quantum domain.

As an example of the physical effect of radiation reaction, we consider a simple harmonic oscillator whose equation of motion, with the force of radiation reaction included, is

$$\ddot{x} + \omega_0^2 x = \tau\dddot{x}. \tag{14.64}$$

Equation (14.64) has solutions of the form $e^{i\lambda t}$ where λ may now be complex, since it satisfies

$$\lambda^2 = \omega_0^2 + i\tau\lambda^3. \tag{14.65}$$

This cubic equation has three solutions in general, but if we assume $\omega_0\tau \ll 1$ so that $\lambda = \pm\omega_0 + \varepsilon$, where ε is small enough so that its square and cube are negligible, we find from this equation,

$$\lambda \approx \omega_0 + \frac{i\tau\omega_0^2}{2}.$$

The imaginary part in λ means the motion is now damped. This implies that a Fourier analysis of the motion is not a pure harmonic of frequency ω_0, but contains a spectrum of frequencies. Consider then an oscillator which starts oscillating at $t = 0$. The spectrum of frequencies $\chi(\omega)$ inherent in it is

$$\chi(\omega) = \int_0^\infty x(t)e^{-i\omega t}\,dt,$$

where for $t > 0$,

$$x(t) = x_0 e^{i\omega_0 t - \tau\omega_0^2 t/2} \tag{14.66}$$

This oscillating charge gives rise to an electromagnetic field with the same time variation as in Eq. (14.66). The Fourier transform of the field components will then contain the factor $\chi(\omega)$ and the square of these components, that define the intensity distribution $I(\omega)$, will contain the factor $|\chi(\omega)|^2$. Hence we find that

$$I(\omega) \propto \frac{1}{(\omega - \omega_0)^2 + (\tau^2\omega_0^4/4)}. \tag{14.67}$$

The radiation thereby represents not a single line but the frequency distribution given by (14.67). This radiation-damped oscillator crudely models the so-called *natural breadth of spectral lines.*

PROBLEMS

1. The transformation between $\boldsymbol{r}^*$ and $\boldsymbol{r}'$ given by

$$\boldsymbol{r}^* = \boldsymbol{r}' - \boldsymbol{r}_0\left(t - \frac{|\boldsymbol{r} - \boldsymbol{r}'|}{c}\right)$$

($\boldsymbol{r}$ and t are passive parameters) is used in Section 14.3 to integrate the δ-function in Eq. (14.3). As stated in that section, show that $dv^* = (1 - \boldsymbol{n}\cdot\boldsymbol{\beta})\,dv'$. Try a one-dimensional version of the transformation first, if the three-dimensional version gets bogged down.

2. Show that the integral representation of Eq. (14.8) leads to the explicit form of (14.7) for the Lienard-Wiechert potentials.

3. The coordinates of a charge q that moves in the x, y-plane are the functions of time: $x = x_0 e^{-t^2/a^2}$, $y = y_0 e^{-t^2/b^2}$. Find $dI(\omega)/d\Omega$ the frequency distribution per steradian of the radiated intensity.
4. Verify Eq. (14.28) for the Poynting vector in the radiation zone.
5. Discuss the polarization properties of the synchrotron radiation.
6. Plane waves with a given propagation direction but *random* polarization scatter from an electron by the Thomson scattering mechanism. Show that the angular distribution of the scattered radiation is as $(1 + \cos^2\theta)$, with θ the angle between the radiation and the incident propagation direction.
7. Calculate the Poynting vector for radiation emitted by two point charges, each of magnitude $q/2$ that move at constant (nonrelativistic) velocities on a circle of radius a, at opposite ends of a diameter. Discuss similarly the radiation if the two point charges were divided into four, each of magnitude $q/4$ uniformly spaced on the circle, and again moving uniformly. What happens to the total radiated energy if the charges are further subdivided into $2N$ uniformly spaced charges of magnitude $q/2N$, and $N \to \infty$?
8. A particle moves along the z-axis with position $z(t) = a\cos\omega t$. Show that $d\overline{W}/d\Omega$, the angular distribution of the time average radiated energy, is

$$\frac{d\overline{W}}{d\Omega} = \frac{q^2 c\beta^4(4 + \beta^2\cos^2\theta)\sin^2\theta}{32\pi a^2(1 - \beta^2\cos^2\theta)^{7/2}}.$$

and that the average power per steradian radiated in the lth harmonic is

$$\frac{dP_l}{d\Omega} = \frac{q^2 c\beta^2 l^2}{2\pi a^2}\tan^2\theta J_l^2(l\beta\cos\theta).$$

REFERENCES

Batygin, V. V., and I. N. Toptygin, *Problems in Electrodynamics*, Academic Press, New York (1964).

This is a résumé of electromagnetic theory; it contains 741 problems and worked solutions, and gives many examples of the emission of radiation in collisions.

Clemmow, P. C., and J. P. Dougherty, *Electrodynamics of Particles and Plasmas*, Addison-Wesley, Reading, Mass. (1969).

The first part of this book treats the electrodynamics of point currents, and includes a clear presentation of the classical Lorentz evaluation of the self force on a point particle.

Landau, L. D., and E. Lifshitz, *The Classical Theory of Fields*, Addison-Wesley, Reading, Mass. (1962).

In the sections on point particles, this book presents a thoroughly relativistic viewpoint.

Monaghan, J. J., "The Heaviside-Feynman expression for the fields of an accelerated dipole," *J. Phys.* **A1**, 112 (1968).

As a preliminary to deriving the fields of a point dipole, Monaghan derives the Heaviside-Feynman expression for the fields of a point charge. The derivation in this text borrows from his.

Rohrlich, F., *Classical Charged Particles*, Addison-Wesley, Reading, Mass. (1965).

A clear presentation of the classical electrodynamics of point particles, and a clarification of the much-vexed subjects of the Lorentz transformation properties of the field energy and field momentum of a particle.

Schott, G. A., *Electromagnetic Radiation and the Mechanical Reactions Arising from It*, Cambridge University Press, Cambridge, England (1912).

This 60-year-old work contains much information on the radiation of particles in orbits, and includes the first calculation of what is now known as synchroton radiation.

15
PERFECT CONDUCTORS AND TIME HARMONIC FIELDS

15.1 INTRODUCTION

In this chapter we discuss boundary value problems with so-called *perfect conductors* in time harmonic fields. Perfect conductors are an idealization of real conductors but we begin with them since their associated boundary conditions are particularly simple. In Chapter 17 the discussion is extended to include imperfect conductors and dielectrics. Boundary value problems with conducting boundaries are, roughly, of two kinds. Problems of the first kind are variously called *scattering*, *diffraction*, *forced oscillation*, or *inhomogeneous* problems; although the names vary, the physics of all these problems is identical. Similarly, problems of the second kind go by the names of *eigenvalue*, *free oscillation*, or *homogeneous* problems.

A problem of the first kind, an *inhomogeneous* one, is the analog of the inhomogeneous problem in electrostatics. Typically, a wave is incident on some conducting body and induces currents which are, however, unknown in advance. These currents produce a field which, together with the incident field, makes up a total field which satisfies the boundary conditions. Since the currents are unknown *a priori*, so is the total field, and there arises the circular situation which typifies boundary value problems: the field depends on the distribution of currents and the distribution of currents depends on the field, and neither is known at the beginning.

In a problem of the second kind, a *homogeneous* problem, there is no external field. However, both $\boldsymbol{E}$ and $\boldsymbol{B}$ satisfy a wave equation and it is generally true that any wave equation (for the vibration of a string, for the displacement of a drumhead, for the Schrödinger wave function ψ, etc.) has solutions describing free or *natural* oscillations of systems which are somehow bound or confined. Moreover, such solutions frequently exist only for certain *natural* or *allowed* frequencies, or frequency *eigenvalues*, and corresponding natural *modes* or *eigenfunctions*. If then the electromagnetic field is confined, either by enclosing it completely with a conducting box (cavity resonator), or partially with hollow cylinders (wave guides), we might expect that only certain frequencies and field configurations will be allowed, and this is indeed the case. There is also an analogy to electrostatics in which the homogeneous problem for a conductor was to find a surface distribution of *charge* which generated a field satisfying the boundary condition that the conductor be an equipotential. For time harmonic fields, given a volume bounded by a conductor, the homogeneous problem is that of finding possible distributions of *current* on the interior surface which generate fields satisfying the boundary conditions. In contrast to the unique solution of the electrostatic case, however, there is in general an infinite number of solutions, each having a different frequency and different current and field distribution or *mode*.

There are fewer examples of soluble problems for time harmonic fields than there were for electrostatics, for which there were few enough indeed. In the present case, some of the electrostatic techniques, such as the method of images, do not apply or apply only infrequently. Furthermore, there are techniques, such as conformal mapping in two dimensions, that do not apply at all. One technique that carries over from electrostatics is that of expanding the field components in sets of orthogonal functions with undetermined coefficients and then determining the coefficients by imposing boundary conditions. But even in those few cases (e.g., for the sphere or cylinder) for which all the coefficients in such a solution can be determined, there are often further practical difficulties in computation. As in the summation problem for harmonic fields, three parameters frequently characterize boundary value problems: size of the object, wavelength, distance to the field point. A series solution, even if it can be found, is rarely practical for numerical computation for all relative values of these parameters. For example, the series solution usually converges slowly in the short wavelength limit and it must then be summed partially, or transformed, to make it practical.

A second important technique for solving boundary value problems with conductors is the use of integral equations. These are potentially powerful, although quite sophisticated mathematical techniques are frequently necessary in dealing with them. In addition, there are various approximate, i.e., perturbational, variational, and numerical methods, which we shall not discuss in detail. An extensive treatment of them is given in Morse and Feshbach (B).

One of the approximate methods we do discuss briefly for the inhomogeneous problem is the Kirchhoff-Huygens theory of *diffraction*. This is of considerable importance in optics and is directed toward calculating the scattering of light from objects or apertures that are much larger than the wavelength of the light. The deviation of the light from linear or geometrical propagation is for historical reasons called *diffraction*,† but, as we have indicated, the physics of diffraction is identical to the physics of electromagnetic scattering in any other inhomogeneous problem.

15.2 BOUNDARY CONDITIONS

Before discussing boundary conditions for time harmonic fields, it is worth first recalling the similar question of boundary conditions in electrostatics. There we had the two equations $\nabla \times \boldsymbol{E} = 0$ and $\nabla \cdot \boldsymbol{E} = 4\pi\rho$. The first led to the condition

† For a discussion of the semantics involved in the use of the word *diffraction* see Van de Hulst (B).

that the tangential component of $\boldsymbol{E}$ was zero at the boundary of the conductor; the second led to the condition that the normal component of $\boldsymbol{E}$ was 4π times the surface charge density. Although each of these conditions is derived from a Maxwell equation they have quite different status in one respect. For, apart from Maxwell's equations, there is a purely mathematical problem: What boundary conditions need to be prescribed on a surface to determine a unique solution for the field inside? In discussing the Dirichlet and Neumann problems, we concluded that the boundary condition on the tangential component of $\boldsymbol{E}$, which is essentially a statement that the surface is an equipotential, will suffice, and it is this condition that is used in electrostatics. The second condition on the normal component of $\boldsymbol{E}$, which of course is still *valid*, has then a different character. It is used *after* the problem is solved to determine the charge density. There is an analogous situation for the time harmonic fields. Here also, as was shown in Section 13.1, there are two independent Maxwell equations,

$$\nabla \times \boldsymbol{E}_\omega = ik\boldsymbol{B}_\omega \tag{15.1}$$

$$\nabla \times \boldsymbol{B}_\omega = -ik\boldsymbol{E}_\omega + \frac{4\pi}{c}\boldsymbol{J}_\omega . \tag{15.2}$$

Each of these equations yields a boundary condition. It will turn out, however, that for the perfect conductors, discussed below, only the first boundary condition, on the tangential component of $\boldsymbol{E}_\omega$, need be applied to determine the solution. The second condition ,on the tangential component of $\boldsymbol{B}_\omega$, serves to determine the surface current density once the fields have been determined from the first condition.

Many conductors obey *Ohm's law* which linearly relates current density $\boldsymbol{J}$ and field $\boldsymbol{E}$ by the *conductivity* σ,

$$\boldsymbol{J} = \sigma\boldsymbol{E}.$$

Roughly speaking, perfect conductors are those for which σ is infinite. Now the only truly sound way of discussing "infinite" conductivity is as a limit, i.e., by first discussing problems with finite σ and *then* letting σ become infinite. The case of finite σ will in fact be discussed in Chapter 16. For the moment, however, we shall argue more loosely to get immediately to the case of perfect conductivity. Consider a conducting medium that obeys Ohm's law, and suppose that σ becomes infinite. Then it must be that $\boldsymbol{E}_\omega$ vanishes in the interior, since otherwise there would be infinite currents in the volume and, associated with them, an unphysical infinite energy. This argument does not, of course, preclude the existence of currents in an infinitesimal thickness near the surface, since an infinite current density in a sheet of infinitesimal thickness need not correspond to an infinite energy. Moreover, the concept of surface currents is consistent with our general picture that conductors in applied fields must have currents induced in them; if these currents cannot exist throughout the volume, the only place for them to exist is on the surface. The concept of *surface currents* in the limit of infinite conductivity will be confirmed in Chapter 16. There we shall find that time harmonic fields in a medium of conductivity σ penetrate the conductor to a distance of the order of the so-called *skin depth* δ, given by $\delta = c/(2\pi\sigma\omega)^{1/2}$. This is very small for many practical cases of interest, and in the limit of infinite σ it becomes zero.

Consider now the boundary conditions associated with the two curl equations (15.1) and (15.2). We shall first derive these conditions *in general* and only then assume perfect conductivity. Using Stokes theorem on (15.1), we find

$$\oint \boldsymbol{E}_\omega \cdot d\boldsymbol{l} = ik \int_S \boldsymbol{B}_\omega \cdot d\boldsymbol{s}. \tag{15.3}$$

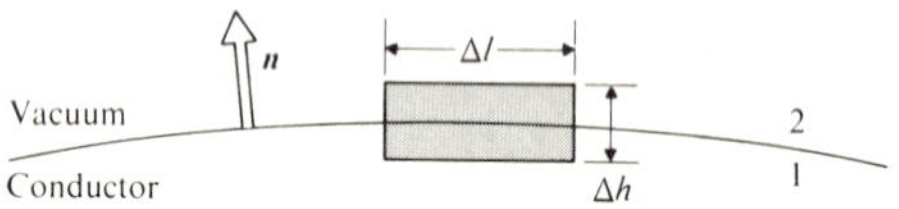

Fig. 15.1 Boundary condition on the tangential component of $\boldsymbol{E}_\omega$.

We apply this to a small rectangular loop of length Δl and height Δh which straddles the boundary between free space and conductor (or other medium) as shown in Fig. 15.1. The left-hand side of (15.3) becomes approximately $(E_{\omega t,1} - E_{\omega t,2})\Delta l$, where $E_{\omega t,1}$ and $E_{\omega t,2}$ are tangential components of $\boldsymbol{E}_\omega$, i.e., their components along Δl. The right-hand side of (15.3) is, however, of the order $|\boldsymbol{B}_\omega|\Delta l\Delta h$. Assuming that B_ω is not singular,† we conclude that, as $\Delta h \to 0$,

$$\boldsymbol{E}_{\omega t,1} = \boldsymbol{E}_{\omega t,2}. \tag{15.4}$$

The vector notation here emphasizes that there are *two* tangential components, corresponding to the two orientations of the rectangle. Since $\boldsymbol{E}_{\omega t}$ can be written as $\boldsymbol{E}_\omega \times \boldsymbol{n}$, where $\boldsymbol{n}$ is the unit normal shown, the condition (15.4) is frequently written in the vector form

$$(\boldsymbol{E}_{\omega 1} - \boldsymbol{E}_{\omega 2}) \times \boldsymbol{n} = 0. \tag{15.5}$$

A boundary condition can be derived from (15.2) in exactly the same way. If both $\boldsymbol{E}_\omega$ and $\boldsymbol{J}_\omega$ on the right-hand side of (15.2) are *nonsingular*, we conclude that

$$(\boldsymbol{B}_{\omega 1} - \boldsymbol{B}_{\omega 2}) \times \boldsymbol{n} = 0. \tag{15.6}$$

We now reconsider these two boundary conditions for the case of perfect conductors. We know for these that $\boldsymbol{E}_{\omega 1}$ vanishes so that on dropping the sub-

† It is not always possible to assume that $\boldsymbol{B}_\omega$ (and $\boldsymbol{E}_\omega$) are nonsingular. They will, in fact, be singular in the region of sharp edges or points, in which case the questions of boundary conditions must be reconsidered. We exclude such cases here. For a discussion of this question, that of the so-called *edge conditions*, see Jones (R).

script 2, (15.4) becomes

$$E_{\omega t} = 0. \qquad \text{Perfect conductor} \tag{15.7}$$

The boundary condition (15.6) must however be reexamined for perfect conductors, because in this case there *are* singular surface currents, contrary to the assumption implicit in its derivation. We do not need to work out the boundary condition anew, since it can be derived by borrowing from the discussion in Chapter 8 for the transition of $\boldsymbol{B}$ across a *current sheet* in magnetostatics. The difference between the magnetostatic case and the time harmonic case is essentially embodied in the term $ik\boldsymbol{E}_\omega$ on the right-hand side of Eq. (15.2). If $\boldsymbol{E}_\omega$ is nonsingular, this term plays no role when the equation is applied to the usual small rectangle straddling the boundary and the results for the magnetostatic case can be applied. Then we find for the case of perfect conductivity, where the fields vanish in the interior,

$$\boldsymbol{n} \times \boldsymbol{B}_\omega = \frac{4\pi}{c} \boldsymbol{J}_s, \qquad \text{Perfect conductor} \tag{15.8}$$

where $\boldsymbol{J}_s$ is the surface current density and $\boldsymbol{B}_\omega$ is the field just *outside* the conductor. To repeat, this equation determines $\boldsymbol{J}_s$ *after* the boundary value problem is solved. Note that the other Maxwell equations, for $\nabla \cdot \boldsymbol{B}_\omega$ and $\nabla \cdot \boldsymbol{E}_\omega$ also have boundary conditions associated with them. These are not independent conditions and are automatically satisfied by any correct solution; they nontheless may be useful. For example, the condition derived from $\nabla \cdot \boldsymbol{E}_\omega = 4\pi\rho_\omega$ determines the *surface charge density* σ_ω.

15.3 PLANE REFLECTORS

In this section we ease into more complicated boundary value problems by discussing the scattering, or as it is more usually called, the *reflection*, of waves from plane conducting surfaces. These waves, it will be recalled from Section 11.8 satisfy the Helmholtz equation

$$\begin{aligned} (\nabla^2 + k^2)\boldsymbol{E}_\omega &= 0 \\ (\nabla^2 + k^2)\boldsymbol{B}_\omega &= 0 \end{aligned} \tag{15.9}$$

and have the transversality properties discussed in that section. In the simplest reflection problem, a plane wave is incident *normally* on a plane conductor. If, as in Fig. 15.2, $x = 0$ is the plane face of the conductor and a wave whose electric vector is $E_0 e^{-ikx}\hat{\boldsymbol{y}}$ is incident from above, then the boundary value problem is solved by adding to this a reflected wave similarly polarized and of equal and opposite amplitude. The resultant total field $\boldsymbol{E}_\omega$ is

$$\boldsymbol{E}_\omega = E_0(e^{-ikx} - e^{ikx})\hat{\boldsymbol{y}}$$

and this obviously satisfies the Helmholtz equation (15.9) as well as the boundary condition (15.7) that the tangential component of $\boldsymbol{E}_\omega$ vanish on the surface. Incidentially, this example illustrates the remark above that the vanishing of the tangential component of $\boldsymbol{E}$ does indeed determine the solution.

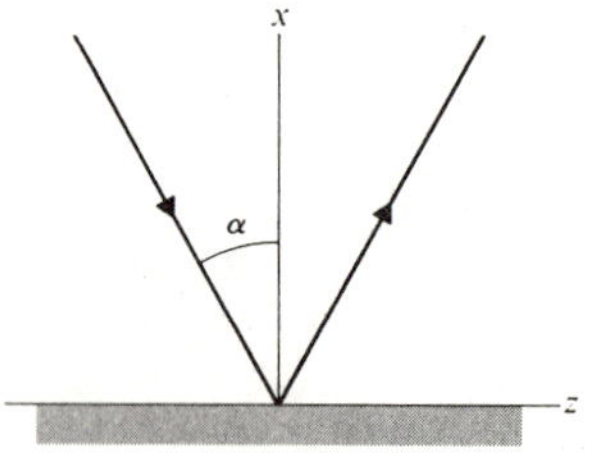

Fig. 15.2 Plane wave incident on plane conductor at angle α.

More generally, suppose a wave is incident from other than the normal direction, at some angle α as in Fig. 15.2. Let the propagation vector $\boldsymbol{k}$ be in the x, z-plane, and the electric vector be in the y-direction. Then $\boldsymbol{k} = k(-\hat{\boldsymbol{x}} \cos \alpha + \hat{\boldsymbol{z}} \sin \alpha)$, and the incident electric field $\boldsymbol{E}_\omega^i$ is

$$\boldsymbol{E}_\omega^i = E_0 e^{ik(-x \cos \alpha + z \sin \alpha)} \hat{\boldsymbol{y}}. \tag{15.10}$$

We would perhaps expect a reflected wave $\boldsymbol{E}_\omega^r$ for which the angle of reflection equals the angle of incidence, and so try

$$\boldsymbol{E}_\omega^r = E_0' e^{ik(x \cos \alpha + z \sin \alpha)} \hat{\boldsymbol{y}}. \tag{15.11}$$

The total field $\boldsymbol{E}_\omega$ is

$$\boldsymbol{E}_\omega = \boldsymbol{E}_\omega^i + \boldsymbol{E}_\omega^r$$

and, since $\boldsymbol{E}_\omega$ is already tangential to the surface, condition (15.7) then becomes

$$\boldsymbol{E}_\omega = 0, \qquad x = 0.$$

From Eqs. (15.10) and (15.11), this implies that $E_0 = -E_0'$, and

$$\boldsymbol{E}_\omega = -2iE_0\{e^{ikz \sin \alpha}\} \sin(kx \cos \alpha) \hat{\boldsymbol{y}}. \tag{15.12}$$

This expression has an interesting interpretation. The first part, in braces, looks like a wave traveling along the z-axis with wave number k_g, and associated wavelength λ_g,

$$k_g = k \sin \alpha = \frac{2\pi}{\lambda_g}. \tag{15.13}$$

Remembering that the usual, or as we may call it, the *free space wave length* λ associated with a plane wave is $\lambda = 2\pi/k$, we can also write

$$\lambda_g = \frac{\lambda}{\sin \alpha}.$$

Thus λ_g is always *greater* than the free space wavelength. By the same token, the velocity of this wave is $v_g = c/\sin\alpha$ which is always greater than c. Although the first term in (15.12) resembles a plane wave propagating in the z-direction, the other spatially varying factor, $\sin(kx\cos\alpha)$, shows that this is not the usual plane wave with constant amplitude across any plane perpendicular to the propagation direction. The expression (15.12) for $\boldsymbol{E}_\omega$ satisfies Maxwell's equations and the boundary condition at the conducting surface $x = 0$; it can then be interpreted as a solution of the boundary value problem of the propagation of a harmonic wave in the presence of a perfect conductor. Note that this solution is one for which the electric field is in the y-direction only and hence perpendicular or *transverse* to the direction of propagation, considered to be in the z-direction. The magnetic field however has components in *both* the y- and z-directions.

Let us extend the problem. Suppose that another conducting plane is added, say at $x = a$, in addition to the one at $x = 0$. Then Eq. (15.12) for $\boldsymbol{E}_\omega$ does not, as it stands, satisfy the boundary conditions at $x = a$, but it can be made to do so for certain values of $\cos\alpha$. Thus if

$$\sin(ka\cos\alpha) = 0$$

which implies that

$$k\cos\alpha = \frac{m\pi}{a}, \qquad m = 1, 2\ldots \tag{15.14}$$

then $\boldsymbol{E}_\omega$ will vanish at $x = a$ as well as at $x = 0$. We will have backed into the solution of the problem of a wave propagating between two conducting planes, a configuration that is sometimes called a *parallel plate wave guide*. Equation (15.14) implies a condition on the propagation constant k_g. Solving this equation for $\cos\alpha$ and (15.13) for $\sin\alpha$ and using $\sin^2\alpha + \cos^2\alpha = 1$, we find that k_g is determined by

$$k_g^2 = k^2 - \frac{m^2\pi^2}{a^2}. \tag{15.15}$$

This equation has an important interpretation. The z-dependence of $\boldsymbol{E}_\omega$ is given by e^{ik_gz}. If k_g is *real*, this dependence corresponds to wave-like motion along the z-axis with a guide wavelength λ_g given by $2\pi/k_g$. If, however, k_g is *imaginary*, then e^{ik_gz} describes an exponential attenuation, or unbounded growth, of $\boldsymbol{E}_\omega$—behavior not of a periodic or wave-like kind. The lowest frequency for which k_g will be real is found by putting $m = 1$ in (15.15) and requiring that the right-hand side be positive, i.e., requiring that $k = \omega/c > \pi/a$. The frequency $\pi c/a$, *below* which k_g is imaginary and hence wave-like propagation ceases, is called the *cut-off frequency*. The choice $m = 1$ implies a particular field configuration. For example, from Eq. (15.12), we see that $\boldsymbol{E}_\omega$ varies as $\sin\pi x/a$, and, with $\boldsymbol{E}_\omega$ fixed, $\boldsymbol{B}_\omega$ can be determined from Maxwell's equations. One finds, on arbitrarily renormalizing so that $B_{\omega z}$ has amplitude B_0, that

$$E_{\omega y} = \frac{iB_0 ka}{\pi} \sin \frac{\pi x}{a} e^{ik_g z},$$

$$B_{\omega x} = -\frac{iB_0 k_g a}{\pi} \sin \frac{\pi x}{a} e^{ik_g z}, \tag{15.16}$$

$$B_{\omega z} = B_0 \cos \frac{\pi x}{a} e^{ik_g z}.$$

Equations (15.16) define the field made corresponding to $m = 1$ in Eq. (15.15); there are other modes, guide wavelengths, and cut-off frequencies, corresponding to the other possible choices for m in Eq. (15.15). We would emphasize, however, that there are also other *kinds* of modes than the TE mode we have just discussed. The treatment above is meant to be illuminating rather than systematic. For example, there are TM modes, in which the *magnetic field* $\boldsymbol{B}$ is *transverse* to the direction of propagation. And there is also a TEM mode in which, as for a plane wave, *both* $\boldsymbol{E}$ and $\boldsymbol{B}$ are transverse to the direction of propagation. It is easy to see that such a mode is possible. For suppose that a plane wave propagates in the z-direction. Now imagine that two conducting planes are added, perpendicular to the direction of polarization. The plane wave still constitutes a solution to propagation between these planes since it satisfies the Helmholtz equation and the boundary condition on tangential $\boldsymbol{E}$. One says that this parallel plate wave guide supports a TEM mode. In Section 15.5 we shall discuss more generally the question of what kinds of guiding wave geometries permit a TEM mode.

Having backed into the solution (15.16) of the problem of wave propagation between parallel planes, we could, with a little more backing, find a solution to a superficially more complicated problem. If we add two conducting planes at $y = 0$ and $y = b$, so that they form, along with the other two, a hollow rectangular cylinder, then (15.16) represents one solution for the problem of propagation in such a hollow cylinder or *wave guide*. Namely, this solution still satisfies the boundary condition that the tangential component of $\boldsymbol{E}_\omega$ be zero at $x = 0$ and $x = a$, but since $\boldsymbol{E}_\omega$ is in the y-direction with no tangential component at the planes $y = 0$ and $y = b$, it *also* satisfies the boundary condition that tangential $\boldsymbol{E}_\omega$ be zero at these planes.

This solution is far from the most general one, however, and at this point we shall stop the present serendipitous development and discuss the problem of wave guides more systematically. The point of the present treatment has been to bring out as simply as possible the most important features of that problem as, for example, the existence of a guide wavelength, of a cut-off frequency, and of the field modes.

15.4 RECTANGULAR WAVE GUIDES

We now discuss the propagation of waves in a cylindrical wave guide of rectangular cross-section, with the cylinder axis parallel to the z-axis. Leaning on the results of the last section, we assume that the z-dependence of any of the field components is as $e^{ik_g z}$, to correspond to a wave traveling down the guide; the wave number k_g is to be determined. We then write

$$\begin{aligned} \boldsymbol{E}_\omega(x, y, z) &= \tilde{\boldsymbol{E}}(x, y)e^{ik_g z} \\ \boldsymbol{B}_\omega(x, y, z) &= \tilde{\boldsymbol{B}}(x, y)e^{ik_g z} \end{aligned} \tag{15.17}$$

and the two curl equations (15.1) and (15.2) become equations for the functions $\tilde{\boldsymbol{E}}$ and $\tilde{\boldsymbol{B}}$

$$\begin{aligned} &\frac{\partial \tilde{E}_z}{\partial y} - ik_g\tilde{E}_y = ik\tilde{B}_x, &\qquad &\frac{\partial \tilde{B}_z}{\partial y} - ik_g\tilde{B}_y = -ik\tilde{E}_x, \\ &-\frac{\partial \tilde{E}_z}{\partial x} + ik_g\tilde{E}_x = ik\tilde{B}_y, &\qquad &-\frac{\partial \tilde{B}_z}{\partial x} + ik_g\tilde{B}_x = -ik\tilde{E}_y, \\ &\frac{\partial \tilde{E}_y}{\partial x} - \frac{\partial \tilde{E}_x}{\partial y} = ik\tilde{B}_z, &\qquad &\frac{\partial \tilde{B}_y}{\partial x} - \frac{\partial \tilde{B}_x}{\partial y} = -ik\tilde{E}_z. \end{aligned} \tag{15.18}$$

From these equations, the transverse components $\tilde{E}_x$, $\tilde{E}_y$, $\tilde{B}_x$, $\tilde{B}_y$ can be expressed in terms of $\tilde{B}_z$ and $\tilde{E}_z$. For example, $\tilde{B}_x$ is found by eliminating $\tilde{E}_y$ between the first and fifth of Eqs. (15.18), and similarly for the other components. One finds

$$\begin{aligned} \tilde{B}_x &= -\frac{i}{k^2 - k_g^2}\left(k\frac{\partial \tilde{E}_z}{\partial y} - k_g\frac{\partial \tilde{B}_z}{\partial x}\right). \\ \tilde{B}_y &= \frac{i}{k^2 - k_g^2}\left(k\frac{\partial \tilde{E}_z}{\partial x} + k_g\frac{\partial \tilde{B}_z}{\partial y}\right). \\ \tilde{E}_x &= \frac{i}{k^2 - k_g^2}\left(k\frac{\partial \tilde{B}_z}{\partial y} + k_g\frac{\partial \tilde{E}_z}{\partial x}\right). \\ \tilde{E}_y &= -\frac{i}{k^2 - k_g^2}\left(k\frac{\partial \tilde{B}_z}{\partial x} - k_g\frac{\partial \tilde{E}_z}{\partial y}\right). \end{aligned} \tag{15.19}$$

These equations open up the possibility of separating the field configurations into transverse electric (TE) and transverse magnetic (TM) modes. First we can solve Eqs. (15.19) under the condition that $\tilde{E}_z = 0$ (TE mode), and then that $\tilde{B}_z = 0$ (TM mode), and consider that the general solution is a superposition of these two cases.

We start with the TM case, for which $\tilde{B}_z = 0$, and concentrate on finding $\tilde{E}_z$. Now $E_{\omega z}$ satisfies the Helmholtz equation (11.49) whence it follows that $\tilde{E}_z$ satisfies

$$\left(\frac{\partial^2}{\partial x^2} + \frac{\partial^2}{\partial y^2} - k_g^2 + k^2\right)\tilde{E}_z = 0. \tag{15.20}$$

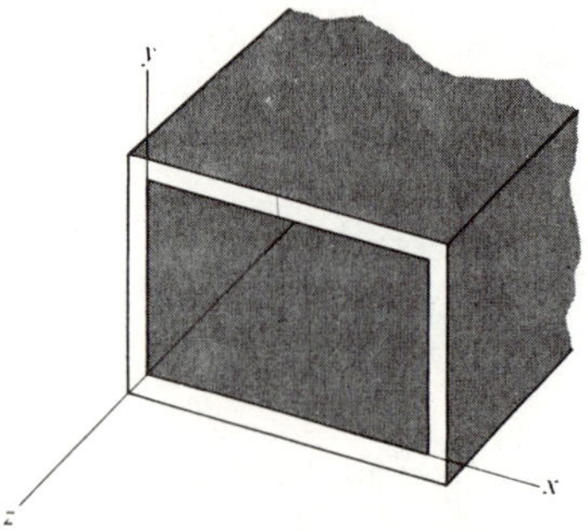

Fig. 15.3 Rectangular wave guide.

Let the hollow interior of the guide be the space $0 < x < a$ and $0 < y < b$, as shown in Fig. 15.3. Now $\tilde{E}_z$ is a tangential component of the field on the guide walls and it must vanish there. We discuss the separated solutions of Eq. (15.20) for $\tilde{E}_z$ in Appendix D. These solutions can be written as a product of one of the functions $\sin k_x x$, $\cos k_x x$, with one of the functions $\sin k_y y$, $\cos k_y y$ provided that

$$k_g^2 = k^2 - k_x^2 - k_y^2. \tag{15.21}$$

Of the four possibilities of product functions, we cannot admit any containing $\cos k_x x$ or $\cos k_y y$ since there is no choice of k_x or k_y which would then make $\tilde{E}_z$ vanish at $x = 0$ and $y = 0$. The only possibility left is the product of the two sines:

$$\tilde{E}_z = E_0 \sin(k_x x) \sin(k_y y). \tag{15.22}$$

For $\tilde{E}_z$ to vanish at $x = a$ and $y = b$ we must take

$$k_x = \frac{m\pi}{a}, \qquad k_y = \frac{n\pi}{b}, \qquad m = 1, 2 \ldots, \qquad n = 1, 2 \ldots. \tag{15.23}$$

The wave number k_g is then determined from (15.21) as

$$k_g^2 = k^2 - \pi^2\left(\frac{m^2}{a^2} + \frac{n^2}{b^2}\right). \tag{15.24}$$

This equation is obviously a generalization of Eq. (15.15). As in the discussion of that equation, we see that the field mode corresponding to a given m and n is *propagated* down the guide, or *attenuated*, according as k_g^2 is *greater* or *less* than zero. The former case will hold if

$$\omega > \omega_{mn}, \tag{15.25}$$

where

$$\omega_{mn}^2 = c^2\pi^2\left(\frac{m^2}{a^2} + \frac{n^2}{b^2}\right) \tag{15.26}$$

ω_{mn} is called the *cut-off frequency* for the given mode. With $\tilde{E}_z$ fixed, we can calculate the transverse components of the field from (15.19) and (15.22). We find that.

$$
\begin{aligned}
E_{\omega x} &= \frac{iE_0 k_g k_x}{k^2 - k_g^2} \cos(k_x x) \sin(k_y y) e^{ik_g z}, \\
E_{\omega y} &= \frac{iE_0 k_g k_y}{k^2 - k_g^2} \sin(k_x x) \cos(k_y y) e^{ik_g z}, \\
E_{\omega z} &= E_0 \sin(k_x x) \sin(k_y y) e^{ik_g z}, \\
B_{\omega x} &= -\frac{k}{k_g} E_{\omega y}, \\
B_{\omega y} &= \frac{k}{k_g} E_{\omega x}, \\
B_{\omega z} &= 0.
\end{aligned}
\tag{15.27}
$$

Note incidentally that although the boundary condition has only been applied to $E_{\omega z}$, both $E_{\omega x}$ and $E_{\omega y}$ also turn out to vanish on the planes on which they are tangential. The field configurations (15.27), for which the *magnetic* field is *transverse* to the direction of propagation, are labeled by the indices m and n and are called for brevity $(\mathrm{TM})_{mn}$ modes.

Now we take up the TE case, for which $\tilde{E}_z$ is zero by definition. Then we focus on $\tilde{B}_z$, which satisfies an equation like (15.20), viz.,

$$
\left(\frac{\partial^2}{\partial x^2} + \frac{\partial^2}{\partial y^2} - k_g^2 + k^2\right)\tilde{B}_z = 0. \tag{15.28}
$$

This also has solutions that are products of $\sin(k_x x)$ or $\cos(k_x x)$ with $\sin(k_y y)$ or $\cos(k_y y)$. Now, from (15.19), the field component $\tilde{E}_y$, a tangential component at $x = 0$ and $x = a$, is proportional to $\partial\tilde{B}_z/\partial x$, and $\tilde{E}_x$, tangential at $y = 0$ and $y = b$, is proportional to $\partial\tilde{B}_z/\partial y$. It is then easy to see that the boundary condition on tangential $\boldsymbol{E}$ requires that we choose

$$
\tilde{B}_z = B_0 \cos(k_x x) \cos(k_y y), \tag{15.29}
$$

where k_x and k_y are again given by (15.23), and satisfy (15.24). With $\tilde{B}_z$ thereby fixed,† we can calculate the field components from (15.19). We find

† With $\tilde{B}_z$ given by (15.29) we see that $B_{\omega z}$ satisfies the condition $\partial B_{\omega z}/\partial n = 0$ on the wave guide walls. It can be shown that this condition, derivable from $\nabla \cdot \boldsymbol{B}_\omega = 0$, and Eqs. (15.18), holds for guides of any cross-section and it is frequently used to determine TM modes. Since $\nabla \cdot \boldsymbol{B}_\omega = 0$ is, however, not independent of the curl equations, this condition is equivalent to the vanishing of tangential $\boldsymbol{E}_\omega$, which was derived from the curl equation (15.11). See Problem 15.8.

$$E_{\omega x} = -\frac{iB_0 kk_y}{k^2 - k_g^2} \cos(k_x x) \sin(k_y y) e^{ik_g z},$$

$$E_{\omega y} = \frac{iB_0 kk_x}{k^2 - k_g^2} \sin(k_x x) \cos(k_y y) e^{ik_g z},$$

$$E_{\omega z} = 0, \tag{15.30}$$

$$B_{\omega x} = \frac{k_g}{k} E_{\omega y},$$

$$B_{\omega y} = -\frac{k_g}{k} E_{\omega x},$$

$$B_{\omega z} = B_0 \cos(k_x x) \cos(k_y y) e^{ik_g z}.$$

The TE modes are also doubly infinite; the *mn*th mode is frequently abbreviated $(\mathrm{TE})_{mn}$. We see that the mode found inadvertently in the discussion of parallel plate wave guides, and given by Eq. (15.16), is in fact the $(\mathrm{TE})_{10}$ mode. For rectangular guides for which $a > b$, this mode has the lowest cut-off frequency of all modes. The dimensions of rectangular guides are frequently chosen so that this is the only propagating mode for the frequency ω in question.

15.5 CIRCULAR AND OTHER WAVE GUIDES

In this section, we shall discuss somewhat generally the propagation of waves in wave guides. The term wave guide usually means a hollow conducting cylinder, but we shall use it to include so-called *coaxial lines*, which consist of a hollow right circular cylinder plus a metallic inner conductor.

A TEM mode of propagation existed in the parallel plate wave guide of Section 15.3 but did not do so in the rectangular guides of Section 15.4. The first general point we want to discuss is the condition under which such a TEM mode can exist. In this discussion, we shall assume that the cylinder axis is the z-axis and that propagation is in the z-direction so that Eqs. (15.18) apply. We consider them under the TEM assumption that both $\tilde{E}_z$ and $\tilde{B}_z$ are zero. The first equation becomes $-k_g \tilde{E}_y = k\tilde{B}_x$ and the fifth becomes $-k\tilde{E}_y = k_g \tilde{B}_x$, showing that

$$k_g = k.$$

There is then no cut-off wavelength for a TEM mode.

We shall now show that such a mode cannot exist in *hollow* wave guides, i.e., wave guides whose cross-section is a closed curve and which contain no inner conductor. Consider $\tilde{E}_x(x, y)$ and $\tilde{E}_y(x, y)$; they constitute a *transverse two-dimensional field* $\tilde{\boldsymbol{E}}(x, y)$. The curl of this field is in the z-direction and vanishes if $\tilde{\boldsymbol{B}}_z$ is zero, as it is for a TEM mode. Then $\tilde{\boldsymbol{E}}$ can be derived from a potential: $\tilde{\boldsymbol{E}} = -\nabla\Phi(x, y)$. Now from $\nabla \cdot \boldsymbol{E}_\omega = 0$, it follows that if $E_{\omega z}$ is zero (TEM mode), then $\nabla \cdot \tilde{E} = 0$. As a consequence, Φ satisfies the two-dimensional Laplace equation. The boundary condition for the wave guide problem is that the tangential component of E_ω be zero on the boundary of the conductor, or in the language of electrostatics, that the conductor be an equipotential at some potential Φ_0. In discussing the interior Dirichlet problem in electrostatics, we concluded that the unique solution to such a problem was a constant: $\Phi = \Phi_0$ in the interior. But this solution corresponds to a vanishing field, $\tilde{E}_x = \tilde{E}_y = 0$. There can then be no TEM mode in a hollow wave guide. Of course, the above argument does not apply to the parallel plate wave guide, whose cross-section is not a closed curve, and, as we have seen, a TEM mode can propagate in such a guide. Nor does it apply to a coaxial line in which a TEM mode can also exist.

The general hollow wave guide can then support only TE or TM modes. We analyze these now by generalizing Eqs. (15.19) for the rectangular guide, i.e., by deriving general formulas for the transverse components in terms of $E_{\omega z}$ and $B_{\omega z}$. Setting $E_{\omega z}$ equal to zero defines TE modes and setting $B_{\omega z}$ to zero defines TM modes.† For a cylindrical wave guide of arbitrary cross-section, with cylinder elements parallel to the z-axis, we assume as before that the z-dependence of all field components is as $e^{ik_g z}$

$$\begin{aligned} \boldsymbol{E}_\omega(\boldsymbol{r}) &= \tilde{\boldsymbol{E}}(\boldsymbol{\rho})e^{ik_g z} \\ \boldsymbol{B}_\omega(\boldsymbol{r}) &= \tilde{\boldsymbol{B}}(\boldsymbol{\rho})e^{ik_g z}, \end{aligned} \tag{15.31}$$

where $\boldsymbol{\rho}$ is a vector in a plane perpendicular to the z-axis. Then we decompose the fields $\boldsymbol{E}_\omega$ and $\boldsymbol{B}_\omega$ into *longitudinal* components, $\boldsymbol{E}_{\omega z}$ and $\boldsymbol{B}_{\omega z}$, and *perpendicular* components. $\boldsymbol{E}_{\omega\perp}$ and $\boldsymbol{B}_{\omega\perp}$:

$$\begin{aligned} \boldsymbol{E}_\omega &= \boldsymbol{E}_{\omega z} + \boldsymbol{E}_{\omega\perp} \\ \boldsymbol{B}_\omega &= \boldsymbol{B}_{\omega z} + \boldsymbol{B}_{\omega\perp}. \end{aligned} \tag{15.32}$$

We also divide the nabla-operator ∇ similarly,

$$\nabla = \nabla_\perp + \nabla_z. \tag{15.33}$$

The Maxwell equation (15.1) for $\nabla \times \boldsymbol{E}_\omega$ becomes

$$(\nabla_\perp + \nabla_z) \times (\boldsymbol{E}_{\omega\perp} + \boldsymbol{E}_{\omega z}) = ik(\boldsymbol{B}_{\omega\perp} + \boldsymbol{B}_{\omega z}). \tag{15.34}$$

On expanding this equation, noting that $\nabla_z \times \boldsymbol{E}_{\omega z}$ is zero and that $\nabla_\perp \times \boldsymbol{E}_{\omega\perp}$ is in the z-direction, we find that

$$\nabla_z \times \boldsymbol{E}_{\omega\perp} + \nabla_\perp \times \boldsymbol{E}_{\omega z} = ik\boldsymbol{B}_{\omega\perp}. \tag{15.35}$$

In the same manner, we obtain from the Maxwell equation for $\nabla \times \boldsymbol{B}_\omega$

$$\nabla_z \times \boldsymbol{B}_{\omega\perp} + \nabla_\perp \times \boldsymbol{B}_{\omega z} = -ik\boldsymbol{E}_{\omega\perp}.$$

† It can be shown that there is no loss of generality in this procedure and that an arbitrary field can be expanded as a sum of TE and TM modes. See p. 243 of Jones (R).

Substituting $\boldsymbol{B}_{\omega\perp}$ from the first of these equations into the second, using

$$(\partial^2/\partial z^2 + k_g^2)E_{\omega z} = 0,$$

and similarly substituting $\boldsymbol{E}_{\omega\perp}$ from the second into the first, we find formulas† relating the transverse components of $\tilde{\boldsymbol{E}}$, $\tilde{\boldsymbol{B}}$ to the longitudinal ones:

$$\tilde{\boldsymbol{E}}_\perp = \frac{i}{k^2 - k_g^2}(k\boldsymbol{\nabla}_\perp \times \tilde{\boldsymbol{B}}_z + k_g\boldsymbol{\nabla}_\perp \tilde{E}_z) \tag{15.36}$$

$$\tilde{\boldsymbol{B}}_\perp = \frac{i}{k^2 - k_g^2}(k_g\boldsymbol{\nabla}_\perp \tilde{B}_z - k\boldsymbol{\nabla}_\perp \times \tilde{\boldsymbol{E}}_z). \tag{15.37}$$

We now specialize the discussion to the circular wave guide. Consider for example the TM case, $\tilde{B}_z = 0$, and focus on $\tilde{E}_z$, which satisfies an equation like (15.20) except that it is now obviously more convenient to express the two-dimensional Laplacian in cylindrical coordinates,

$$\frac{1}{\rho}\frac{\partial}{\partial\rho}\left(\rho\frac{\partial\tilde{E}_z}{\partial\rho}\right) + \frac{1}{\rho^2}\frac{\partial^2\tilde{E}_z}{\partial\varphi^2} + (k^2 - k_g^2)\tilde{E}_z = 0. \tag{15.38}$$

This has the form of the two-dimensional Helmholtz equation that is discussed in Appendix D. The solution that is finite at $\rho = 0$ is

$$\tilde{E}_z = E_0 J_n(\gamma\rho)\begin{cases}\sin n\varphi \\ \cos n\varphi\end{cases} \tag{15.39}$$

where

$$\gamma^2 = k^2 - k_g^2.$$

For $\tilde{E}_z$ to vanish at $\rho = a$, γa must be one of the roots of the Bessel function J_n. For a given order n, there are an infinite number of such roots, that can be arranged in a sequence of increasing magnitude, labeled by the index l. We denote these roots by x_{nl}; they satisfy

$$J_n(x_{nl}) = 0.$$

Then we can label γ similarly

$$\gamma_{nl} = x_{nl}/a. \tag{15.40}$$

The propagation constant k_g is determined from

$$k_g^2 = k^2 - \frac{x_{nl}{}^2}{a^2}. \tag{15.41}$$

† Equations (15.19) for Cartesian coordinates are, of course, special cases of these equations.

Once again propagation in a given mode, defined by the indices n,l, will be possible only for frequencies greater than the cut-off frequency ω_{nl},

$$\omega_{nl} = \frac{cx_{nl}}{a}. \tag{15.42}$$

The lowest cut-off frequency for all modes is that for $n = 0$, $l = 1$, corresponding to the root x_{01} of the zeroth order Bessel function: $x_{01} \approx 2.405$. With $\tilde{E}_z$ determined, the field components can be calculated from (15.36), and (15.37). Omitting for simplicity the subscripts n, l on γ and k_g, we find:

$$\begin{aligned}
E_{\omega\rho} &= \frac{iE_0 k_g \gamma}{k^2 - k_g^2} J_n'(\gamma\rho) \begin{Bmatrix} \sin n\varphi \\ \cos n\varphi \end{Bmatrix} e^{ik_g z}, \\
E_{\omega\varphi} &= \frac{iE_0 k_g n}{k^2 - k_g^2} \begin{Bmatrix} \cos n\varphi \\ -\sin n\varphi \end{Bmatrix} \frac{J_n(\gamma\rho)}{\rho} e^{ik_g z}, \\
E_{\omega z} &= E_0 J_n(\gamma\rho) \begin{Bmatrix} \sin n\varphi \\ \cos n\varphi \end{Bmatrix} e^{ik_g z},
\end{aligned} \tag{15.43}$$

$$B_{\omega\rho} = -\frac{k}{k_g} E_{\omega\varphi}, \qquad B_{\omega\varphi} = \frac{k}{k_g} E_{\omega\rho}, \qquad B_{\omega z} = 0.$$

Note from these equations that a given propagation constant k_g is the same for the two *different* mode configurations that vary as $\sin n\varphi$ or $\cos n\varphi$. This phenomenon, of a single eigenvalue being associated with more than one eigenfunction, is called *degeneracy*.

The TE case is handled in much the same way. First consider $\tilde{B}_z$, which satisfies an equation like (15.38). The solution can be written as a product of a Bessel function and $\sin n\varphi$ or $\cos n\varphi$ as in Eq. (15.39):

$$\tilde{B}_z = B_0 J_n(\gamma\rho) \begin{cases} \sin n\varphi \\ \cos n\varphi \end{cases}$$

However, the only tangential component of $\tilde{\boldsymbol{E}}$ is $\tilde{E}_\varphi$, which from (15.36) is

$$\tilde{E}_\varphi = \frac{ik}{k^2 - k_g^2} \frac{\partial \tilde{B}_z}{\partial \rho}.$$

The condition that $\tilde{E}_\varphi$ vanish at $\rho = a$ is now one that involves the zeros of the *derivatives* of the Bessel functions. With this exception, the development is so similar to that for the TE case that we leave the details to a problem.

15.6 RESONANT CAVITIES

Another important problem involving time harmonic fields and conducting boundaries is that of the *resonant cavity*, which is a region of space bounded by conducting walls. A section of wave guide whose ends are closed off by a conducto might typify a resonant cavity. The mathematical problem associated with such a cavity is that of finding solutions of the time harmonic Maxwell equations that satisfy the boundary condition (15.7), $\boldsymbol{E}_{\omega t} = 0$, on all the walls. It will turn out that these conditions can usually be satisfied only for certain values of the frequency ω. These are called the *natural* or *resonant frequencies* of the cavity, and they are analogous to the natural frequencies in other oscillation problems, for example, those of a plucked string or a vibrating drumhead.

Since resonant cavities are so closely related to wave guides, it is convenient to begin by borrowing some of the results from the discussion above on wave guides. Consider the TE field components, Eq. (15.30), for the rectangular wave guide. These represent a solution of Maxwell's equations satisfying the boundary conditions on the four walls of the guide provided that (15.24) is satisfied. Now suppose that the guide is terminated by conducting walls at $z = 0$ and $z = d$. Both $E_{\omega x}$ and $E_{\omega y}$ are tangential to these walls and must vanish on them. To effect this, we first observe that all the field components in Eq. (15.38) satisfy Maxwell's equations equally if the factor $e^{ik_g z}$ is replaced by $e^{-ik_g z}$, or by $\sin k_g z$ or $\cos k_g z$. With the choice $\sin k_g z$, the boundary condition on $E_{\omega x}$ and $E_{\omega y}$ is satisfied automatically at $z = 0$, and can be satisfied at $z = d$ as well by taking

$$k_g = \frac{p\pi}{d}, \tag{15.44}$$

where p is an integer. Then Eq. (15.24) for $k^2 = \omega^2/c^2$ becomes one for the *natural frequencies* ω_{mnp} of the cavity,

$$\frac{\omega_{mnp}^2}{c^2} = \pi^2\left(\frac{m^2}{a^2} + \frac{n^2}{b^2} + \frac{p^2}{d^2}\right). \tag{15.45}$$

With the functional form of $E_{\omega x}$ and $E_{\omega y}$ defined, we can calculate the remaining field components easily. We find

$$\begin{aligned}
E_{\omega x} &= -\frac{ikB_0}{\lambda^2}\left(\frac{n\pi}{b}\right)\cos\frac{m\pi x}{a}\sin\frac{n\pi y}{b}\sin\frac{p\pi z}{d},\\
E_{\omega y} &= \frac{ikB_0}{\lambda^2}\left(\frac{m\pi}{a}\right)\sin\frac{m\pi x}{a}\cos\frac{n\pi y}{b}\sin\frac{p\pi z}{d},\\
E_{\omega z} &= 0,\\
B_{\omega x} &= -\frac{B_0}{\lambda^2}\left(\frac{m\pi}{a}\right)\left(\frac{p\pi}{d}\right)\sin\frac{m\pi x}{a}\cos\frac{n\pi y}{b}\cos\frac{p\pi z}{d},\\
B_{\omega y} &= -\frac{B_0}{\lambda^2}\left(\frac{n\pi}{b}\right)\left(\frac{p\pi}{d}\right)\cos\frac{m\pi x}{a}\sin\frac{n\pi y}{b}\cos\frac{p\pi z}{d},\\
B_{\omega z} &= B_0\cos\frac{m\pi x}{a}\cos\frac{n\pi y}{b}\sin\frac{p\pi z}{d},
\end{aligned} \tag{15.46}$$

where

$$\lambda^2 = \left(\frac{m\pi}{a}\right)^2 + \left(\frac{n\pi}{b}\right)^2.$$

The solutions (15.27) for the TM modes can, of course, be modified in a similar way. One again finds that the resonant frequencies are given by Eq. (15.43), but now for a given mode—a given m,n,p—the field configuration is different. Again this is the phenomenon of *mode degeneracy*, encountered first in circular wave guides. It is clear that the modes for the cylindrical wave guide can be modified in much the same manner as above to provide the solutions for a circular resonant cavity. The details are left for a problem.

The perfect conductors we discuss do not have any ohmic losses associated with them. These losses are important practically, in defining the rate at which a given mode will decay. This problem, that of the so-called Q of the cavity is discussed later, when imperfect conductors are considered.

15.7 SCATTERING FROM A CIRCULAR CYLINDER

We have been discussing homogeneous or eigenvalue problems. The other general type of problem is the *inhomogeneous* or *scattering* problem, to which we turn now.

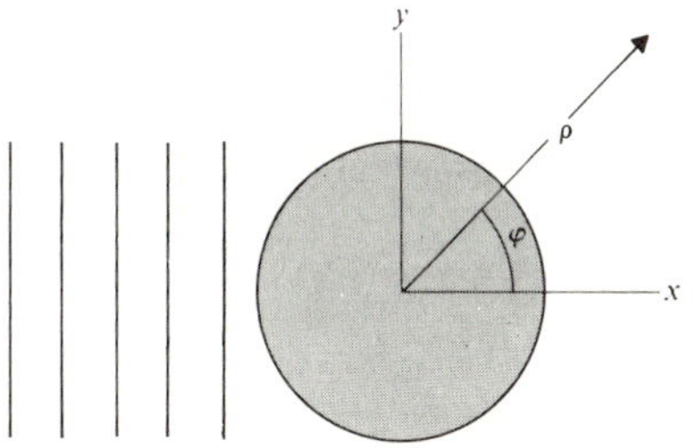

Fig. 15.4 Scattering of plane wave by circular cylinder.

One example of a scattering problem that can be solved by superposition of elementary solutions is that of a plane wave incident normally on an infinitely long, perfectly conducting circular cylinder. The same problem is discussed by means of integral equations in the next section. Figure 15.4 shows the cylinder, with axis along the z-axis; the plane wave is incident from $-\infty$ along the x-axis. Consider first the polarization of the incident wave. A plane wave of *arbitrary* polarization can be considered as a linear combination of one polarized along the z-axis and one polarized along the y-axis. In effect then, the most general case is treated by solving independently the scattering problem for polarization along the y- and z-axes. Polarization along the z-axis will lead to currents on the cylinder in the z-direction only; it is then easy to see that $B_{\omega z}$ will be zero for the scattered wave. Such a cylindrical wave, for which $B_{\omega z} = 0$ but $E_{\omega z} \neq 0$, is frequently called an *E-polarized wave.* Similarly, polarization in the y-direction leads only to transverse currents $J_{\omega x}$ and $J_{\omega y}$ and hence to a field for which $E_{\omega z} = 0$ but $B_{\omega z} \neq 0$. Such a field is called *B-polarized.*

Consider first the E-polarized case. The incident field, $E^i_{\omega z}$, of unit amplitude, is

$$E^i_{\omega z} = e^{ikx}.$$

The scattered wave $E^s_{\omega z}$ must satisfy the Helmholtz equation and the so-called *radiation condition,* i.e., it must correspond to an *outgoing* wave at infinity. The solutions of the Helmholtz equation in cylindrical coordinates involve Bessel and Neumann functions. From their asymptotic forms, given in Appendix D, we conclude, much as we did in Section 13.6, that the appropriate linear combination to satisfy the radiation condition is the Hankel function

$$H_m(k\rho) = J_m(k\rho) + iN_m(k\rho).$$

For ρ large, these contain the typical outgoing cylindrical wave factor $e^{ik\rho}/\sqrt{\rho}$:

$$H_m(k\rho) \to \sqrt{\frac{2}{\pi k\rho}}\, e^{i[k\rho-((2m+1)\pi/4)]}, \qquad \rho \to \infty. \tag{15.47}$$

We can then take as the scattered field

$$E^s_{\omega z} = \sum_{m=-\infty}^{\infty} A_m H_m(k\rho) e^{im\varphi} \tag{15.48}$$

The total field $E_{\omega z}$ is the sum of the incident and scattered fields, so with $x = \rho \cos \varphi$,

$$E_{\omega z} = e^{ik\rho \cos\varphi} + \sum_{m=-\infty}^{\infty} A_m H_m(k\rho) e^{im\varphi}. \tag{15.49}$$

Now, $E_{\omega z}$ is the tangential component of E_ω, and it must vanish on the cylinder. To apply this condition, we expand $e^{ik\rho\cos\varphi}$, which, of course, satisfies the Helmholtz equation, in the elementary solutions of that equation in cylindrical coordinates. These elementary solutions are $e^{im\varphi}$ times a Bessel or Neumann function of order m. Since $e^{ik\rho\cos\varphi}$ is finite for $\rho = 0$, the Neumann functions cannot enter into its expansion so we can write

$$e^{ik\rho\cos\varphi} = \sum_{m=-\infty}^{\infty} C_m J_m(k\rho) e^{im\varphi}, \tag{15.50}$$

where the C_m are to be determined. Multiplying (15.50) by $e^{-in\varphi}$, integrating over φ, and using the integral representation (D.28) for the Bessel function, we find that C_m is just i^m and thus have the very useful *plane wave expansion*:

$$e^{ik\rho\cos\varphi} = \sum_{m=-\infty}^{\infty} i^m J_m(k\rho) e^{im\varphi}. \tag{15.51}$$

With this expansion in (15.49), the condition $E_{\omega z} = 0$ at $\rho = a$ yields the coefficients A_m, and we find for the E-polarized case that

$$E_{\omega z} = e^{ik\rho\cos\varphi} - \sum_{m=-\infty}^{\infty} i^m \frac{J_m(ka)}{H_m(ka)} H_m(k\rho)e^{im\varphi} \tag{15.52}$$

Similarly, for a B-polarized wave, the incident field is

$$B^i_{\omega z} = e^{ik\rho\cos\varphi}, \tag{15.53}$$

and the scattered field $B^s_{\omega z}$ can be expanded in a series like (15.48). The boundary condition on the total tangential $\mathbf{E}_\omega$ is now the condition that

$$E_{\omega\varphi} = \frac{1}{k}\frac{\partial B_{\omega z}}{\partial\rho} = 0,$$

and this fixes the coefficients in the expansion of $B^s_{\omega z}$. We find

$$B_{\omega z} = e^{ik\rho\cos\varphi} - \sum_{m=-\infty}^{\infty} i^m \frac{J'_m(ka)}{H'_m(ka)} H_m(k\rho)e^{im\varphi}, \tag{15.54}$$

where the prime means derivative with respect to the argument.

With these results, we calculate some quantities of physical interest and first consider the angular distribution of the scattered energy at large distances. For this we need the Poynting vector in the radiation zone, i.e., for $k\rho \gg 1$. From (15.47) and (15.48) we find, for the case of E-polarization:

$$E^s_{\omega z} \simeq \sqrt{\frac{2}{\pi k\rho}}\, e^{ik\rho} \sum_{m=-\infty}^{\infty} A_m e^{i[m\varphi - ((2m+1)\pi/4)]}. \tag{15.55}$$

Moreover, it was found in Section 13.6 that $B_{\omega\theta} = -E_{\omega z}$ in the radiation zone. Thus the time average Poynting vector is radially outward and the magnitude $\bar{S}$ is given by

$$\bar{S}(\varphi) = \frac{c}{8\pi}\left|E_{\omega z}\right|^2 = \frac{c}{4\pi^2 k\rho} \sum_{m=-\infty}^{\infty}\sum_{m'=-\infty}^{\infty} A_m A^*_{m'} e^{i(m-m')\varphi} e^{-(i\pi/2)(m-m')}. \tag{15.56}$$

The total power scattered per unit length of the cylinder, P_s, is

$$P_s = \rho\int_0^{2\pi} \bar{S}(\varphi)\, d\varphi. \tag{15.57}$$

In characterizing the power scattered by an object it is convenient to normalize by defining a quantity called the *cross-section*. Since the cross-section will be useful later, we define it rather generally; it assumes a slightly different form for scattering from infinite cylinders. The scattered power P_s in *three dimensions* is defined to be the amount of energy per unit time that is radiated out through a large closed surface; it is given by the Poynting vector integrated over a large *sphere.* The incident power per unit area, the energy/(cm^2 × sec) in the incident plane wave, is just S_i, the magnitude of the Poynting vector for the incident wave. The cross-section for scattering, σ_s, is then defined for three-dimensional scattering by

$$\sigma_s = \frac{P_s}{S_i}. \tag{15.58}$$

Obviously σ_s has the dimensions of an area, and it is as if energy were extracted from this area of the plane wave front and scattered.

For an infinite cylinder, the total power scattered is infinite and this definition must be modified. We then introduce the power scattered *per unit length* as the energy/sec radiated through a cylindrical area of unit height and large radius and define the cross-section σ_{sc} for scattering from the cylinder by

$$\sigma_{sc} = \frac{P_s\,(\text{unit length})}{S_i}. \tag{15.59}$$

Since σ_{sc} has dimensions of length rather than area, it is sometimes called the *scattering width* instead of the scattering cross-section. We defer to the more popular, if less accurate, usage and call it the cross-section.

To return to E-polarized scattering by a cylinder, the cross-section σ_{sc} is easily found to be, for $\boldsymbol{E}$-polarization,

$$\sigma_{sc} = \frac{4}{k} \sum_{m=-\infty}^{\infty} \frac{J_m^2\,(ka)}{J_m^2(ka) + N_m^2(ka)}. \tag{15.60}$$

For B-polarization, it is

$$\sigma_{sc} = \frac{4}{k} \sum_{m=-\infty}^{\infty} \frac{J_m'^2(ka)}{J_m'^2(ka) + N_m'^2(ka)}. \tag{15.61}$$

For these equations to be useful, we must be able to evaluate them numerically. The technique of this evaluation depends mainly on the size of ka. For ka small, the small argument approximations for the Bessel functions can be used and only a few terms in the sum over m may suffice. But for the opposite case, for example, in the scattering of light from a cylinder of macroscopic size, ka may well be very large and hundreds or thousands of terms in the sum need to be taken. The expressions as they stand may then be quite useless, even for evaluation by computer. In this case, various techniques too specialized to discuss here must be used for summing the series. We consider then only the relatively simple long wavelength limit. From equation (15.60) it is easy to see that only $m = 0$ need be included in the sum and the Poynting vector $\bar{S}$ and cross-section σ_{sc} turn out to be, for E-polarization

$$\left.\begin{aligned}\bar{S}(\varphi) &\to \frac{c}{16k\rho\, ln^2\,(1/ka)},\\ \sigma_{sc} &\to \frac{\pi^2}{k\, ln^2\left(\dfrac{1}{ka}\right)},\end{aligned}\right\}\quad ka \to 0. \tag{15.62}$$

The scattering is independent of angle in this approximation. For the B-polarized case, one finds similarly

$$\left.\begin{aligned}\bar{S}(\varphi) &\to \frac{ca}{64\rho}(ka)^3\,(1 - 2\cos\varphi)^2,\\ \sigma_{sc} &\to \frac{3}{4}\pi^2 a\,(ka)^3\end{aligned}\right\}\quad ka \to 0. \tag{15.63}$$

15.8 USE OF INTEGRAL EQUATIONS FOR SCATTERING FROM AN ARBITRARY CYLINDER

In this section we use an integral equation to formulate the problem of scattering of an E-polarized wave by a cylinder, not necessarily circular in cross-section. The case of a B-polarized wave can be treated equally, but we shall not discuss it here. The virtue of integral equations applied to scattering problems is that, like their electrostatic analogs discussed in Section 4.7, they hold for bodies of any shape and for incident fields of any kind. They can also be solved numerically and thus provide a method for calculating the scattering from a much larger class of geometries than the circular cylinder and sphere. In fact, for the special case of scattering by a circular cylinder, we shall show that the integral equation we derive can also be solved analytically and reproduces, more conveniently in some respects, the results of the last section.

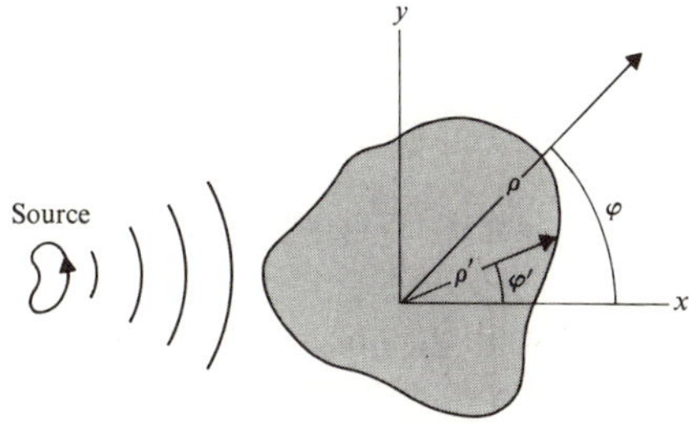

Fig. 15.5 Scattering of E-polarized wave from cylinder of arbitrary cross section.

Consider then the scattering of an incident wave $E^i_{\omega z}$ polarized in the z-direction (but not necessarily plane) from the cylinder shown in cross-section in Fig. 15.5. Let $\boldsymbol{\rho}$ be a vector in a cylindrical coordinate system to the field point and $\boldsymbol{\rho}'$ a vector to a point on the surface of the cylinder. The incident wave $E^i_{\omega z}(\boldsymbol{\rho})$

is assumed to be some known function. The surface current† J_S that is induced will be in the z-direction (out of the paper) and can be considered as a function of φ', the angle associated with $\boldsymbol{\rho}'$. As in the last section, the *total* field $E_{\omega z}$ is the incident field plus the scattered field $E_{\omega z}^s$ due to these currents. We can write an expression for this latter field using Eq. (13.46) for the vector potential $A_{\omega z}$ of a unit line source. Now $E_{\omega z} = ikA_{\omega z}$, so that $E_{\omega z}$ for a unit line source along the z-axis is $-(k\pi/c)H_0(k\rho)$. For the present problem then, the scattered field $E_{\omega z}^s$ is

$$E_{\omega z}^s = -\frac{k\pi}{c}\int J_S(\varphi')H_0(k|\boldsymbol{\rho} - \boldsymbol{\rho}'|)\,dl', \tag{15.64}$$

where dl' is an element of length along the perimeter of the cylinder. Then

$$E_{\omega z}(\boldsymbol{\rho}) = E_{\omega z}^i(\boldsymbol{\rho}) - \frac{k\pi}{c}\int J_S(\varphi')H_0(k|\boldsymbol{\rho} - \boldsymbol{\rho}'|)\,dl'. \tag{15.65}$$

We shall focus on J_S as the unknown quantity to be determined. Letting $\boldsymbol{\rho}$ in this equation become a vector $\boldsymbol{\rho}_s$ to the surface, we obtain the basic integral equation for determining this quantity:

$$0 = E_{\omega z}^i(\boldsymbol{\rho}_s) - \frac{k\pi}{c}\int J_S(\varphi')H_0(k|\boldsymbol{\rho}_s - \boldsymbol{\rho}'|)\,dl'. \tag{15.66}$$

The kernel of this equation is weakly singular, but this by itself presents no serious difficulties.

The first application of Eq. (15.66) will be to the problem treated in the last section, the scattering of an E-polarized plane wave by a *circular* cylinder; the results of that section will be reproduced. First expand $J_S(\varphi')$ in the complete set $e^{il\varphi'}$ to find

$$J_S(\varphi') = \sum_{l=-\infty}^{\infty} D_l e^{il\varphi'}. \tag{15.67}$$

Then use the expansion (13.55) for $H_0(k|\boldsymbol{\rho}_s - \boldsymbol{\rho}'|)$, which becomes, since both $\boldsymbol{\rho}_s$ and $\boldsymbol{\rho}'$ have magnitude a,

$$H_0(k|\boldsymbol{\rho}_s - \boldsymbol{\rho}'|) = \sum_{m=-\infty}^{\infty} e^{im(\varphi-\varphi')}J_m(ka)H_m(ka).$$

We substitute this into (15.66), do the integrals over φ' using the orthogonality of the set $e^{im\varphi}$, and recall that the incident wave $E_{\omega z}^i$ is now the plane wave $e^{ikx} = e^{ik\rho\cos\varphi}$. Equation (15.65) becomes, with $dl' = a\,d\varphi'$,

$$0 = e^{ika\cos\varphi} - \frac{2\pi^2 ka}{c}\sum_{l=-\infty}^{\infty} D_l e^{il\varphi}J_l(ka)H_l(ka).$$

† We use the capital subscript S in this section to avoid confusion with J_s as the notation for a Bessel function.

Expanding $e^{ika\cos\varphi}$ by (15.51) and equating coefficients of $e^{il\varphi}$, we find

$$D_l = \frac{i^l c}{2\pi^2 ka H_l(ka)},$$

whence

$$J_S(\varphi') = \frac{c}{2\pi^2 ka} \sum_{l=-\infty}^{\infty} \frac{i^l e^{il\varphi'}}{H_l(ka)}. \tag{15.68}$$

This result can also be derived from the solution of the last section, and the equation $\boldsymbol{J}_S = (c/4\pi)\boldsymbol{n} \times \boldsymbol{B}_\omega$; conversely, the results of that section can be derived from (15.68). Note incidentally that the reason that the integral equation (15.66) can be solved as readily as it has been is due to the convenient expansion of Eq. (13.55) for the kernel. Similar expansions do not in general exist for the kernels of the integral equations that occur in other problems of diffraction. These equations can however be solved numerically on digital computers. A discussion of techniques for so doing is given in Harrington (R).

The integral equation (15.66) can also be used to derive an interesting result, the *optical theorem*, which relates the amplitude for scattering in the forward direction to the total scattering cross-section. We derive this now, for the case of the E-polarized wave and, for simplicity, for the circular cylinder, although more general derivations are possible.

The scattering amplitude is defined in terms of $E^s_{\omega z}$ of Eq. (15.64). To find the form of $E^s_{\omega z}$ at large distances, the asymptotic form of $H_0(k|\boldsymbol{\rho} - \boldsymbol{\rho}'|)$ for large $|\boldsymbol{\rho} - \boldsymbol{\rho}'|$ can be used as well as the approximation $\rho \gg \rho'$, much as in the derivation of Eq. (13.58). Then $E^s_{\omega z}$ becomes, with $dl' = a\,d\varphi'$

$$E^s_{\omega z} = -e^{i(k\rho - \pi/4)} \sqrt{\frac{2\pi k a^2}{\rho c^2}} \int_0^{2\pi} J_S(\varphi') e^{-ika\cos(\varphi' - \varphi)}\,d\varphi'.$$

The integral in this term is called the *scattering amplitude* $f(\varphi)$,

$$f(\varphi) = \int_0^{2\pi} J_S(\varphi') e^{-ika\cos(\varphi' - \varphi)}\,d\varphi'. \tag{15.69}$$

As above let P_S represent the power radiated (per unit length) through a cylindrical surface S_0 of large radius ρ

$$P_S = \frac{c}{8\pi} \operatorname{Re}\colon \int_{S_0} E^{*s}_{\omega z}(\rho, \varphi')(-B^s_{\omega\rho}(\rho, \varphi'))\rho\,d\varphi'.$$

But this is equal to the power radiated out from the cylindrical *conducting surface C*:

$$P_S = \frac{c}{8\pi} \operatorname{Re}\colon \int_C E^{*s}_{\omega z}(a, \varphi')(-B^s_{\omega\rho}(a, \varphi'))a\,d\varphi'.$$

Now from Maxwell equations, $\partial E^s_{\omega z}/\partial\rho = -ikB^s_{\omega\varphi}$, so we have, omitting the arguments (a, φ') of the field variables for simplicity,

$$P_S = \frac{c}{8\pi}\mathrm{Re}\colon \int_C \frac{1}{ik} E^{*s}_{\omega z}\left(\frac{\partial}{\partial\rho} E^s_{\omega z}\right) a\, d\varphi'. \tag{15.70}$$

In general, $E_{\omega z} = E^i_{\omega z} + E^s_{\omega z}$ and, at the surface of the cylinder, $E^{*s}_{\omega z} = -E^{*i}_{\omega z}$. Thus (15.70) becomes

$$P_S = -\frac{c}{8\pi}\mathrm{Re}\colon \int_C \frac{1}{ik} E^{*i}_{\omega z}\frac{\partial}{\partial\rho}(E_{\omega z} - E^i_{\omega z}) a\, d\varphi'.$$

But

$$\int_C E^{*i}_{\omega z}\frac{\partial}{\partial\rho}\left(E^i_{\omega z}\right) d\varphi' = 0$$

so we have, using $\partial E_{\omega z}/\partial\rho = -ikB_{\omega\varphi} = -(4\pi/c)ikJ_S$,

$$P_S = \tfrac{1}{2}\mathrm{Re}\colon \int_C e^{-ika\cos\varphi} J_S(\varphi') a\, d\varphi.$$

Since the incident Poynting vector S_i is just $c/8\pi$, we have

$$\sigma_{sc} = \frac{4\pi}{c}\mathrm{Re}\colon \int_C e^{-ika\cos\varphi'} J_S(\varphi') a\, d\varphi'. \tag{15.71}$$

Comparing this result with the definition (15.69) of the scattering amplitude $f(\varphi')$, we have the *optical theorem*,

$$\sigma_{sc} = \frac{4\pi a}{c}\mathrm{Re}\colon f(0). \tag{15.72}$$

15.9 DIFFRACTION BY A HALF-PLANE

There are relatively few problems that can be solved exactly, and of these the majority involve geometries for which the Helmholtz equation separates. The cylinder we have just discussed is one example; another is the sphere. Since a conducting strip can be thought of as an elliptic cylinder of infinite eccentricity, the problem of scattering by such a two-dimensional strip can be solved by using two-dimensional elliptic coordinates and the related *Mathieu functions.* Similarly the problem of a conducting disk can be solved, although not easily and not compactly, by using the separated solutions appropriate to spheroidal coordinates. And there are only a few others, references to which can be found in Bouwkamp (R).

There is considerable interest in still another soluble problem that is important physically and that was one of the first to be solved by other than the separation of variables technique. This is the problem of *scattering* (or as it usually is called, *diffraction*) of a plane wave by a perfectly conducting half-plane. This problem was first solved by Arnold Sommerfeld in 1896, by a technique† that relied heavily on the theory of functions of a complex variable. Since then, various other methods of treating it have been devised. It has been solved by reducing it to an integral equation of a special kind (Clemmow (R)), by introducing new coordinates in which the wave equation separates (van Bladel (B)), and in fact by a variant of the separation of variables that involves multi-valued solutions (Morse and Feshbach (B)). The specifics of these methods are frequently quite complicated and often involve advanced techniques in the theory of functions, so we shall not present any of them in detail. Since the problem is so important, however, we shall sketch briefly the method of solution using the integral equation, and discuss some of the results. These are important in their own right and also in putting later approximate solutions in proper perspective.

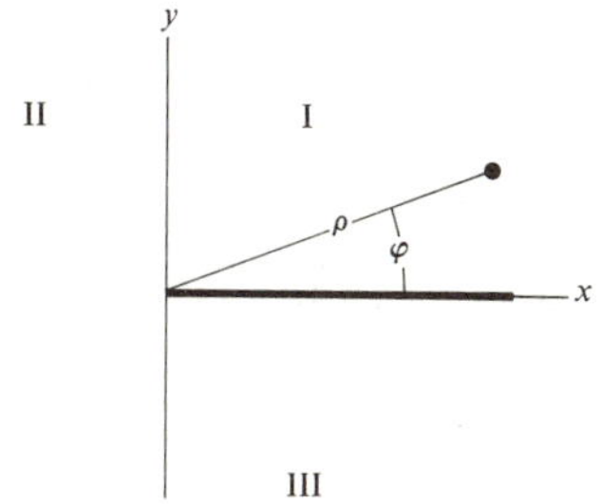

Fig. 15.6 Diffraction by the perfectly conducting half-plane $y = 0, x > 0$.

Consider then a perfectly conducting half-plane, which can be idealized, in the way discussed below, to be of zero thickness. It can be taken, as in Fig. 15.6, to coincide with the half-plane $y = 0$, $x > 0$. We suppose that waves are incident from above, i.e., from $y > 0$. The problem can be solved for a variety of incident waves; for example, for plane waves of either polarization incident at an arbitrary angle or for an incident wave generated by a line source parallel to the z-axis. We shall discuss a special case of these, but one which contains the essential features of more general solutions: diffraction of a z-polarized plane wave at normal incidence, i.e., of a wave that moves *down* along the y-axis and is given by

$$E^i_{\omega z} = e^{-iky}.$$

Any physical conductor is, of course, of finite thickness; we discuss now the idealization to zero thickness. According to the general ideas discussed previously, currents will be induced on the conductor and, in fact, on both sides of it. In the limit of small (as compared to a wavelength) thickness, the currents on the top and

† The solution was discussed in Sommerfeld, A., *Optics*, Academic Press, New York (1954).

bottom of the conductor will essentially overlie one another and the field produced at a point in space due to any narrow strip of the conductor will be due to the *sum* of the currents on either side of the strip. We can lump the currents on the two sides and consider their sum simply as *the* current. Physically, the incident wave can be thought of as producing a *current sheet* in the plane $y = 0$, $x > 0$. The boundary condition is that tangential $\boldsymbol{E}$ vanish on the surface of the conductor, but since the top and bottom surfaces are infinitesimally separated, this becomes the condition that tangential $\boldsymbol{E}$ vanish for $y = 0, x > 0$. This condition is expressed in general by the integral equation (15.66), which is easily adapted to the present case. Thus the incident wave evaluated over the surface is unity, and $\boldsymbol{\rho}_s$ and $\boldsymbol{\rho}'$ become x and x', respectively, so we have

$$0 = 1 - \frac{k\pi}{c}\int_0^\infty J_S(x')H_0(k|x - x'|)dx', \qquad x > 0. \tag{15.73}$$

This modest looking equation for the unknown current is in fact quite a difficult one to solve. It differs in an apparently trivial way from a kind of integral equation which is readily solved. Namely, integral equations like (15.73) with *difference* kernels, i.e., kernels $K(x, x')$ which are a function only of $x - x'$ and for which the variables x and x' vary from $-\infty$ to ∞, can be solved in a straightforward way by using Fourier transforms. But the fact that the variable x in (15.73) ranges only from zero to infinity makes a profound difference and consequently (15.73) can only be solved by using advanced techniques of so-called *generalized Fourier analysis*.

Although we shall not present the details of the solution of (15.73), we can make plausible one form that the solution takes. As we have indicated, the scattered or diffracted wave is that from a z-polarized current sheet in the plane $y = 0$. Putting the problem of diffraction aside for the moment, consider quite generally the form of the field that such a current sheet might produce. Take the simplest case first, a uniform sheet of z-directed current extending from $x = -\infty$ to $+\infty$. The electric field this produces will be the upward plane wave e^{+iky} above the sheet (taking unit amplitude for convenience) and the downward plane wave e^{-iky} below. We can write this as

$$E_{\omega z} = e^{\pm iky},$$

where the upper (lower) sign is associated with y greater (less) than zero. The associated magnetic field is $B_{\omega x} = \pm e^{\pm iky}$ so that the magnitude of the surface current density is $c/2\pi$. More generally now, imagine the current sheet that would generate a plane wave traveling upward in the x, y-plane at an angle α above the sheet and downward at an angle $-\alpha$ below the sheet, i.e., one for which $E_{\omega z}$ is, with the same convention for upper and lower signs,

$$E_{\omega z} = e^{ik\rho\cos(\varphi \mp \alpha)}. \tag{15.74}$$

The corresponding magnetic field $B_{\omega x}$ will be $\pm \sin\alpha\, e^{ik\rho(\cos\varphi \mp \alpha)}$ and the surface current J_S then depends harmonically on x by means of

$$J_S = \frac{c}{2\pi} \sin\alpha\, e^{ikx\cos\alpha}. \tag{15.75}$$

In a nutshell, the argument that follows is this: Eq. (15.75) for the current associated with the plane wave (15.74) resembles one Fourier component in a Fourier integral. Since summing over all such components generates, by the Fourier integral theorem, an *arbitrary surface current*, we argue that an analogous sum over the associated plane waves (15.74) will represent an *arbitrary field*. Consider then the representation by a Fourier integral of an arbitrary surface current $J_S(x)$:

$$J_S(x) = \frac{1}{2\pi}\int_{-\infty}^{\infty} f(\xi) e^{i\xi x}\, d\xi. \tag{15.76}$$

With $\xi = k\cos\alpha$ this becomes

$$J_S(x) = -\frac{k}{2\pi}\int_C f(k\cos\alpha)\sin\alpha\, e^{ikx\cos\alpha}\, d\alpha, \tag{15.77}$$

where the contour C remains to be specified. It is clear that this contour must be one in the complex α plane since $\xi/k = \cos\alpha$ must range from $-\infty$ to ∞, and this is possible only if α assumes complex values. Consider now the contour C in the α-plane that will reproduce the necessary range of ξ from $-\infty$ to ∞. We assert that it is the one shown in Fig. 15.7. Thus over the first leg of the integration,

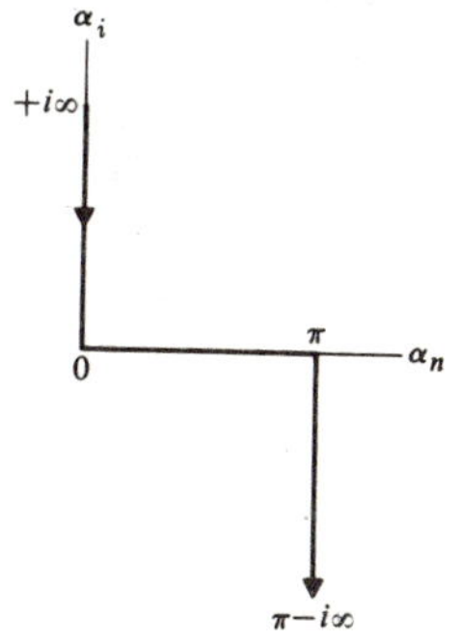

Fig. 15.7 The integration contour C in the α-plane.

from $i\infty$ to zero, $\cos\alpha$ ranges from $+\infty$ to 1; over the leg from 0 to π, it varies from 1 to -1; and over the third leg, from -1 to $-\infty$. Hence the contour C reproduces the essential range of ξ; and, except for inconsequential factors, the integral (15.77) over C is the Fourier integral, which can represent an arbitrary surface current. As we have seen, each harmonic component $e^{ikx\cos\alpha}$ of (15.75) can be thought of as being associated with the plane wave field (15.74). We can then think of the superposition current integral (15.82) as being associated with an electric field that is a superposition of plane wave fields. Consequently it is plausible to assume that the field $E^s_{\omega z}$ due to a general current sheet can be represented by

$$E^s_{\omega z} = \int_C P(\cos\alpha) e^{ik\rho\cos(\varphi\mp\alpha)}\, d\alpha, \tag{15.78}$$

where $P(\cos\alpha)$ is some weighting function that depends on the current. Equation (15.78) represents $E^s_{\omega z}$ as a *spectrum of plane waves.* However, not all the waves in that spectrum are ordinary plane waves: those that correspond to complex α are so-called *evanescent* or *decaying waves.*

Equation (15.78) is then a possible form for the diffracted field. The question of finding $P(\cos\alpha)$ so that the integral equation (15.73) is satisfied is beyond the scope of this book and we shall merely quote the simple result

$$P(\cos\alpha) = \frac{i}{2\pi}\frac{\sqrt{1-\cos\alpha}}{\cos\alpha}.$$

The total field $E_{\omega z}$ can then be written

$$E_{\omega z} = e^{-iky} + \frac{i}{2\pi}\int_C \frac{\sqrt{1-\cos\alpha}}{\cos\alpha} e^{ik\rho\cos(\varphi\mp\alpha)}\, d\alpha. \tag{15.79}$$

Once given this solution, the evaluation of the integral in it is still a task that leans heavily on the theory of functions of a complex variable and we shall merely present the results without derivation.† The exact evaluation of (15.79) is given in terms of the complex Fresnel integral $F(a)$ defined by

$$F(a) = \int_a^\infty e^{i\xi^2}\, d\xi. \tag{15.80}$$

The total field $E_{\omega z}$ (*including* the incident wave) is

$$E_{\omega z} = \frac{e^{-i\pi/4}}{\sqrt{\pi}}\{e^{-iky}F(-\sqrt{2k\rho}\cos[\tfrac{1}{2}\varphi - \tfrac{1}{4}\pi]) - e^{iky}F(-\sqrt{2k\rho}\cos[\tfrac{1}{2}\varphi + \tfrac{1}{4}\pi])\}. \tag{15.81}$$

Although the incident field does not appear explicitly in (15.81), it can be shown to be there implicitly. By using approximate expansions of the Fresnel integral, the

† For details, see Born and Wolf (*R*).

solution (15.81) can be evaluated in the limits $k\rho \gg 1$ and $k\rho \ll 1$. We consider the former case. In this limit of very short wavelength, the field is, not unexpectedly, related to the *geometrical optical field* E^g defined by

$$E^g = \begin{cases} e^{-iky} - e^{+iky}, & \text{in I}: 0 < \varphi < \pi/2; \\ e^{-iky}, & \text{in II}: \pi/2 < \varphi < 3\pi/2; \\ 0, & \text{in III}: 3\pi/2 < \varphi < 2\pi. \end{cases} \tag{15.82}$$

This field E^g is the one that would be produced if geometrical optics were valid, i.e., if each point on the plane wave front were associated with a ray perpendicular to it, and if, for that ray, the angle of incidence equaled the angle of reflection. In region I (reflection region) the field consists of the incident e^{-iky} and reflected wave $-e^{iky}$. There is, of course, no reflection in region II (illuminated region), and only the incident wave e^{-iky} is present there. Finally in region III (shadow region) geometrical optics would predict no field at all.

In terms of E^g, the total field for $k\rho \gg 1$ is given as a sum of it and a diffracted field E^d:

$$E_{\omega z} = E^g + E^d, \tag{15.83}$$

where

$$E^d \simeq \frac{e^{i\pi/4}}{\sqrt{\pi}} \frac{\sin \varphi/2}{\cos \varphi} \frac{e^{ik\rho}}{\sqrt{k\rho}}. \tag{15.84}$$

The approximation implied by (15.84) is valid except near the angles $\varphi = \pi/2$ and $\varphi = 3\pi/2$. Away from these, it has an interesting interpretation. The wave E^d appears to be two dimensional, $e^{ik\rho}/\sqrt{\rho}$, with an amplitude factor $(\sin \varphi/2)/(\cos \varphi/2)$, as if it were produced by some line source at $\rho = 0$. This equivalent line source cannot be a simple cylindrically symmetric one, since it contains the φ-dependent angular factor. It is more akin to the field due to a two-dimensional multipole. We can understand this result more intuitively by looking at the expression for the induced current or the surface. From (15.81) the magnetic field, and hence the induced current density J_S, can be found. It turns out to be

$$J_S \simeq \frac{c}{2\pi} + 0((kx)^{-3/2}), \qquad kx \gg 1;$$

$$J_S \simeq \frac{c}{2(\pi)^{3/2}} e^{-i\pi/4} \left\{ \frac{i}{\sqrt{kx}} + 2\sqrt{kx} \right\} e^{ikx}, \qquad kx \ll 1.$$

For x small, near the edge of the conductor, the current density is singular as $1/\sqrt{x}$. This being so, it is not surprising that the diffracted wave appears to be produced by a line source at the edge of the conductor.

15.10 DIFFRACTION BY AN APERTURE

The half-plane problem is perhaps the simplest of planar diffraction problems. More general problems of diffraction by an aperture of arbitrary shape in a conducting plane can generally be solved only approximately. In this section we discuss some of the approximate techniques. Consider then the perfectly conducting screen of Fig. 15.8 with an aperture of arbitrary shape. A wave is incident from the left and we want to calculate the resultant field.

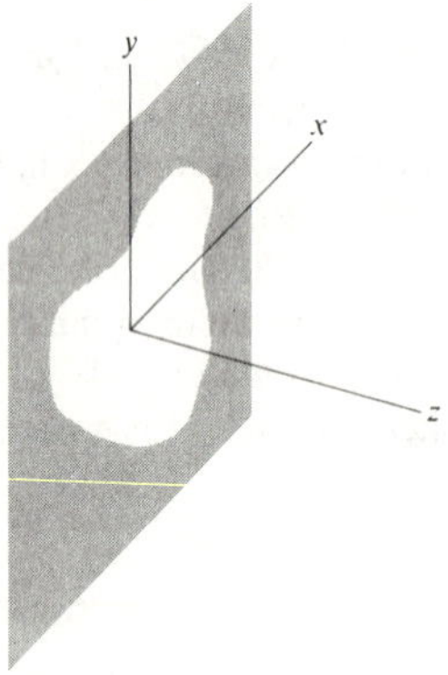

Fig. 15.8 Aperture in perfectly conducting screen.

For certain special shapes of the aperture, the solution to the half-plane problem can sometimes be exploited. Thus, suppose the aperture were a slit of width $2d$. We can consider that the slit is generated by two conducting half-planes, for which $z = 0$ and whose edges, parallel to the y-axis, are at $x = d$ and $x = -d$. If a plane wave is incident on this geometry, we can think of the problem as a *composite* one, similar to those discussed in electrostatics in Section 4.6. Namely, either of the conducting planes (say, plane I) can be considered to be irradiated by the incident wave *and* by the wave generated by plane II. The wave generated by the current on plane II is, as we have seen, approximately that of a line current singularity at its edge. Approximately then, plane I has incident on it the plane wave *plus* a cylindrical wave due to an effective line source (of unknown strength) generated by plane II. Now the solutions of the half-plane problem are known for irradiation by an incident plane wave, and by a line current of known strength. Using these solutions, we can set up the consistency conditions that determine the strength of the effective line sources of planes I and II and so solve approximately† the problem of the slit that is wide compared to a wavelength.

As another example, take the diffraction of a plane wave by a large (with

† Karp, S. N. and A. Russek, "Diffraction by a Wide Slit," *J. Appl. Phys.* **27**, 886 (1956).

respect to the wavelength) circular hole in a plane conductor. We anticipate that there will be a current singularity near the edge of the hole. Moreover, for large holes, this edge will be approximately straight over distances of the order of a wavelength, and one would expect the current singularity to be approximately the same as for the half-plane. Working along these lines, Braunbek† has exploited the solution of the half-plane problem to solve the problem of scattering by a circular hole in a conducting screen.

An approximate theory of diffraction from an aperture, important in optics, is the *Kirchoff theory* and we discuss it now. The theory is based on an integral formula similar to the electrostatic integral identity of Eq. (5.23). Like that identity its derivation begins with Green's (first) theorem of Eq. (5.18), relating volume and surface integrals of functions $\psi(\mathbf{r}')$ and $\chi(\mathbf{r}')$ (it is convenient here to prime ∇^2 to show that it acts on $\mathbf{r}'$)

$$\int_V (\psi\nabla'^2\chi - \chi\nabla'^2\psi)\,dv' = \int_S \left(\psi\frac{\partial\chi}{\partial n} - \chi\frac{\partial\psi}{\partial n}\right) ds'. \tag{15.85}$$

Now let F be any of the six Cartesian components of a time harmonic electromagnetic field whose time-separated amplitudes are $\mathbf{E}_\omega$, $\mathbf{B}_\omega$. We suppose that the field is produced by current sources that are localized in some portion of space. Consider a volume which does *not* contain any of these sources and let this be the volume V, bounded by the surface S as in Fig. 15.9, to which Eq. (15.85) applies.

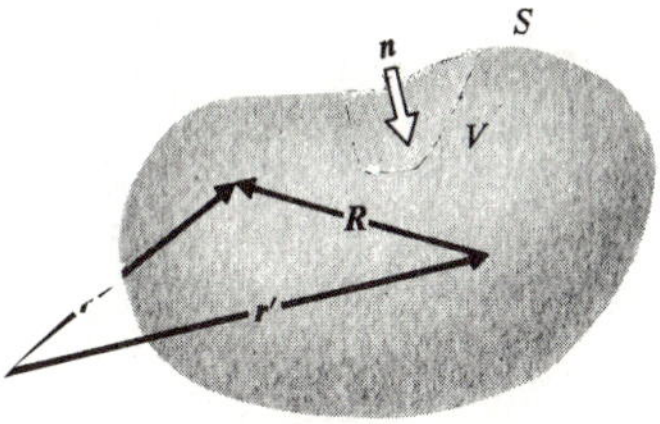

Fig. 15.9 For the application of Green's theorem to diffraction problems.

As the figure indicates, it is now convenient to let $\mathbf{n}$ be the *inward* drawn normal, which fact introduces a minus sign on one side of Eq. (15.85). Now in V we choose χ to be one of the rectangular field components

$$\chi \equiv F$$

so that

$$(\nabla'^2 + k^2)F = 0. \tag{15.86}$$

† See Bouwkamp (R) for further details and for references.

For ψ we choose

$$\psi = \frac{e^{ik|\boldsymbol{r}-\boldsymbol{r}'|}}{|\boldsymbol{r}-\boldsymbol{r}'|}$$

and so we have

$$(\nabla'^2 + k^2)\psi = -4\pi\delta(\boldsymbol{r}' - \boldsymbol{r}). \tag{15.87}$$

Putting (15.86) and (15.87) into (15.85) and including the minus sign just mentioned yields the basic integral identity

$$F(\boldsymbol{r}) = -\frac{1}{4\pi}\int_S \left(\frac{e^{ik|\boldsymbol{r}-\boldsymbol{r}'|}}{|\boldsymbol{r}-\boldsymbol{r}'|}\frac{\partial F}{\partial n} - F\frac{\partial}{\partial n}\frac{e^{ik|\boldsymbol{r}-\boldsymbol{r}'|}}{|\boldsymbol{r}-\boldsymbol{r}'|}\right) ds'. \tag{15.88}$$

This identity is of the same nature as the electrostatic one, to which it reduces for $k \to 0$: we refer the reader to the discussion of Section 5.3.

To apply (15.88) to the problem of Fig. 15.8, let the surface† S be the plane $z = 0$ so that the volume V is the half-space $z > 0$. Now from (15.88) F can be calculated for $z > 0$ if F and $\partial F/\partial n$ are known on the surface, much as in the electrostatic case. But in general F and $\partial F/\partial n$ are *not* known *a priori*: As with Laplace's equation, the theory of the Helmholtz equation shows that its solution throughout a volume is fixed by the values of F *or* $\partial F/\partial n$ on the surface. *One* of these quantities may then be assumed known *a priori*, but the other must be determined.

Despite these qualifications, the Kirchhoff theory makes the basic assumptions that F and $\partial F/\partial n$ vanish everywhere on the plane $z = 0$ except in the aperture, and that their values in the aperture are those for the incident wave. Now it is conceivable, although it would *a priori* appear quite fortuitous, that these values of F and $\partial F/\partial n$ are the correct ones. This is a small straw to grasp at, but even it is unsubstantial. It can be shown that the basic Kirchhoff assumption is self-contradictory.

Despite these objections, there is one point in favor of the theory: it seems to work, at least for problems in which the aperture size is much larger than the wavelength, as is the case for light. Moreover, it often holds surprisingly well even for aperture sizes only a few, or several, wavelengths in dimension. A critical discussion of the significance of the theory is to be found in the books by Baker and Copson, and the review articles of Bouwkamp and of Wolf, all cited in the references at the end of the chapter.

The theory is partially justified by experiment and we now consider some of its details. Suppose the incident wave F^i corresponds to propagation in the

† More precisely, we should include a very large hemispherical surface extending to the right as part of S, but since such a hemisphere has no effect, we shall neglect it from the beginning.

z-direction, but with (for generality) an amplitude which may depend on x and y. That is, suppose $F^i(x', y', z')$ is of the form

$$F^i(x', y', z') = A(x', y')e^{ikz'}.$$

Then the values of F and $\partial F/\partial n$ that go into the integral (15.88) are

$$F(x', y', 0) = A(x', y')$$

and

$$\frac{\partial F}{\partial n} = \left.\frac{\partial F}{\partial z'}\right|_{z'=0} = ikA(x', y').$$

Moreover, according to the Kirchhoff assumption, the integration goes only over the aperture. With $R = |\boldsymbol{r} - \boldsymbol{r}'|$ we have,

$$\frac{\partial}{\partial n}\frac{e^{ikR}}{R} = -\left(ik - \frac{1}{R}\right)\frac{z}{R}\frac{e^{ikR}}{R}$$

and Eq. (15.88) becomes

$$F(\boldsymbol{r}) = -\frac{1}{4\pi}\int_{\text{Aperture}} A(x', y')\frac{e^{ikR}}{R}\left[ik + \left(ik - \frac{1}{R}\right)\frac{z}{R}\right] ds'. \tag{15.89}$$

This is the basic formula; the problem is now that of evaluating the integral in it. This integral is very similar to the superposition integral in the summation problem with time harmonic currents. As in that case, there are three parameters on which the integral depends: wavelength λ, typical aperture dimension d, and distance r from aperture to observation point. Again as in that case, the integral can rarely be done exactly, and approximations are made that depend on the relative values of the three parameters. Two important approximations are those for *Fraunhofer diffraction* and *Fresnel diffraction.* The former is the analog of the radiation zone approximation, i.e., it makes the assumptions,

$$r \gg d, \qquad kr \gg 1.$$

In this case

$$\frac{e^{ik|\boldsymbol{r}-\boldsymbol{r}'|}}{|\boldsymbol{r} - \boldsymbol{r}'|} \simeq \frac{e^{ikr}}{r}e^{-i\boldsymbol{k}\cdot\boldsymbol{r}'},$$

where $\boldsymbol{k}$ is a vector of magnitude k in the direction of observation. We assume observation near the forward direction, so that $z/R \approx 1$ and use $ik \gg 1/R$. Then (15.89) becomes the well-known *Fraunhofer diffraction formula*

$$F(\boldsymbol{r}) = -\frac{ik}{2\pi}\frac{e^{ikr}}{r}\int_{\text{Aperture}} A(x', y')e^{-i\boldsymbol{k}\cdot\boldsymbol{r}'} ds'. \tag{15.90}$$

In principle, F is any of the three rectangular components of $\boldsymbol{E}_\omega$ or $\boldsymbol{B}_\omega$. In applications, however, the polarization properties of the radiation are usually glossed over; the intensity $I(P)$ of radiation at a point, defined as the time average energy/sec passing the point (measured, say, by the blackening of a photographic plate) is usually taken simply as being proportional to $|F|^2$.

As an example, we consider the Fraunhofer diffraction by a circular aperture, of radius a. We take $\boldsymbol{k}$ in the y, z-plane making an angle θ with the z-axis, $\boldsymbol{k} = k(\hat{\boldsymbol{y}} \sin\theta + \hat{\boldsymbol{z}} \cos\theta)$, and use $\boldsymbol{r}' = r'(\hat{\boldsymbol{x}} \cos\varphi + \hat{\boldsymbol{y}} \sin\varphi)$. Then $I(P)$ becomes $I(\theta)$, where

$$I(\theta) \propto |F|^2 = C\left|\int_0^a r'\,dr' \int_0^{2\pi} e^{-ikr'\sin\theta\sin\varphi}\,d\varphi\right|^2.$$

and C is a proportionality constant. From Eq. (D.28), we see that the integral over φ yields a Bessel function of order zero, namely, $2\pi J_0(kr'\sin\theta)$, and then the integral over r' can be done using

$$\int_0^x x' J_0(x')\,dx' = xJ_1(x).$$

We find thus the formula for the intensity in Fraunhofer diffraction from a circular aperture

$$I(\theta) = C(2\pi a^2)^2\left(\frac{J_1(ka\sin\theta)}{ka\sin\theta}\right)^2.$$

In *Fresnel diffraction*, the observation point is near the edges of the shadow that would be predicted by geometrical optics. The expansion of the factor $e^{ik|\boldsymbol{r}-\boldsymbol{r}'|}/|\boldsymbol{r}-\boldsymbol{r}'|$ is rather delicate for the Fresnel case. Oversimplifying a bit, let us assume that the coordinates x, y of the observation point are of the order d, but that $z \gg d$. Then

$$|\boldsymbol{r}-\boldsymbol{r}'| \equiv ((x-x')^2 + (y-y')^2 + z^2)^{1/2} \approx z\left(1 + \frac{(x-x')^2}{2z} + \frac{(y-y')^2}{2z} + \cdots\right).$$

By changing variables to $\xi = (x-x')/\sqrt{2}$ or $\xi = (y-y')/\sqrt{2}$ one can see that the integrals important for the evaluation of Fresnel diffraction patterns will involve the *Fresnel integral*

$$\int_a^\infty e^{i\xi^2}\,d\xi.$$

These remarks scarcely do justice, however, to the somewhat complicated subject of Fresnel diffraction and we refer the reader to standard treatises, e.g., Born and Wolf (R) for a further discussion.

PROBLEMS

1. A wave guide of right triangular cross-section is bounded by conducting walls at $x = a$, $y = 0$, $x = y$. Find the cut-off frequencies and field modes. [*Hint*: Superpose degenerate modes for the square guide, which modes satisfy the boundary condition of vanishing tangential $\boldsymbol{E}$ at $x = a$ and $y = 0$, in such a manner as to satisfy the boundary condition at $x = y$ as well.]

2. A unit line source of current is parallel to an infinite conducting circular cylinder of radius a, a distance d from its axis. Find the Poynting vector of the scattered radiation at large distances.

3. Verify from the series solution (15.52) that the current on an infinite cylinder has the value (15.68) derived from the integral equation solution. [*Hint*: What is a Wronskian?]

4. Derive Eq. (15.54) for the expansion coefficients of the field in the scattering of a B-polarized wave from a cylinder.

5. Show that the boundary value problem of a dipole over an infinite plane conductor can be solved by using images. Can the problem of a dipole external to a conducting sphere also be solved by images, as it can in electrostatics?

6. The optical theorem of Eq. (15.72) holds for cylinders of arbitrary cross-sections. Verify it for a circular cylinder by using in it the explicit results (15.60) for σ_{sc} and (15.68) for J_S.

7. Verify Eqs. (15.62) and (15.63) for the scattering cross-sections of a circular cylinder in the longwave length limit.

8. Use Eq. (15.37) to show that for TE modes in a wave guide the condition of vanishing tangential $\boldsymbol{E}_\omega$ is equivalent to the vanishing of $\partial B_{\omega z}/\partial n$ on the guide walls.

9. Discuss the TM modes in a guide of circular cross-section along the lines of the discussion of TE modes in the text, finding the propagation constants and the field components for each mode.

10. A coaxial line consists of a solid inner conducting cylinder of radius a, concentric with a circularly cylindrical outer conductor of inner radius b. Given that there is a TEM mode possible for which (cylindrical coordinates ρ, φ, z) $E_{\omega\rho} = (E_0/\rho)e^{ikgz}$, find all other field components and find k_g.

11. Show from Eq. (15.68) that scattering of an E-polarized wave in the long wavelength limit leads to a current distribution on the cylinder that is independent of φ. Find the current distribution for scattering of a B-polarized wave, and find its angular distribution in the long wavelength limit.

12. Calculate the surface current on the walls of a rectangular wave guide in which the TE_{10} mode is propagating. Sketch the lines of current flow.

13. Find from (15.90) the Fraunhofer diffraction pattern for a plane wave incident normally on a rectangular aperture in a plane conducting screen.

REFERENCES

Baker, B. B., and E. T. Copson, *The Mathematical Theory of Huygen's Principle*, Oxford, Clarendon Press (1950).

This book contains a rigorous mathematical statement of Huygen's principle and a discussion of the consistency, or rather inconsistency, of the Kirchhoff approximation.

Borgnis, F. E., and C. H. Papas, *Electromagnetic Waveguides and Resonators*, Vol. XVI *Encyclopedia of Physics*, Springer, Berlin (1968).

A standard and thorough discussion of wave guides and resonators of all kinds.

Born, M., and E. Wolf, *Principles of Optics*, Pergamon, New York (1959).

There is substantial detail on the classical theory of diffraction, and there are many examples of the evaluation of Fresnel and Fraunhofer patterns.

Bouwkamp, C. J., *Diffraction Theory, Reports on Progress in Physics,* XVIII, 35 (1954).

A cogent critical review of the by now century-old theory of diffraction. There is a bibliography of over 500 papers.

Clemmow, P. C., *The Plane Wave Spectrum Representation of Electromagnetic Fields*, Pergamon, Oxford (1965).

Our discussion of the diffraction by a half-plane is patterned after the treatment in this lucid work, where complete details may be found. Also see Born and Wolf above.

Harrington, R. F., *Field Computation by Moment Methods*, Macmillan, New York (1968).

Among other useful topics, the numerical solution of the integral equations for scattering problems is discussed.

Jones, D. S., *The Theory of Electromagnetism*, Macmillan, New York (1964).

There is a wide spectrum of time harmonic boundary value problems in Jones. Section 8.7 of the book discusses the transformation of the formal series solution for the problem of scattering by a cylinder to a form that is practical for computation in the short wave length limit.

Wolf, E., Some Recent Research on Diffraction of Light, in *Symposium on Modern Optics*, Polytechnic Press, Brooklyn (1967).

16
TIME HARMONIC FIELDS IN MATTER

16.1 INTRODUCTION

We have thus far bypassed the problem of fields inside matter, but it is clear that to discuss many phenomena in dielectrics and conductors, something must be known about them. Ultimately, the detailed theory of such fields is a complicated one that really is the province of quantum mechanics, statistical physics, etc. There is, however, a phenomenological and classical theory that explains many of the gross features of the experimental results in terms of simple models, and it is this we discuss. We shall treat only time harmonic fields; these are important in their own right and can, of course, be considered as one Fourier component of an arbitrary time-varying field. In Chapter 17, we synthesize the time harmonic results to discuss the propagation of an arbitrarily time dependent field in matter.

Our basic viewpoint is this: all matter contains charged particles—free electrons, bound electrons, ions, etc. An electromagnetic wave incident on a sample of matter sets the charges into motion and it becomes, in effect, a spatial distribution of currents. The different natures of the currents, continuous or discrete, oscillatory or damped, etc., translate themselves into different macroscopic properties and determine whether the matter is conducting or insulating, transparent or opaque, singly or double refracting, etc. These various macroscopic effects of the currents can, in turn, be subsumed under one heading: given certain approximations, a time harmonic wave of frequency ω can be shown to propagate in matter with an (in general) complex propagation constant k' that is different from the free space $k = \omega/c$.

The rest of this chapter is essentially an elaboration on this last sentence. In Section 16.2, we show that, if matter contains a *continuous* current distribution that satisfies a generalized Ohm's law (current proportional to field), the propagation constant is indeed changed. The next section describes simple classical models of *point* currents in various kinds of matter: conductors, dielectrics, plasmas. Then we assume that an *ensemble* of such point currents effectively constitutes a continuous distribution. This assumption is reasonable and obvious for conductors and plasmas, but it is not obviously correct for dielectrics. In the last two sections, we look at dielectrics from a more detailed point of view and try to make plausible some of the previous assumptions about them.

16.2 PROPAGATION CONSTANT OF A CURRENT-BEARING MEDIUM

We begin by considering the interior of a region in which a *continuous* time harmonic current $\boldsymbol{J}_\omega(\boldsymbol{r})$ flows. Taking the curl of the two Maxwell time harmonic curl equations (13.4a, b), it is straightforward to show, assuming there is no net charge ($\nabla \cdot \boldsymbol{E}_\omega = 0$), that the rectangular components of $\boldsymbol{E}_\omega$ and $\boldsymbol{B}_\omega$ satisfy

$$(\nabla^2 + k^2)\boldsymbol{E}_\omega = -\frac{i4\pi k\boldsymbol{J}_\omega}{c}$$
$$(\nabla^2 + k^2)\boldsymbol{B}_\omega = -\frac{4\pi\nabla \times \boldsymbol{J}_\omega}{c}. \tag{16.1}$$

Now we make the basic assumption, which might be called a *generalized Ohm's Law*, that $\boldsymbol{J}_\omega$ is proportional to $\boldsymbol{E}_\omega$ with an arbitrary proportionality constant Γ†

$$\boldsymbol{J}_\omega = \Gamma\boldsymbol{E}_\omega. \tag{16.2}$$

The constant Γ is quite general: it may be real, imaginary, or complex, corresponding to currents that are in phase, out of phase, or partially in phase with the driving field. It will, in general, be frequency dependent.

With the assumption (16.2), we have

$$\left(\nabla^2 + k^2 + \frac{i4\pi k\Gamma}{c}\right)\boldsymbol{E}_\omega = 0$$
$$\left(\nabla^2 + k^2 + \frac{i4\pi k\Gamma}{c}\right)\boldsymbol{B}_\omega = 0. \tag{16.3}$$

This is the result anticipated in the introduction: in a medium described by (16.2), the *fields propagate with a wave vector* of magnitude k', given by

$$k'^2 = k^2 + \frac{i4\pi k\Gamma}{c}. \tag{16.4}$$

The result (16.4) may be expressed in terms of a *complex index of refraction*, *n*, defined by

$$n = \frac{k'}{k}, \tag{16.5}$$

whence

$$n^2 = 1 + \frac{i4\pi\Gamma}{ck}. \tag{16.6}$$

---

† The assumption that Γ is a simple scalar implies that conditions in the medium are isotropic. This is frequently not a good assumption. If, for example, there is an applied magnetic field, the point currents will be affected anisotropically, and Eq. (16.2) will not hold. See Problem 2 at the end of this chapter.

In general, Γ is complex,

$$\Gamma = \Gamma_r + i\Gamma_i.$$

so that k' will also have a real and imaginary part,

$$k' = a + i\beta,$$

where

$$\alpha^2 = \frac{k^2 - \dfrac{4\pi k\Gamma_i}{c} + \sqrt{\left(k^2 - \dfrac{4\pi k\Gamma_i}{c}\right)^2 + \dfrac{4\pi k\Gamma_r}{c}}}{2}$$

$$\beta^2 = \frac{-\left(k^2 - \dfrac{4\pi k\Gamma_i}{c}\right) + \sqrt{\left(k^2 - \dfrac{4\pi k\Gamma_i}{c}\right)^2 + \left(\dfrac{4\pi k\Gamma_r}{c}\right)^2}}{2}. \qquad (16.7)$$

Assuming that the result of (16.4) applies to real matter, the widely-differing responses of various kinds of matter to electromagnetic fields can be qualitatively explained in terms of the various possibilities for Γ. If, for example, Γ is real, as it will be for conductors, the propagation constant will contain both a real and imaginary part; the latter leads to an attenuation of the field in a conductor, as experiment verifies. If, on the other hand, Γ is pure imaginary, as it often is for dielectrics, the propagation constant will be real, yielding a real index of refraction. Further, Γ may be complex, as it is in lossy dielectrics; or it may be real for some frequencies and imaginary for others, as will be seen in plasmas.

The problem now is to find the conditions under which (16.4) applies to real matter. We have assumed tacitly in (16.2) that Γ is a constant, i.e., characteristic of the material and independent of position in the medium. Then Eq. (16.2) is one between a current $\boldsymbol{J}_\omega$, defined everywhere in the medium, and $\boldsymbol{E}_\omega$, similarly defined. There is no difficulty in principle in interpreting this equation for conductors in which we can assume that $\boldsymbol{J}$ exists wherever $\boldsymbol{E}$ exists. For dielectrics, on the other hand, the currents are confined to atoms or molecules that are, say, localized at sites of a crystal lattice. Or the currents may be those in the atoms or molecules of a dilute gas or liquid so that, throughout most of space, there is no current. Equation (16.2) cannot hold exactly, since there may be a field $\boldsymbol{E}_\omega$ at some point between the molecules and yet $\boldsymbol{J}_\omega$ is necessarily zero there. If, for example, we think of the idealized case of point dipoles, then almost everywhere in space there is no current, almost everywhere $(\nabla^2 + k^2)\boldsymbol{E}_\omega = 0$, and, at first sight, it is not clear how (16.2) applies. However, we shall find that this equation can make sense in a certain approximation, even for dielectrics.

16.3 THE CURRENTS IN MATTER

Setting aside modern solid-state physics, we now adopt classical models of matter in which the current is due to the motion of single charged particles. This motion is, in turn, determined by the classical forces acting on the particle. The models of force are then central and we shall consider four different ones: a plasma, a conductor, a dielectric, and a lossy dielectric.

Consider a charged particle (electron) of mass m and charge e in a medium in which an electric field $\boldsymbol{E} = \boldsymbol{E}_0 e^{-i\omega t}$ is applied. The particle will be driven by this applied force subject to whatever other forces may act. First, it may be that the distribution of charges is dilute enough that there is essentially no other force acting on any electron. The electrons and the assumed positive background of charge then form a *plasma*, for which the equation of motion of the electron is

$$m\ddot{\boldsymbol{r}} = e\boldsymbol{E}_0 e^{-i\omega t} \qquad \text{Plasma.}$$

An example of a plasma is the *ionosphere* which contains electrons that are essentially free in a more or less uniform background of positive ions.

More generally, in a *conductor* there will be some force on the electron over and above the applied driving force. For example, its acceleration will be impeded by the collisions it makes with the impurities or with the vibrating lattice ions. Phenomenologically, it turns out to be reasonable to assume that the effects of these collisions are described by a *damping force* proportional to the velocity $\boldsymbol{v}$; we write this as $m\gamma\boldsymbol{v}$. The equation of motion is then

$$m\ddot{\boldsymbol{r}} + m\gamma\dot{\boldsymbol{r}} = e\boldsymbol{E}_0 e^{-i\omega t}, \qquad \text{Conductor}$$

that of a damped harmonic oscillator.

For a *pure dielectric*, we assume a model in which there is no damping but rather that the electron is bound to an origin with a certain natural frequency. This frequency is a crude counterpart of the natural frequencies (derivable in principle from quantum mechanics) of electrons bound in atoms. If we write the restoring force involving this frequency as $-m\omega_0^2\boldsymbol{r}$, the equation of motion of the electron is

$$m\ddot{\boldsymbol{r}} + m\omega_0^2\boldsymbol{r} = e\boldsymbol{E}_0 e^{-i\omega t} \qquad \text{Dielectric.}$$

Finally, we consider a *lossy dielectric* for which the equation of motion of an electron has *both* a damping term *and* a harmonic restoring force. It is

$$m\ddot{\boldsymbol{r}} + m\gamma\dot{\boldsymbol{r}} + m\omega_0^2\boldsymbol{r} = e\boldsymbol{E}_0 e^{-i\omega t} \qquad \text{Lossy dielectric.} \tag{16.8}$$

The motion of the electrons described by any of the last four equations constitutes a point current, and to discuss the effect of these currents we shall need the steady-state solutions of these equations. Consider Eq. (16.8), which represents the most general of the four models of point current; the results for the other models can then be found by setting γ, or ω_0, or both to zero. The steady-state solution of this equation is found by writing†

$$\boldsymbol{r} = \boldsymbol{r}_\omega e^{-i\omega t} \tag{16.9}$$

† The usual convention about taking the real part of complex quantities is understood.

whence

$$\boldsymbol{r}_\omega = \frac{e\boldsymbol{E}_0}{m(\omega_0^2 - \omega^2 - i\omega\gamma)}. \tag{16.10}$$

From $\boldsymbol{v}_\omega = -i\omega\boldsymbol{r}_\omega$, we have

$$\boldsymbol{v}_\omega = \frac{-i\omega e\boldsymbol{E}_0}{m(\omega_0^2 - \omega^2 - i\omega\gamma)}. \tag{16.11}$$

16.4 HEURISTIC THEORY OF PROPAGATION IN MATTER

A set of electrons, each described by one of the equations on the preceding page, constitutes an ensemble of point currents. In conductors and plasmas, it is reasonably clear that such an ensemble approximates a continuous current. Each point current corresponds to an electron trajectory. Think of a little bundle of such neighboring trajectories that pass normally through a small area A. Assuming that the velocity at this surface of each of the particles is approximately the same and that in some average sense, the trajectories pierce this surface randomly, we can speak of an *average* current density $\boldsymbol{J}_\omega$. If the number of particles per unit volume in the neighborhood of the surface is N, then in this *average* sense

$$\boldsymbol{J}_\omega = \mathrm{N}e\boldsymbol{v}_\omega. \tag{16.12}$$

As we have remarked in the introduction, it is not clear that the smoothing or averaging process implied by Eq. (16.12) is valid for the localized currents in dielectrics. Nonetheless, we shall in this section apply this assumption to dielectrics for the sake of a formal unity and because it leads to standard formulas for the index of refraction. We investigate this assumption more closely in later sections. Using (16.11) for $\boldsymbol{v}_\omega$ in (16.12) and identifying the $\boldsymbol{E}_0$ in it with the $\boldsymbol{E}_\omega$ in the generalized Ohm's law (16.2), we find for Γ

$$\Gamma = -\frac{iN\omega e^2}{m(\omega_0^2 - \omega^2 - i\omega\gamma)}. \tag{16.13}$$

We apply (16.13) first to plasmas, for which ω_0 and γ are both zero, so that we obtain from (16.4)

$$k'^2 = k^2\left(1 - \frac{4\pi Ne^2}{m\omega^2}\right). \tag{16.14}$$

This can be written as

$$k'^2 = k^2\left(1 - \frac{\omega_p^2}{\omega^2}\right), \tag{16.15}$$

where the *plasma frequency* ω_p is defined by

$$\omega_p^2 = \frac{4\pi Ne^2}{m}.$$

We see that for a wave with frequency ω, such that $\omega > \omega_p$, the propagation constant k' is real and that such a wave can propagate through the plasma. For frequencies smaller than ω_p, however, the index of refraction becomes imaginary and a wave of such frequency will be damped. This phenomenon is important in the study of the properties of the ionosphere. If a wave of a given frequency is incident on the ionospheric plasma, in which N is a function of height, the reflection properties of the ionosphere will change at that height at which $\omega^2 = \omega_p^2$. By measuring the times that electromagnetic pulses of various frequencies take to return to the transmitter, one can map out the electron density in the ionosphere.

In conductors the natural frequency ω_0 in (16.13) is zero so that we have

$$\Gamma = \frac{i\omega e^2 N}{m(\omega^2 + i\omega\gamma)}. \tag{16.16}$$

Now it is found experimentally that the damping constant γ, which obviously has dimensions of frequency, is typically of the order of 10^{17} sec^{-1}. This means that, for frequencies ω less than this, ω^2 in the denominator of Eq. (16.16) is negligible and approximately

$$\Gamma = \frac{e^2 N}{m\omega\gamma}. \tag{16.17}$$

Then the quantity Γ is purely real; for the present case of conductors, this quantity is traditionally called σ, the *conductivity*. The relation (16.2) becomes *Ohm's law*, in its usual form

$$\boldsymbol{J}_\omega = \sigma \boldsymbol{E}_\omega.$$

We have effectively derived this law, a special case of the postulated generalized Ohm's law, on the basis of a simple model. This crude derivation embodies *Drude's model of conductivity*.

In terms of the conventional notation σ, the propagation constant k' is given by

$$k'^2 = \frac{\omega^2}{c^2} + \frac{i4\pi\sigma\omega}{c^2}. \tag{16.18}$$

Consider now the orders of magnitude of the two terms on the right-hand side of (16.18). For good conductors like metals, σ which has dimensions sec^{-1}, is typically of order 10^{17} sec^{-1}. This means that $4\pi\sigma\omega/c^2$ in (16.18) is much larger than ω^2/c^2 for all wavelengths down to infrared wavelengths. For such cases, we can drop $(\omega/c)^2$ from the right-hand side. Tracing through the derivation of (16.18) shows that this approximation is equivalent to neglecting the displacement current in a metal with respect to the conduction current. With $\sqrt{i} = (1 + i)/\sqrt{2}$. we now have

$$k' = (1 + i)/\delta, \tag{16.19}$$

where the *skin depth* δ is defined by

$$\delta = \frac{c}{\sqrt{2\pi\sigma\omega}}. \tag{16.20}$$

The fields $\boldsymbol{E}_\omega$ and $\boldsymbol{B}_\omega$ now satisfy

$$(\nabla^2 + k'^2)\begin{pmatrix} \boldsymbol{E}_\omega \\ \boldsymbol{B}_\omega \end{pmatrix} = 0 \tag{16.21}$$

in a conductor. To consider briefly some features of the solutions of these equations, we take the simplest possible geometry wherein the conductor occupies the half-space $z > 0$ and assume that the field components do not depend on x and y. Letting F_ω denote any of the Cartesian components of $\boldsymbol{E}_\omega$ or $\boldsymbol{B}_\omega$, we have

$$\frac{d^2F_\omega}{dz^2} + k'^2F_\omega = 0.$$

The solution of this equation for a wave propagating into the conductor is

$$F_\omega = e^{ik'z} = e^{iz/\delta}e^{-z/\delta}. \tag{16.22}$$

The term $e^{-z/\delta}$ shows that this wave, of unit amplitude at $z = 0$, falls off to an amplitude of $1/e$ in a distance δ. The fields, and hence the currents, are then effectively confined to a thin skin near the surface, of order of magnitude δ. Although we have derived these results for an infinite plane boundary, it is clear they hold for a curved boundary provided that the radii of curvature are large compared to the penetration depth. The fact of zero-depth of field penetration in the limit of perfect conductivity is one that was anticipated in Chapter 15 on perfect conductors.

For dielectrics, we shall simply apply Eq. (16.12) formally; a closer discussion of the properties of dielectrics is given in the next two sections. From (16.4) and (16.13), we have

$$k'^2 = k^2\left(1 + \frac{4\pi Ne^2}{m(\omega_0^2 - \omega^2 - i\omega\gamma)}\right). \tag{16.23}$$

If not all electrons are bound with the same frequency, but there are N_l electrons per unit volume having frequency ω_l and damping constant γ_l, the same assumptions that led to the last equation yield

$$k'^2 = k^2\left(1 + \frac{4\pi e^2}{m}\sum_l \frac{N_l}{\omega_l^2 - \omega^2 - i\omega\gamma_l}\right). \tag{16.24}$$

The index of refraction defined by (16.5) is then

$$n^2 = 1 + \frac{4\pi e^2}{m}\sum_l \frac{N_l}{\omega_l^2 - \omega^2 - i\omega\gamma_l}. \tag{16.25}$$

This last formula is a standard one in the theory of dielectrics, although it is usually derived somewhat differently. Although the derivation as it stands is dubious, (16.25) does seem to provide a reasonable quasi-empirical formula that fits the properties of many dielectrics.

16.5 ONE-DIMENSIONAL DIELECTRIC MODELS

In this section, we consider more detailed models, mainly one-dimensional, of dielectrics, both to clarify the physics that is involved and to provide some theoretical justification for the assumptions of the last section.

The general picture of a dielectric is that of an ensemble of discrete elements that can be made to radiate under the action of an incident field. In natural crystalline dielectrics the elements are atoms or molecules; in the *artificial* dielectrics that are used in microwave applications the elements are generally metallic radiators. These elements may be arrayed in periodic lattices or, as in gaseous or liquid dielectrics, more or less at random. For natural dielectrics, the induced currents are sometimes called *polarization currents* since they are due to the motion of electrons that are bound to the atoms or molecules, just as *polarization charge* is a displacement of bound charge.

In principle, the propagation properties of a periodic dielectric could be calculated, given the lattice structure and the details of the individual scatterers. The problem is similar to the so-called *band structure problem* of solid-state theory, which is that of calculating the energy bands of periodic atomic lattices. However, the electromagnetic problem is more difficult because of the vectorial nature of the electromagnetic field. It is then important to point out that the *existence* of the propagation constant can be shown to be a consequence of quite general arguments. That is, the general theory of wave propagation in a periodic lattice shows that the wave amplitude must assume a special form which can be considered to describe propagation with a constant $\boldsymbol{k}'$ different from the free space constant. Consider, for example, such an amplitude ψ, which may correspond to a Schrödinger amplitude in the presence of a lattice of potentials, or the pressure amplitude of an acoustic wave in a lattice of sound scatterers, or one component of a time harmonic electric field in a lattice of dipoles. The general theory we have mentioned shows that, for a three-dimensional lattice, ψ has the form

$$\psi(\boldsymbol{r}) = e^{i\boldsymbol{k}'\cdot\boldsymbol{r}}u_{k'}(\boldsymbol{r}), \tag{16.26}$$

where k' is the *propagation constant* and $u_{k'}$ is a function with the *same periodicity as the lattice.* This result is known as *Floquet's theorem.*†

† This theorem can perhaps be made plausible by the following argument. Suppose, for definiteness, that ψ were a Schrödinger amplitude whose squared magnitude represented the position probability of a particle. For a periodic lattice, we would expect this probability to be periodic and this will be the case if the amplitude itself is the product of a periodic function factor $e^{i\boldsymbol{k}'\cdot\boldsymbol{r}}$ (which disappears on taking the absolute magnitude). Floquet's theorem is proven in many references. See, for example, Collin (R).

The proof of the existence of $\boldsymbol{k}'$ is one matter; the problem of actually calculating $\boldsymbol{k}'$ for a given lattice is quite another. And the problem of calculating $\boldsymbol{k}'$ for a random collection of scatterers (as in a gaseous dielectric) is even more difficult since, in that case, even the positions of the scatterers are not known, except statistically.

Happily, we are not concerned with the details of such calculations; our aim is simply to illustrate the physics of the refractive index by simple models. To this end, we consider first a one-dimensional model. To motivate our discussion, consider the problem of wave propagation in the lattice shown in Fig. 16.1. This represents a simple cubic array, with lattice constant a, of dipolar radiating elements oriented in the x-direction. Suppose now that the dipoles in each of the planes parallel to the x-axis are divided into two, that each of the two resultant dipoles are positioned somewhere in the plane, and that these two are further divided and repositioned, etc., much as continuous dipolar distributions were generated in electrostatics. The plane becomes a kind of one-dimensional dipolar current sheet and the array of sheets then models the original lattice for propagation perpendicular to them, along the z-axis.

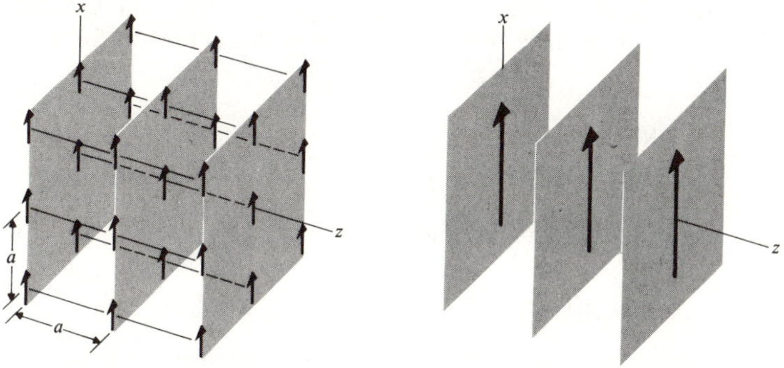

Fig. 16.1 Simple cubic lattice and idealized one-dimensional model.

We examine the propagation of the fields in this idealized lattice. Since the current is in the x-direction, we shall consider only the x-component of $\boldsymbol{E}$. For convenience, *we drop the subscripts and simply write* E *for* $E_{\omega x}$ *and* J *for* $J_{\omega x}$. Then, from Eq. (16.1), E satisfies

$$\left(\frac{d^2}{dz^2} + k^2\right) E(z) = -\frac{i4\pi k J(z)}{c}. \tag{16.27}$$

Suppose that, in a given current sheet, J is proportional to E according to Eq. (16.2). For the present case of lossless dielectrics, we have from (16.13) for Γ

$$\Gamma = -\frac{iN\omega e^2}{m(\omega_0^2 - \omega^2)}.$$

Recognizing that the current $J(z)$ exists only in the planes $z = 0, \pm a, \pm 2a, \ldots$, we can write it as

$$J(z) = \Gamma E(z) a \sum_{m=-\infty}^{\infty} \delta(z - ma)$$

so that Eq. (16.27) becomes†

$$\left(\frac{d^2}{dz^2} + k^2\right) E = \beta E \sum_{m=-\infty}^{\infty} \delta(z - ma), \tag{16.28}$$

where

$$\beta = -\frac{4\pi k a}{c} \frac{N\omega e^2}{m(\omega_0^2 - \omega^2)}. \tag{16.29}$$

Let the region between 0 and a be denoted by $\oplus$ and that between $-a$ and 0 by $\ominus$. By Floquet's theorem,

$$E(z) = e^{ik'z} u(z), \tag{16.30}$$

where $u(z)$ is periodic of period a. Since in $\oplus$ $E(z)$ satisfies $d^2E/dz^2 + k^2E = 0$, the solution in that region is $Ae^{ikz} + Be^{-ikz}$. We recast this in Floquet form:

$$E_+(z) = e^{ik'z}(Ae^{i(k-k')z} + Be^{-i(k+k')z}). \tag{16.31}$$

Comparing Eqs. (16.30) and (16.31), we have

$$u_+(z) = Ae^{i(k-k')z} + Be^{-i(k+k')z}, \qquad 0 < z < a.$$

The solution u_- in $\ominus$ is, from the periodicity of u, the same function of $z + a$ that u_+ is of z. In $\ominus$ then

$$E_-(z) = e^{ik'z}(Ae^{i(k-k')(z+a)} + Be^{-i(k+k')(z+a)}), \qquad -a < z < 0. \tag{16.32}$$

As a first boundary condition, we require that E be continuous through the origin

$$E_+(0) = E_-(0). \tag{16.33}$$

We integrate Eq. (16.28) over the δ-function at the origin to get the second boundary condition

$$\left.\frac{dE_+}{dz}\right|_0 - \left.\frac{dE_-}{dz}\right|_0 = \beta E(0). \tag{16.34}$$

Applying these two conditions, we find that

$$A + B = Ae^{i(k-k')a} + Be^{-i(k+k')a}$$

$$ik(1 - e^{i(k-k')a})A - ik(1 - e^{-i(k+k')a})B = \beta(A + B).$$

† The analogous equation (model) for the propagation of Schrödinger waves in a periodic lattice of δ-function potentials is called the *Krönig-Penney model.*

Setting the determinant of these equations to zero gives, after some algebra, a transcendental equation that determines k'

$$\cos k'a = \cos ka + \frac{\beta a}{2} \frac{\sin ka}{ka}. \tag{16.35}$$

This equation for k' as a function of $\omega = kc$ is the simplest example of all the equations which exist in principle for more complicated lattices but which cannot be calculated in practice.

The form of (16.35) in the limit of weak scattering ($k' \approx k$) is interesting. Expanding it to lowest order in β, we find that

$$k' = k - \frac{\beta}{2ka}. \tag{16.36}$$

With β from (16.29), this becomes

$$k' = k + \frac{2\pi N k e^2}{m(\omega_0^2 - \omega^2)}$$

which is just the value yielded by Eq. (16.23) for $k' \approx k$ and $\gamma = 0$.

There is another illuminating way of calculating the refractive index of the one-dimensional model of current sheets. To introduce it, first consider the problem of a single such sheet at the origin. In its presence, $E(z)$ satisfies

$$\left(\frac{d^2}{dz^2} + k\right) E(z) = \beta E \delta(z).$$

If a wave e^{ikz} is incident on the sheet from the left, the (idealized) dipoles in the sheet will be set into forced vibration. In addition to the incident wave e^{ikz}, there will then be a reflected wave Re^{-ikz} to the left of the sheet and a transmitted wave Te^{ikz} traveling to the right,

$$E(z) = e^{ikz} + Re^{-ikz}, \qquad z < 0;$$
$$E(z) = Te^{ikz}, \qquad z > 0.$$

Applying the two boundary conditions (16.33) and (16.34), we find that the reflection and transmission coefficients are

$$R = \frac{\beta}{2ik - \beta}$$
$$T = \frac{2ik}{2ik - \beta}. \tag{16.37}$$

Consider now a finite one-dimensional array consisting of $n + 1$ current sheets, at the origin, and at $z = a, 2a, \ldots, na$. Let a wave e^{ikz} be incident from the left on the first sheet. It will produce a transmitted wave Te^{ikz} in the region $0 < z < a$. This transmitted wave, on passing through the second sheet, will also be multiplied by T so the wave amplitude for $a < z < 2a$ will be $T^2 e^{ikz}$. We here neglect the reflected wave which returns from the second sheet and is reflected back again from the first. Each time the wave passes through a current sheet, its amplitude is multiplied by T, so that after the last sheet the transmitted wave will be approximately $T^{n+1} e^{ikz}$. From Eq. (16.37), with $n + 1 \approx n$, we have

$$T^{n+1} = \left(1 + \frac{i\beta}{2k}\right)^{-n-1} \approx \left(1 - \frac{in\beta}{2k}\right) \approx e^{-(in\beta/2k)}.$$

The wave at na, i.e., just to the right of the $(n + 1)$th sheet can then be written as

$$e^{ikna} e^{-(in\beta/2k)} = e^{i(k-(\beta/2ka))na}. \tag{16.38}$$

Suppose that we could consider that the propagation constant for this lattice were k'. An incident wave, e^{ikz}, of unit amplitude at $z = 0$ would then become the wave $e^{ik'na}$ after the $(n + 1)$th scatterer. Comparing this result with Eq. (16.38), we find that $k' = k - (\beta/2ka)$ in this approximation—which is just the result (16.36).

There is a moral to be drawn here. Consider a model of a random dielectric consisting of $n + 1$ scatterers that are distributed in any manner over a length na so that *only on the average* is the spacing a. The present derivation of the approximate formula (16.36) does not use the fact that the potentials are uniformly spaced but only that there are $n + 1$ scatterers occupying a length na. In this limit of weak *scattering, a continuous or smeared-out distribution gives the same result as a discrete one.* We made a similar assumption in applying (16.12) to dielectrics in three dimensions and the present results perhaps provide some small justification for it.

16.6 THREE-DIMENSIONAL DIELECTRIC MODELS

The gap between the one-dimensional dielectric models of the last section and three-dimensional models that involve the vectorial electromagnetic field is a large one. In the first part of this section, we partially bridge it by discussing scattering models of what might be called *scalar dielectrics* in a spirit similar to the one used in our discussion of the one-dimensional scattering models. That is, we assume an ensemble of scatterers for a scalar wave amplitude ψ that may represent the pressure amplitude in an acoustic wave, a Schrödinger probability amplitude, etc. An incident wave falling on the scatterers induces radiation from each; the superposition of incident and induced fields appears to be one that propagates through the medium with a propagation constant k' different from the free space $k = \omega/c$.

In free space, we assume that the amplitude ψ satisfies the Helmholtz equation $(\nabla^2 + k^2)\psi = 0$. If a plane wave e^{ikz} is incident on one of the scatterers of the ensemble, we shall assume that it generates a spherical wave ae^{ikr}/r (S-wave scattering) at a distance r. The quantity a is called the *scattering amplitude.*

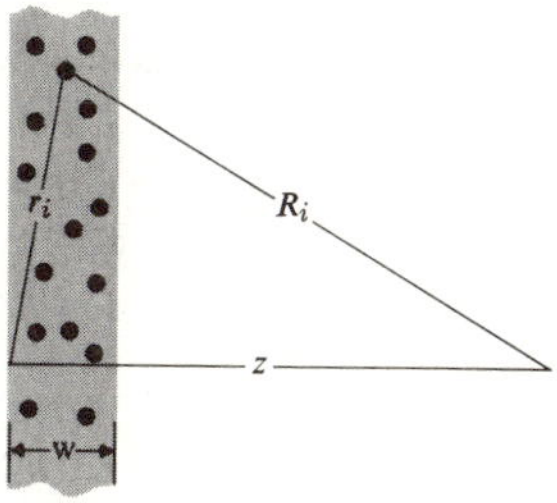

Fig. 16.2 Slab-shaped ensemble of scatterers.

We shall discuss two different derivations for the index of an ensemble of such scatterers. In the first, consider a slab-shaped ensemble that has a wave e^{ikz} incident from the left, as shown in Fig. 16.2. We assume that the scattering is weak, i.e., that the wave incident on any scatterer is essentially the incident wave. We neglect thereby those waves which impinge on a scatterer having been previously scattered, much as we neglected the reflected and re-reflected waves in the one-dimensional model of the last section. With this assumption, the scatterer at $\boldsymbol{r}_i$ in Fig. 16.2 emits a wave whose magnitude at the field point z is ae^{ikR_i}/R_i. The total wave amplitude at z is

$$\psi = e^{ikz} + a\sum_i \frac{e^{ikR_i}}{R_i}.$$

We now suppose that the wavelength is long compared to the interparticle spacing so that the difference between e^{ikR_i} and e^{ikR_j} is small if i and j refer to neighboring scatterers. Then the sum can be replaced by an integral. If R is the distance from the scatterers in a small volume dv to the field point, the wave amplitude at z, including the incident wave, is

$$\psi = e^{ikz} + a\int \frac{e^{ikR}}{R} N\, dv,$$

where N is the (uniform) density of scatterers. Suppose that the thickness w of the slab is small enough that for the great bulk of the scatterers the distance r to the scatterers in dv is essentially a cylindrical coordinate, so that $dv = 2\pi wrdr$. From $r^2 + z^2 = R^2$, we have $r\,dr = R\,dR$ and $dv = 2\pi wR\,dR$. Then

$$\psi = e^{ikz} + 2\pi Naw \int_z^\infty e^{ikR}\, dR. \tag{16.39}$$

The integral diverges at the upper limit, but we argue that this divergence comes from the unphysical assumption of a uniform density of scatterers that extends to infinity. One can, for example, imagine the density falling off at infinity as $e^{-\varepsilon|z|}$, where ε is arbitrarily small, and this makes the contribution of the upper limit vanish. Neglecting the upper limit then, Eq. (16.39) gives for the field at $z = w$

$$\psi = e^{ikw} + \frac{i2\pi Naw}{k} e^{ikw} \approx e^{ikw(1+(2\pi Na/k^2))} \tag{16.40}$$

If the incident field e^{ikz}, which has unit amplitude at the origin, had propagated through the slab with propagation vector k', the field at $z = w$ would be

$$\psi = e^{ik'w}. \tag{16.41}$$

Comparing (16.40) with (16.41), we have

$$k' = k\left(1 + \frac{2\pi Na}{k^2}\right), \tag{16.42}$$

the analog of the one-dimensional result (16.36). The weakness and looseness of this derivation are as obvious to the author as to the reader, but it does give a feeling for the physics that is involved.

For the second derivation of (16.41), we again consider point scatterers located at $\boldsymbol{r}_1, \boldsymbol{r}_2, \ldots, \boldsymbol{r}_n$, but do not assume in this instance that the ensemble necessarily forms a slab-shaped region. We again imagine the elements scattering under the influence of an incident wave e^{ikz} but do not require that the scattering be weak. More realistically, we suppose that the scattered wave produced by the ith particle is proportional to the scattering amplitude a and to the total wave amplitude $\psi(\boldsymbol{r}_i)$, at that point. This amplitude is that due to the incident wave plus that due to the other scatterers. We assume, however, that the scatterers are far enough apart that the spherical wave emitted by one is essentially a plane wave when it reaches a neighboring scatterer. This assumption is required by the use of the scattering amplitude a, since it is defined with respect to an incident plane wave. With these assumptions, the amplitude $\psi(\boldsymbol{r})$ at some field point $\boldsymbol{r}$ is

$$\psi(\boldsymbol{r}) = e^{ikz} + a \sum_i \psi(\boldsymbol{r}_i) \frac{e^{ik|\boldsymbol{r}-\boldsymbol{r}_i|}}{|\boldsymbol{r}-\boldsymbol{r}_i|}. \tag{16.43}$$

Again we shall replace the sum in (16.43) by an integral using the assumption that the discrete distribution can be replaced by a smeared out or continuous one. This might correspond to a model in which the atoms are not fixed but are free to move throughout the volume with some probability distribution. Then $\boldsymbol{r}_i$ is replaced by a continuous variable $\boldsymbol{r}'$ and the contribution of the atom at $\boldsymbol{r}_i$ by the contribution of the $N(\boldsymbol{r}')\,dv'$ atoms in the volume dv' The amplitude at $\boldsymbol{r}$ becomes

$$\psi(\boldsymbol{r}) = e^{ikz} + a \int \frac{N(\boldsymbol{r}')\psi(\boldsymbol{r}')e^{ik|\boldsymbol{r}-\boldsymbol{r}'|}}{|\boldsymbol{r}-\boldsymbol{r}'|}\,dv'. \tag{16.44}$$

Operating on Eq. (16.44) with $(\nabla^2 + k^2)$ yields a δ-function in the integrand according to Eq. (13.14) and the resulting equation can be written as

$$\left(\nabla^2 + k^2\left(1 + \frac{4\pi Na}{k^2}\right)\right)\psi = 0.$$

Effectively, it is an equation for the propagation of a wave with wave number k' given by

$$k'^2 = k^2\left(1 + \frac{4\pi Na}{k^2}\right). \tag{16.45}$$

If, in fact, the scattering is weak ($4\pi Na \ll k^2$), this equation becomes approximately Eq. (16.42). In this limit, then, the two more or less distinct derivations, one of which assumes weak scattering and one of which assumes the replacement of a discrete distribution by a continuous one, are equivalent.

The general scattering picture above is equally valid in principle as a physical description of the origin of the refractive index for electromagnetic waves. But in detail the picture is more complicated, partly because the electromagnetic field is a vectorial one and partly (and concomitantly) because the scattered fields of the individual elements are more complex. As we have seen, the field of a time harmonic dipole is made up of three parts: the quasi-static, induction, and radiation fields. Realistically, the field that acts on any scatterer, the *effective field*, is (as it was for the scalar case) the incident field plus that due to all the other scatterers. But the complex structure of the dipole field makes the field of the other scatterers even more difficult to calculate than for the scalar case where it was already intractable. The calculation of the index of refraction from first principles has then all the difficulties of calculating the dielectric constant from first principles in electrostatics and of similarly calculating the index of refraction for scalar waves. And even this is under the assumption that the individual scatterers are pure dipoles, an assumption that is not always verified.

The heuristic theory of currents in dielectrics in Section 16.4 yielded Eq. (16.25), a standard formula. This formula is, however, usually derived in terms of a time dependent polarization, and the derivation is then different from the present one. We discuss the conventional theory now so that we can relate the present treatment to it.

Our model of a dielectric has been that of an ensemble of charged particles, each bound to some origin by harmonic forces. If the particle whose position is $\boldsymbol{r}$ is displaced time harmonically with respect to that origin, it not only constitutes a point current, but also constitutes an instantaneous dipole moment $\boldsymbol{p}_\omega$

$$\boldsymbol{p}_\omega = e\boldsymbol{r}_\omega. \tag{16.46}$$

If we use Eq. (16.10) for $\boldsymbol{r}_\omega$, then

$$\boldsymbol{p}_\omega = \frac{e^2\boldsymbol{E}_0}{m(\omega_0^2 - \omega^2 - i\omega\gamma)}. \tag{16.47}$$

In this equation, $\boldsymbol{E}_0$ is, of course, the total field acting on a dipole and, for a single dipole in space, it is just the applied field $\boldsymbol{E}_a$. But for a dipole in a lattice, $\boldsymbol{E}_0$ must be taken to be the applied field *plus* the field due to all the other dipoles and, as it was in electrostatics, this field is impossible to calculate in general. One can (and people frequently do) multiply (16.47) by N, the particle density, to get an equation with $\boldsymbol{P}_\omega$, the *volume density* of dipole moment on the left-hand side. The equation which results is meaningless, since the $\boldsymbol{E}_0$ that appears in it is not well defined.

There is one case when $\boldsymbol{E}_0$ is approximately known. For very dilute dielectrics, $\boldsymbol{E}_0$ is obviously just the applied field $\boldsymbol{E}_a$. We have then (we take the case $\gamma = 0$)

$$\boldsymbol{P}_\omega = \frac{e^2 \boldsymbol{E}_a N}{m(\omega_0^2 - \omega^2)}. \tag{16.48}$$

Consider (16.48) in the low frequency limit, $\omega \to 0$. Then $\boldsymbol{P}_\omega$ just becomes the static polarization density $\boldsymbol{P}$ discussed in Chapter 6. From the equation

$$\boldsymbol{P} = \chi_e \bar{\boldsymbol{E}} = \chi_e(\boldsymbol{E}_a + \boldsymbol{E}_d) \approx \chi_e \boldsymbol{E}_a,$$

we have $\chi_e = e^2 N/(m(\omega_0^2 - \omega^2))$. This equation really makes sense only for $\omega = 0$, but for formal reasons, we have kept the explicit ω dependence. Since the dielectric constant ε is given by $\varepsilon = 1 + 4\pi\chi_e$, we have

$$\varepsilon = 1 + \frac{4\pi e^2 N}{m(\omega_0^2 - \omega^2)}.$$

Comparing this with Eq. (16.25) for n^2, in which we take $\gamma_l = 0$ and one frequency ω_0, we find that

$$\varepsilon = n^2. \tag{16.49}$$

This is a well-known equation usually called *Maxwell's relation.* Our derivation of it is not the usual one. More commonly, one starts from the set of Maxwell equations that involves $\boldsymbol{D}$ and $\boldsymbol{H}$ and, by formal manipulations and the assumption of certain so-called *constitutive relations* between $\boldsymbol{D}$ and $\boldsymbol{E}$, and $\boldsymbol{H}$ and $\boldsymbol{B}$, arrives at (16.49). But this formal theory throws no light on the underlying physics. The interpretation of the Maxwell relation is rather peculiar in that the dielectric constant refers to a static field; the index of refraction necessarily refers to a time harmonic one. The meaning of this equation then is that it is an approximation valid for those low frequencies for which the frequency dependence of the index is small. In fact, Eq. (16.49) is satisfied by some substances, but it is far from universally correct.

There is a generalization of (16.49) that is known as the *Lorenz-Lorentz relation.* It can be derived by assuming that the effective field that acts on an individual time-harmonic dipole can be calculated much as the effective field acting on a static dipole was calculated in Section 6.6. But it is difficult to see how the static calculation, which is limited to lattice sites of a cubic lattice and takes into account only the near or quasi-static part of the electric dipole field, is relevant to the present more general problem.

There is one fundamental problem that remains. We have considered above the fields produced by time harmonic sources that are *external* to matter and shown that these fields propagate inside matter with a modified propagation constant. We have not, however, considered the problem of calculating the fields of sources (charges and currents) that are embedded *in* the matter itself. We recall that the similar problem in electrostatics had a relatively simple solution. The field of sources embedded in an (infinite) dielectric medium was the same as the field produced in free space, but was divided everywhere by the dielectric constant. Unfortunately, there is no real reason for this same result to apply to time harmonic dipolar matter since the structure of the time harmonic dipolar field is quite different from the static one. If Maxwell's relationship or its Lorenz-Lorentz generalization were generally valid, they would provide an answer, assuming that the dielectric constant in them still describes the effective reduction of the strength of charges in time harmonic dipolar media. There is no sound physical basis for making this assumption, and indeed one knows that Maxwell's relation is not even generally correct; nonetheless, the assumption is frequently made for lack of a better one. The problem of calculating the electrical properties of a realistic dielectric from first principles is thus still an open one.

PROBLEMS

1. Prove Eq. (16.1), relating $\boldsymbol{E}_\omega$ and $\boldsymbol{B}_\omega$ to their source $\boldsymbol{J}_\omega$.

2. As a generalization of $\boldsymbol{J}_\omega = \Gamma \boldsymbol{E}_\omega$ consider an *anisotropic medium* in which $\boldsymbol{J}_\omega$ is related to $\boldsymbol{E}_\omega$ by the *dyadic of generalized conductivity* $\boldsymbol{\Gamma} = \Gamma_{xx}\hat{\boldsymbol{x}}\hat{\boldsymbol{x}} + \Gamma_{xy}\hat{\boldsymbol{x}}\hat{\boldsymbol{y}} + \cdots \Gamma_{zz}\hat{\boldsymbol{z}}\hat{\boldsymbol{z}}$, according to the law $\boldsymbol{J}_\omega = \boldsymbol{\Gamma} \cdot \boldsymbol{E}_\omega$. Show that

$$\nabla^2 \boldsymbol{E}_\omega + k^2 \boldsymbol{E}_\omega + \frac{ik4\pi\boldsymbol{\Gamma} \cdot \boldsymbol{E}_\omega}{c} = 0$$

$$\nabla^2 \boldsymbol{B}_\omega + k^2 \boldsymbol{B}_\omega + \frac{ik4\pi\boldsymbol{\Gamma} \cdot \boldsymbol{B}_\omega}{c} = 0.$$

These equations are the basis for the study of *double refraction* or *birefringence.* If, for example, one assumes a plane wave solution of the form $\boldsymbol{E}_\omega = \boldsymbol{A}e^{i\boldsymbol{k}' \cdot \boldsymbol{r}}$ with unknown propagation constant $\boldsymbol{k}'$ the equation for $\boldsymbol{E}_\omega$ becomes a homogeneous set for the components A_x, A_y, A_z. The vanishing of the determinant of this set can be shown to yield *two* physically meaningful values of the *magnitude* of $\boldsymbol{k}$ for a given *direction* of propagation.

3. An important case of an anisotropic medium is a dilute plasma in an applied magnetic field (the ionosphere is an example), for which the equation of motion of an electron is

$$m(d\boldsymbol{v}/dt) = e\boldsymbol{E} + (e\boldsymbol{v} \times \boldsymbol{B}/c).$$

Suppose that $\boldsymbol{B}$ is a static field in the z-direction, $\boldsymbol{B} = B_0\hat{z}$, and that $\boldsymbol{E}$ is time harmonic, $\boldsymbol{E} = \boldsymbol{E}_\omega e^{-i\omega t}$, so that we can assume $\boldsymbol{v} = \boldsymbol{v}_\omega e^{-i\omega t}$. Solve the equation of motion for the steady state to find, with N as the number density of electrons and $\omega_c = eB_0/mc$,

$$J_{\omega x} = Nev_{\omega x} = \frac{ie^2 N\omega}{m(\omega^2 - \omega_c^2)}\left(E_{\omega x} + \frac{i\omega_c}{\omega} E_{\omega y}\right)$$

$$J_{\omega y} = Nev_{\omega y} = \frac{ie^2 N\omega}{m(\omega^2 - \omega_c^2)}\left(-\frac{i\omega_c}{\omega} E_{\omega x} + E_{\omega y}\right)$$

$$J_{\omega z} = Nev_{\omega z} = \frac{ie^2 N}{m\omega} E_{\omega z}.$$

4. Read off from the results of Problem 3 above the components of the conductivity dyadic $\mathbf{\Gamma}$ defined by $\boldsymbol{J}_\omega = \mathbf{\Gamma} \cdot \boldsymbol{E}_\omega$. Use these components and the result of Problem 2 to show that a wave propagating along the z-direction (the direction of the magnetic field) has the two values of the propagation constant

$$k'^2 = k^2\left(1 - \frac{4\pi N e^2}{m\omega(\omega \pm \omega_c)}\right).$$

5. Following the general outline of the last problem, find the propagation constant for a wave propagating in a plasma at an angle θ to the applied magnetic field. The resulting formula is a special case of the *Appleton-Hartree formula*, important in the theory of ionospheric propagation. See Clemmow and Dougherty (B) for more detail.

6. Because of the periodicities involved in it, Eq. (16.28) for the one-dimensional dielectric model can be usefully treated by using Fourier series. With $E(z) = e^{ik'z}u(z)$ the periodic $u(z)$ can be expanded as $u(z) = \sum_{l=-\infty}^{\infty} A_l e^{il2\pi z/a}$. Recognizing that $\sum_{m=-\infty}^{\infty} \delta(z - ma)$ is a periodic function, show that the relation between k' and k (the dispersion relation) is given by

$$1 = \frac{\beta}{a} \sum_{l=-\infty}^{\infty} \frac{1}{k^2 - (k' + 2\pi l/a)^2}.$$

Verify that this form is equivalent to (16.35).

REFERENCES

Brillouin, L., *Wave Propagation in Periodic Structures*, 2d ed, Dover, New York (1953).

The lucidity of Brillouin's work has made it a classic. It is still one of the best treatments of the subject although the waves referred to in the title are not electromagnetic ones.

Collin, R. C., *Field Theory of Guided Waves*, McGraw-Hill, New York (1960).

Chapter 12 is a clear treatment of artificial dielectrics, i.e., of wave propagation in man-made periodic structures, but the general viewpoint is equally applicable to natural crystalline dielectrics.

Ewald, P. P., "Crystal Optics for Visible Light and X Rays," *Rev. Mod. Phys.* **37**, 46 (1965).

Ewald was a pioneer in relating the macroscopic property of refraction to the radiation of scattered waves from dipoles in a lattice, at a time (1911) when the very concepts of a spatial lattice and the periodic structure of crystals were still highly tentative. This review article lucidly summarizes the physics involved in the work of a lifetime.

Twersky, V., "Multiple Scattering of Electromagnetic Waves by Arbitrary Configurations," *J. Math. Phys.* **8,** 589, (1967).

Twersky has written widely on the subject of multiple scattering of waves from an ensemble of scatterers, and the conditions under which such an ensemble acts as a continuous medium, with a propagation constant modified from the free space one. The paper referred to here is rather formal, mainly because of the vectorial nature of the electromagnetic field, but it has reference to earlier papers on scalar scattering that are very illuminating physically. One such is:

Twersky, V., "Multiple Scattering of Waves and Optical Phenomena," *J. Opt. Soc. America,* **52,** #2, 145 (1962).

17 PROPAGATION IN IMPERFECT CONDUCTORS AND DIELECTRICS

In this chapter, we discuss problems that are connected with the fact that a harmonic wave in a medium propagates with a frequency dependent propagation constant. For a single unbounded medium, one such problem is that of the change of shape of a wave packet, or pulse, as it propagates. In space containing two different media with boundaries, there is the problem of the penetration into, and reflection from, the second medium of a wave that originates in the first, and of waves that may be guided along the boundary.

17.1 WAVES IN DISPERSIVE MEDIA

We have seen in Chapter 16 that a time harmonic plane wave of frequency ω propagates in a material medium with a frequency dependent propagation constant k' that is different from the free space constant $k = \omega/c$. To put it differently, the velocity of a harmonic wave in matter varies with the frequency ω. This phenomenon is called *dispersion* and a medium that exhibits it is said to be dispersive.† There are other examples of what may be considered as dispersive media. Thus, in a wave guide, a given mode propagates down the guide with a propagation constant $k_g(\omega)$ whose frequency dependence is different, in general, from that in free space. Now the harmonic time dependence of such modes, or of plane waves, entails a time variation that is the same in the remote past as it will be in the distant future. Such idealized time dependence does not exist in nature. Any electromagnetic field is generated at some time and ceases to exist at another. An important question then is that of the propagation of an electromagnetic pulse with *arbitrary* time (and space) dependences. We discuss this now.

We begin by reviewing the problem of a pulse, or wave of a general shape, in free space. This is almost trivial since such a pulse travels with unchanged shape at velocity c. For example, for propagation along the z-axis, any rectangular component of $\boldsymbol{E}$ and $\boldsymbol{B}$, call it F, satisfies the wave equation

$$\frac{\partial^2 F}{\partial z^2} - \frac{1}{c^2}\frac{\partial^2 F}{\partial t^2} = 0. \tag{17.1}$$

† The origins of the terms dispersive and dispersion are to be found in the fact that if a beam of white light is incident on a refracting medium (a prism, for example), the beam will be spread out or *dispersed* into rays of different color on emerging from that medium.

The general solution of (17.1) is, as we have found,

$$F(z, t) = f(z - ct) + g(z + ct), \tag{17.2}$$

where f and g are arbitrary functions. Equation (17.2) represents a wave of arbitrary shape traveling to the right with velocity c and another arbitrary one traveling to the left. It will be useful here to derive this result anew by Fourier analysis. Writing

$$F(z, t) = \int_{-\infty}^{\infty} F_\omega(z)e^{-i\omega t}\,d\omega, \tag{17.3}$$

we see from (17.1) that F_ω satisfies, with $k = \omega/c$,

$$\frac{d^2F_\omega}{dz^2} + k^2F_\omega = 0. \tag{17.4}$$

The general solution of (17.4) is

$$F_\omega = \frac{1}{c}(Ae^{ikz} + Be^{-ikz}), \tag{17.5}$$

where A and B are to be considered as functions of ω (or of k) and where the factor $1/c$ is introduced for later convenience. Putting (17.5) into (17.3) and changing the integral to one over k instead of ω yields

$$F(z, t) = \int_{-\infty}^{\infty} (A(k)e^{ik(z-ct)} + B(k)e^{-ik(z+ct)})\,dk. \tag{17.6}$$

This result is, of course, simply the representation by Fourier integrals of the two general functions $f(x - ct)$ and $g(x + ct)$ of Eq. (17.2).

An arbitrary pulse $F(z, t)$ in a dispersive medium can be Fourier analyzed as in (17.3). Then F_ω satisfies (17.4) except that in it k must now be considered a general function of ω: $k = k(\omega)$. With this understanding, F_ω is given by a linear combination of $e^{ik(\omega)z}$ and $e^{-ik(\omega)z}$. We put such a combination back into (17.3), consider that the ω vs. k relation is inverted so that ω is a function of k: $\omega = \omega(k)$, and integrate over k. Then $F(z, t)$ can be written as

$$F(z, t) = \int_{-\infty}^{\infty} (C(k)e^{i(kz-\omega(k)t)} + D(k)e^{-i(kz+\omega(k)t)})\,dk. \tag{17.7}$$

In general, there is now a *qualitative* difference between (17.6) and (17.7). The latter equation becomes the former if $\omega = kc$, whereupon the time dependence becomes again universal function of the combinations $z - ct$ and $z + ct$. But for more general $\omega(k)$ dependence, each of the two *wave packets* (or *pulses*) in (17.7) are now functions of z and t as *independent* variables, and therefore these packets do not travel with unchanged shape. Physically, we may say that the pulses in

(17.7) are synthesized from harmonic plane wave components, each of which travels with a different velocity, so that a given superposition of these components at one time will become a different superposition at later times.

We now extend the above discussion by considering an initial value problem. Given the shape, i.e., the spatial dependence of a pulse at $t = 0$, what is the later space and time dependence? For simplicity, we shall consider only a pulse $F(z, t)$ traveling to the right and we suppose that it has a certain functional form at $t = 0$:

$$F(z, 0) = h(z).$$

For nondispersive media, the solution is trivial. That function which satisfies (17.1), which represents a wave to the right, and which reduces to $h(z)$ at $t = 0$, is

$$F(z, t) = h(z - ct). \tag{17.8}$$

For a dispersive medium, we must use (17.7). For a wave traveling to the right, $D(k) = 0$. At $t = 0$

$$F(z, 0) = h(z) = \int_{-\infty}^{\infty} C(k)e^{ikz}\,dk, \tag{17.9}$$

whence

$$C(k) = \frac{1}{2\pi}\int_{-\infty}^{\infty} e^{-ikz'}h(z')\,dz'. \tag{17.10}$$

substituting this value for $C(k)$ into (17.7), we have

$$\begin{aligned} F(z, t) &= \int_{-\infty}^{\infty} C(k)e^{i(kz-\omega(k)t)}\,dk \\ &= \frac{1}{2\pi}\int_{-\infty}^{\infty}\int_{-\infty}^{\infty} h(z')e^{-i(kz'-kz+\omega(k)t)}dz'\,dk. \end{aligned} \tag{17.11}$$

For a nondispersive medium for which $\omega = kc$, the integral on k yields $2\pi\delta(z' - z + ct)$, and the result (17.8) is reproduced. But for a dispersive medium, the integral over k cannot now be done explicitly and there is not much one can say about (17.11) in general.

We note, however, that there is a functional dependence of ω and k beyond the one in free space which permits a pulse to propagate unchanged in shape. If there were a medium in which ω were proportional to k, but with a proportionality constant c' *different* from c so that $\omega = kc'$, then a pulse would obviously propagate with velocity c'. And if somewhat more generally, ω were a general linear function of k, that is, $\omega = \omega_0 + kc'$, essentially the same result would be obtained: The constant ω_0 would simply introduce a phase factor. If, for a general dispersive medium, it were possible to *approximate* the $\omega(k)$ relation by a linear function, a pulse would move unchanged in shape but with a velocity different from c. Such an approximate linear relation does, in fact, frequently hold. Thus, in the amplitude function $C(k)$ in Eq. (17.10), not all wave numbers appear with the same weight. Frequently, $C(k)$ peaks at some value k_0, and has a certain width around k_0. In that case, for t not too large, only values of ω in the neighborhood of $\omega(k_0)$ will contribute to the integral and $\omega(k)$ can be expanded about the point k_0, keeping only the linear term in the expansion,

$$\omega(k) = \omega(k_0) + (k - k_0)\left.\frac{d\omega}{dk}\right|_{k=k_0} + \cdots. \tag{17.12}$$

The quantity

$$\left.\frac{d\omega}{dk}\right|_{k=k_0}$$

has dimensions of velocity and is defined to be the *group velocity* v_g

$$v_g = \left.\frac{d\omega}{dk}\right|_{k=k_0}. \tag{17.13}$$

Then

$$\omega(k) = \omega(k_0) + (k - k_0)v_g + \cdots. \tag{17.14}$$

On putting (17.14) into (17.11), we have

$$F(z, t) = e^{i\phi t} \int C(k)e^{ik(z - v_g t)}\, dk, \tag{17.15}$$

where the phase factor ϕ is

$$\phi = k_0 v_g - \omega(k_0).$$

The integral in (17.15) represents $h(z - v_g t)$ so that

$$F(z, t) = e^{i\phi(t)}h(z - v_g t). \tag{17.16}$$

The phase factor ϕ is often inconsequential. Frequently, for example, one is interested in the absolute magnitude of $F(z, t)$, in which case the phase factor disappears. Aside from the phase factor, Eq. (17.16) represents a pulse which, as a superposition of a group of plane waves, moves to the right with velocity v_g and this is the origin of the name *group velocity*.

We can help clarify the nature of the approximation in (17.16) with an exact example. Let the initial pulse be

$$h(z) = e^{-z^2/2a^2}.$$

Then $C(k)$ is

$$C(k) = \frac{1}{2\pi}\int_{-\infty}^{\infty} e^{-z^2/2a^2 - ikz}\, dz = \frac{a}{\sqrt{2\pi}}e^{-k^2a^2/2}. \tag{17.17}$$

For the $\omega(k)$ dependence, we will assume a form which is contrived to permit an exact integration that can be compared with the approximate form above. Namely, we take

$$\omega(k) = \Omega\left(1 - \frac{(k-K)^2b^2}{2}\right). \tag{17.18}$$

Then the group velocity is, since $C(k)$ peaks at $k = 0$,

$$v_g = \Omega K b^2.$$

With (17.17) and (17.18) in (17.11), we have

$$F(z,t) = \frac{a}{\sqrt{2\pi}} \int_{-\infty}^{\infty} e^{-(k^2a^2/2) + ikz - i\Omega(1-((k-K)^2b^2/2))t}\, dk.$$

The integral involves a quadratic form in an exponent and so can be done. We find that

$$F(z,t) = \frac{a e^{i\Omega t(1-(K^2b^2/2))}}{(a^2 - i\Omega t b^2)^{1/2}} e^{-((z-K\Omega t b^2)^2/2(a^2 - i\Omega t b^2))}. \tag{17.19}$$

This expression reduces correctly to $e^{-z^2/2a^2}$ for $t = 0$, and it confirms the qualitative features of the above discussion of group velocity. For sharply peaked distributions in k (i.e., for a large) and/or for short times ($a \gg \Omega t b^2$), the pulse is given by $e^{i\phi t}e^{-(z-v_g t)^2/2a^2}$ as (17.16) predicts. Later, however, beginning roughly at a time $T \sim a^2/\Omega b^2$ the amplitude of the pulse begins to diminish and the pulse broadens. The time T is larger for larger a (broader pulse in space) or as b gets smaller, i.e., as the dispersion properties of the medium become less marked. Essentially similar results hold for any distribution characterized by a width a.

17.2 ATTENUATION IN IMPERFECT CONDUCTORS

In Chapter 15, we assumed that boundary value problems with perfect conductors were good approximations to analogous problems with imperfect (but good) conductors. The skin effect discussed in Section 16.4 confirms these results in one sense: in imperfect condtuctors (at sufficiently high frequencies) the current is often a surface current to first approximation since its penetration depth is usually small compared to the other transverse dimensions over which it flows. On the other hand, even this small penetration of current and field in matter has an important qualitative effect. The electric field can now do work on the charged particles in the matter and, by various collisional processes, this work may be transformed into heat. This is the mechanism of *resistive* or *Joule heating* and is, of course, the physical origin of the damping term in the Drude model of Section 16.4. This damping has the practical consequence that energy is extracted from any wave that may exist in the vicinity of the conductor, thereby weakening or attenuating the wave. We now consider this attenuation in more detail.

We shall use the results for perfect conductors as a kind of zero order approximation, in which the effect of finite conductivity is taken to be a small perturbation on the results for infinite conductivity. Consider first the space part of the time harmonic fields at the surface of the conductor. For a boundary value problem with *perfect* conductivity, there was only a tangential (parallel) component $\boldsymbol{B}_\omega^{\|}$ of $\boldsymbol{B}_\omega$, and a normal component of $\boldsymbol{E}_\omega$. Consider now the analogous surface fields for a geometrically similar boundary value problem with imperfect conductors; for example, the field at a point in air close to an inner metallic surface of a cavity or wave guide. We make the reasonable assumption that this air field is unchanged to first approximation or, in other words, that it is almost independent of the fact that the current penetration in the imperfect conductor is small rather than of strictly zero depth. Let a coordinate l measure distance normally into the conductor, so that the surface of the conductor at the point considered is the plane $l = 0$, and the region $l > 0$ is conducting. Let $\boldsymbol{B}_{\omega p}^{\|}(0)$ be the tangential field that would exist just to the left of $l = 0$ for *perfect* conductivity. If the similar field with the *imperfect* conductor is $\boldsymbol{B}_\omega^{\|}(0)$, our approximation is

$$\boldsymbol{B}_\omega^{\|}(0) \approx \boldsymbol{B}_{\omega p}^{\|}(0). \tag{17.20}$$

With this assumption, we can find $\boldsymbol{B}_\omega^{\|}$ in the imperfect conductor since it is known to vary there as $e^{l(i-1)/\delta}$ and is continuous through the origin. We assume then that

$$\boldsymbol{B}_\omega^{\|}(l) = \boldsymbol{B}_{\omega p}^{\|}(0)e^{l(i-1)/\delta}. \tag{17.21}$$

From (17.21), other field components in the conductor can be calculated. It is $\boldsymbol{E}_\omega^{\|}$, the component of $\boldsymbol{E}_\omega$ parallel to the surface, that is of most interest here. From $\nabla \times \boldsymbol{B}_\omega = 4\pi\sigma\boldsymbol{E}_\omega/c - (i\omega/c)\boldsymbol{E}_\omega$, we find, for the case of usual interest, $\sigma \gg \omega$ (displacement current negligible with respect to the conduction current), that $\boldsymbol{E}_\omega^{\|}$ is *perpendicular*† to $\boldsymbol{B}_\omega^{\|}$ and, with $\boldsymbol{n}$ the normal to the conductor, is given by

$$\boldsymbol{E}_\omega^{\|}(l) = \frac{c(1-i)}{4\pi\sigma\delta}(\boldsymbol{n} \times \boldsymbol{B}_{\omega p}^{\|}(0))e^{l(i-1)/\delta}. \tag{17.22}$$

The importance of $\boldsymbol{E}_\omega^{\|}(l)$ arises in that it is needed to calculate the energy dissipated in the metal. This energy loss can be calculated from either of two viewpoints. First, we see that for the present imperfectly conducting case, a Poynting vector can be formed from (17.21) and (17.22) and it is directed into the metal. This implies a flux into the metal of energy which is dissipated as Joule heat. This dissipated energy is extracted from the energy resident in the fields of the wave guide mode; hence the fields will be attenuated as the wave travels down the guide. The second viewpoint for calculating Joule heating begins with the fact that the rate per unit volume at which a field $\boldsymbol{E}$ does work on a current $\boldsymbol{J}$ is $\boldsymbol{E} \cdot \boldsymbol{J}$. For time harmonic fields the *average* rate of such work which is expended in Joule heat

† The superscript $\|$ on both $\boldsymbol{B}_\omega^{\|}$ and $\boldsymbol{E}_\omega^{\|}$ does not, of course, imply that these vectors are parallel to each other, but merely that they are both parallel to the surface.

is easily seen to be $\frac{1}{2}\,\mathrm{Re}\colon (\boldsymbol{E}_\omega \cdot \boldsymbol{J}_\omega^*)$. Using $\boldsymbol{J}_\omega = \sigma \boldsymbol{E}_\omega$, we can write for $d\bar{P}/dv$, the time average power loss per unit *volume* due to Joule heating,

$$\frac{d\bar{P}}{dv} = \frac{\sigma}{2}|\boldsymbol{E}_\omega|^2. \tag{17.23}$$

In the present approximation $\boldsymbol{E}_\omega$ is taken to be $\boldsymbol{E}_\omega^{||}$ since the currents are assumed to flow parallel to the surface.

In detail then, from the first viewpoint, a time average Poynting vector of magnitude $\bar{S}_n$ is directed normally into the conductor and at the surface of the conductor is given by

$$\bar{S}_n = \frac{c^2}{32\pi^2\sigma\delta}|\boldsymbol{B}_{\omega p}^{||}(0)|^2. \tag{17.24}$$

It is more convenient to express $\bar{S}_n$ in terms of the current $\boldsymbol{J}_s$ using the fact that

$$\frac{4\pi|\boldsymbol{J}_s|}{c} = |\boldsymbol{B}_{\omega p}^{||}(0)|.$$

Writing $d\bar{P}/da$ as the average rate of power dissipated per unit *area* of surface, we have

$$\frac{d\bar{P}}{da} = \bar{S}_n = \frac{|\boldsymbol{J}_s|^2}{2\sigma\delta}. \tag{17.25}$$

To derive this result from the second viewpoint, we have

$$\frac{d\bar{P}}{da} = \int \frac{d\bar{P}}{dv}\, dl.$$

Using (17.23), we obtain

$$\frac{d\bar{P}}{da} = \frac{\sigma}{2}\int_0^\infty |\boldsymbol{E}_\omega^{||}(l)|^2\, dl = \frac{|\boldsymbol{B}_{\omega p}^{||}(0)|^2}{16\pi^2\sigma\delta^2}\int_0^\infty e^{-2l/\delta}\, dl = \frac{|\boldsymbol{J}_s|^2}{2\sigma\delta},$$

which is in agreement with (17.25).

As an application of these results, we consider attentuation in a rectangular guide. Let $\bar{P}_g$ be the average power flowing down the guide, i.e., the energy per second passing through some cross-sectional plane in the guide. Because of the attenuation, this power will decrease exponentially so that if the guide axis is the z-axis,

$$\bar{P}_g = P_{g0}e^{-2\alpha z}$$

or

$$\alpha = -\frac{d\bar{P}_g/dz}{2\bar{P}_g}. \tag{17.26}$$

To illustrate the calculation of α, consider the commonly used TE_{10} mode, although the same principles apply to other modes. For this mode, we have from Eq. (15.16), in the coordinates of Fig. 15.3 for the only nonvanishing field components

$$E_{\omega y} = \frac{iB_0 ka}{\pi} \sin \frac{\pi x}{a} e^{ik_g z},$$

$$B_{\omega x} = -\frac{k_g}{k} E_{\omega y}, \tag{17.27}$$

$$B_{\omega z} = B_0 \cos \frac{\pi x}{a} e^{ik_g z},$$

where

$$k_g^2 = k^2 - \frac{\pi^2}{a^2}.$$

We calculate $\bar{P}_g$ in Eq. (17.26) in terms of the magnitude $\bar{S}_g$ of the Poynting vector for energy flow down the guide:

$$\bar{P}_g = \int \bar{S}_g ds,$$

where the integration goes over the guide cross-section. From

$$\bar{S}_g = (c/8\pi)\,\mathrm{Re}:(\boldsymbol{E}_\omega \times \boldsymbol{B}_\omega^*),$$

we have

$$\bar{S}_g = \frac{ck_g ka^2 B_0^2}{8\pi^3} \sin^2 \frac{\pi x}{a}.$$

It follows that $\bar{P}_g$ is

$$\bar{P}_g = \frac{ck_g ka^3 b B_0^2}{16\pi^3}. \tag{17.28}$$

To calculate $d\bar{P}_g/dz$, we must integrate $d\bar{P}/da$ of Eq. (17.25), the power loss per *unit area*, over the perimeter of a cross-section of the guide to give power loss per *unit length*:

$$\frac{d\bar{P}_g}{dz} = -\int_{\substack{\text{Guide}\\ \text{walls}}} \frac{d\bar{P}}{da}\, d\xi, \tag{17.29}$$

where $d\xi$ is an element of length along the perimeter. For calculating $d\bar{P}/da$, the current on the side wall of the guide is, from $(4\pi/c)\boldsymbol{J}_s = \boldsymbol{n} \times \boldsymbol{B}_\omega$,

$$\boldsymbol{J}_s|_{x=0} = -\hat{\boldsymbol{y}} \frac{cB_0}{4\pi} e^{ik_g z}$$

and the current on the bottom wall is

$$\boldsymbol{J}_s|_{y=0} = \frac{cB_0}{4\pi}\left(\hat{\boldsymbol{x}}\cos\frac{\pi x}{a} + \hat{\boldsymbol{z}}\frac{ik_g a}{\pi}\sin\frac{\pi x}{a}\right)e^{ik_g z}.$$

With these expressions in (17.25) and (17.29), we find readily that

$$\frac{d\bar{P}_g}{dz} = -\frac{c^2B_0^2}{16\pi^2\sigma\delta}\left(b + \frac{k^2a^3}{2\pi^2}\right).$$

Then from (17.26), the attenuation constant is

$$\alpha = \frac{c\pi\left(b + \dfrac{k^2a^3}{2\pi^2}\right)}{2\sigma\delta k_g k a^3 b}.$$

For the usual wave guide for which $a \approx b$, the behavior of α as a function of frequency ω involves first a singularity in the attenuation near the cut-off frequency $\omega_e = \pi c/a$ due to the factor k_g in the denominator. There is then a shallow minimum when ω is a few times ω_c and a subsequent slow increase of α with ω. A typical value for α at microwave frequencies is of the order of a few hundred meters.

A second important application of the formulas above is to estimate the rate of ohmic loss in cavities. In the perfect cavities discussed in Section 15.6, any given mode will oscillate indefinitely. But in a real cavity, the oscillation will damp gradually because of heating loss in the walls. The rapidity of this damping is characterized by a figure of merit called Q. If the mode has frequency ω_0, then Q is defined by

$$Q = \omega_0 \frac{\text{Energy stored in cavity}}{\text{Energy lost per second in walls}}. \tag{17.30}$$

The Q's of cavities can be calculated much as the attenuation of wave guides was calculated above.

17.3 REFLECTION AND REFRACTION AT A DIELECTRIC INTERFACE

The important problem of the reflection and refraction of a wave at the interface of two media of different dielectric constants is a boundary value problem in which the physics is identical in principle to that involved in time harmonic problems involving conductors. If, for example, one medium is vacuum and a plane wave is incident on a second (dielectric) medium, the incident harmonic plane wave generates oscillating time harmonic dipoles (or dipolar currents) that produce a field of their own. The strength of the currents is, however, not known in advance and therein lies the essence of problem. Although this problem is a boundary value problem, it is especially simple because of the great symmetry and simple geometry. It turns out that it is solved by adding to the incident plane waves only two other plane waves, a *reflected* and a *transmitted* (or refracted) one.

The geometry is shown in Fig. 17.1 with the plane $z = 0$ taken as the interface between the two media, labeled 1 and 2.

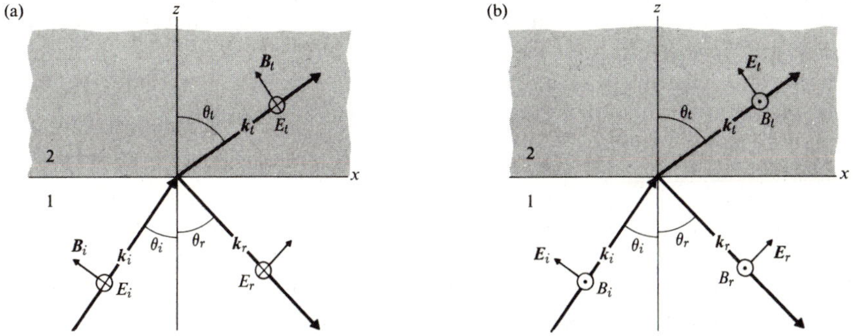

Fig. 17.1 Reflection and refraction at a dielectric interface. (a) Electric vector perpendicular to plane of incidence. (b) Electric vector parallel to plane of incidence.

The electric field, incident from below, is the space part of a time harmonic plane wave and would be written $\boldsymbol{E}_\omega^i$ with the usual convention. For simplicity, however, we shall *omit the subscript ω throughout this section* and write, for example

$$\boldsymbol{E}_\omega^i \equiv \boldsymbol{E}^i.$$

In terms of the *vector* propagation constant, $\boldsymbol{k}_i$, the incident fields are

$$\boldsymbol{E}_i = \boldsymbol{A}e^{i\boldsymbol{k}_i\cdot\boldsymbol{r}}, \qquad \boldsymbol{B}_i = (\boldsymbol{k}_i/k)\times\boldsymbol{E}_i \qquad \text{Incident.} \tag{17.31}$$

Here $\boldsymbol{B}_i$ is derived from the assumed $\boldsymbol{E}_i$ by $\nabla \times \boldsymbol{E}_i = ik\boldsymbol{B}_i$. The normal $\hat{\boldsymbol{z}}$ to the plane $z = 0$ and the vector $\boldsymbol{k}_i$ define a *plane of incidence* which can, without loss of generality, be taken to be the x, z-plane as we have done in Fig. 17.1. We now postulate the existence of two other plane waves and will show that these suffice to solve the boundary value problem. These *reflected* and *transmitted* waves, amplitudes $\boldsymbol{E}_r$ and $\boldsymbol{E}_t$, respectively, are

$$\boldsymbol{E}_r = \boldsymbol{R}e^{i\boldsymbol{k}_r\cdot\boldsymbol{r}}, \quad \boldsymbol{B} = (\boldsymbol{k}_r/k)\times\boldsymbol{E}_r \qquad \text{Reflected} \tag{17.32}$$

and

$$\boldsymbol{E}_t = \boldsymbol{T}e^{i\boldsymbol{k}_t\cdot\boldsymbol{r}}, \quad \boldsymbol{B}_t = (\boldsymbol{k}_t/k)\times\boldsymbol{E}_t \qquad \text{Transmitted.} \tag{17.33}$$

The vectors $\boldsymbol{k}_r$ and $\boldsymbol{k}_t$ must, for the moment, be considered arbitrary in direction, for although $\boldsymbol{k}_i$ is in the x, z-plane, we cannot assume the same *a priori* for $\boldsymbol{k}_r$ and $\boldsymbol{k}_t$. The magnitudes $k_i \equiv |\boldsymbol{k}_i|$ etc. of the wave vectors are, with n_1 and n_2 the indices of refraction of the two media and $k = \omega/c$,

$$k_i = k_r = n_1 k$$
$$k_t = n_2 k. \tag{17.34}$$

Consider now the boundary conditions which state that the tangential components of $\boldsymbol{E}$ and $\boldsymbol{B}$ be continuous. These must be satisfied in two steps. First, if the tangential components of the three fields (17.31), (17.32), and (17.33) are to be matched at $z = 0$, it is clear that the spatial dependence given by the exponents must be identical. However, this is a necessary but not a sufficient condition. Secondly, the vector coefficients $\boldsymbol{A}$, $\boldsymbol{R}$, and $\boldsymbol{T}$ must be determined. The first condition, that the spatial variation of the three fields be identical at $z = 0$, is

$$(\boldsymbol{k}_i \cdot \boldsymbol{r})_{z=0} = (\boldsymbol{k}_r \cdot \boldsymbol{r})_{z=0} = (\boldsymbol{k}_t \cdot \boldsymbol{r})_{z=0}. \tag{17.35}$$

The first equality in Eq. (17.35) yields $k_{ix}x = k_{rx}x + k_{ry}y$. For this condition to hold for all x and y, we must have $k_{ry} = 0$, showing that $\boldsymbol{k}_r$ lies in the plane of incidence and also that $k_{ix} = k_{rx}$. Similarly, from the second equality in (17.35), $\boldsymbol{k}_t$ must also lie in this plane, so that $\boldsymbol{k}_i$, $\boldsymbol{k}_r$, $\boldsymbol{k}_t$ are all coplanar. Moreover, from the geometry of Fig. 17.1, we have from $k_{ix} = k_{rx}$ that $k_i \sin\theta_i = k_r \sin\theta_r$. Or, since $k_i = k_r$,

$$\sin\theta_i = \sin\theta_r\,; \tag{17.36}$$

the *angle of incidence equals the angle of reflection*. Similarly, the equality of k_{ix} and k_{tx} yields

$$k_i \sin\theta_i = k_t \sin\theta_t$$

or

$$\frac{\sin\theta_i}{\sin\theta_t} = \frac{n_2}{n_1}. \tag{17.37}$$

Equation (17.37) is known as *Snell's law* of refraction.

The conditions (17.36) and (17.37) are quite general ones that are independent of the detailed vectorial nature of the wave field; they hold, for example, for the reflection and refraction of scalar waves. These conditions by themselves do not then guarantee the continuity of tangential $\boldsymbol{E}$ and $\boldsymbol{B}$ across the boundary. To satisfy these conditions, we must specify more closely the polarization of the fields. It is convenient to consider the general case of arbitrary incident polarization as a linear combination of a wave with polarization perpendicular to the plane of incidence and one with polarization parallel to the plane of incidence. The reflected and transmitted waves will then be similarly polarized. These two cases are sketched as parts (a) and (b) of Fig. 17.1.

Consider first the case of $\boldsymbol{E}$ perpendicular to the plane of incidence, i.e., in the y-direction. The vector coefficients $\boldsymbol{A}$, $\boldsymbol{R}$, $\boldsymbol{T}$ become scalar ones, with subscript $_\perp$ to denote this case.

$$\begin{aligned} E_{i\perp} &= A_\perp e^{i\boldsymbol{k}_i \cdot \boldsymbol{r}}, \\ E_{r\perp} &= R_\perp e^{i\boldsymbol{k}_r \cdot \boldsymbol{r}}, \\ E_{t\perp} &= T_\perp e^{i\boldsymbol{k}_t \cdot \boldsymbol{r}}. \end{aligned} \tag{17.38}$$

Then from the continuity of tangential $\boldsymbol{E}$ at the boundary,

$$A_\perp + R_\perp = T_\perp. \tag{17.39}$$

The condition on tangential $\boldsymbol{B}$ becomes, with $B_{ix} = -A_\perp n_1 \cos\theta_i$, $B_{rx} = R_\perp n_1 \cos\theta_r$, and $B_{tx} = -T_\perp n_2 \cos\theta_t$:

$$n_1(A_\perp - R_\perp)\cos\theta_i = n_2 T_\perp \cos\theta_t. \tag{17.40}$$

Solving (17.39) and (17.40) for the ratios $R_\perp/A_\perp$ and $T_\perp/A_\perp$, we find, using Snell's law, that

$$\begin{aligned} \frac{R_\perp}{A_\perp} &= \frac{1 - \dfrac{\tan\theta_i}{\tan\theta_r}}{1 + \dfrac{\tan\theta_i}{\tan\theta_r}} = -\frac{\sin(\theta_i - \theta_t)}{\sin(\theta_i + \theta_t)} \\ \frac{T_\perp}{A_\perp} &= \frac{2}{1 + \dfrac{\tan\theta_i}{\tan\theta_t}} = \frac{2\cos\theta_i \sin\theta_t}{\sin(\theta_i + \theta_t)}. \end{aligned} \tag{17.41}$$

For the second case, incident wave polarized parallel to the plane of incidence, we use $A_{||} = |\boldsymbol{A}|$ in (17.31) and, similarly $R_{||} = |\boldsymbol{R}_{||}|$ and $T_{||} = |\boldsymbol{T}|$ in (17.32) and (17.33). The boundary conditions yield

$$\cos\theta_i(A_{||} - R_{||}) = \cos\theta_t T_{||}$$

$$n_1(A_{||} + R_{||}) = n_2 T_{||}.$$

These equations lead to the results

$$\begin{aligned} \frac{R_{||}}{A_{||}} &= \frac{\tan(\theta_i - \theta_t)}{\tan(\theta_i + \theta_t)} \\ \frac{T_{||}}{A_{||}} &= \frac{2\cos\theta_i \sin\theta_t}{\sin(\theta_i + \theta_t)\cos(\theta_i - \theta_t)}. \end{aligned} \tag{17.42}$$

There are two phenomena worthy of note in connection with the above discussion. Consider the case of polarization in the plane of incidence. We see from Eq. (17.42) that $R_{||}/A_{||}$ will be zero for $\theta_i + \theta_t = \pi/2$. Putting this condition into Snell's law, using $\sin\theta_t = \sin(\pi/2 - \theta_i) = \cos\theta_i$, we see that the angle of incidence θ_B (called the *Brewster angle*) for which this happens is defined by

$$\tan\theta_B = \frac{n_2}{n_1}. \tag{17.43}$$

If a wave with arbitrary polarization is incident on a dielectric interface, it can be considered to be a linear combination of a wave polarized parallel to, and a wave polarized perpendicular to, the plane of incidence. At the Brewster angle, the parallel component will not be reflected so that the reflected wave will be plane polarized in a plane perpendicular to the plane of incidence. This effect can then be made the basis of a device for polarizing an unpolarized beam of radiation.

The second phenomenon is that of *total internal reflection.* In either (a) or (b) of Fig. 17.1, suppose that the index n_1 is greater than n_2. Then, from Snell's law, $\sin\theta_t = (n_1/n_2)\sin\theta_i$ so that θ_t is always greater than θ_i. There will then be some value of θ_i, call it θ_{int}, for which $\theta_t = \pi/2$; this angle is defined by

$$\sin\theta_{\text{int}} = \frac{n_2}{n_1}.$$

Since, in general,

$$\boldsymbol{E}_t = \boldsymbol{T}e^{ikn_2(x\sin\theta_t + z\cos\theta t)}, \tag{17.44}$$

for $\theta_t = \pi/2$ there is no wave in the second medium; the z-dependence vanishes. Now if $\theta_i > \theta_{\text{int}}$, then $\sin\theta_t$ is larger than unity from Snell's law and, as a consequence, $\cos\theta_t$ is imaginary:

$$\cos\theta_t = \sqrt{1 - \left(\frac{n_1}{n_2}\right)^2 \sin^2\theta_i} = i\sqrt{\sin^2\theta_i\left(\frac{n_1}{n_2}\right)^2 - 1}.$$

Equation (17.44) becomes

$$\boldsymbol{E}_t = \boldsymbol{T}e^{-kn_2z\sqrt{(n_1/n_2)^2\sin^2\theta_i - 1}}e^{ikn_2x\sin\theta_t}.$$

This corresponds to a wave which is exponentially attentuated as a function of z, and which propagates as a function of x with propagation constant $kn_2\sin\theta_t$. Such a wave is the prototype of a *surface wave* discussed in more detail in the next section.

17.4 SURFACE WAVES IN DIELECTRICS

Time harmonic waves in a dielectric propagate with wave number $k' \neq \omega/c$ and if we had to deal only with unbounded dielectrics, nothing much more need be said. The surface dividing two dielectrics introduces more interest, however. In discussing the problem of a plane wave incident on a plane interface separating two dielectrics, we found that, under certain conditions, the fields attenuated exponentially in the medium of the smaller index, but propagated as a *surface wave* along the interface, with a real wave number. The plane interface is just one example of geometries that can guide or support surface waves. Given the proper dimensions, dielectric constant, etc., surface waves can exist at the boundaries of many cylindrical geometries. Such cylinders can then act as dielectric wave guides, propagating energy down them much as do metallic guides. Just as there are circularly cylindrical wave guides, so can circularly cylindrical dielectric rods act as wave guides, and much as metallic parallel plates guide waves, so can a dielectric slab.

As the results of Section 17.3 indicate a surface wave can be thought of as one for which the wave vector is complex. For example, an ordinary time harmonic plane wave amplitude ψ propagating parallel to the x, z-plane is given by

$$\psi = e^{i(k_x x + k_z z)}$$

and satisfies $(\nabla^2 + k^2)\psi = 0$, where k_x and k_z are assumed real and $k_x^2 + k_z^2 = k^2$. But suppose now that k_x or k_z were imaginary, say, $k_x = i\gamma$. Then an equally good solution of $(\nabla^2 + k^2)\psi = 0$, describing a *surface wave* propagating in the z-direction, is

$$\psi = e^{-\gamma x + ik_z z},$$

where $-\gamma^2 + k_z^2 = k^2$. Surface waves bear many resemblances to bound states in quantum mechanics. This is not surprising since the governing equations are essentially the same. The potential $V(r)$ of quantum mechanics has its analog in dielectric matter in $n^2(r) - 1$, the deviation of the square of the index from unity.

Although it is easy enough to show the mathematical existence of surface waves, the question of whether they exist in a given dielectric geometry is one that has to be analyzed for each possible dielectric wave guide. The analysis of such guides is similar to that for metallic ones in many respects, such as the existence of guide wavelengths, modes, etc., so we shall not pursue it here at any length. Rather, we shall take the simplest example, one mode in a plane parallel dielectric slab wave guide, to typify the general theory.

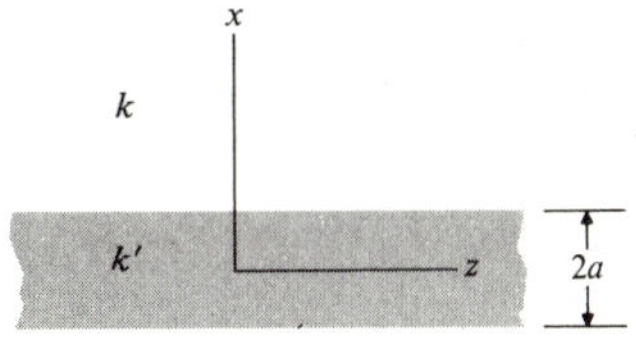

Fig. 17.2 Dielectric slab supporting surface wave.

Consider then the dielectric slab, infinite in the y-direction and of thickness $2a$, shown in Fig. 17.2. Let the propagation constants be k and k' outside and inside the slab, respectively, and assume that there is no y-dependence of the field components. Each rectangular component of $\boldsymbol{E}_\omega$ and $\boldsymbol{B}_\omega$ satisfies the Helmholtz equation

$$(\nabla^2 + k^2)\begin{pmatrix} \boldsymbol{E}_\omega \\ \boldsymbol{B}_\omega \end{pmatrix} = 0$$

outside the slab and satisfies the similar expression, with k replaced by k', inside the slab. Guided by the results of the last section, we assume that the z-dependence of all field components is the same, as $e^{ik_g z}$, inside and outside the slab. With the assumption of no y-dependence, we have then

$$\begin{aligned} \boldsymbol{E}_\omega(x, z) &= \tilde{\boldsymbol{E}}(x)e^{ik_g z} \\ \boldsymbol{B}_\omega(x, z) &= \tilde{\boldsymbol{B}}(x)e^{ik_g z}. \end{aligned} \tag{17.45}$$

The form (17.45) has the consequence that the field configurations divide into TE modes ($E_{\omega z} = 0$) and TM modes ($B_{\omega z} = 0$) just as in the discussion of metallic guides in Section 15.4 and as exemplified in Eqs. (15.19).

For the sake of an example, we choose to discuss a TM mode. On the assumption of no y-dependence, we see from Eqs. (15.19) that the only nonvanishing field components are $E_{\omega x}$, $E_{\omega z}$, $B_{\omega y}$. Consider the tangential component $B_{\omega y}$. With (17.45), we have

$$\frac{d^2\tilde{B}_y}{dx^2} + (k'^2 - k_g^2)\tilde{B}_y = 0, \qquad |x| < a; \tag{17.46}$$

$$\frac{d^2\tilde{B}_y}{dx^2} + (k^2 - k_g^2)\tilde{B}_y = 0, \qquad |x| > a. \tag{17.47}$$

The form of the solution of (17.46) for $|x| < a$ depends, of course, on whether $k'^2 - k_g^2$ is greater or less than zero. Taking the former case for definiteness, the two independent solutions of (17.46) are $\cos(k'^2 - k_g^2)^{1/2}x$ and $\sin(k'^2 - k_g^2)^{1/2}x$, even and odd, respectively, about $x = 0$. This evenness and oddness constitute a further subclassification of the TM modes. We treat the two cases separately and take the even case first:

$$\tilde{B}_y = C\cos(k'^2 - k_g^2)^{1/2}x.$$

For the even solution of (17.47) outside the slab, we can take, since we anticipate exponential decay of the field,

$$\tilde{B}_y = De^{-\sqrt{k_g^2 - k^2}|x|}, \qquad |x| > a.$$

The continuity of $B_{\omega y}$ at $x = \pm a$ yields

$$C\cos\sqrt{k'^2 - k_g^2}a = De^{-\sqrt{k_g^2 - k^2}a}. \tag{17.48}$$

The other boundary condition is on $E_{\omega z} = (i/k\,n^2)(\partial B_{\omega y}/\partial x)$, and the condition that this be continuous at $x = \pm a$ is

$$-C\sqrt{k'^2 - k_g^2}\sin\sqrt{k'^2 - k_g^2}a = -Dn^2\sqrt{k_g^2 - k^2}\,e^{-\sqrt{k_g^2 - k^2}a}. \tag{17.49}$$

Dividing (17.49) by (17.48), the transcendental equation determining k_g in terms of k, a, and the index $n = k'/k$, is

$$\sqrt{n^2k^2 - k_g^2}\tan\sqrt{n^2k^2 - k_g^2}a = n^2\sqrt{k_g^2 - k^2}. \tag{17.50}$$

With k_g given by (17.50), the solution is now completely determined.

For the odd solution in the slab,

$$\tilde{B}_y = C'\sin(k'^2 - k_g^2)^{1/2}x,$$

we must assume an odd solution in the exterior as well and so take

$$\tilde{B}_y = \pm D' e^{-\sqrt{k_g^2 - k^2}|x|},$$

where the $\pm$ sign refers to $x \gtrless a$. Much as above, the transcendental equation determining k_g is

$$\sqrt{n^2k^2 - k_g^2} \cot \sqrt{n^2k^2 - k_g^2}\,a = -\sqrt{k_g^2 - k^2}. \tag{17.51}$$

The pattern of solution for other kinds of geometries is similar to the above except that there appear the solutions of the Helmholtz equation in the appropriate coordinates; for example, Bessel functions and cylindrical harmonics if one is dealing with a cylindrical dielectric rod.

PROBLEMS

1. An electromagnetic plane wave in a medium I with a propagation constant k_1 is incident normally on a dielectric slab of thickness d, extending between $x = 0$ and $x = d$. The propagation constant of the slab is k_2 and it is backed by a perfect conductor at $x = d$. The electric field amplitude in I is $E_0 e^{ikx}$ and the similarly polarized reflected wave has amplitude $rE_0 e^{-ikx}$. Find r and show that $|r| = 1$, as would be expected.

2. A plane dielectric slab of thickness d, propagation constant k_2, separates two infinite media I and III, with propagation constants k_1 and k_3, respectively. A wave with electric field amplitude $E_0 e^{ik_1x}$ is incident normally from the left. A similarly polarized reflected wave $rE_0 e^{-ik_1x}$ is thereby generated in I and a transmitted wave $tE_0 e^{ik_3x}$ is generated in III. Show that

$$r = \frac{r_{12} + r_{23}e^{2ik_2d}}{1 + r_{12}r_{23}e^{2ik_2d}}$$

$$t = \frac{(1 + r_{12})(1 + r_{23})e^{i(k_2 - k_3)d}}{1 + r_{12}r_{23}e^{2ik_2d}},$$

where $r_{ij} = (k_i - k_j)/(k_i + k_j)$.

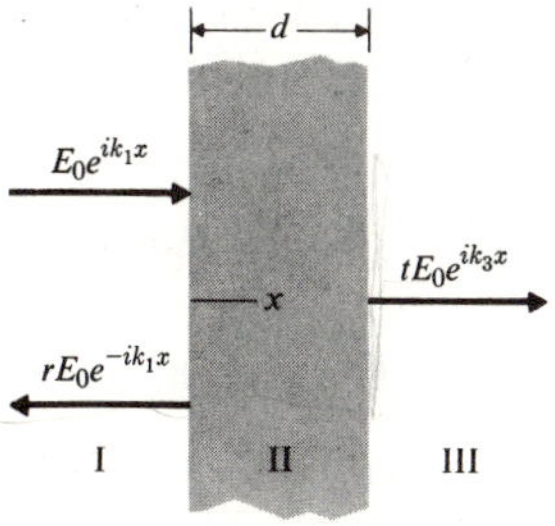

3. In Problem 2 show that if $k_2^2 = k_1k_3$ and if the wavelength $\lambda_2 = 2\pi/k_2$ in the second medium is related to the slab thickness by $d = \lambda_2/4$, there will be no reflected wave ($r = 0$). Such a dielectric slab is called a *quarter wave plate*.

4. In Problem 2, the power *transmission* coefficient T, which represents the power per unit area transmitted toward the right, is defined by $T = |t|^2$. In the same way, the power *reflection* coefficient R is $R = |r|^2$. Show that $R + T = 1$.

5. The boundary value problem of reflection from a slab, stated in Problem 2, can be usefully formulated by an integral equation. Taking $k_1 = k_3$ for simplicity and letting $E(x)$ be the (transversely polarized) field at any point this equation is

$$E(x) = E_0 e^{ik_1 x} + \frac{1}{2ik_1} \int_{\text{Slab}} (k_1^2 - k_2^2) e^{ik_1|x-x'|} E(x')\, dx',$$

where the Green function $G(x, x') = (1/2ik)e^{ik|x-x'|}$ satisfies

$$(d^2/dx^2 + k^2)G(x, x') = \delta(x - x').$$

Verify the correctness of this integral equation by showing that $E(x)$ defined by it: (a) satisfies $(d^2/dx^2 + k_1^2)E(x) = 0$ in I and III; (b) satisfies $(d^2/dx^2 + k_2^2)E(x) = 0$ in II; (c) has the correct asymptotic form for $|x|$ large; (d) satisfies the boundary conditions. Note that this integral equation is valid even for nonuniform slabs, for which k_2 is a function of x.

6. The integral equation of Problem 5 is useful in that it readily yields approximate solutions. Suppose for example that $k_1 \approx k_2$ so the scattering due to the slab is "small." Then the approximation $E(x) \approx E_0 e^{ik_1 x}$ can be used in the integral as a *first iterate*, or, as it is sometimes called, the *first Born approximation.* Show that in this approximation

$$r = \frac{k_1 - k_2}{2k_1}(1 - e^{2ik_1 d})$$

and compare with the exact value. Find t.

7. Show that the attenuation constant α for the TEM mode in coaxial line (cf. Problem 10 in Chapter 15) is

$$\alpha = \frac{c}{8\pi\sigma\delta} \frac{(1/a + 1/b)}{\ln b/a}.$$

8. Find the attenuation constant for the TE_{11} mode in rectangular guide.

9. An infinite dielectric cylinder of propagation constant k' and in the geometry of Fig. 15.4, scatters a plane wave with electric vector $\hat{z}E_0 e^{ikz}$. Show that the wave scattered by the cylinder is

$$E_{sc} = \hat{z}E_0 \sum_{m=-\infty}^{\infty} B_m H_m(k\rho) e^{im\varphi},$$

where H_m is a Hankel function and B_m is given by (a prime denotes differentiation)

$$B_m = \frac{i^m(k'J_m(ka)J_m'(k'a) - kJ_m(k'a)J_m'(ka))}{kH_m'(ka)J_m(k'a) - k'H_m(ka)J_m'(k'a)}.$$

Calculate similarly the scattered field if the polarization of the incident plane wave is normal to the cylinder axis.

10. Discuss the TE modes in the dielectric slab of Fig. 17.2. Show that the only nonvanishing field components are B_x, B_z, E_y. The modes again divide into those even and those odd about $x = 0$. For the even modes, find the equation analogous to Eq. (17.51) that determines k_g and find also all the field components inside and outside the slab.
11. A TEM mode is possible in a parallel plate wave guide (cf. Section 3 in Chapter 15). Suppose a dielectric slab, with faces parallel to the wave guide plates and thickness *less* than the plate separation is inserted in such a guide. Is a TEM mode now possible? Is a TEM mode possible if the slab completely fills the guide?
12. Find the Q (Eq. 17.50) of a rectangular cavity for the mode with field components given by Eq. (15.46).

REFERENCES

Harrington, R. F., *Time Harmonic Electromagnetic Fields,* McGraw-Hill, New York (1961).

A more complete discussion of the dielectric slab wave guide is to be found here, as well as a treatment of guided waves in other dielectric geometries.

Sommerfeld, A., *Optics*, Academic Press, New York (1954).

Sommerfeld discusses the propagation of a pulse in a medium with a more realistic law of dispersion than the one we have used. A related discussion is found in:

Stratton, J. A., *Electromagnetic Theory*, McGraw-Hill, New York (1941).

APPENDIXES

APPENDIX A
CONVERSION TABLE FOR GAUSSIAN ⇄ MKS UNITS

The following table gives the relations between the units of common quantities in the rationalized MKS system and in the Gaussian (CGS) system used in this book.

Read "equals," with the numerical factor of column three, between the second and fourth columns. For example, one newton equals 10^5 dynes.

$$\hat{3} = 2.9979250 \pm .0000010$$

Quantity	Rationalized MKS		Gaussian
Length	1 meter (m)	10^2	centimeter (cm)
Mass	1 kilogram (kg)	10^3	gram (g)
Time	1 second (sec)	1	second (sec)
Force	1 newton	10^5	dyne
Work / Energy	1 joule	10^7	erg
Power	1 watt	10^7	erg sec^{-1}
Charge	1 coulomb	$\hat{3} \times 10^9$	statcoulomb
Current	1 ampere	$\hat{3} \times 10^9$	statampere
Current density	1 amp m^{-2}	$\hat{3} \times 10^5$	statamp cm^{-2}
Electric field	1 volt m^{-1}	$1/\hat{3} \times 10^{-4}$	statvolt cm^{-1}
Potential	1 volt	$1/\hat{3} \times 10^{-2}$	statvolt
Conductivity	1 mho m^{-1}	$\hat{3}^2 \times 10^9$	sec^{-1}
Resistance	1 ohm	$1/\hat{3}^2 \times 10^{-11}$	sec cm^{-1}
Capacitance	1 farad	$\hat{3}^2 \times 10^{11}$	cm

APPENDIX B
DELTA FUNCTIONS

The first point to grasp about δ-functions is that they are not functions. They can be considered on occasion as a notational *shorthand*, or as the *limit* of a *sequence of functions*, or as a *mnemonic* for certain integration formulas, but they are not functions in the sense that $\sin x$ or e^{-x} are functions.

Their first guise in electromagnetic theory is perhaps this: We can write a general expression for the potential Φ of any charge distribution by using the potential for a charge element and the principle of superposition. If the charge distribution is partially continuous with density $\rho(\boldsymbol{r}')$, and partially discrete with charges q_i at $\boldsymbol{r}_i$, then the expression for the potential at $\boldsymbol{r}$ consists of an integral and a sum:

$$\Phi(\boldsymbol{r}) = \int \frac{\rho(\boldsymbol{r}')\,dv'}{|\boldsymbol{r} - \boldsymbol{r}'|} + \sum_i \frac{q_i}{|\boldsymbol{r} - \boldsymbol{r}_i|}. \tag{B.1}$$

Now the integral and the sum are more closely akin than they may look: the same physics, Coulomb's law and superposition, is in both of them. It is then convenient to avoid the redundancy of writing both of them by generalizing the distribution $\rho(\boldsymbol{r}')$ so that it includes sharply localized integrals, i.e., point charges as a special case. We can then write the expression (B.1) in terms of the integral only, and this makes a very convenient shorthand.

To elaborate on this, consider one-dimensional charge distributions that are functions, say, of the x-coordinate. Suppose that we have such a distribution that is partially continuous and partially discrete. We can think of the discrete part, the one-dimensional "point charges," as consisting of thin charge slabs at $x_1, x_2, \ldots, x_n$. The continuous part is described by a density distribution which is some function of x. How can we describe the "point charges" as well in terms of a density distribution? Clearly the density distribution that describes a unit "*point*" source at, say, the origin must be a highly singular function. It must be nonvanishing essentially only at the origin, and moreover its integral must be unity. As a function of x, it is not an ordinary function, but if we call such a kind of "function" $\delta(x)$ it must have the properties

$$\begin{gathered} \delta(x) = 0, \qquad x \neq 0, \\ \int \delta(x)\,dx = 1. \end{gathered} \tag{B.2}$$

Moreover, since it is such a sharp function, if it appears in an integrand with any other function $f(x)$ which is *regular* at the origin, then the contribution to the integral will come only from $x = 0$. Then we can evaluate $f(x)$ in the integrand at the origin and get the basic defining equation of the δ-function,

$$\int f(x)\delta(x)\,dx = f(0). \tag{B.3}$$

With this definition, a source of strength q at $x = a$ is represented by $q\delta(x - a)$, where the generalized defining equation is

$$\int f(x)\delta(x - a)\,dx = f(a). \tag{B.4}$$

The extension to three dimensions is then straightforward. If there is a point charge q at $\boldsymbol{r} = \boldsymbol{r}_1$, i.e., at $x = x_1$, $y = y_1$, $z = z_1$, we can represent it by

$$\rho(x, y, z) = q\delta(x - x_1)\delta(y - y_1)\delta(z - z_1)$$

or, as we shall write more succinctly,

$$\rho(\boldsymbol{r}) = q\delta(\boldsymbol{r} - \boldsymbol{r}_1).$$

A sum of charges with charge q_i at $\boldsymbol{r}_i$ is then represented by

$$\rho(\boldsymbol{r}) = \sum_i q_i\delta(\boldsymbol{r} - \boldsymbol{r}_i). \tag{B.5}$$

Given this last expression, we can write the general expression for the potential Φ of a charge distribution of *any* kind as

$$\Phi(\boldsymbol{r}) = \int \frac{\rho(\boldsymbol{r}')}{|\boldsymbol{r} - \boldsymbol{r}'|}\,dv'. \tag{B.6}$$

for it is easy to verify that this formula, combined with the representation of point charges given by Eq. (B.5), just leads back to Eq. (B.1).

From the above point of view, the δ-function is defined by its integral properties, and the only real requirement on it is that the representation of a point charge by a δ-function must, when substituted in (B.6), lead back to a known correct formula. We see immediately, then, how to represent δ-functions in coordinates other than Cartesian. For example, in spherical coordinates r, θ, φ, $\delta(\boldsymbol{r} - \boldsymbol{r}_1)$ is represented by

$$\delta(\boldsymbol{r} - \boldsymbol{r}_1) = \frac{1}{r^2 \sin\theta}\delta(r - r_1)\delta(\theta - \theta_1)\delta(\varphi - \varphi_1).$$

The factor $1/r^2 \sin\theta$ is, of course, inserted here because in spherical coordinates the volume element is $r^2 \sin\theta\, d\theta\, d\varphi\, dr$.

As we have presented it up to now, the δ-function is a notational shorthand, and the integral properties of the function essentially define it. But as happens so often in mathematics, once the concept has been defined, it invites generalization. For example, in the expression (B.6) we may want to integrate by parts, or change variables, or expand in some series. The question then arises as to how to treat the δ-function part of the distribution function in these processes. Can we generalize the concept of the δ-function to make it close enough to the ordinary concept of a function so that at least some of the processes of analysis are defined for it? The answer is a limited yes. Although we cannot represent the δ-function as an ordinary function, we *can* represent it as the *limit of a sequence* of ordinary functions.

To elaborate on this, we shall go from the specific to the general. Consider for example the function

$$\frac{\sin \lambda x}{\pi x}.$$

This is a function which has the value λ/π at the origin, which integrates to unity independently of the value of λ, and, which as a function of x, becomes sharper as λ increases. In the limit when λ tends to infinity, it is then *infinite* at the origin, of *zero width* and with *unit integral*. In this limit, we can consider it to be a representation of the δ-function, and so can write

$$\delta(x) = \lim_{\lambda \to \infty} \frac{\sin \lambda x}{\pi x}.$$

This specific example is only one of an infinite number of similar representations. Start with any function $g(x)$ that has an absolute maximum at the origin, that drops off at least as $1/|x|$ for x large, and whose integral from $-\infty$ to ∞ is unity,

$$\int_{-\infty}^{\infty} g(x)\, dx = 1.$$

Take the function $\lambda g(\lambda x)$. This function has unit integral, by the last equation. Moreover, from the assumption on the behavior for large x

$$\lim_{\lambda \to \infty} \lambda g(\lambda x) = 0, \qquad x \neq 0.$$

For $x = 0$, on the other hand, the function is $\lambda g(0)$ which becomes infinite in the limit $\lambda \to \infty$. In this limit then, the function $\lambda g(\lambda x)$ satisfies all the requirements for the δ-function, and so we write

$$\delta(x) = \lim_{\lambda \to \infty} \lambda g(\lambda x).$$

As a rather popular special example of this general class, consider the function

$$g(x) = \frac{1}{\pi(x^2 + 1)}.$$

The prescription just given leads to the formula

$$\delta(x) = \lim_{\lambda \to \infty} \frac{\lambda}{\pi(\lambda^2 x^2 + 1)}.$$

This representation is usually written in a slightly different way, obtained by letting $\lambda = 1/\varepsilon$. Then

$$\delta(x) = \lim_{\varepsilon \to 0} \frac{\varepsilon}{\pi(x^2 + \varepsilon^2)};$$

Either from representations like these or from the basic definition (B.3), we can derive various useful formulas that enable us to treat δ-functions almost like ordinary functions. We list some of these formulas here. In them, a prime on a function means differentiation with respect to its variable.

$$\delta(ax - b) = 1/|a|\delta(x - b/a)$$

$$\delta(x^2 - a^2) = 1/2a[\delta(x - a) + \delta(x + a)]$$

$$\int \delta(x - a)\delta(x - b)\, dx = \delta(a - b) \tag{B.7}$$

$$\int f(x)\delta'(x)\, dx = -f'(0)$$

$$\int g(x)\delta(f(x))\, dx = \left. \frac{g(x)}{|df/dx|} \right|_{f(x)=0}$$

In the discussion above of δ-functions, they are presented as essentially a mathematical device invented to fulfil a pragmatic end; we put them in the mathematical formalism for our own purposes, and there they stay. What is surprising, however, is that once we are acquainted with them, we recognize them in rather unexpected places, sometimes associated with otherwise innocent mathematics. Thus, consider a complete orthonormal set of functions in one dimension; call them $\varphi_1(x), \varphi_2(x), \ldots, \varphi_n(x) \ldots$ Then any reasonable function $f(x)$ can be expanded in terms of this set,

$$f(x) = \sum_n a_n \varphi_n(x), \tag{B.8}$$

and the coefficients a_n are given by

$$a_n = \int f(x')\varphi_n^*(x')\, dx'$$

so that

$$f(x) = \sum_n \left[\int f(x')\varphi_n^*(x')\, dx' \right] \varphi_n(x).$$

If we assume we can interchange the summation and integration in this expression, it becomes

$$f(x) = \int f(x')\{\sum_n \varphi_n^*(x')\varphi_n(x)\}\, dx. \tag{B.9}$$

This is a somewhat remarkable equation and the quantity in braces, which is a function of x and x', is a remarkable function. In general, if $f(x')$ is integrated on the right-hand side with an *arbitrary* function of x and x' in the braces, we will not get the function $f(x)$ as a result. We will get $f(x)$ *only* if the function in braces has the basic integral property of the δ-function. For, with the tentative identification

$$\delta(x - x') = \sum_n \varphi_n^*(x')\varphi_n(x), \tag{B.10}$$

we can do the integration in Eq. (B.9) using Eq. (B.4) and indeed get back $f(x)$. Moreover, it is plausible that the sum in Eq. (B.10) has the other qualitative properties of the δ-function. For example, for $x = x'$, each term of the sum is $|\varphi_n(x)|^2$, a positive quantity. For this case then the sum tends to be large and, in fact, is generally infinite. For $x \neq x'$, on the other hand, the terms in the sum are sometimes positive and sometimes negative; there tends to be considerable cancellation, and the sum is small. Thus the sum has the basic quantitative and qualitative behavior of the δ-function and we are justified in keeping the provisional identification of Eq. (B.10).

With the general formula (B.10), we see that there are as many "representations" of the δ-function as there are orthonormal sets. For example, take the set

$$\frac{1}{\sqrt{2\pi}} e^{im\varphi}, \qquad m = 0, \pm 1, \pm 2, \ldots,$$

where φ is defined over the region between 0 and 2π, and the set is orthonormal there. Then we have

$$\delta(\varphi - \varphi') = \frac{1}{2\pi} \sum_{m=-\infty}^{\infty} e^{im(\varphi - \varphi')}.$$

There are analogous representations when in the set φ_n, the index n is not a discrete but is a continuous one; in this case, n is replaced by a variable k and the sum in (B.8) becomes an integral

$$f(x) = \int a(k)\varphi_k(x)\, dk. \tag{B.11}$$

In exactly the same way, however, we find that

$$\delta(x - x') = \int \varphi_k^*(x)\varphi_k(x')\, dk.$$

For example if the functions $\varphi_k(x)$ are

$$\varphi_k(x) = e^{ikx}/\sqrt{2\pi}$$

so that (B.11) is essentially the Fourier integral representation, we have

$$\delta(x - x') = (1/2\pi)\int e^{ik(x-x')}\,dk$$

or, with $x - x' = y$,

$$\delta(y) = (1/2\pi)\int e^{iky}\,dk. \tag{B.12}$$

The analogous equation for the three-dimensional function

$$\delta(\boldsymbol{r}) \equiv \delta(x)\delta(y)\delta(z)$$

is

$$\delta(\boldsymbol{r}) = (1/(2\pi)^3)\int e^{i\boldsymbol{k}\cdot\boldsymbol{r}}\,d\boldsymbol{k}, \tag{B.13}$$

where $d\boldsymbol{k} = dk_x\,dk_y\,dk_z$.

There is another important way in which the δ-function emerges from apparently innocent mathematics. Consider $\nabla^2(1/r)$, the operation of the Laplacian on the reciprocal of the spherical coordinate r. For r different from zero, the function $1/r$ is perfectly regular, and on using the expression for ∇^2 in spherical coordinates, we have

$$\nabla^2\left(\frac{1}{r}\right) = \frac{1}{r^2}\frac{d}{dr}\left(r^2\frac{d}{dr}\left(\frac{1}{r}\right)\right) = 0, \qquad r \neq 0.$$

At $r = 0$, however, we cannot differentiate since $1/r$ is singular there; thus $\nabla^2(1/r)$ is not defined at the origin. However, it is frequently the case in mathematics that functions which are not defined at a point because they are singular there can be integrated over. Consider, for example, the functions $\log x$ or $1/\sqrt{x}$ which are infinite, i.e., nondefined at the origin, but which still make sense in an integration through the origin. What is usually done in these cases is to redefine the integral, perhaps by first excluding a small region near the origin, and then go to the limit when the region shrinks to a point. In short, a common prescription is to redefine the integral in the principal value sense. It is then interesting to ask whether, although $\nabla^2(1/r)$ is infinite and therefore undefined at the origin, its *integral* over a small region containing the origin can be sensibly defined. Consider the integral of $\nabla^2(1/r)$ over a small sphere S centered at the origin. If the volume of the sphere is V we have by the divergence theorem

$$\int_V \nabla^2(1/r)\,dv = \int_V \nabla\cdot\nabla(1/r)\,dv = \int_S \nabla(1/r)\cdot d\boldsymbol{s} = -4\pi.$$

In this way we can give meaning to $\nabla^2(1/r)$ when it is integrated. Now consider a function $F(x, y, z)$, which is regular at the origin and which appears in the integral

$$\int F(x, y, z)\nabla^2(1/r)\,dv;$$

the integrand is a strange one: it vanishes everywhere except at the origin. We can then evaluate $F(x, y, z)$ at this point to get

$$\int F(x, y, z)\nabla^2(1/r)\,dv = F(0, 0, 0)\int \nabla^2(1/r)\,dv = -4\pi F(0, 0, 0).$$

But we recognize here the basic defining equation of the three-dimensional δ-function so we can subsume the discussion above in the equation

$$\nabla^2(1/r) = -4\pi\delta(\boldsymbol{r}). \tag{B.14}$$

APPENDIX C
NOTES ON VECTORS AND DYADICS

In this appendix, we briefly review a few topics in vector and dyadic analysis that are important in electromagnetic theory.

DIVERGENCE AND CURL

For defining the divergence at a point, consider a small volume ΔV of arbitrary shape which encloses the point, and which is bounded by a closed surface ΔS. The divergence of $\boldsymbol{A}$, written $\nabla \cdot \boldsymbol{A}$ or div $\boldsymbol{A}$, is defined as the limit of the flux of $\boldsymbol{A}$ through the surface ΔS, divided by the magnitude of the volume ΔV, as the surface shrinks down onto the point. Thus, with $\boldsymbol{ds}$ an outwardly drawn vector element of area, we have

$$\nabla \cdot \boldsymbol{A} = \lim_{\Delta V \to 0} \left(\int_{\Delta S} \boldsymbol{A} \cdot \boldsymbol{ds} / \Delta V \right). \tag{C.1}$$

Since the curl† of $\boldsymbol{A}$, written curl $\boldsymbol{A}$ or $\nabla \times \boldsymbol{A}$, is a vector, a prescription for calculating it at a point implies a prescription for calculating its component along an arbitrary direction. Such a direction may be thought of as an arrow associated with the point. Imagine a small plane closed curve C, of area ΔS, that contains the point, and thereby encloses the shaft, of the arrow; let the plane of C be perpendicular to the arrow. Define a positive sense of the traversal of C, such that the relation of this sense to the direction of the arrow is the same as the relation of the turning of a right-handed screw to its direction of advance. Then the component of $\nabla \times \boldsymbol{A}$ in the given direction is defined as the limit of the line integral, in the positive sense, of the component of A tangential to C, divided by the area ΔS, as the curve shrinks down onto the point. If d symbolizes the direction indicated by the arrow, then

$$(\nabla \times \boldsymbol{A})_d = \lim_{\Delta S \to 0} \left(\oint_C \boldsymbol{A} \cdot \boldsymbol{dl} / \Delta S \right). \tag{C.2}$$

† In the older literature, particularly the German, another name for curl $\boldsymbol{A}$ is rot $\boldsymbol{A}$, where *rot* derives from *rotation*.

DIVERGENCE THEOREM AND STOKES' THEOREM

Two important theorems closely related to the above definitions are the *divergence theorem* (also known as Gauss' theorem) and *Stokes' theorem.* They are proven in vector analysis; we merely record their statements here. For the divergence theorem, consider a finite volume V bounded by a closed surface S, for which the outwardly directed element of surface area is $\boldsymbol{ds}$. Then the theorem is

$$\int_V \nabla \cdot \boldsymbol{A}\, dv = \int_S \boldsymbol{A} \cdot \boldsymbol{ds}. \tag{C.3}$$

Note the relation of this formula to the definition of divergence in (C.1). The definition is recovered by applying the formula to an infinitesimal volume ΔV.

For Stokes' theorem, consider a closed curve C, a directed element of length $\boldsymbol{dl}$ on it, and a surface S that spans (contains) the curve. Then the theorem is

$$\int_S (\nabla \times \boldsymbol{A}) \cdot \boldsymbol{ds} = \int_C \boldsymbol{A} \cdot \boldsymbol{dl}, \tag{C.4}$$

where the sense of traversal of the curve and the direction of the positive normal to the surface are related as for a right-handed system. Again, the definition (C.2) of curl is recovered by applying Stokes' theorem to an infinitesimal surface.

DIFFERENTIAL OPERATIONS IN VARIOUS COORDINATE SYSTEMS

The definitions of divergence and curl above are quite general and independent of any specific set of coordinates. It is shown in vector analysis, that these definitions are equivalent to the representation of divergence and curl as *differential operators*, and, in applications, the latter are usually more important than the general definitions. The *form* of these differential operators of course differs from coordinate system to coordinate system; some of the forms are listed on page 386 for reference. Included as well are the forms of two other common operators: the gradient of a scalar function Φ, written as $\nabla\Phi$ or grad Φ, and the divergence of such a gradient, $\nabla \cdot \nabla\Phi$, also called the Laplacian, $\nabla^2\Phi$.

THE FORMULA $\nabla \times (\nabla \times \boldsymbol{A}) = \nabla\nabla \cdot \boldsymbol{A} - \nabla^2\boldsymbol{A}$

An expression that occurs frequently is that for the curl of the curl of a vector, $\nabla \times (\nabla \times \boldsymbol{A})$. It often appears in an equation in which it is related to a source function $\boldsymbol{S}$, as

$$\nabla \times (\nabla \times \boldsymbol{A}) = \boldsymbol{S}. \tag{C.5}$$

This equation represents three partial differential equations, in variables that depend on the choice of the coordinate system. For an arbitrary system, these

Cartesian coordinates (x, y, z)	Cylindrical coordinates (ρ, φ, z)	Spherical coordinates (r, θ, φ)
Gradient: $\nabla\psi$ or grad ψ		
$\hat{\boldsymbol{x}}\frac{\partial\psi}{\partial x}+\hat{\boldsymbol{y}}\frac{\partial\psi}{\partial y}+\hat{\boldsymbol{z}}\frac{\partial\psi}{\partial z}$	$\hat{\boldsymbol{\rho}}\frac{\partial\psi}{\partial\rho}+\hat{\boldsymbol{\varphi}}\frac{1}{\rho}\frac{\partial\psi}{\partial\varphi}+\hat{\boldsymbol{z}}\frac{\partial\psi}{\partial z}$	$\hat{\boldsymbol{r}}\frac{\partial\psi}{\partial r}+\hat{\boldsymbol{\theta}}\frac{1}{r}\frac{\partial\psi}{\partial\theta}+\hat{\boldsymbol{\varphi}}\frac{1}{r\sin\theta}\frac{\partial\psi}{\partial\varphi}$
Divergence: $\nabla\cdot\boldsymbol{A}$ or div $\boldsymbol{A}$		
$\frac{\partial A_x}{\partial x}+\frac{\partial A_y}{\partial y}+\frac{\partial A_z}{\partial z}$	$\frac{1}{\rho}\frac{\partial(\rho A_\rho)}{\partial\rho}+\frac{1}{\rho}\frac{\partial A_\varphi}{\partial\varphi}+\frac{\partial A_z}{\partial z}$	$\frac{1}{r^2}\frac{\partial(r^2A_r)}{\partial r}+\frac{1}{\mathrm{r}\sin\theta}\frac{\partial(\sin\theta A_\theta)}{\partial\theta}+\frac{1}{r\sin\theta}\frac{\partial A_\varphi}{\partial\varphi}$
Curl: $\nabla\times\boldsymbol{A}$ or curl $\boldsymbol{A}$		
$\hat{\boldsymbol{x}}\left(\frac{\partial A_z}{\partial y}-\frac{\partial A_y}{\partial z}\right)+\hat{\boldsymbol{y}}\left(\frac{\partial A_x}{\partial z}-\frac{\partial A_z}{\partial x}\right)+\hat{\boldsymbol{z}}\left(\frac{\partial A_y}{\partial x}-\frac{\partial A_x}{\partial y}\right)$	$\hat{\boldsymbol{\rho}}\left(\frac{1}{\rho}\frac{\partial A_z}{\partial\varphi}-\frac{\partial A_\varphi}{\partial z}\right)+\hat{\boldsymbol{\varphi}}\left(\frac{\partial A_\rho}{\partial z}-\frac{\partial A_z}{\partial\rho}\right)+\hat{\boldsymbol{z}}\frac{1}{\rho}\left(\frac{\partial(\rho A_\varphi)}{\partial\rho}-\frac{\partial A_\rho}{\partial\varphi}\right)$	$\hat{\boldsymbol{r}}\frac{1}{r\sin\theta}\left(\frac{\partial(\sin\theta\,A_\varphi)}{\partial\theta}-\frac{\partial A_\theta}{\partial\varphi}\right)+\hat{\boldsymbol{\theta}}\frac{1}{r\sin\theta}\frac{\partial A_r}{\partial\varphi}-\frac{1}{r}\frac{\partial(rA_\varphi)}{\partial r}+\hat{\boldsymbol{\varphi}}\frac{1}{r}\left(\frac{\partial(rA_\theta)}{\partial r}-\frac{\partial A_r}{\partial\theta}\right)$
Laplacian: $\nabla^2\psi=\nabla\cdot\nabla\psi$		
$\frac{\partial^2\psi}{\partial x^2}+\frac{\partial^2\psi}{\partial y^2}+\frac{\partial^2\psi}{\partial z^2}$	$\frac{1}{\rho}\frac{\partial}{\partial\rho}\left(\rho\frac{\partial\psi}{\partial\rho}\right)+\frac{1}{\rho^2}\frac{\partial^2\psi}{\partial\varphi^2}+\frac{\partial^2\psi}{\partial z^2}$	$\frac{1}{r^2}\frac{\partial}{\partial r}\left(r^2\frac{\partial\psi}{\partial r}\right)+\frac{1}{r^2\sin\theta}\frac{\partial}{\partial\theta}\left(\sin\theta\frac{\partial\psi}{\partial\theta}\right)+\frac{1}{r^2\sin^2\theta}\frac{\partial^2\psi}{\partial\varphi^2}$

equations are rather complicated. In spherical coordinates, for example, if we take the θ component of both sides of this equation, we find that the left-hand side involves not only A_θ, but A_r and A_φ as well and similarly for the r or φ component. In general, then, Eq. (C.5) stands for a *coupled* set of three partial differential equations.

It happens that this set has a particularly simple appearance in rectangular coordinates. It is easy to verify that, for example, the x-component of (C.5) can be written as

$$\frac{\partial}{\partial x}(\nabla \cdot \boldsymbol{A}) - \nabla^2 A_x = S_x. \tag{C.6}$$

With similar equations for the other two components, all three can be written at once, in a kind of shorthand, as

$$\nabla(\nabla \cdot \boldsymbol{A}) - \nabla^2 \boldsymbol{A} = \boldsymbol{S}. \tag{C.7}$$

The only meaning in this last equation, and concomitantly the only meaning in the hitherto undefined *Laplacian of a vector*, $\nabla^2 \boldsymbol{A}$, is as a mnemonic for three rectangular equations. But Eq. (C.7) is, alas, indecently suggestive. It looks like a vector equation that invites being broken into components in *arbitrary* coordinates in some such way as $\nabla_r(\nabla \cdot \boldsymbol{A}) - \nabla^2 A_r = S_r$. Many have done just that, to their ultimate sorrow (it is not correct); there is nothing in the derivation of (C.7), based as it is on Cartesian coordinates, to imply that it should be correct.

One reason that it is particularly useful to use rectangular coordinates in (C.5) is that frequently one deals with fields whose divergence is zero. In this case, Eq. (C.5) uncouples and reduces in rectangular coordinates to *three separate equations, one for each rectangular component of* $\boldsymbol{A}$.

HELMHOLTZ'S THEOREM

There is an important theorem and a corollary on vector fields, both of which are sometimes called *Helmholtz's theorem*, that we discuss now. Consider a vector function $\boldsymbol{C}(\boldsymbol{r})$ which is properly regular and bounded at infinity. What we shall choose to call Helmholtz's theorem is this: The function $\boldsymbol{C}$ can be written as the sum of *two* vector functions, for one of which the divergence is identically zero and for the other of which the curl is identically zero. In short, the decomposition

$$\boldsymbol{C}(\boldsymbol{r}) = \boldsymbol{D}(\boldsymbol{r}) + \boldsymbol{F}(\boldsymbol{r}) \tag{C.8}$$

is always possible, where

$$\nabla \cdot \boldsymbol{D} = 0, \qquad \nabla \times \boldsymbol{F} = 0. \tag{C.9}$$

The proof consists of actually exhibiting $\boldsymbol{D}$ and $\boldsymbol{F}$ in terms of $\boldsymbol{C}$. We begin by writing

$\boldsymbol{D}$ and $\boldsymbol{F}$ in terms of two other functions φ and $\boldsymbol{\chi}$ (potentials):

$$\boldsymbol{D} = \nabla \times \boldsymbol{\chi} \tag{C.10}$$

$$\boldsymbol{F} = -\nabla\varphi. \tag{C.11}$$

Then Eqs. (C.9) are satisfied automatically. Taking the divergence of (C.11) and the curl of (C.10), we have

$$\nabla^2\varphi = -\nabla \cdot \boldsymbol{C} \tag{C.12}$$

$$\nabla \times (\nabla \times \boldsymbol{\chi}) = \nabla \times \boldsymbol{C}. \tag{C.13}$$

The first of these equations is like Poisson's equation (2.15), in which the source term ρ is replaced by $\nabla \cdot \boldsymbol{C}/4\pi$. The second is like the magnetostatic Eq. (7.23) for the vector potential $\boldsymbol{A}$, in which $4\pi\boldsymbol{J}/c$ is replaced by $\nabla \times \boldsymbol{C}$. Then Eqs. (C.12, C.13) can be solved by similar superposition integrals, namely,

$$\varphi = \frac{1}{4\pi}\int \frac{\nabla' \cdot \boldsymbol{C}(\boldsymbol{r}')}{|\boldsymbol{r} - \boldsymbol{r}'|}\, dv',$$

$$\boldsymbol{\chi} = \frac{1}{4\pi}\int \frac{\nabla' \times \boldsymbol{C}(\boldsymbol{r}')}{|\boldsymbol{r} - \boldsymbol{r}'|}\, dv' \qquad \text{(Cartesian coordinates).}$$

Since $\boldsymbol{D}$ and $\boldsymbol{F}$ can now be found from φ and χ, we have exhibited the decomposition claimed by Helmholtz's theorem and thereby proved it. It is a corollary of this theorem that a vector function is determined if its curl and divergence are known everywhere; this is, in fact, the result usually referred to in the text as Helmholtz's theorem. Note finally that the formula $\boldsymbol{F} = -\nabla\varphi$ resembles that for the calculation of the electric field from the potential. Such a field, when produced by a point source, is *longitudinal* with respect to the radius vector from the charge to the field point. Similarly $\boldsymbol{D} = \nabla \times \chi$ is like that for calculating the field $\boldsymbol{B}$ from the vector potential. In this case, the field is *transverse* to the radius vector from a field producing element to the field point. For this reason $\boldsymbol{F}$ is sometimes called the *longitudinal* part of $\boldsymbol{C}$, and $\boldsymbol{D}$, the *transverse* part.

GENERAL FORMULAS

In the following $\boldsymbol{A}, \boldsymbol{B}, \boldsymbol{C}, \boldsymbol{D}$ are vector fields and φ and ψ are scalar fields.

$$\boldsymbol{A} \cdot \boldsymbol{B} \times \boldsymbol{C} = \boldsymbol{B} \cdot \boldsymbol{C} \times \boldsymbol{A} = \boldsymbol{C} \cdot \boldsymbol{A} \times \boldsymbol{B}$$

$$\boldsymbol{A} \times (\boldsymbol{B} \times \boldsymbol{C}) = (\boldsymbol{A} \cdot \boldsymbol{C})\boldsymbol{B} - (\boldsymbol{A} \cdot \boldsymbol{B})\boldsymbol{C}$$

$$\boldsymbol{A} \times (\boldsymbol{B} \times \boldsymbol{C}) + \boldsymbol{B} \times (\boldsymbol{C} \times \boldsymbol{A}) + \boldsymbol{C} \times (\boldsymbol{A} \times \boldsymbol{B}) = 0$$

$$(\boldsymbol{A} \times \boldsymbol{B}) \cdot (\boldsymbol{C} \times \boldsymbol{D}) = \boldsymbol{A} \cdot [\boldsymbol{B} \times (\boldsymbol{C} \times \boldsymbol{D})] = (\boldsymbol{A} \cdot \boldsymbol{C})(\boldsymbol{B} \cdot \boldsymbol{D}) - (\boldsymbol{A} \cdot \boldsymbol{D})(\boldsymbol{B} \cdot \boldsymbol{C})$$

$$(\boldsymbol{A} \times \boldsymbol{B}) \times (\boldsymbol{C} \times \boldsymbol{D}) = (\boldsymbol{A} \times \boldsymbol{B} \cdot \boldsymbol{D})\boldsymbol{C} - (\boldsymbol{A} \times \boldsymbol{B} \cdot \boldsymbol{C})\boldsymbol{D}$$

$$
\begin{aligned}
\nabla(\varphi\psi) &= \varphi\nabla\psi + \psi\nabla\varphi \\
\nabla\cdot(\varphi \boldsymbol{A}) &= \boldsymbol{A}\cdot\nabla\varphi + \varphi\nabla\cdot\boldsymbol{A} \\
\nabla\times(\varphi\boldsymbol{A}) &= \varphi\nabla\times\boldsymbol{A} - \boldsymbol{A}\times\nabla\varphi \\
\nabla\cdot(\boldsymbol{A}\times\boldsymbol{B}) &= \boldsymbol{B}\cdot(\nabla\times\boldsymbol{A}) - \boldsymbol{A}\cdot(\nabla\times\boldsymbol{B}) \\
\nabla\times(\boldsymbol{A}\times\boldsymbol{B}) &= \boldsymbol{A}(\nabla\cdot\boldsymbol{B}) - \boldsymbol{B}(\nabla\cdot\boldsymbol{A}) + (\boldsymbol{B}\cdot\nabla)\boldsymbol{A} - (\boldsymbol{A}\cdot\nabla)\boldsymbol{B} \\
\nabla(\boldsymbol{A}\cdot\boldsymbol{B}) &= \boldsymbol{A}\times(\nabla\times\boldsymbol{B}) + \boldsymbol{B}\times(\nabla\times\boldsymbol{A}) + (\boldsymbol{B}\cdot\nabla)\boldsymbol{A} + (\boldsymbol{A}\cdot\nabla)\boldsymbol{B} \\
\nabla\times\nabla\varphi &= 0 \\
\nabla\cdot(\nabla\times\boldsymbol{A}) &= 0
\end{aligned}
\tag{C.14}
$$

If V is a volume bounded by a closed surface S, with $\boldsymbol{ds}$ positive outward from the enclosed volume, then

$$
\begin{aligned}
\int_V (\nabla\cdot\boldsymbol{A})\,dv &= \oint_S \boldsymbol{A}\cdot\boldsymbol{ds}, \qquad \text{(Divergence theorem)} \\
\int_V (\nabla\varphi)\,dv &= \int_S \varphi\,\boldsymbol{ds}, \\
\int_V (\nabla\times\boldsymbol{A})\,dv &= \int_S (\boldsymbol{ds}\times\boldsymbol{A}).
\end{aligned}
\tag{C.15}
$$

If S is an open surface bounded by a contour C, and the relation of the contour line element $\boldsymbol{dl}$ and the normal to the surface $\boldsymbol{ds}$ is a right-handed one, then

$$
\begin{aligned}
\int_S (\nabla\times\boldsymbol{A})\cdot\boldsymbol{ds} &= \int_C \boldsymbol{A}\cdot\boldsymbol{dl}, \qquad \text{(Stokes' theorem)} \\
\int_S (\boldsymbol{ds}\times\nabla\varphi) &= \int_C \varphi\,\boldsymbol{dl}.
\end{aligned}
\tag{C.16}
$$

DYADS AND DYADICS

Frequently in physics one has to deal with two vectors that are somehow associated. For example, one vector may be a function of the other. Thus, consider an elastic body, a point of which is defined by the position vector $\boldsymbol{R}$. If the body is strained, the point moves to some other point $\boldsymbol{R} + \boldsymbol{S}$. The strained state of the body is thereby characterized by the vectors $\boldsymbol{R}$ and $\boldsymbol{S}$, where $\boldsymbol{S}$ is a function of $\boldsymbol{R}$. To show that two such vectors are associated, we may write them adjacently. We *define* the dyad $\mathscr{D}$ as such an entity, i.e., as the two vectors standing side by side:

$$\mathscr{D} = \boldsymbol{RS}. \tag{C.17}$$

Strained bodies aside, we can think of $\boldsymbol{R}$ and $\boldsymbol{S}$ as two arbitrary vectors, and of (C.17) as a general definition of a dyad.

This definition is an empty one without rules for operating with it. We shall discuss here only the simplest of these, focusing on the needs of the text. Given a vector $\boldsymbol{F}$, it is useful to define two dot products, $\mathscr{D}\cdot\boldsymbol{F}$ and $\boldsymbol{F}\cdot\mathscr{D}$, and a cross product, $\mathscr{D}\times\boldsymbol{F}$, by

$$\begin{aligned} \mathscr{D}\cdot\boldsymbol{F} &= \boldsymbol{R}(\boldsymbol{S}\cdot\boldsymbol{F}), \\ \boldsymbol{F}\cdot\mathscr{D} &= (\boldsymbol{F}\cdot\boldsymbol{R})\boldsymbol{S}, \\ \mathscr{D}\times\boldsymbol{F} &= \boldsymbol{R}(\boldsymbol{S}\times\boldsymbol{F}). \end{aligned} \tag{C.18}$$

The dot products yield vectors and the cross product yields another dyad.

A *dyadic* is a linear combination of dyads. The dyad (C.17) can be written in a useful expanded form, as a *dyadic*. This form is

$$\begin{aligned} \mathscr{D} = {} & R_xS_x\hat{\boldsymbol{x}}\hat{\boldsymbol{x}} + R_xS_y\hat{\boldsymbol{x}}\hat{\boldsymbol{y}} + R_xS_z\hat{\boldsymbol{x}}\hat{\boldsymbol{z}} + R_yS_x\hat{\boldsymbol{y}}\hat{\boldsymbol{x}} + R_yS_y\hat{\boldsymbol{y}}\hat{\boldsymbol{y}} + R_yS_z\hat{\boldsymbol{y}}\hat{\boldsymbol{z}} \\ & + R_zS_x\hat{\boldsymbol{z}}\hat{\boldsymbol{x}} + R_zS_y\hat{\boldsymbol{z}}\hat{\boldsymbol{y}} + R_zS_z\hat{\boldsymbol{z}}\hat{\boldsymbol{z}}. \end{aligned} \tag{C.19}$$

It is easy to verify that the expanded form on the right-hand side is equivalent to the left-hand side as far as the operations we have defined, assuming the distributive law. A special dyadic is the unit dyadic $\mathscr{I}$ defined by

$$\mathscr{I} = \hat{\boldsymbol{x}}\hat{\boldsymbol{x}} + \hat{\boldsymbol{y}}\hat{\boldsymbol{y}} + \hat{\boldsymbol{z}}\hat{\boldsymbol{z}}. \tag{C.20}$$

The dot or cross product of any vector with the unit dyadic just yields the vector itself.

The expanded form (C.19), or a slight generalization thereof, frequently crops up in another way in physics. Suppose that a vector $\boldsymbol{A}$ is a linear function of $\boldsymbol{B}$, by which we mean that each component of $\boldsymbol{A}$ is a linear function of each of the components of $\boldsymbol{B}$,

$$\begin{aligned} A_x &= \mathscr{L}_{xx}B_x + \mathscr{L}_{xy}B_y + \mathscr{L}_{xz}B_z, \\ A_y &= \mathscr{L}_{yx}B_x + \mathscr{L}_{yy}B_y + \mathscr{L}_{yz}B_z, \\ A_z &= \mathscr{L}_{zx}B_x + \mathscr{L}_{zy}B_y + \mathscr{L}_{zz}B_z. \end{aligned} \tag{C.21}$$

This equation can be written as the dyadic equation

$$\boldsymbol{A} = \mathscr{L}\cdot\boldsymbol{B}, \tag{C.22}$$

where the dyadic $\mathscr{L}$ is formed on the pattern of (C.19), i.e., as $\mathscr{L}_{xx}\hat{\boldsymbol{x}}\hat{\boldsymbol{x}} + \mathscr{L}_{xy}\hat{\boldsymbol{x}}\hat{\boldsymbol{y}} + \cdots$. There is clearly a close relationship between dyadics and second rank tensors. The essential difference is one of emphasis: with tensors, the transformation properties under rotation are usually paramount, whereas with dyadics, it is the algebraic properties in a given frame.

APPENDIX D
SOLUTION OF LAPLACE AND HELMHOLTZ EQUATIONS BY SEPARATION OF VARIABLES

In this Appendix, we summarily discuss a common method for finding solutions of the Laplace equation, $\nabla^2\Phi = 0$, and of the Helmholtz equation, $(\nabla^2 + k^2)\Phi = 0$. The method is that of *separation of variables.* Laplace's equation is, of course, a special case of the Helmholtz equation and we might discuss it as such. But it is clearer, if slightly repetitious, to go from the special to the general case. We shall consider Cartesian, cylindrical, and spherical coordinates. These systems are the most common ones, but there are many others, such as ellipsoidal, parabolic, bispherical, etc., for which the same general technique applies; a good treatment is given in Morse and Feshbach (B). Our discussion will be unbalanced in that we shall concentrate on the examples that are especially relevant for electromagnetic theory.

The first basic idea of the method of separation of variables is to seek an elementary solution of a partial differential equation in several variables as a *product of functions of one variable.* This elementary solution will involve a parameter known as the *separation constant.* By summing or integrating over a spectrum of values of this parameter, we can sometimes find a solution that is flexible enough to satisfy rather general boundary conditions. The separation constant is, *a priori*, completely arbitrary: it can be positive, negative, or even complex, and the nature of the solutions can depend critically on how it is chosen. Different boundary conditions may demand different assumptions. Moreover, it frequently happens that, with the sign of the constant fixed, the boundary conditions dictate that only certain values (*characteristic values* or *eigenvalues*) of the constant correspond to acceptable solutions (*characteristic functions* or *eigenfunctions*). These remarks are exemplified at various points in this Appendix and in the text.

CARTESIAN COORDINATES (x, y, z)

We start with the two-dimensional *Laplace equation*,

$$\frac{\partial^2\Phi}{\partial x^2} + \frac{\partial^2\Phi}{\partial y^2} = 0, \tag{D.1}$$

and seek an elementary solution as a product of $X(x)$, a function of x only, and $Y(y)$, a function of y only:

$$\Phi(x, y) = X(x)Y(y). \tag{D.2}$$

Substituting (D.2) into (D.1) and dividing the result by XY yields

$$-\frac{X''}{X} = \frac{Y''}{Y}, \tag{D.3}$$

where a prime on a function means differentiation with respect to its variable. In this equation, a function of x on the left-hand side equals a function of y on the right-hand side. The only way that a function of one variable can equal a function of another variable for all values of the variables is for both functions to be constant. We can therefore equate both sides of (D.3) to a *separation constant*. If we arbitrarily choose the separation constant to be a number α^2 which is positive or zero, we have the two equations

$$X'' + \alpha^2 X = 0,$$

$$Y'' - \alpha^2 Y = 0.$$

For $\alpha \neq 0$ this choice thus leads to trigonometric functions or complex exponentials for X, such as $e^{\pm i\alpha x}$, and to hyperbolic or exponential functions, such as $e^{\pm \alpha y}$, for Y. We can then take as the elementary separated solution

$$\Phi_\alpha = e^{\pm i\alpha x} e^{\pm \alpha y}. \tag{D.4}$$

Taking appropriate linear combinations of these solutions we can form, for example, the function $(A \sin \alpha x + B \cos \alpha x)(C \sinh \alpha y + D \cosh \alpha y)$ which is quoted in Section 5.2 as a solution of the Laplace equation. In addition to (D.4), there is, for $\alpha = 0$, a solution for X that is a linear function of x, and similarly for Y. Moreover, since (D.4) is a solution for *any* value of α, a sum or integral over a spectrum of values of α will remain a solution. By assuming the opposite sign for α^2 (replacing α by $i\alpha$), we find that the hyperbolic functions become trigonometric and vice versa. There is then no paucity of solutions, but this is neither here nor there. In solving partial differential equations, it is the fitting of solutions to the particular boundary conditions that is much more onerous than finding a multiplicity of general solutions in the first place.

The solution of *Laplace's equation in three dimensions*,

$$\frac{\partial^2\Phi}{\partial x^2} + \frac{\partial^2\Phi}{\partial y^2} + \frac{\partial^2\Phi}{\partial z^2} = 0,$$

proceeds in much the same way. We write

$$\Phi = X(x)Y(y)Z(z)$$

and find that the equation becomes

$$-\frac{X''}{X} = \frac{Y''}{Y} + \frac{Z''}{Z}.$$

By the same arguments as before, we can equate both sides of this equation to a constant α^2 to find

$$X'' + \alpha^2 X = 0, \tag{D.5}$$

and

$$-\frac{Y''}{Y} = \frac{Z''}{Z} - \alpha^2. \tag{D.6}$$

The solutions for X are then $\sin \alpha x$ and $\cos \alpha x$ or, equivalently, $e^{i\alpha x}$ and $e^{-i\alpha x}$. If we equate both sides of (D.6) to another constant β^2, we obtain

$$Y'' + \beta^2 Y = 0$$

$$Z'' - (\alpha^2 + \beta^2)Z = 0,$$

and the elementary separated solutions can be taken to be

$$\Phi_{\alpha\beta} = e^{\pm i\alpha x} e^{\pm i\beta y} e^{\pm\sqrt{\alpha^2+\beta^2}z}. \tag{D.7}$$

Again, by summing or integrating over α and β, by writing the exponentials in terms of trigonometric and hyperbolic functions, or by permuting the variables x, y, and z, we generate a large spectrum of solutions.

The technique of solution of the *Helmholtz equation* is so similar to that for Laplace's that we shall merely record the answer. The two-dimensional version is not important in electromagnetic theory. The basic solution of the three-dimensional one,

$$\left(\frac{\partial^2}{\partial x^2} + \frac{\partial^2}{\partial y^2} + \frac{\partial^2}{\partial z^2} + k^2\right)\psi = 0, \tag{D.8}$$

is

$$\psi_{\alpha\beta} = e^{\pm i\alpha x} e^{\pm i\beta y} e^{\pm i\sqrt{k^2-\alpha^2-\beta^2}z}. \tag{D.9}$$

The solution is thus sinusoidal or exponential in the z-direction depending on whether k^2 is greater or less than $\alpha^2 + \beta^2$. This consideration is important in the theory of wave guides.

CYLINDRICAL COORDINATES (ρ, φ, z)

Two-dimensional solutions

We discuss first the two-dimensional solutions of *Laplace's equation*, i.e., those without z-dependence, which therefore correspond to geometries that are infinite in the z-direction. For this case, the equation is

$$\rho\frac{\partial}{\partial\rho}\left(\rho\frac{\partial\Phi}{\partial\rho}\right) + \frac{\partial^2\Phi}{\partial\varphi^2} = 0. \tag{D.10}$$

We seek a solution of the form

$$\Phi(\rho, \varphi) = R(\rho)F(\varphi), \tag{D.11}$$

and find that (D.11) becomes

$$\frac{\rho}{R}\frac{d}{d\rho}\left(\rho\frac{dR}{d\rho}\right) = -\frac{1}{F}\frac{d^2F}{d\varphi^2}. \tag{D.12}$$

In this equation, a function of ρ now equals a function of φ, and the usual argument of the method applies. We equate both sides of (D.12) to a separation constant; this is arbitrarily chosen as positive and is written as p^2. We have then

$$\frac{d^2F}{d\varphi^2} + p^2F = 0 \tag{D.13}$$

and

$$\rho\frac{d}{d\rho}\left(\rho\frac{dR}{d\rho}\right) - p^2R = 0. \tag{D.14}$$

The solutions of Eq. (D.13) will be most conveniently taken as the trigonometric functions $\sin p\varphi$ and $\cos p\varphi$. In general, p is arbitrary, but there is an important case in which it is quantized. Suppose that there is a problem in which the whole range of φ between 0 and 2π comes into consideration, as for a circular cylinder. If we require that Φ be continuous, and in particular that as φ approaches 2π, Φ approaches its value at zero, we must have

$$\cos 2\pi p = \cos 0 = 1$$

$$\sin 2\pi p = \sin 0 = 0.$$

These equations can be satisfied by taking p to be zero or a positive† integer n. We then label F with a subscript n and write the solutions of Eq. (D.13) as

$$F_n = \begin{cases} \cos n\varphi \\ \sin n\varphi \end{cases} \qquad n = 0, 1, 2. \tag{D.15}$$

An expansion in these functions is just a Fourier series. Now p is set equal to n in Eq. (D.14) and the function R is also labeled with a subscript. It is then easy to verify that its solutions are just powers ρ^n and ρ^{-n}, for $n \neq 0$, and a logarithmic term for $n = 0$:

$$R_0 = \ln \rho, \qquad R_n = \rho^{\pm n}. \tag{D.16}$$

Taking linear combinations with arbitrary coefficients of the elementary solutions

† This is no restriction, since the solutions for negative integers are not linearly independent of those for positive.

(D.15) and (D.16), we find that the general solutions of (D.10), subject to the conditions mentioned, is

$$\Phi = A \ln \rho + \sum_{n=1}^{\infty} (B_n \rho^n + C_n \rho^{-n})(D_n \cos n\varphi + E_n \sin n\varphi). \tag{D.17}$$

It is essentially this solution that emerged from the discussion of the summation problem for two-dimensional charge distributions.

This last and common result far from exhausts the solutions that are implicit in the method of separation of variables. For example, for $p = 0$ in (D.13), there is also a solution linear in φ which has not been considered. And if φ ranges only over part of the domain from zero to 2π, as it might in problems involving wedge or sector-shaped cylinders, there is no need to quanitize p. Also, if we had chosen the separation constant with the opposite sign, for instance, $-p^2$ instead of p^2, the F functions would be $e^{p\varphi}$ and $e^{-p\varphi}$ and the solutions for R would be modified. These solutions are not especially appropriate for boundary conditions at fixed ρ for all φ but can be used for boundary conditions at fixed φ for all ρ. For example, the problem of a two-dimensional line charge parallel to a dielectric wedge can be solved in terms of them. The details are given in Smythe (B).

Helmholtz's equation, in two-dimensional cylindrical coordinates is

$$\frac{1}{\rho}\frac{\partial}{\partial \rho}\left(\rho \frac{\partial \psi}{\partial \rho}\right) + \frac{1}{\rho^2}\frac{\partial^2 \psi}{\partial \varphi^2} + k^2 \psi = 0. \tag{D.18}$$

The separation assumption $\psi = R(\rho)F(\varphi)$ yields

$$\frac{\rho}{R}\frac{d}{d\rho}\left(\rho \frac{dR}{d\rho}\right) + k^2\rho^2 = -\frac{1}{F}\frac{d^2F}{d\varphi^2}. \tag{D.19}$$

We take the separation constant to be p^2, as for the Laplace equation. The solutions for the equation in F are then conveniently taken to be complex exponentials

$$F = e^{\pm ip\varphi}.$$

The equation for R becomes, with $x = k\rho$,

$$\frac{d^2R}{dx^2} + \frac{1}{x}\frac{dR}{dx} + \left(1 - \frac{p^2}{x^2}\right)R = 0. \tag{D.20}$$

This equation is called *Bessel's equation*. It is a standard one in higher analysis, and we shall shortly summarize some of its properties. For the moment, we merely remark that the most common case of interest is usually that when the solutions in φ are continuous. This quantizes the parameter p, as before, to be only an integer or zero. We write

$$F_n = e^{\pm in\varphi}, \qquad n = 0, 1, 2 \ldots. \tag{D.21}$$

In this case, the two independent solutions of Eq. (D.20) can be taken as the so-called *Bessel function* J_n or *Neumann function* N_n, both of which we shall define shortly. In terms of them, the separated solutions of (D.18), subject to the continuity condition on φ, are

$$\psi_n = \begin{Bmatrix} J_n(kr) \\ N_n(kr) \end{Bmatrix} e^{\pm in\varphi}, \qquad n = 0, 1, 2, \ldots. \tag{D.22}$$

The Bessel equation (D.20) is important beyond its present context, so we shall briefly list some of the properties of the functions that are its solutions. The theory of these functions is extensive, and the literature huge, so that we shall barely be able to scratch the surface here. As a second-order differential equation, Eq. (D.20) has two independent solutions. These can be expressed as power series, and, for p *nonintegral*, they are often taken to be the two Bessel functions of order p and $-p$. These are designated by J_p and J_{-p} and are defined by

$$J_p(x) = \left(\frac{x}{2}\right)^p \sum_{j=0}^{\infty} \frac{(-)^j}{j!\Gamma(j+p+1)}\left(\frac{x}{2}\right)^{2j} \tag{D.23}$$

and

$$J_{-p}(x) = \left(\frac{x}{2}\right)^{-p} \sum_{j=0}^{\infty} \frac{(-)^j}{j!\Gamma(j-p+1)}\left(\frac{x}{2}\right)^{2j}. \tag{D.24}$$

where $\Gamma(x)$ is the *gamma-function* of x. Thus J_p is regular and J_{-p} is singular at the origin. It turns out that when p is integral, J_p is not linearly independent of J_{-p}. A second linearly independent solution can be found and it is frequently taken to be the *Neumann function* N_p defined in general by

$$N_p(x) = \frac{J_p(x)\cos p\pi - J_{-p}(x)}{\sin p\pi}. \tag{D.25}$$

Then for p integral *or* nonintegral the basic pair of solutions can be taken to be J_p and N_p. Further, it is often convenient to take *linear combinations* of J_p and N_p as the basic pair of solutions. Two such combinations of importance are the *Hankel functions* $H_p^{(1)}$ and $H_p^{(2)}$:

$$H_p^{(1)} = J_p + iN_p, \qquad H_p^{(2)} = J_p - iN_p. \tag{D.26}$$

It is frequently useful to know the asymptotic or limiting values of the various functions defined above. We list them here.

For $x \ll 1$

$$J_p(x) \to (1/\Gamma(p+1))(x/2)^p,$$

$$N_0(x) \to (2/\pi)(\ln(x/2) + 0{\cdot}5772\cdots), \tag{D.27a}$$

$$N_p(x) \to -(\Gamma(p)/\pi)(2/x)^p, \qquad p > 0.$$

For $x \gg 1, p$

$$J_p(x) \to \sqrt{2/\pi x}\cos(x - p\pi/2 - \pi/4),$$
$$N_p(x) \to \sqrt{2/\pi x}\sin(x - p\pi/2 - \pi/4), \qquad \text{(D.27b)}$$
$$H_p^{(1,2)} \to \sqrt{2/\pi x}\, e^{\pm i(x - p\pi/2 - \pi/4)}.$$

Important properties of these functions are the recursion relations. If F_p stands for either J_p, N_p, or H_p, it can be shown that

$$F_{p-1}(x) + F_{p+1}(x) = \frac{2p}{x}F_p(x)$$

$$F_{p-1}(x) - F_{p+1}(x) = 2\frac{dF_p}{dx}$$

Useful integral representations are:

$$J_p(x) = \frac{1}{2\pi i^m}\int_0^{2\pi} e^{i(x\cos\varphi + m\varphi)}\,d\varphi.$$

$$H_0^{(1)}(x) = -\frac{i}{\pi}\int_{-\infty}^{\infty}\frac{e^{i\sqrt{x^2+s^2}}}{\sqrt{x^2+s^2}}\,ds. \qquad \text{(D.28)}$$

Solutions with *z*-dependence

Bessel functions also emerge from the solution of both the Laplace and Helmholtz equations in the general case in cylindrical coordinates when there is z-dependence. We discuss this now, beginning with the Laplace equation,

$$\frac{1}{\rho}\frac{\partial}{\partial\rho}\left(\rho\frac{\partial\Phi}{\partial\rho}\right) + \frac{1}{\rho^2}\frac{\partial^2\Phi}{\partial\varphi^2} + \frac{\partial^2\Phi}{\partial z^2} = 0. \qquad \text{(D.30)}$$

Writing

$$\Phi = R(\rho)F(\varphi)Z(z),$$

we find as before that $(1/F)(d^2F/d\varphi^2)$ must be equated to a separation constant. Under the same assumptions about continuity of solution as a function of φ, we conclude that the φ dependence is again given by Eq. (D.21). With the φ-dependence separated off, Eq. (D.30) becomes

$$\frac{1}{\rho R}\frac{d}{d\rho}\left(\rho\frac{dR}{d\rho}\right) - \frac{n^2}{\rho^2} = -\frac{1}{Z}\frac{d^2Z}{dz^2}. \qquad \text{(D.31)}$$

A function of z equals a function of ρ, and once more we must choose a separation constant. If there is periodicity in the z-direction the constant can be taken of a sign to give trigonometric functions of z as solutions. Alternatively, it can be chosen to yield exponential or hyperbolic functions. We begin with the latter case, for which the z-equation becomes, with separation constant $-\lambda^2$,

$$\frac{d^2Z}{dz^2} - \lambda^2 Z = 0.$$

Here the solutions are $e^{\pm\lambda z}$ or $\sinh \lambda z$ and $\cosh \lambda z$. With the two separation constants disposed of, the equation for $R(\rho)$ is now unique:

$$\frac{d^2R}{d\rho^2} + \frac{1}{\rho}\frac{dR}{d\rho} + \left(\lambda^2 - \frac{n^2}{\rho^2}\right)R = 0. \tag{D.32}$$

This equation, nominally a function of ρ, λ, and n separately, is in fact a function only of $\lambda\rho$ and n as the substitution $x = \lambda\rho$ immediately shows. Hence (D.32) becomes

$$\frac{d^2R}{dx^2} + \frac{1}{x}\frac{dR}{dx} + \left(1 - \frac{n^2}{x^2}\right)R = 0. \tag{D.33}$$

Thus the solution of the R-equation is once again a Bessel function. The elementary separated solution of Eq. (D.30) can then be taken to be

$$\Phi_{n\lambda} = \begin{Bmatrix} J_n(\lambda\rho) \\ N_n(\lambda\rho) \end{Bmatrix} e^{\pm\lambda z} e^{\pm in\varphi}, \qquad n = 0, 1, 2 \ldots . \tag{D.34}$$

Choosing the opposite sign of the separation constant in the equation for z is equivalent to replacing λ by an imaginary quantity $i\gamma$ in Eq. (D.34). The z-dependence is now as $e^{\pm i\gamma z}$, and the Bessel functions become functions of imaginary argument. As a result, the solutions are

$$\Phi_{n\gamma} = \begin{Bmatrix} J_n(i\gamma\rho) \\ N_n(i\gamma\rho) \end{Bmatrix} e^{\pm i\gamma z} e^{\pm in\varphi}. \tag{D.35}$$

These functions of imaginary argument are frequently called *modified* or *hyperbolic Bessel functions*. In the present case of imaginary arguments the basic pair of solutions of (D.33) is often taken to be not J_n and N_n, but two functions I_n and K_n defined by

$$I_n(x) = i^{-n} J_n(ix)$$

$$K_n(x) = (\pi/2) i^{n+1} H_n^{(1)}(ix).$$

These two functions are convenient in that their asymptotic forms have a certain symmetry. For large x, I_n approaches $e^x/\sqrt{2\pi x}$ and K_n approaches $\sqrt{\pi/2x}\, e^{-x}$.

We now discuss the solution of the *Helmholtz equation*,

$$\frac{1}{\rho}\frac{\partial}{\partial\rho}\left(\rho\frac{\partial\psi}{\partial\rho}\right) + \frac{1}{\rho^2}\frac{\partial^2\psi}{\partial\varphi^2} + \frac{\partial^2\psi}{\partial z^2} + k^2\psi = 0. \tag{D.36}$$

We shall be rather summary since the main features of the separated solutions have already been discussed. The dependence of the solutions on φ will be given by Eq. (D.21) under the assumption of continuity in φ that was already discussed. As for the Laplace equation, the dependence on z will be either as $e^{\pm \gamma z}$ or as $e^{\pm i\gamma z}$, depending on the sign of a separation constant. For the Helmholtz equation, which is, of course, derived from the wave equation, the most common case of interest is when the dependence is as $e^{\pm i\gamma z}$, since this corresponds to wave-like motion along the z-axis. With this choice, the dependence on ρ is unambiguous, and turns out to be a solution of Bessel's equation. If we arbitrarily take the two linearly independent solutions of this equation to be J_n and N_n, the elementary separated solution of the Helmholtz equation is:

$$\psi_{n\gamma} = \begin{Bmatrix} J_n(\sqrt{k^2 - \gamma^2}\rho) \\ N_n(\sqrt{k^2 - \gamma^2}\rho) \end{Bmatrix} e^{\pm i\gamma z \pm in\varphi}. \tag{D.37}$$

SPHERICAL COORDINATES (r, θ, φ)

We turn to the Laplace and Helmholtz equations in spherical coordinates. It will be convenient to start with the more general Helmholtz equation and discuss the dependence of the solution on the angles θ and φ. The same dependence will then obtain for the Laplace equation; only the functions of r will differ.

The Helmholtz equation is

$$\frac{1}{r^2}\frac{\partial}{\partial r}\left(r^2 \frac{\partial \psi}{\partial r}\right) + \frac{1}{r^2 \sin\theta}\frac{\partial}{\partial \theta}\left(\sin\theta \frac{\partial \psi}{\partial \theta}\right) + \frac{1}{r^2 \sin^2\theta}\frac{\partial^2 \psi}{\partial \varphi^2} + k^2\psi = 0. \tag{D.38}$$

With the assumed separated solution

$$\psi = R(r)P(\theta)F(\varphi),$$

we are led to solutions for F of the form $e^{\pm i\gamma\varphi}$, where γ^2 is a separation constant. With the familiar assumption of continuity in φ, the parameter γ must be zero or an integer. Traditionally this integer is called m, the exponent is taken with the *positive* sign only, and m is then allowed to take on negative integral values:

$$F(\varphi) \to e^{im\varphi}, \qquad m = 0, \pm 1, \pm 2, \ldots.$$

The equations for $R(r)$ and $P(\theta)$ become, with λ^2 a separation constant,

$$r^2 \frac{d^2R}{dr^2} + 2r\frac{dR}{dr} + (k^2r^2 - \lambda^2)R = 0 \tag{D.39}$$

and

$$\frac{1}{\sin\theta}\frac{d}{d\theta}\left(\sin\theta \frac{dP}{d\theta}\right) + \left(\lambda^2 - \frac{m^2}{\sin^2\theta}\right)P = 0. \tag{D.40}$$

With the substitution $x = \cos\theta$, we have for (D.40)

$$\frac{d}{dx}\left((1 - x^2)\frac{dP}{dx}\right) + \left(\lambda^2 - \frac{m^2}{1 - x^2}\right)P = 0. \tag{D.41}$$

Consider first the case $m = 0$ for which

$$\frac{d}{dx}\left((1 - x^2)\frac{dP}{dx}\right) + \lambda^2 P = 0.$$

We summarize the results of the solution of this equation. If a power-series solution is tried, it turns out that generally the series represents a function infinite at $x = \pm 1$. But if all coefficients beyond some fixed one in the formally infinite series are zero, i.e., if the series really is a polynomial, the function is finite at $x = \pm 1$. The condition that the series break off is that $\lambda^2 = l(l + 1)$, where l is zero or a positive integer. Usually, but not always, the physics of the problem requires that the solution be finite at θ = zero and π, i.e., at $x = \pm 1$. In this case, applying this condition generates a sequence of polynomial solutions, one for each value of l; these are the *Legendre polynomials* $P_l(x)$. The first few of them are:

$$P_0(x) = 1$$

$$P_1(x) = x$$

$$P_2(x) = \tfrac{1}{2}(3x^2 - 1)$$

$$P_3(x) = \tfrac{1}{2}(5x^3 - 3x)$$

$$P_4(x) = \tfrac{1}{8}(35x^4 - 30x^2 + 3).$$

A general definition of $P_l(x)$ is comprised in *Rodrigues' formula*:

$$P_l(x) = \frac{1}{2^l l!}\frac{d^l}{dx^l}(x^2 - 1)^l. \tag{D.42}$$

Useful recursion relations can be derived from this formula. Among them are:

$$\frac{dP_{l+1}}{dx} - \frac{dP_{l-1}}{dx} - (2l + 1)P_l = 0,$$

$$(l + 1)P_{l+1} - (2l + 1)xP_l + lP_{l-1} = 0, \tag{D.43}$$

$$(x^2 - 1)\frac{dP_l}{dx} - lxP_l + lP_{l-1} = 0.$$

As a solution of a *Sturm-Liouville problem*, the Legendre polynomials satisfy an orthogonality relation which turns out to be

$$\int_{-1}^{1} P_{l'}(x)P_l(x)\,dx = \frac{2}{2l + 1}\delta_{ll'}. \tag{D.44}$$

Moreover, the functions form a complete set for the expansion of a function of x over the interval -1 to 1 (or of $\cos\theta$ over the interval 0 to π). The coefficients are then determined in the usual way by the help of (D.44).

The general case of Eq. (D.41), with $m \neq 0$, is similar to that for $m = 0$. Again, for finite solutions at $x = \pm 1$, λ^2 must equal $l(l+1)$. It turns out that, for a given l, the values of m are no longer *arbitrary* integers, but are limited to the values

$$-l, -(l-1), -(l-2)\cdots 0, \cdots (l-1), l.$$

For a given l, there is a different solution for each allowed value of m. The functions are labeled by two indices l and m, are called *associated Legendre polynomials* (or *functions*) and are written P_l^m. They are found, for positive m, to be related to the ordinary Legendre polynomials by

$$P_l^m(x) = \frac{(-)^m}{2^l l!}(1-x^2)^{m/2}\frac{d^{l+m}}{dx^{l+m}}(x^2-1)^l. \tag{D.45}$$

For negative m, they are defined by

$$P_l^{-m}(x) = (-)^m\frac{(l-m)!}{(l+m)!}P_l^m(x). \tag{D.46}$$

For a given m, the associated Legendre polynomials form an orthogonal set on the interval $-1 \leqslant x \leqslant 1$. The orthogonality relation reads

$$\int_{-1}^{1} P_{l'}^m(x)P_l^m(x)\,dx = \frac{2}{2l+1}\frac{(l+m)!}{(l-m)!}\delta_{ll'}. \tag{D.47}$$

Equation (D.44) is a special case of this, since $P_l^0(x) = P_l(x)$.

An arbitrary function of θ $(0 < \theta < \pi)$ can be expanded in terms of the $P_l^m(\cos\theta)$ and an arbitrary function of φ $(0 < \varphi < 2\pi)$, in terms of the functions $e^{im\varphi}$. The product of these two functions $P_l^m(\cos\theta)e^{im\varphi}$ then forms a complete set, when doubly summed over l and m, for expansion of an arbitrary function of θ and φ. These product functions, when normalized in a certain way, are called *spherical harmonics*, and are designated by $Y_{lm}(\theta, \varphi)$. There are various conventions for the normalization. The one we adopt is

$$Y_{lm}(\theta, \varphi) = \sqrt{\frac{2l+1}{4\pi}\frac{(l-m)!}{(l+m)!}}\,P_l^m(\cos\theta)e^{im\varphi}, \tag{D.48}$$

which implies that

$$\int_0^{2\pi} d\varphi \int_0^{\pi} Y_{l'm'}^* Y_{lm}\sin\theta\,d\theta = \delta_{ll'}\delta_{mm'}. \tag{D.49}$$

From Eq. (D.46), we find

$$Y_{l,-m}(\theta, \varphi) = (-)^m Y_{lm}^*(\theta, \varphi). \tag{D.50}$$

This equation defines the spherical harmonics with negative m in terms of those with positive m. The first few spherical harmonics, with positive m, are

$$Y_{00} = \frac{1}{\sqrt{4\pi}}$$

$$Y_{11} = -\sqrt{\frac{3}{8\pi}} \sin\theta\, e^{i\varphi}, \qquad Y_{10} = \sqrt{\frac{3}{4\pi}} \cos\theta, \tag{D.51}$$

$$Y_{22} = \sqrt{\frac{15}{32\pi}} \sin^2\theta\, e^{2i\varphi},\ Y_{21} = -\sqrt{\frac{15}{8\pi}} \sin\theta \cos\theta\, e^{i\varphi},\ Y_{20} = \sqrt{\frac{5}{16\pi}}\ (3\cos^2\theta - 1).$$

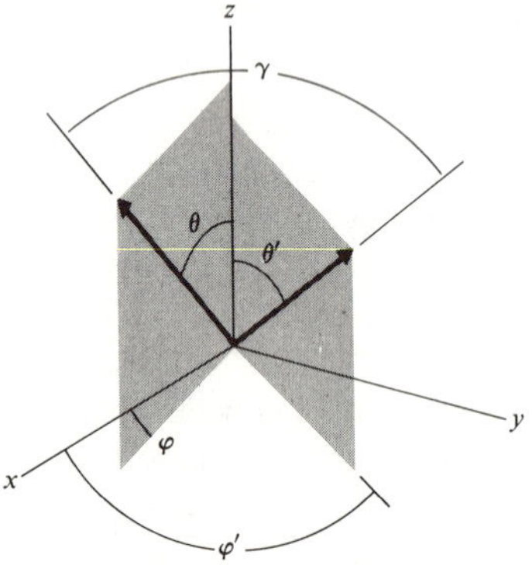

Fig. D.1 Angles involved in the addition theorem.

An important application of spherical harmonics is in the *addition theorem*, which we quote here. Consider two directions in space characterized by the angles θ, φ, and θ', φ', respectively, as shown in Fig. D.1. The angle γ between those two directions is a function of θ, φ, and θ', φ', and so then is $P_l(\cos\gamma)$. We can imagine $P_l(\cos\gamma)$, as a function of all four variables, expanded in terms of a sum over l, m, l', m' of the products $Y_{lm}(\theta, \varphi)\, Y_{l'm'}(\theta', \varphi')$. It turns out that this expansion is in fact simpler than one might have guessed, in that it contains terms with only $l = l'$ and $m = m'$. Explicitly, it is

$$P_l(\cos\gamma) = \frac{4\pi}{2l+1} \sum_{m=-l}^{l} Y_{lm}(\theta, \varphi) Y_{lm}^*(\theta', \varphi'). \tag{D.52}$$

It is this formula that is called the *addition theorem of spherical harmonics*.

So much for the angular dependence of solutions in spherical coordinates; we return to the radial equations. We first consider the *Laplace* equation; the radial equation is just (D.39) with k set equal to zero, with $\lambda^2 = l(l + 1)$, and with a subscript on R:

$$r^2\frac{d^2R_l}{dr^2} + 2r\frac{dR_l}{dr} - l(l+1)R_l = 0.$$

It is easy to verify that the two solutions of this equation are

$$R_l = \begin{cases} r^l \\ r^{-(l+1)} \end{cases}.$$

With this result we can put together the general solution (subject to all the specific choices of separation constants) of the Laplace equation:

$$\Phi(r, \theta, \varphi) = \sum_{l,m} (A_{lm}r^l + B_{lm}r^{-(l+1)})Y_{lm}(\theta, \varphi). \tag{D.53}$$

It is just this solution that emerged from the discussion of the summation problem in electrostatics.

Now we consider the radial equation (D.39) for the Helmholtz equation, again with $\lambda^2 = l(l+1)$ and a subscript on R. With the substitution

$$R_l(r) = S_l(r)/(kr)^{1/2},$$

it is readily found that

$$r^2\frac{d^2S_l}{dr^2} + r\frac{dS_l}{dr} + ((kr)^2 - (l+\tfrac{1}{2})^2)S_l = 0.$$

The solution of this equation is a Bessel or Neumann-function of half integral order, so that $R_l(r)$ is:

$$R_l(r) = \begin{cases} J_{l+1/2}(kr)/(kr)^{1/2} \\ N_{l+1/2}(kr)/(kr)^{1/2} \end{cases}$$

or any of the linear combinations of J_l and N_l that have already been discussed. It is now useful to define functions j_l, n_l, and h_l that are generally called *spherical Bessel functions* and that are related to the Bessel, Neumann, and Hankel functions by

$$\begin{aligned} j_l(x) &= \sqrt{\frac{\pi}{2x}}J_{l+1/2}(x), \\ n_l(x) &= \sqrt{\frac{\pi}{2x}}N_{l+1/2}(x), \\ h_l^{(1,2)}(x) &= \sqrt{\frac{\pi}{2x}}(J_{l+1/2}(x) \pm iN_{l+1/2}(x)). \end{aligned} \tag{D.54}$$

Arbitrarily choosing j_l and n_l rather than some other pair as the basic solution, the general solution of the Helmholtz equation is:

$$\psi(r, \theta, \varphi) = \sum_{l,m} (A_{lm} j_l(kr) + B_{lm} n_l(kr)) Y_{lm}(\theta, \varphi).$$

We tabulate here some of the properties of spherical Bessel functions. They are expressible in terms of elementary functions and the first few are:

$$\begin{aligned}
&j_0(x) = \frac{\sin x}{x}, \qquad n_0(x) = -\frac{\cos x}{x}, \qquad h_0^{(1)}(x) = \frac{e^{ix}}{ix};\\
&j_1(x) = \frac{\sin x}{x^2} - \frac{\cos x}{x}, \qquad n_1(x) = -\frac{\cos x}{x^2} - \frac{\sin x}{x},\\
&h_1^{(1)}(x) = -\frac{e^{ix}}{x}\left(1 + \frac{i}{x}\right);\\
&j_2(x) = \left(\frac{3}{x^3} - \frac{1}{x}\right)\sin x - \frac{3\cos x}{x^2},\\
&n_2(x) = -\left(\frac{3}{x^3} - \frac{1}{x}\right)\cos x - \frac{3\sin x}{x^2},\\
&h_2^{(1)}(x) = \frac{ie^{ix}}{x}\left(1 + \frac{3i}{x} - \frac{3}{x^2}\right).
\end{aligned} \tag{D.55}$$

Their limiting form for *small* x ($x \ll l$) is

$$\begin{aligned}
j_l(x) &\to \frac{x^l}{1.3.5\cdots(2l+1)}\\
n_l(x) &\to -\frac{1.1.3.5\cdots(2l-1)}{x^{l+1}}.
\end{aligned} \tag{D.56}$$

For large x, ($x \gg l$) the asymptotic forms are

$$\begin{aligned}
j_l(x) &\to \frac{1}{x}\sin\left(x - \frac{l\pi}{2}\right),\\
n_l(x) &\to -\frac{1}{x}\cos\left(x - \frac{l\pi}{2}\right),\\
h_l^{(1)}(x) &\to (-i)^{l+1}\frac{e^{ix}}{x}.
\end{aligned} \tag{D.57}$$

If $\gamma_l(x)$ stands for any of these functions, then

$$\frac{2l+1}{x}\gamma_l(x) = \gamma_{l-1}(x) + \gamma_{l+1}(x)$$

$$\frac{d\gamma_l}{dx} = \frac{1}{2l+1}[l\gamma_{l-1}(x) - (l+1)\gamma_{l+1}(x)]. \tag{D.58}$$

BIBLIOGRAPHY

The following books are general ones that are relevant to, or cited in, more than one chapter.

Abraham, M., and R. Becker, *Electricity and Magnetism*, Blackie, London (1937).

Abramovitz, Milton, and Irene A. Stegun, *Handbook of Mathematical Functions*, Dover, New York (1965).

Born, M., and E. Wolf, *Principles of Optics*, Pergamon, New York (1959).

Butkov, E., *Mathematical Physics*, Addison-Wesley, Reading, Mass. (1968).

Clemmow, P. C., *The Plane Wave Spectrum Representation of Electromagnetic Fields*, Pergamon, Oxford (1966).

Clemmow, P. C., and J. Dougherty, *Electrodynamics of Particles and Plasmas*, Addison-Wesley, Reading, Mass. (1969).

Harnwell, B. P., *Principles of Electricity and Electromagnetism*, 2nd ed., McGraw-Hill, New York (1949).

Harrington, R. E., *Time Harmonic Electromagnetic Fields*, McGraw-Hill, New York (1961).

Iwanenko, D., and A. Sokolow, *Klassische Feldtheorie*, Academie-Verlag, Berlin (1953).

Jackson, J. D., *Classical Electrodynamics*, Wiley, New York (1962).

Jeans, J. H., *Mathematical Theory of Electricity and Magnetism*, 5th ed., Cambridge University Press (1948).

Jeffreys, H., and B. S., *Methods of Mathematical Physics*, 3rd ed., Cambridge University Press (1956).

King, R. W. P., and T. T. Wu, *Scattering and Diffraction of Waves*, Harvard University Press, Cambridge, Mass. (1959).

Landau, L. D., and E. N. Lifshitz, *Classical Theory of Fields*, Addison-Wesley, Reading, Mass. (1951).

Landau, L. D., and E. N. Lifshitz, *Electrodynamics of Continuous Media*, Addison-Wesley, Reading, Mass. (1960).

Lorentz, H. A., *Theory of Electrons*, 2nd ed. Dover, New York (1952).

Mason, M., and W. Weaver, *The Electromagnetic Field*, University of Chicago Press (1929); Dover, New York (1952).

Maxwell, J. C., *Treatise on Electricity and Magnetism*, 3rd ed., Dover, New York (1954).

Morse, P. M., and H. Feshbach, *Methods of Theoretical Physics*, McGraw-Hill, New York (1953).

Panofsky, W. K. H., and M. Phillips, *Classical Electricity and Magnetism*, Addison-Wesley, Reading, Mass. (1955).

Purcell, E. M., *Electricity and Magnetism*, McGraw-Hill, New York (1968).

Rohrlich, F., *Classical Charged Particles*, Addison-Wesley, Reading, Mass. (1965).

Schelkunoff, S. A., *Electromagnetic Waves*, Van Nostrand, New York (1943).

Schott, G. A., *Electromagnetic Radiation*, Cambridge University Press (1912).

Smythe, W. R., *Static and Dynamic Electricity*, 2nd ed., McGraw-Hill, New York (1950).

Sommerfeld, A., *Electrodynamics*, Academic, New York (1952).

Stratton, J. A., *Electromagnetic Theory*, McGraw-Hill, New York (1941).

van Bladel, J., *Electromagnetic Fields*, McGraw-Hill, New York (1964).

Van de Hulst, H. C., *Light Scattering by Small Particles*, Wiley, New York (1957).

Weber, E., *Electromagnetic Fields, Theory and Applications*, Vol. 1, *Mapping of Fields*, Wiley, New York (1950).

Wendt, G., *Statische Felder und Stationare Strome*, in *Handbuch der Physik*, Vol. 16, *Elektrische Felder und Wellen*, Springer, Berlin (1958).

Whittaker, E. T., *History of the Theories of Aether and Electricity*, Nelson, London (1951, 1953).

INDEX

INDEX

A CATALOG OF SELECTED

DOVER BOOKS

IN SCIENCE AND MATHEMATICS